TWO DECADES OF ADULT T-CELL LEUKEMIA AND HTLV-I RESEARCH

Gann Monograph on Cancer Research

The "Gann Monograph on Cancer Research" series is promoted by the Japanese Cancer Association. This semiannual series of monographs was initiated in 1966 by the late Dr. Tomizo Yoshida (1903–1973) and is now published jointly by Japan Scientific Societies Press, Tokyo and S. Karger AG, Basel. Each volume consists of collected contributions on current topics in cancer problems and allied research fields. The planning for each volume is done by the Monograph Committee of the Japanese Cancer Association, with the final approval of the Board of Directors. It is hoped that the series will serve as an important source of information in the field of cancer research.

The publication of these monographs owes much to the financial support given by the late Professor Kazushige Higuchi, the Jikei University School of Medicine.

Japanese Cancer Association

JAPANESE CANCER ASSOCIATION
Gann Monograph on Cancer Research No. 50

TWO DECADES OF ADULT T-CELL LEUKEMIA AND HTLV-I RESEARCH

Edited by
KAZUO SUGAMURA, TAKASHI UCHIYAMA, MASAO MATSUOKA
and MARI KANNAGI

JAPAN SCIENTIFIC SOCIETIES PRESS
Tokyo

KARGER
Basel·Freiburg·Paris·London·New York·Bangalore
Bangkok·Singapore·Tokyo·Sydney

January 2003

Published jointly by
JAPAN SCIENTIFIC SOCIETIES PRESS
2-10 Hongo, 6-chome, Bunkyo-ku, Tokyo 113-0033, Japan
ISBN 4-7622-3003-0

and

S. KARGER AG
P.O. Box, CH-4009 Basel, Switzerland
ISBN 3-8055-7543-2

Sole distribution rights outside Japan granted to S. KARGER AG, Basel.

PREFACE

In the early 1980s, tremendous waves of interest regarding human retrovirology blossomed with the discovery of human T-cell leukemia virus type I (HTLV-I) and human immunodeficiency virus (HIV). From the time HTLV-I was demonstrated to be etiologically related to adult T-cell leukemia (ATL), a new era of human viral leukemogenesis was born in place of the long history of animal models.

In April 2001, the Japanese Cancer Association was privileged to host a symposium in Kyoto, Japan on ATL entitled "Two Decades since the Discovery of the Viral Pathogenesis of ATL." This timely symposium was intended to promote the advancement of basic and clinical studies on HTLV-I and ATL. During the last two decades, HTLV-I was established to be the causative virus for not only ATL but also HTLV-I-associated myelopathy/tropical spastic paraparesis (HAM/TSP). Although HTLV-I was molecularly characterized to harbor a variety of oncogenic properties in its Tax protein, the mechanisms of development of ATL and HAM/TSP have not yet been fully resolved. Epidemiological studies of HTLV-I led to the discovery of HTLV-I-endemic areas in the world, namely southwest Japan, the Caribbean region and some areas of Africa. The subsequent disclosure that the main route of HTLV-I transmission is through mother-to-child via breast feeding enabled the development of appropriate intervention strategies that led to successful reduction of HTLV-I prevalence in the HTLV-I endemic areas in Japan. This achievement turned out to be a milestone in the preventive trials for human leukemia.

Since ATL is known to occur in a small proportion of HTLV-I carriers 40 to 50 years after HTLV-I infection, critical events following this infection are thought to be required for ATL leukemogenesis. To establish effective therapeutic measures for ATL, we have to scrutinize the critical events that accompany ATL development.

The papers presented at the memorial symposium in Kyoto included exciting advancements in pathogenesis and protective epidemiology as well as descriptive historical profiles of HTLV-I and ATL. For this monograph, we selected not only papers presented there, but papers of several eminent individuals who have extensively researched HTLV-I and ATL throughout the world. We hope the publication will provide new insights into the oncogenic mechanism mediated by HTLV-I and the development of preventive and therapeutic strategies of ATL. We would further like to extend our appreciation to all the contributors to this Gann Monograph.

January 2003 The Editors

CONTENTS

GENERAL REVIEW OR MESSAGE FOR THE NEXT GENERATION

A SYNOPSIS OF THE ORIGIN OF ATLV/HTLV

Yorio HINUMA

*Emeritus Professor of Kyoto University**

It has been two decades since we reported an unequivocal etiological link between human retrovirus infection and adult T-cell leukemia (ATL) (*1*). On this occasion, it is meaningful to track the history of human retrovirus studies to not only document the exciting findings in this field, but particularly to assist the next generation in discovering as yet unknown virus-related diseases.

During an ATL conference chaired by Dr. Masao Hanaoka of the Institute of Virus Research in Tokyo held on June 5th, 1980, I strongly contemplated the possible viral etiology of ATL, having convinced myself that antigen detection by the immunofluorescence method used for more than fifteen years in my previous work with Epstein-Barr virus (EBV), was a possible avenue to follow. A most notable feature revealed at the conference was the ATL-associated clustering of the birthplace of patients in the southern part of Japan. This fact continued to linger in my mind.

Immediately after the conference in Tokyo I commenced a series of studies which on November 24th, 1980 culminated in Kinya Nagata, one of my research fellows in the Institute for Virus Research of Kyoto University showing me a brilliant fluorescence under a microscope (Fig. 1). The fluorescence was generated by combining the sera from ATL patients with MT-1 cells (an ATL cell line) using the indirect immunofluorescence method. A combination of sera from healthy individuals with the same cells did not produce any fluorescence. We tentatively called the antigen detectable in MT-1 cells as ATLA (ATL cell-associated antigen) (*1*). We found that ATLA strongly reacted with the sera from all 44 patients with ATL in Japan. Interestingly, 26 out of 100 healthy adults in the ATL endemic area had antibodies against ATLA. In contrast, only a few (2/105) healthy adults from the nonendemic areas had antibodies to ATLA. In parallel studies, in collaboration with Dr. Masuyo Nakai who had expertise in retrovirus electroscopic analysis, numerous type C virus particles were seen in MT-1 cells after treatment with 5-iodo-2′-deoxyuridime (IdUrd) (Fig. 2) (*1*). This also appeared to profoundly increase ATLA expression. The MT-1 cell line, established by Dr. Isao Miyoshi (*4*), was the only ATL cell line available at that time. We did find that ATLA neither associated with other human lymphoblastoid cell lines including T cell lines, B cell lines and non-T and non-B cell lines, nor with herpesviruses including EBV, herpes simplex virus, varicella-zoster virus, human cytomegalovirus, herpesvirus saimiri and Marek disease virus. For the first time, a strong case for the etiological relationship of the C-type virus particles with ATL was convincingly demonstrated (*1*).

In 1981, we found and reported that the C-type virus particles were able to enter and transform normal human T-cells during co-cultivation of cord blood lymphocytes

* Konoe-cho, Yoshida, Sakyo-ku, Kyoto 606-8501, Japan

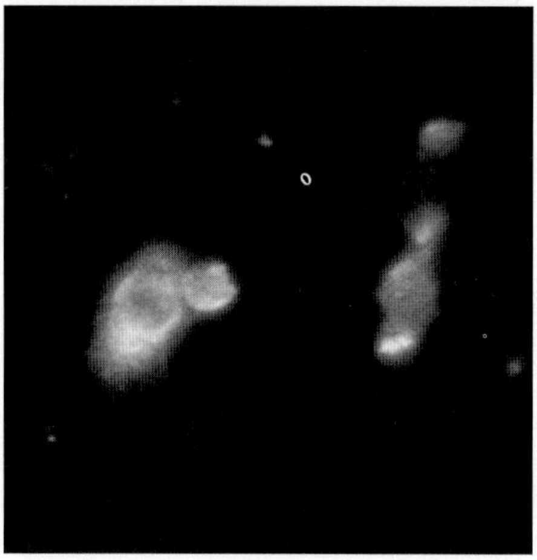

FIG. 1. Detection of ATLA in MT-1 cells by indirect immunofluorescence

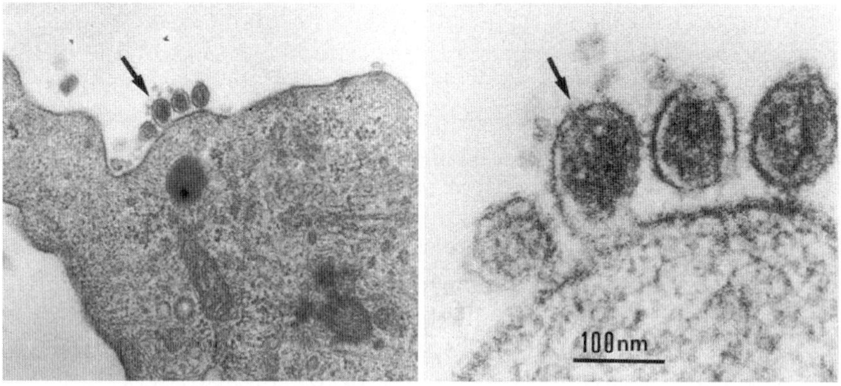

FIG. 2. Thin-section electron micrographs of C-type virus particles in pelleted MT-1 cells

and fresh ATL cells. The resultant cell was transformed into a high virus-producer cell line which we called MT-2 (*5*). This observation established the basis of the transmission of and transformation by the C-type virus on human T-cells. This was the first *in vitro* evidence that the C-type virus associated with ATL bears the transformation capacity for human T-cells. Furthermore, based on the biochemical conclusion that the virus produced by MT-2 is a novel human retrovirus, and on the detection of proviral DNA in the genome of peripheral cells of patients with ATL, we proposed to call this virus ATL virus (ATLV) in 1982 (*10*).

ATL was first described as a new type of T-cell malignancy by Dr. Kiyoshi Takatsuki and co-workers of Kyoto University in 1977 (*7*). It was characterized by an adult onset, subacute or chronic leukemia, with a rapidly progressive terminal course and a common occurrence of lymphoadenopathy, hepatosplenomegaly, and frequent skin involvement. Unlike the long history of animal retrovirus studies, there was a scarcity of information on human retroviruses. The first description of retroviruses associated with human origin officially came in 1979, when van der Loo *et al.* reported the presence of C-type virus particles in skin biopsies and lymph nodes of seven patients with mycosis fungoides and in two patients with Sezary's syndrome (*8*). The virus particles were found mostly in Langerhans cells in the skin and interdigitating cells of lymph nodes but not in lymphoma cells. The authors did not, however, isolate the virus or determine whether the virus was etiologically associated with these neoplasias.

In 1980, a US group led by Dr. Robert Gallo reported the presence of a C-type retrovirus in two human T-cell lines, HUT102 and CTCL-3, established from the lymph nodes of a 26-year-old black male patient with mycosis fungoides in the USA, as well as in fresh peripheral blood lymphocytes obtained from the same patient (*6*). They called the virus human cutaneous T-cell lymphoma virus with the abbreviation HTLV. In 1982, they reported the results of a serological study of HTLV in which it was shown that only one patient with mycosis fungoides and one patient with Sezary's syndrome out of 23 patients with various T-cell malignancies (not specified) had serum antibodies to the HTLV p24 core protein. However, no evidence was provided to depict a causal relationship between HTLV and mycosis fungoides or Sezary's syndrome or any other human malignancies for that matter (*3*). Subsequently, the same group reported that a high proportion of Japanese ATL patients (12 out of 13) had antibodies to p24 of HTLV. To this end, molecular analyses of the viral genome of ATLV and HTLV identified that both viruses, ATLV and HTLV, are the same human retrovirus (*9*).

An explosive investigation on ATLV/HTLV and ATL ensued and extensive reviews documented these early trials. To describe these later events after these early studies is beyond the scope of this article. A critical issue I must emphasize at present is that no means to treat and/or cure ATL has been established thus far, despite a monumental amount of research and many, many trials.

Finally, I would like to conclude the natural history of ATLV/HTLV by recollecting the early facts of the geographical clustering of ATL and ATLV antibodies which had strongly attracted me. The serological study told us that 26% of 100 healthy adults over 20 years of age in ATL-endemic areas also had anti-ATLA, but only a few of the 100 persons in nonendemic areas had anti-ATLA. It was noteworthy that of 13 persons with anti-ATLA in the nonendemic area, 10 had been born in an endemic area: 7 of 8 patients with ATL, one of two patients with malignant T-cell lymphoma, one patient with hypertension, and one of two healthy persons (*1*). The results of this screening test indicated that a considerable proportion of adults living or born in an endemic area have antibody to ATLA and that most, if not all, patients having ATL have anti-ATLA, whereas only a few of the adults from ATL-nonendemic areas have the antibody. Information on the geographic biases of ATLA/HTLV infection to endemic areas was enough of a hint to consider the origin and dynamic movement of Japanese races. My view on this is summarized elsewhere (*2*).

I have devoted much of my time to studies of virus and virus-mediated diseases and am still keen to participate on such exciting matters in the future.

Acknowledgments

I am grateful to Dr. Kazuo Sugamura for help in the preparation of this manuscript.

REFERENCES

1. Hinuma, Y., Nagata, K., Hanaoka, M., Nakai, M., Matsumoto, T., Kinoshita, K. I., Shirakawa, S., and Miyoshi, I. Adult T-cell leukemia: antigen in an ATL cell line and detection of antibodies to the antigen in human sera. *Proc. Natl. Acad. Sci. USA*, **78**, 6476–6480 (1981).
2. Hinuma, Y. Natural history of the retrovirus associated with a human leukemia. *BioAssay*, **3**, 205–209 (1985).
3. Kalyanaraman, V. S., Sarngadharan, M. G., Nakao, Y., Ito, Y., Aoki, T., and Gallo, R. C. Natural antibodies to the structural core protein (p24) of the human T-cell leukemia (lymphoma) retrovirus found in sera of leukemia patients in Japan. *Proc. Natl. Acad. Sci. USA*, **79**, 1653–1657 (1982).
4. Miyoshi, I., Kubonishi, I., Sumida, M., Hiraki, S., Tsubota, T., Kimura, I., Miyamoto, K., and Sato, J. A novel T-cell line derived from adult T-cell leukemia. *Gann*, **71**, 155–156 (1980).
5. Miyoshi, I., Kubonishi, I., Yoshimoto, S., Akagi, T., Ohtsuki, Y., Shiraishi, Y., Nagata, K., and Hinuma, Y. Type C virus particles in a cord T-cell line derived by co-cultivating normal human cord leukocytes and human leukaemic T cells. *Nature*, **294**, 770–771 (1981).
6. Poiesz, B. J., Ruscetti, F. W., Gazdar, A. F., Bunn, P. A., Minna, J. D., and Gallo, R. C. Detection and isolation of type C retrovirus particles from fresh and cultured lymphocytes of a patient with cutaneous T-cell lymphoma. *Proc. Natl. Acad. Sci. USA*, **77**, 7415–7419 (1980).
7. Uchiyama, T., Yodoi, J., Sagawa, K., Takatsuki, K., and Uchino, H. Adult T-cell leukemia: clinical and hematologic features of 16 cases. *Blood*, **50**, 481–492 (1977).
8. van der Loo, E. M., van Muijen, G. N. P., van Vloten, W. A., Beens, W., Scheffer, E., and Meijer, C. J. L. M. C-type virus-like particles specifically localized in Langerhans cells and related cells of skin and lymphnodes of patients with mycosis fungoides and Sezary's syndrome. *Virchows Arch. B Cell Pathol. Incl. Mol. Pathol.*, **31**, 193–203 (1979).
9. Watanabe, T., Seiki, M., and Yoshida, M. HTLV type I (U. S. isolate) and ATLV (Japanese isolate) are the same species of human retrovirus. *Virology*, **133**, 238–241 (1984).
10. Yoshida, M., Miyoshi, I., and Hinuma, Y. Isolation and characterization of retrovirus from cell lines of human adult T-cell leukemia and its implication in the disease. *Proc. Natl. Acad. Sci. USA*, **79**, 2031–2035 (1982).

Gann Monograph on Cancer Research 50, 2003

HUMAN LEUKEMIA VIRUS—A GODSEND AT LONG LAST

Isao MIYOSHI

*Department of Medicine, Kochi Medical School**

Around the time adult T-cell leukemia (ATL) was proposed as a new disease by Takatsuki and his colleagues (Uchiyama *et al.*, *Blood*, **50**, 481, 1977), all efforts to detect human leukemia virus had failed and many investigators were giving up the idea of viral etiology of human leukemia. After working on murine leukemia viruses, I joined the study on ATL in 1978 by coculturing ATL cells with human cord leukocytes. I added the cord leukocytes in the hope that they would possibly stimulate the growth of ATL cells that had been degenerating *in vitro*. This coculture method gave rise to an ATL cell line, MT-1 (Miyoshi *et al.*, *Gann*, **71**, 155, 1980) and, unexpectedly, a cord T-cell line, MT-2 (Miyoshi *et al.*, *Gann*, **72**, 978, 1981). To our surprise, both cell lines were persistently infected with type-C virus (Hinuma *et al.*, *Proc. Natl. Acad. Sci. USA*, **78**, 6476, 1981; Miyoshi *et al.*, *Nature*, **294**, 770, 1981), later designated human T-cell leukemia virus type I (HTLV-I). MT-2 was established by transformation of cord T-cells by HTLV-I released from ATL cells during coculture. Thus, we were fortunate enough to be able to isolate HTLV-I and to demonstrate its T-cell transforming capacity at the same time. Presumably, cytokines such as interleukin-2 (IL-2) were produced during coculture and supported the proliferation of T-cells. At about this time in the United States, Gallo and his associates succeeded in growing an HTLV-I-infected T-cell line with the use of T-cell growth factor now known as IL-2 from a patient with cutaneous T-cell lymphoma (Poiesz *et al.*, *Proc. Natl. Acad. Sci. USA*, **77**, 7415, 1980). These culture techniques laid the foundation for the subsequent detection of human immunodeficiency virus (Barré-Sinoussi *et al.*, *Science*, **220**, 868, 1983).

Shortly after the detection of HTLV-I, we were excited to find that many Old World monkeys including Japanese macaques were infected with HTLV-I-like virus, now called simian T-cell leukemia virus type I (STLV-I) (Miyoshi *et al.*, *Lancet*, **ii**, 658, 1982; *Int. J. Cancer*, **32**, 333, 1983). HTLV-I and STLV-I appeared to be identical in serological and ultrastructural aspects and were shown to be more than 90% homologous by nucleotide sequence analysis. It is conspicuous that HTLV-I is prevalent among Japanese people and so is STLV-I among Japanese monkeys. The monkey virus is likely to have been transmitted to humans in ancient times but when and how this event occurred remains unknown. My hypothesis is that Jomon people, ancestors of modern Japanese, who lived on the Japan Islands about 2,500 to 16,000 years ago, became infected with the monkey virus through eating virus-carrying monkeys. This does not exclude the possibility that some of them had already been infected with HTLV-I when they first appeared on the Japan Islands. The Jomon people made their living by hunting, fishing, and collecting nuts and fruit. It is to be noted that

* Kohasu, Okoh-cho, Nankoku, Kochi 783-8505, Japan

monkey bones were excavated from the Jomon remains in various regions of Japan. Even today monkey meat is sold at butcher shops in certain parts of Africa, and there are pet lovers who keep monkeys as their companions. These and other people should be warned against the possible risk of STLV-I infection from monkeys.

We were then intrigued by the finding that rabbits were highly susceptible to HTLV-I infection, in contrast to abortive or inconsistent infection of mice and rats with HTLV-I. The rabbit model allowed us to perform serological and virological analyses on blood-borne and milk-borne transmission of HTLV-I; HTLV-I infection through blood transfusion or breast feeding was found to be preventable by passive immunization of rabbits with anti-HTLV-I IgG prepared from pooled plasma from seropositive healthy blood donors (Miyoshi *et al.*, *Gann Monogr.*, **39**, 195, 1992). Similarly, passive immunization protected Japanese monkeys from HTLV-I infection (Murata *et al.*, *Leukemia*, **10**, 1971, 1996). In Japan all blood donors have been screened for HTLV-I antibody since 1986; any seropositive blood components are totally discarded. In some virus-endemic regions of Japan, seropositive pregnant mothers are being advised to switch to bottle feeding. Nevertheless, intrapartum transmission of HTLV-I is still possible (Sawada *et al.*, *JAMA*, **273**, 284, 1995). Our passive immunization experiments indicate that cell-to-cell infection of HTLV-I can be blocked by anti-HTLV-I IgG with a high neutralizing antibody titer against vesicular stomatitis virus bearing envelope antigens of HTLV-I. Therefore, its clinical application would be warranted for the prevention of intrapartum and milk-borne transmission of HTLV-I from mother to infant, especially in situations where bottle feeding is not feasible for some reason.

DISCOVERY OF ADULT T-CELL LEUKEMIA: REMINISCENCES

Kiyoshi TAKATSUKI

*Professor Emeritus, Kumamoto University**

In the history of medicine the cycle involving the discovery of a new disease, the study of its etiology, and discussions of countermeasures must have been repeated many times. However, a cycle frequently lasts for several centuries and involves many twists and turns; hence, the details of many such past cycles are unknown. In recent years, however, the cycle of discovering, reporting and recognizing a new disease and then, most importantly, defining its etiology and considering the preventive measures has proceeded rapidly, reflecting the general trend of rapid changes in social phenomena. One example is adult T-cell leukemia (ATL). It was around the year 1973 that we came to recognize the existence of ATL, previously an unknown disease. Here, I would like to give a retrospective view concerning the background of our studies on ATL.

Many clinicians in Japan probably feel that the descriptions of diseases in the literature from abroad differ from the features of the diseases observed in their practices in Japan. We can mention many examples in the field of hematology. One of these is that in Japan the incidence of chronic lymphocytic leukemia is quite low, being only 2% of all hematologic malignancies. What differs is not only the incidence but also the detailed symptoms and signs of diseases. This vague feeling that an autochthonous pathology probably exists in Japan may be considered as the key factor in the background of our study of ATL.

The second major factor is the recent progress made in immunology. Having an interest in immunoglobulin abnormalities, I studied multiple myeloma and related diseases in the days when even the word 'immunoglobulin' was not yet in use. The central theme of immunology at that time was to identify the structure of antibodies. Immunology advanced rapidly, and myeloma was defined as a B-cell malignancy. Therefore, it was quite natural to enlarge the objectives of our clinical studies to all lymphoproliferative diseases.

At this stage of our research in Kyoto, joint studies with young researchers were initiated. This was the third and most important element of the background of our ATL study. I became acquainted with Drs. Takashi Uchiyama and Junji Yodoi, who are currently both working in Kyoto, during their postgraduate training. Dr. Yodoi prepared an antiserum against human thymocytes and invited us to direct our attention to T-cells. In the course of examining patients with various lymphoproliferative disorders, we arrived at the conclusion that ATL was a disease which had not been described anywhere before. It was recognized that T-cell malignancy had a relatively high incidence among Japanese adults and that most of the patients with this disease were from Kyushu. This discovery was made from our bedside observations rather than from laboratory work. Thereafter, Dr. Uchiyama went to the National Cancer

* Present Address: Kitano Hospital, 2-4-20 Ohgimachi, Kita-ku, Osaka 530-8480, Japan

Institute, Bethesda to work with Dr. Thomas A. Waldmann and raised a monoclonal antibody, called 'Tac' antibody, which later played an important role in our ATL studies. Dr. Yodoi has developed a new field of research by identifying a novel cytokine, ATL-derived factor (ADF), which has proved to be important in redox regulation. At the end of 1981, I moved from Kyoto to Kumamoto, which is located in the middle of Kyushu. Our studies advanced remarkably there, due mainly to the efforts of excellent young co-workers including Drs. Kazunari Yamaguchi, Toshio Hattori, and Masao Matsuoka.

This discovery of ATL ushered in some dramatic developments in oncology, virology and, unexpectedly, neurology and other fields of medicine. When we reported a series of 13 patients with ATL in 1976 on the occasion of the 16th International Congress of Hematology in Kyoto, it was stated that 'attempt to elucidate leukemogenesis in this disease should be directed towards exploring the genetic background and a possible viral involvement.' HTLV (human T-cell leukemia virus), the pathogen of ATL, was first reported by Dr. Robert C. Gallo and his co-workers in Bethesda in 1980 and 1981. They isolated HTLV from cultured cells taken from a patient with an aggressive variant of mycosis fungoides and another with Sezary syndrome. Although both patients were said to have cutaneous T-cell lymphoma, they had some unusual features which, in retrospect, link them to the clinical entity now called ATL.

In Japan, co-culturing of ATL cells with umbilical cord lymphocytes was first done successfully by Dr. Isao Miyoshi's group in Okayama, who obtained the cell line MT-1. Dr. Yorio Hinuma and his co-workers in Kyoto demonstrated that ATL patients have antibodies against presumed viral antigens on MT-1 cells by indirect immunofluorescence. Subsequently, a retrovirus was isolated and called ATLV (adult T-cell leukemia virus). Since Dr. Mitsuaki Yosida and his group in Tokyo showed that HTLV and ATLV are, in fact, identical, the term HTLV-1 (human T-cell lymphotropic virus type 1) has been commonly used.

HTLV-1 is endemic in southwestern Japan, especially Kyushu and Okinawa, in the Caribbean Islands, and in Central Africa. It has also been revealed that there are HTLV-1 carriers in South America, Papua New Guinea, the Solomon Islands, South China, and in other isolated populations in the world, including Ainus in Hokkaido and Aborigines in Australia. These epidemiological studies have been promoted by international collaborations, to which Drs. William Blattner in Bethesda, Guy de Thé in Paris, Kazuo Tajima in Nagoya, Shunro Sonoda in Kagoshima, and many other epidemiologists have made important contributions. Dr. Daniel Catovsky and his colleagues delineated the clinical features of ATL patients originating in the Caribbean Islands.

On the other hand, several diseases have been found to be related to HTLV-1 infection. Drs. Antoinne Gessain and Guy de Thé first reported the association of tropical spastic paraparesis (TSP) and HTLV-1, and Dr. Mitsuhiro Osame and his co-workers in Kagoshima studied features of HTLV-1-associated myelopathy (HAM/TSP) in detail. HTLV-1 uveitis was subsequently described by a study group of ophthalmologists in Kyushu. HTLV-1 may also be associated with bronchopneumonitis, arthritis, polymyositis and other inflammatory conditions. Moreover, it was noted that an immunodeficiency state may be induced by HTLV-1 infection.

In Japan, HTLV-1 carriers have been estimated to be 1.2 million, and more than

700 cases of ATL have been diagnosed each year. The majority of HTLV-1 transmission occurs via one of three routes, all of which require the passage of virus-infected cells. HTLV-1 carrier mothers transmit the virus to newborns mainly through breast milk. Dr. Shigeo Hino and his group in Nagasaki conducted intensive field work regarding the mother-to-infant infection. Carrier mothers in Japan have been instructed to refrain from breast feeding or to modify the feeding procedures to prevent HTLV-1 infection. There is convincing evidence that HTLV-1 can be transmitted from individual to individual by sexual contact, especially from males to females, and through blood transfusions. All blood donated at the Red Cross Blood Centers in Japan was subjected to HTLV-1 antibody testing beginning in November 1986.

Treatment of ATL is the most difficult task. Dr. Masanori Shimoyama in Tokyo has organized a multicenter study group for the chemotherapy of ATL. Many other trials have been reported: deoxycoformycin, α-interferon combined with azidothymidine, and so on. The use of humanized monoclonal antibodies against the interleukin-2 receptor has been challenged by Dr. Waldmann's group. Successful allogeneic bone marrow transplantation for patients with ATL was reported from several institutions.

Before my retirement from Kumamoto University in 1996, I edited a book titled "Adult T-cell Leukaemia" published by Oxford University Press in 1994. Many of the above-mentioned investigators contributed chapters to this book. It may be pertinent here to commemorate the establishment of the International Retrovirology Association, which was announced on the occasion of the 5th International Conference on Human Retrovirology held in Kumamoto on May 11–13, 1992.

Immediately after the discovery of ATL and HTLV-1, the acquired immune deficiency syndrome (AIDS) began to be recognized, and a retrovirus, the human immunodeficiency virus (HIV), was also found to be a causative agent in this disease. The advances made in the worldwide study of AIDS and HIV owe much to the knowledge gained about the relation between ATL and HTLV-1.

[illegible] in 1994 [illegible]

[illegible] HTLV-1 [illegible] and HTLV-1 [illegible]

MOLECULAR BIOLOGY OF HTLV-1: PLEIOTROPIC FUNCTION OF Tax PROTEIN IN CELL REGULATION

Mitsuaki YOSHIDA

*Banyu Tsukuba Research Institute**

The human T cell leukemia virus-1 (HTLV-1) is a retrovirus which causes adult T cell leukemia (ATL) and neurological disorder, the tropical spastic paraparesis (HAM/TSP). The pathogenic factor is highly likely to be the viral protein Tax, which is a key regulator of viral replication. In addition to viral replication, Tax exerts 1) *trans*-activation and -repression of transcription through interaction with various transcription factors and co-activators, 2) dysregulation of cell cycle through inhibition of some tumor suppressor proteins, and 3) accumulation of mutation in cellular genes through suppressing DNA repair enzymes and checkpoint functions. These pleiotropic effects on various cellular targets seem to cooperate in promoting cell proliferation and fixation of genetic abnormalities, ultimately resulting in cell transformation and leukemogenesis.

It has been 20 years since human T cell leukemia virus-1 (HTLV-1) was identified as the first human retrovirus and causative agent of adult T cell leukemia (ATL). Our group has been involved in the molecular biology of HTLV-1 from the initial identification and proof of the retrovirus in the ATL-derived cell line, and up to the present in the molecular characterization and updating the molecular knowledge of its pathogenesis. Here, I would like to summarize our contribution to the understanding of the molecular biology of HTLV-1.

Unlike acute leukemia viruses of animal origin, HTLV-1 has no typical oncogene derived from a cellular gene but has an extra sequence of 1.6 kb (called the pX region) in the 3′ terminal region of its genome (*25*). The pX region codes for several proteins including Tax (*13, 17*), Rex (*14*), p21X-III (*14*), and some other minor proteins (*15*). The *cis*-acting elements in the long terminal repeats of HTLV-1 are uniquely regulated by a *trans*-acting factor, Tax (*4, 26*). Thus, Tax is essential for the effective replication of HTLV-1. Tax also has capacity to immortalize primary T cells (*22*), to transform rodent fibroblasts (*35*), and to induce tumors in transgenic mice expressing Tax (*21*).

Viral involvement in leukemogenesis was originally predicted based on the clonal integration of the proviruses in leukemic cells of ATL patients: every ATL cell contained the proviral genome integrated in its chromosomal DNA, and the integration sites were clonal within individuals indicating that leukemic cells originated from a single infected T cell (*38*). The sites, however, were not common among patients (*24*).

*Tsukuba, Ibaraki 300-2611, Japan

Therefore, the promoter-insertion model of the LTR (long terminal repeat) by the integrated proviruses seems unlikely. Consequently, a *trans*-acting viral factor, Tax, has been proposed for the viral pathogenesis.

Pleiotropic Function of Tax

Tax is a 40 kDa protein mainly localized in the nucleus and originally identified as a *trans*-activator of the viral transcriptional enhancer, the 21-bp sequence in the LTR. This function of Tax was found to activate cellular oncogenes, growth factors and some of their receptors, while some other cellular genes were found to be repressed by this protein. Tax was also found to bind to and inhibit some tumor suppressor proteins. These functions of Tax are apparently the molecular basis of Tax-induced abnormal cell proliferation. More interestingly, Tax was found to have a capacity to induce mutation rate and apoptotic resistance, which might aid in our understanding of leukemogenesis.

Transcriptional Activation

Transcriptional activation by Tax protein is basically achieved by two independent mechanisms: direct binding to enhancer-binding proteins on the DNA such as CREB, NF-κB and SRF, and inhibition or destabilization of NF-κB, such as IκBα.

CREB: In the normal regulation of CRE (cAMP responsive element)-dependent transcription, CRE binding protein (CREB) is phosphorylated by protein kinase A upon stimulation of the cell (*6*), and the phosphorylation allows its binding to a transcriptional activator, CBP (CREB binding protein) (*3*). The CBP thus associated with DNA would acetylate histone through its histone-acetylating activity,

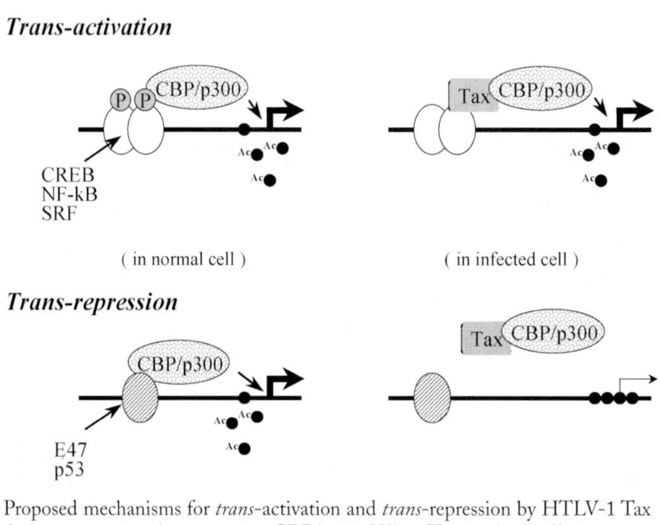

FIG. 1. Proposed mechanisms for *trans*-activation and *trans*-repression by HTLV-1 Tax
Tax binds to transcriptional co-activator, CBP/p300. When Tax has high affinity to a specific transcription factor, the complex binds to enhancer through this factor and activates transcription, but when Tax has no affinity to the transcription factor, Tax in the complex interferes with the interaction of CBP/p300 with the transcription factor.

resulting in opening the nucleosome structure around the 21-bp enhancer sequence and thus leading to activation of transcriptional initiation. Thus, CRE-dependent transcription is controlled in a signal dependent fashion. Tax on the other hand, was found to bind to CBP (*16, 27*); and therefore would be able to bridge between CREB and CBP without any specific phosphorylation of CREB. This complex formation of CREB-Tax-CBP would explain the constitutive activation of the 21-bp enhancer in HTLV-1-infected cells once Tax is expressed (*34*). A major question in this mechanism is that the consensus sequence of the CRE is not *trans*-activated by Tax, and a unique but small flanking sequence upstream of the CRE is required (*5*). This puzzle is still not well solved, but a 21-bp specific conformational change of CREB or direct interaction of Tax to enhancer 21-bp DNA sequence have been discussed.

NF-κB: HTLV-1-infected T cell lines frequently produce various lymphokines such as interleukin-6 (IL-6), granulocyte macrophage-colony stimulating factor (GM-CSF), tumor necrosis factor β (TNFβ) and others (*37*). This over-expression is mediated by activation of a transcription factor family, NF-κB. Coprecipitation of an NF-κB precursor, p105, with Tax was an original discovery of interaction of Tax protein and NF-κB (*7*). This established that Tax binds to NF-κB family proteins including NF-κB p50, p65, c-Rel (*29*) and NF-κB-2 and activates the NF-κB-dependent transcription. The NF-κB proteins bind to p300, therefore, a similar mechanism to activation of CRE-dependent transcription through Tax-CREB interaction has been predicted for the activation of NF-κB-dependent transcription by Tax.

Inactivation of IκB: In unstimulated cells, NF-κB proteins are held in cytoplasm as an inactive complex with IκB (*1*). Stimulatory signals phosphorylate IκB and destabilize it, resulting in nuclear translocation of the active form of NF-κB, thus enhancing NF-κB-dependent transcription. In addition to the Tax binding to NF-κB proteins, its binding to IκBα inducing destabilization of IκBα was demonstrated (*28*). As further mechanisms, the activation of cytokine-inducible kinases, IKKα and IKKβ, were reported to result in constitutive phosphorylation and degradation fol-

	Function	*Direct Target*	*Partner*	*Final Target genes*
Tax →	Trans-activation of transcription	CREB NF-kB, IkB SRF	CBP, p300	HTLV IL-6, IL-2Rα, TGFβ Bcl-x
→	Trans-repression of transcription	CBP, p300	E47 p53	DNA polβ p18ink p53-dependent Bax
→	Inhibition of tumor suppressor	p16ink, p15ink hDlg	Rb/Cdk4 Apc	Myc, PCNA, DNA polα β-Catenin
→	Inhibition of check point, DNA repair & apoptosis	hMad-1 Topo I		hMad-1 DNA polβ, Topo I, PCNA Bcl-x, Bax

FIG. 2. Pleiotropic function of Tax

Four different categories of Tax function and their direct targets and final target genes are summarized.

lowed by nuclear translocation of NF-κB (*18*). Therefore, two independent processes appear to be involved in activation of NF-κB, also indicating that Tax functions in the cytoplasm.

Transcriptional Repression

Tax-mediated *trans*-repression was described for *DNA polymerase β* gene (*9*), then for *p18INK4c* gene, an inhibitor of CDK4, and for *p53*-dependent gene expression. P18INK4c can arrest the cell cycle at G1 phase, but Tax can cancel this effect through down-regulation of p18INK4 expression (*31*). For efficient expression of p18INK4, E47 binds to E-Box in the promoter and the E47 should interact with p300 directly or indirectly. Our data clearly demonstrated that Tax binds to p300 and interferes with its interaction with E47-E-Box protein, thus down-regulating the transcription (*31*).

Trans-repression of p53-dependent transcription: Tax also *trans*-represses the p53-dependent transcription (*20*), which requires interaction of p53 with both the p53-binding site and CBP. Here again it is proposed that Tax binding to CBP interferes with the interaction of CBP with p53 on the DNA (*31*). Whatever the mechanism, the *trans*-repression of p53-dependent transcription is expected to play a role in leukemogenesis, simply because it is a tumor suppressor protein involved frequently in many types of human tumors.

A Mechanism for Trans-activation and -Repression

We now propose a unified principle for the *trans*-activation and -repression: "Tax binds to CBP/p300 in general. When Tax in the Tax-CBP/p300 has high affinity to transcription factors, the interaction would result in *trans*-activation skipping any modification or further requirement, but when Tax in the Tax-CBP/p300 has no affinity to transcription factors, it would result in *trans*-repression." In *trans*-repression, the competing capacity would be greatly affected by the expression level of CBP/p300 and its affinity to specific transcription factors. Thus, the cascade of Tax in transcriptional regulation would be unexpectedly broad in its targets and variable in its extent. However, the full situation seems to be more complex, since histone acetyltransferase-independent *trans*-activation has been described even in the CBP/p300-dependent transcription system (*10*).

Inhibition of Tumor Suppressor Proteins

Tax targets the ankyrin motifs in the IκB (*8*). The structural information that most of the p16INK4a consists of ankyrin motifs suggested it could be the target of Tax binding. Thus, Tax was found to interact with negative regulators of CDK4 (*30*). A typical member of the INK4 family is p16INK4a, which is also known to be a tumor suppressor gene in many types of human tumors. When the negative function of p16INK4a is impaired by Tax, CDK4/6 is activated and phosphorylates retinoblastoma protein (Rb), a tumor suppressor protein, which results in activation of E2F and then activates various genes for DNA synthesis. Through this mechanism, a transcriptional regulator Tax can promote cell cycle progression directly.

p16INK4a and p15INK4b, cell cycle regulators: The binding of Tax to p16INK4a was demonstrated *in vitro* using GST-p16INK4a and also *in vivo* using co-transfection followed by immuno-precipitation (*30*). The binding of Tax to p16INK4a inhibited its negative function on CDK4/6, and thus activated the CDK4. When the cell line U2OS, which lacks *p16INK4a* gene, but retains the wild type Rb, was transfected with p16INK4a, the cell cycle was arrested at G1. Further transfection of Tax rescued the cells from the G1 arrest. The findings indicated that the binding of Tax to p16INK4a was able to activate CDK4/6 and phosphorylated Rb, and then induce G1/S phase transition of the target cells (*30*). Exactly the same results were obtained with p15INK4b, another member of the INK4 family; therefore, both p16INK4a and p15INK4b are functionally knocked out by Tax protein. These two genes are recognized as tumor suppressor genes, *MTS-1* and *-2*, and are frequently deleted in human cancer cells. Therefore, the Tax-induced functional inactivation of these two INK4 family proteins might mimic the effect of deletion of these genes in cancer cells, strongly suggesting involvement of this mechanism in ATL development.

hDlg which associates with APC: Tax also binds to another tumor suppressor protein, a human counterpart of *Dlg*, drosophila tumor suppressor gene. hDlg is a signaling molecule downstream of the Wnt/Frizzle pathway and interacts with the C-terminus of a tumor suppressor protein, APC. Tax was found to bind to the same domain, PDZ domain, of hDlg to which APC binds, therefore, Tax and APC compete with the same site of the hDlg molecule (*32*). In fact, Tax is able to compete with hDlg inducing cell growth.

Enhancement of Mutation Rate

Tax affects genes whose functions are not directly linked to cell proliferation; one is *trans*-repression of DNA polymerase β (*9*), a key enzyme for DNA repair. *Trans*-repression of this gene might result in higher frequency of mutation in host cells. Other possible mechanisms for higher mutation rate have also been proposed: direct inhibition of topoisomerase I by Tax (*33*) and *trans*-activation by Tax of *PCNA* gene (cellular proliferating cell nuclear antigen (*12*)). As expected from these molecular analyses, a mutagenic effect of Tax on host cell chromosomes was directly demonstrated (*19, 23*). Thus, Tax expression and modulation of DNA polymerase β, topoisomerase I, and/or PCNA predispose cells to accumulation of DNA damage and contribute to malignant conversion of infected cells.

In relation to DNA damage and its repair, attention should be focused on the Tax effects on p53. *Trans*-repression of p53-dependent transcription is particularly notable (*20, 34*). It is apparent that the reduction of p53-dependent transcription would reduce the cell cycle checkpoint function of p53, thus leading cells to accumulation of DNA mutations in cells. Continuous disposition of infected cells to such a situation would accelerate malignant transformation and thus progression to ATL. In this respect, HsMAD1, a human homologue of yeast mitotic checkpoint protein MAD1 (*11*) is an exciting target of Tax. HsMAD1 is a component of the mitotic checkpoint system which prevents anaphase and commitment to cellular division until chromosomal alignment is properly completed. Therefore, abrogation of the mitotic checkpoint function of HsMAD1 is likely to link to abnormal cell division and accumulation of chromosomal abnormalities. These abnormal cells are triggered to undergo apoptosis if they are normally regulated. Tax protein, however, is able to prevent apoptosis which arises from suppression of p53 function through activation of NF-κB and *trans*-activation of Bcl-X (*36*) or expression of XIAP (X-chromosome-linked inhibitor of apoptosis) protein. Another possible mechanism is repression of *bax* gene expression (*2*).

In summary, most of the pleiotropic functions of Tax protein cooperate in promoting cell proliferation, accumulation of DNA damage and avoiding apoptosis of abnormal cells infected with HTLV-1. However, it should also be emphasized that Tax alone is apparently not sufficient to explain the final leukemia. It is still necessary to identify second, third or even more molecular events during its progression for a full understanding of ATL.

REFERENCES

1. Baeuerle, P. A. The inducible transcription activator NF-kappa B: regulation by distinct protein subunits. *Biochim. Biophys. Acta*, **1072**, 63–80 (1991).
2. Brauweiler, A., Garrus, J. E., Reed, J. C., and Nyborg, J. K. Repression of bax gene expression by the HTLV-1 Tax protein: implications for suppression of apoptosis in virally infected cells. *Virology*, **231**, 135–140 (1997).
3. Chrivia, J. C., Kwok, R. P., Lamb, N., Hagiwara, M., Montminy, M. R., and Goodman, R. H. Phosphorylated CREB binds specifically to the nuclear protein CBP. *Nature*, **365**, 855–859 (1993).
4. Fujisawa, J., Seiki, M., Kiyokawa, T., and Yoshida, M. Functional activation of the long terminal repeat of human T-cell leukemia virus type I by a trans-acting factor. *Proc. Natl. Acad. Sci. USA*, **82**, 2277–2281 (1985).
5. Fujisawa, J., Toita, M., and Yoshida, M. A unique enhancer element for the trans activator (p40tax) of human T-cell leukemia virus type I that is distinct from cyclic AMP- and 12-O-tetradecanoylphorbol-13-acetate-responsive elements. *J. Virol.*, **63**, 3234–3239 (1989).
6. Gonzalez, G. A., Yamamoto, K. K., Fischer, W. H., Karr, D., Menzel, P., Biggs, W. D., Vale, W. W., and Montminy, M. R. A cluster of phosphorylation sites on the cyclic AMP-regulated nuclear factor CREB predicted by its sequence. *Nature*, **337**, 749–752 (1989).
7. Hirai, H., Fujisawa, J., Suzuki, T., Ueda, K., Muramatsu, M., Tsuboi, A., Arai, N., and Yoshida, M. Transcriptional activator Tax of HTLV-1 binds to the NF-kappa B precursor p105. *Oncogene*, **7**, 1737–1742 (1992).
8. Hirai, H., Suzuki, T., Fujisawa, J., Inoue, J., and Yoshida, M. Tax protein of human T-cell leukemia virus type I binds to the ankyrin motifs of inhibitory factor kappa B and induces nuclear translocation of transcription factor NF-kappa B proteins for transcriptional activation. *Proc. Natl. Acad. Sci. USA*, **91**, 3584–3588 (1994).
9. Jeang, K. T., Widen, S. G., Semmes, O. J. T., and Wilson, S. H. HTLV-I trans-activator protein, tax, is a trans-repressor of the human beta-polymerase gene. *Science*, **247**, 1082–1084 (1990).
10. Jiang, H., Lu, H., Schitz, R., Pise-Masison, C., Ogryzko, V., Nakatani, Y., and Brady,

J. PCAF interacts with tax and stimulates tax transactivation in a histone acetyltransferase-independent manner. *Mol. Cell. Biol.*, **19**, 8136–8145 (1999).

11. Jin, D. Y., Spencer, F., and Jeang, K. T. Human T cell leukemia virus type 1 oncoprotein Tax targets the human mitotic checkpoint protein MAD1. *Cell*, **93**, 81–91 (1998).
12. Kao, S. and Marriott, S. Disruption of nucleotide excision repair by the human T-cell leukemia virus type 1 Tax protein. *J. Virol.*, **73**, 4299–4304 (1999).
13. Kiyokawa, T., Seiki, M., Imagawa, K., Shimizu, F., and Yoshida, M. Identification of a protein (p40x) encoded by a unique sequence pX of human T-cell leukemia virus type I. *Gann*, **75**, 747–751 (1984).
14. Kiyokawa, T., Seiki, M., Iwashita, S., Imagawa, K., Shimizu, F., and Yoshida, M. p27x-III and p21x-III, proteins encoded by the pX sequence of human T-cell leukemia virus type I. *Proc. Natl. Acad. Sci. USA*, **82**, 8359–8363 (1985).
15. Koralnik, I. J., Gessain, A., Klotman, M. E., Lo Monico, A., Berneman, Z. N., and Franchini, G. Protein isoforms encoded by the pX region of human T-cell leukemia/lymphotropic virus type I. *Proc. Natl. Acad. Sci. USA*, **89**, 8813–8817 (1992).
16. Kwok, R. P., Laurance, M. E., Lundblad, J. R., Goldman, P. S., Shih, H., Connor, L. M., Marriott, S. J., and Goodman, R. H. Control of cAMP-regulated enhancers by the viral transactivator Tax through CREB and the co-activator CBP. *Nature*, **380**, 642–646 (1996).
17. Lee, T. H., Coligan, J. E., Sodroski, J. G., Haseltine, W. A., Salahuddin, S. Z., Wong-Staal F., Gallo, R. C., and Essex, M. Antigens encoded by the 3′-terminal region of human T-cell leukemia virus: evidence for a functional gene. *Science*, **226**, 57–61 (1984).
18. Li, X., Murphy, K., Palka, K., Surabhi, R., and Gaynor, R. The human T-cell leukemia virus type-1 Tax protein regulates the activity of the IkappaB kinase complex. *J. Biol. Chem.*, **274**, 34417–34424 (1999).
19. Miyake, H., Suzuki, T., Hirai, H., and Yoshida, M. Trans-activator Tax of human T-cell leukemia virus type 1 enhances mutation frequency of the cellular genome. *Virology*, **253**, 155–161 (1999).
20. Mulloy, J. C., Kislyakova, T., Cereseto, A., Casareto, L., LoMonico, A., Fullen, J., Lorenzi, M. V., Cara, A., Nicot, C., Giam, C., and Franchini, G. Human T-cell lymphotropic/leukemia virus type 1 tax abrogates p53-induced cell cycle arrest and apoptosis through its CREB/ATF functional domain [In Process Citation]. *J. Virol.*, **72**, 8852–8860 (1998).
21. Nerenberg, M., Hinrichs, S. H., Reynolds, R. K., Khoury, G., and Jay, G. The tat gene of human T lymphotropic virus type I induces mesenchymal tumors in transgenic mice. *Science*, **237**, 1324–1329 (1987).
22. Neuveut, C., Low, K. G., Maldarelli, F., Schmitt, I., Majone, F., Grassmann, R., and Jeang, K. T. Human T-cell leukemia virus type 1 Tax and cell cycle progression: role of cyclin D-cdk and p110Rb. *Mol. Cell. Biol.*, **18**, 3620–3632 (1998).
23. Philpott, S. and Buehring, G. Defective DNA repair in cell with human T-cell leukemia/bovine leukemia viruses: role of tax gene. *J. Natl. Cancer Inst.*, **91**, 933–942 (1999).
24. Seiki, M., Eddy, R., Shows, T. B., and Yoshida, M. Nonspecific integration of the HTLV provirus genome into adult T-cell leukaemia cells. *Nature*, **309**, 640–642 (1984).
25. Seiki, M., Hattori, S., Hirayama, Y., and Yoshida, M. Human adult T-cell leukemia virus: complete nucleotide sequence of the provirus genome integrated in leukemia cell

DNA. *Proc. Natl. Acad. Sci. USA*, **80**, 3618–3622 (1983).
26. Sodroski, J. G., Rosen, C. A., and Haseltine, W. A. Trans-acting transcriptional activation of the long terminal repeat of human T lymphotropic viruses in infected cells. *Science*, **225**, 381–385 (1984).
27. Suzuki, T., Fujisawa, J. I., Toita, M., and Yoshida, M. The trans-activator tax of human T-cell leukemia virus type 1 (HTLV-1) interacts with cAMP-responsive element (CRE) binding and CRE modulator proteins that bind to the 21-base-pair enhancer of HTLV-1. *Proc. Natl. Acad. Sci. USA*, **90**, 610–614 (1993).
28. Suzuki, T., Hirai, H., Murakami, T., and Yoshida, M. Tax protein of HTLV-1 destabilizes the complexes of NF-kappa B and I kappa B-alpha and induces nuclear translocation of NF-kappa B for transcriptional activation. *Oncogene*, **10**, 1199–1207 (1995).
29. Suzuki, T., Hirai, H., and Yoshida, M. Tax protein of HTLV-1 interacts with the Rel homology domain of NF-kappa B p65 and c-Rel proteins bound to the NF-kappa B binding site and activates transcription. *Oncogene*, **9**, 3099–3105 (1994).
30. Suzuki, T., Kitao, S., Matsushime, H., and Yoshida, M. HTLV-1 Tax protein interacts with cyclin-dependent kinase inhibitor p16INK4A and counteracts its inhibitory activity towards CDK4. *EMBO J.*, **15**, 1607–1614 (1996).
31. Suzuki, T., Narita, T., Uchida-Toita, M., and Yoshida, M. Down-regulation of the INK4 family of cyclin-dependent kinase inhibitors by tax protein of HTLV-1 through two distinct mechanisms. *Virology*, **259**, 384–391 (1999).
32. Suzuki, T., Ohsugi, Y., Uchida-Toita, M., Akiyama, M., and Yoshida, M. Tax oncoprotein of HTLV-1 binds to the human homologue of Drosophila discs large tumor suppressor protein, hDLG, and perturbs its function in cell growth control. *Oncogene*, **18**, 5967–5972 (1999).
33. Suzuki, T., Uchida-Toita, M., Andoh, T., and Yoshida, M. HTLV-1 tax oncoprotein binds to DNA topoisomerase I and inhibits its catalytic activity. *Virology*, **270**, 291–298 (2000).
34. Suzuki, T., Uchida-Toita, M., and Yoshida, M. Tax protein of HTLV-1 inhibits CBP/p300-mediated transcription by interfering with recruitment of CBP/p300 onto DNA element of E-box or p53 binding site [In Process Citation]. *Oncogene*, **18**, 4137–4143 (1999).
35. Tanaka, A., Takahashi, C., Yamaoka, S., Nosaka, T., Maki, M., and Hatanaka, M. Oncogenic transformation by the tax gene of human T-cell leukemia virus type I *in vitro*. *Proc. Natl. Acad. Sci. USA*, **87**, 1071–1075 (1990).
36. Tsukahara, T., Kannagi, M., Ohashi, T., Kato, H., Arai, M., Nunez, G., Iwanaga, Y., Yamamoto, N., Ohtani, K., and Fujii, M. Induction of Bcl-x(L) expression by human T-cell leukemia virus type 1 Tax through NF-kappaB in apoptosis-resistant T-cell transfectants with Tax. *J. Virol.*, **73**, 7981–7987 (1999).
37. Yoshida, M. HTLV-1 oncoprotein Tax deregulates transcription of cellular genes through multiple mechanisms. *J. Cancer Res. Clin. Oncol.*, **121**, 521–528 (1995).
38. Yoshida, M., Seiki, M., Yamaguchi, K., and Takatsuki, K. Monoclonal integration of human T-cell leukemia provirus in all primary tumors of adult T-cell leukemia suggests causative role of human T-cell leukemia virus in the disease. *Proc. Natl. Acad. Sci. USA*, **81**, 2534–2537 (1984).

ON THE EARLY PERIOD OF THE DISCOVERY OF HUMAN RETROVIRUSES AND LESSONS FOR THE FUTURE

Robert C. GALLO

Institute of Human Virology, University of Maryland Biotechnology Institute[*1] *and Department of Microbiology and Immunology, School of Medicine, University of Maryland*[*2]

Background

I am pleased to provide personal comments on the early history of human retroviruses and the possible lessons that research on them may have for students.

We have known of viruses for a little over one century, and in the earliest decades, of course, they were not characterized. We know them in that period mainly as agents too tiny to see by the standard microscopy of the day, but still often able to produce disease when inoculated into animals in ultrafiltrates derived from extracts of some biological specimens. It is interesting that in retrospect retroviruses were among the earliest known viruses. However, in those early days they were usually described as tumor viruses or oncornaviruses, and when biochemical techniques could be exploited the term RNA tumor viruses came into play, even though it was soon appreciated that this kind of virus also included viruses that apparently never caused animal tumors but instead non-neoplastic diseases. Other names were also used, especially names like types A, B, C and D, that derived from the morphology of the variant forms of these viruses as seen by the technique of electron microscopy. Electron microscopy, of course, greatly accelerated the pace of virological research and was a mainstay of virology laboratories from the middle of the last century until the 1980s when biochemical, molecular biological, and immunological techniques made electron microscopy far less useful and more the specialty of the experts in fine structure.

By the early 1970s the name retrovirus became the standard and official name. This occurred when the late Howard Temin's provirus theory was verified showing that the RNA of these viruses replicated through an integrated DNA form (the provirus), a phenomenon mediated by a special DNA polymerase discovered by Mizutani and Temin and by Baltimore that catalyzed the transcription of the viral RNA genome into DNA upon entry into the cell. The enzyme was eventually called *reverse transcriptase* (*1*, *2*).

From the early part of the 20th century up until about the mid-1970s there was a strong consideration that viruses of a few types, including (perhaps especially so) retroviruses, might cause various human cancers. In response to this the National Cancer Institute in Bethesda established the Virus Cancer Program. There was also a strong suspicion that a human retrovirus (or several of them) might be important candidates for causing some cancers. During most of those years there was also respect for the possibility of the emergence of new and serious unexpected epidemic diseases.

[*1] 15825 Shady Grove Road, Suite 55, Rockville, MD 20850, USA
[*2] 725W Lombard Street, Baltimore, MD 21201, USA

However, insidiously by the late 1970s, attitudes had changed. In the US it seems as if the idea of serious epidemics were thought of as things of historical interest or for the "developing" world only. (It had been a few decades since polio and over 5 decades since the great deadly influenza epidemic.) Evidence for this is illustrated by the fact that some US medical schools terminated departments of microbiology during this period. Similarly, viruses as a cause or co-factors in cancer was an idea that was now dismissed, as evidenced by a Cold Spring Harbor meeting on cancer origins that essentially drew that conclusion. But by this same period, no idea was more suppressed and ridiculed than the notion of human retroviruses. The very idea that they could even exist was fiercely attacked and there were some valid reasons for this, including failure of many scientific groups to find them after a few decades of searching, the known fact that human serum/complement lysed most animal retroviruses, and that in the known animal retrovirus models they replicated so sufficiently that they were readily detectable. However, within a few years (by 1980–81) all three of these bases were shattered:

(1) human retroviruses (human T-cell leukemia viruses, HTLVs) were discovered and linked to a human leukemia and later to neurological disease;
(2) viruses were shown to be involved in about 20% of all human cancer; and
(3) the greatest epidemic in medical history suddenly emerged and ironically was caused by another retrovirus (human immunodeficiency virus, HIV).

The lessons from this for the future appear clear. We need to remember and to teach medical history because epidemic diseases that often disappeared may suddenly and unexpectedly re-emerge, and we should always be vigilant of this possibility. Secondly, we should not always think that highly selected animal models would always predict for disease in an outbred population like humans.

Preparation for the Discoveries of Human Retroviruses

Finding the first human retroviruses was difficult for many reasons. Among the difficulties was the very low replication of HTLV-I and HTLV-II, and the scarcity of the disease (adult T-cell leukemia, ATL) in the US. As we will see later, HIV was a much easier task because of its much greater replication, HTLV-I had proved the existence of human retroviruses, and because of the technology developed for its discovery, HTLV-I provided a clear path to the discovery of other viruses of the subfamily, Retroviridae such as HIV-I.

In 1970, I made the decision to devote a substantial percentage of our laboratory to the finding of human retroviruses. The approach I promoted was based on a few premises: first, that human retroviruses might cause disease with very little virus replication, and that the common animal models were misleading because they were, quite naturally, selected for high disease and its accompanying high virus replication. Consequently, I decided we would need very sensitive techniques to find them. We based our hopes on the enzyme assay for reverse transcriptase (RT). We spent several years (1970–1978) making this assay as sensitive and as specific as possible (*3–10*). It was important that if RT assays were to be used to detect a retrovirus the assay could distinguish RT from each of the cellular DNA polymerases.

The second premise I promoted was that we should not rely only on permanent cell lines. To grow the virus it was generally assumed one would have to transmit it to some animal or human cell line. Because of prior experience, we worried about the confusion that might arise if those cell lines harbored some animal virus. Therefore, I emphasized the need for growing different types of primary human blood cells. Our first hints of a retrovirus came with a myeloid (granulocytic leukemia); so I was particularly concerned with growing these kinds of cells in liquid culture. We searched for growth factors that would enable us to do so, and during those empirical studies we found a growth factor for T-cells that eventually would be known as interleukin-2 or IL-2 (*11–13*). Key contributors to this discovery were my co-workers Doris Morgan and Alan Wu. However, the full development of the discovery was made possible by Frank Ruscetti's work. Like any growth factor we found it was tested on leukemia cells, in this case, of course, T-cells. During this period, a cell line was established from a human T-cell malignancy by A. Gazdar and his co-workers with the help of IL-2 provided by our group. A very dedicated post-doctoral-fellow, Bernie Poiesz, had just joined me, and applying these techniques detected a retrovirus in these cells. However, for me this marked only the beginning of what would be the discovery of HTLV-I. My third premise for claiming human retrovirus was that we must also provide the following: have technology to distinguish any candidate human virus from an animal virus contamination, a problem that had plagued the field (including myself) previously. We must also have immune assays that would enable us to prove the virus was not "a one time event" but existed in the human population to some extent. Further, I recognized that we needed to find the virus in primary fresh tumor cells from this patient and not only in a cell line established previously from a tumor biopsy specimen from him, and finally to provide evidence that this virus caused this leukemia. These objectives were readily and rapidly achieved because our laboratory was so prepared, and later they were applied again to AIDS and HIV. Key contributions in this respect came at this time from Drs. Marjorie Guroff, Carl Saxinger, Marvin Reitz, Phil Markham, M. Sarngadharan and subsequently by Mika Popovic, Flossie Wong-Staal, Veffa Franchini and V. Kalyanaraman, and outside our laboratory from an epidemiological study of William Blattner and virological and molecular studies of Max Essex and William Haseltine. Soon thereafter we obtained a second and a third isolate, and we began to become a little suspicious about their geographic origin and of the race of the patients. We were ready to publish, and our first reports came out in 1980 and 1981 (*14–21*).

Information and Help from Japan

Stimulated by discussion with the late Yohei Ito of Kyoto, my close friend and collaborator, we turned some of our attention to Japan. Ito told us of work by Professor Takatsuki and his co-worker showing clustering of a very special ATL in parts of Japan, especially in its South (*27*). He obtained serum from some patients that we quickly found were positive. It soon became clear that HTLV was endemic in Japan and that the sporadic disease we saw in the US was the same as ATL in Japan. We also knew from our epidemiologist collaborators, in particular William Blattner and his co-workers, and from discussions held in London with Dr. Daniel Catovsky and several other British colleagues, that the same disease was also endemic in some of the Caribbean Islands. Indeed, this meeting in 1981 was a "punctuation mark" in our understanding of the HTLV-I link to this particular kind of T-cell malignancy. The diagnosis of the first US case from whom we isolated HTLV-I was called cutaneous T-cell lymphoma by our clinical collaborators. The meeting with Catovsky and colleagues clarified for us that this disease was identical to Takatsuki's ATL, and explained why most of our patients had a clinical picture not simply of a lymphoma but more frequently a leukemic manifestation. The discussions at the meeting also explained our finding HTLV-I in certain African-Americans. Soon it became clear that sex, blood exposure and mother-to-child were the means of transmission, that ATL (and HTLV-I) were not only endemic in parts of Japan but also among certain African tribes and their descendants (*22–26*).

Ito arranged a meeting near Kyoto with Japanese scientists. We hoped for collaborations. We learned that Japanese scientists, led by Yorio Hinuma, had independently found a retrovirus, and had serological data of its clustering in areas of high frequency of ATL. They would soon report their independent results after our prior publications (*29*).

They called their isolate ATLV. Later, collaborative studies among our group and Yoshida and others proved ATLV and HTLV-I were the same. Of course, I cannot know the details of the contributions to this work from Japan. That is described by others in this journal, but it is nonetheless clear to me that major advances in addition to Hinuma's isolate included: the establishment of HTLV immortalization of T-cells by Miyoshi (*30*), the pioneering molecular biology of Yoshida (*31*), and, of course, Takatsuki's defining the disease, ATL (*27*).

Soon we found a distinct but highly related new retrovirus from a patient with "hairy cell" T-cell leukemia. We called it HTLV-II (*31*). However, HTLV-II has not been linked to any disease in a convincing manner. Shortly after, Antoinne Gessain and Guy de Thé and co-workers in France showed that HTLV-I was also the cause of a serious neurological disease (tropical spastic paraperesis), and this was verified and extended by the important studies of Osame and others in Japan (*33–35*).

What was very difficult about the HTLVs were their discovery and isolation. Linking them to ATL as causative agents was not so difficult for three reasons: (1) retroviruses were well known causes of leukemia of animals; therefore, finding one in association with a human leukemia in and of itself was presumptive evidence; (2) because HTLV-I is a very uncommon human infection and, when found in a particular leukemia with regularity, the likelihood of causation is high; and (3) because molecular studies can be (and were) carried out (chiefly by Yoshida in Japan but also Flossie Wong-Staal and Marvin Reitz in our group) which essentially proved their causative role.

Exactly the opposite is true for HIV and AIDS. As we will see, finding it is (was) not so hard. Showing it to be the causative agent of AIDS was associated with some difficulties.

Legacies of HTLV Discoveries

Soon HTLV research would be greatly overshadowed by the new pandemic of AIDS and the retrovirus—HIV.

Of course, this is because AIDS is so much more of a threat to the world due to its far greater replication, infectivity, and proportion of infected people who succumb to it. However, in no way does this lessen the reality that the early years of the science around HTLV were far more difficult than for HIV. The lasting contributions of HTLV-I research in my mind (other than HTLV-I being the historical first discovered human retrovirus) are likely to be: (1) helping in defining a molecular basis of a leukemia and of a chronic neurological disease; (2) practical usefulness as an agent for immortalizing human T-cells; and (3) helping the discovery of HIV by techniques and by proving the existence of human retroviruses.

HIV/AIDS: the Early History of the Discovery, Rapid Advances, and Controversies

AIDS was recognized for the first time in 1981 by US clinicians and defined as an immune deficiency characterized by a progressive decline in CD4 T-cells, opportunistic infections and often accompanied by a strange and previously rare tumor known as Kaposi sarcoma. US epidemiologists led by the CDC and in particular by its leader on AIDS, James Curran, quickly suggested its probable infectious origin, defined risk groups, and highlighted modes of transmission as by blood, sex, and mother to fetus/infant. However, many theories including non-infectious ones widely circulated. In 1982 Max Essex and I suggested that the most likely cause was a new retrovirus, presuming it would be one highly related to HTLV-I and HTLV-II. These ideas came out of two considerations: (1) Essex's knowledge and experiences with feline leukemia virus (FeLV), e.g. William Jarrett and his group had shown years earlier that variants of FeLV could cause immune suppression; and (2) the experience with HTLV-I such as its T-cell tropisim, modes of transmission, slight capacity to induce immune impairment, and prevalence in Africa. I went further and proposed it would likely be a recombinant virus with an HTLV-I gag-pol region and a novel 3′ region.

There is no doubt that these ideas promoted our work and the experiments soon to begin in France, and there is no doubt that these ideas were the ones that were productive. However, paradoxically though the cause of AIDS was a new retrovirus, it was not one closely related to the HTLVs. There is also no doubt that I held too long to the idea that HIV was likely a member of the HTLV "family."

The technologies used for the frequent isolation of HTLV-I were well exploited for the isolation of HIV by all groups. Again the techniques were sensitive RT assays and culture of the patient's T-cells with IL-2. This led to the first isolation in 1983 of HIV from a patient with lymph gland swelling by the French group led by Luc Montagnier and Jean-Claude Chermann. They later called the virus LAV. The virus was not then shown to be linked to AIDS, and they had difficulties in promoting its growth in permanent cell culture. In the spring of 1984 we published a series of papers (*37–40*) which included the first continuous cell-line propagation of the virus allowing its large-scale production and, therefore, its more detailed characterization. These papers also included the development of a sensitive and specific blood test for the virus that was rapidly brought to nations all over the world after its adaptation by pharmaceutical and biotechnology companies. This test made the blood supplies safe by the beginning of 1985 and allowed the epidemic to be followed by public health officials for the first time. Before this success the disease could only be detected when clinical signs appeared, which we would soon learn was usually 5 to 10 years after infection. Finally, these papers included the claim that the new virus was the cause of AIDS based on frequent virus detection and/or isolation from AIDS patients and risk groups (a total of 48), and larger sero-epidemiological results based on the blood test.

Because our test worked and because some of our isolates could be continuously produced, the patents were awarded to the NIH (US). Montagnier *et al.* had also filed a patent, but this test detected only 17% to 20% of patients with AIDS. Montagnier also reported his success in producing HIV in the cell line called BJAB, a B-cell line. However, this was never borne out, and it is highly unlikely that HIV can be significantly produced by B-cells.

When we published, we suspected that our isolates and the one from France would be the same type. Molecular analysis revealed that to be the case.

A terrible period soon ensued because of a lawsuit over the blood test patents and trial by the press led by a highly biased reporter for a Chicago newspaper who targeted my colleagues and me. This was elevated to sensationalism by the unfounded allegation of misappropriation of the virus due to the finding that one of our many isolates had been accidentally cross contaminated with a virus sent to us from France. Years later we would all learn that the same HIV strain had also been an unknowing contaminant of Montagnier's isolate. In other words, Montagnier believed he was sending us his original isolate, and we believed that to be the case. Instead, he sent us another strain of HIV that contaminated his culture, and its behavior was much different from the properties of the LAV strain of HIV, thus causing confusion. Still later the same strain contaminated isolates of several other labs. The fallout of this was an orchestrated effort to renegotiate a royalty settlement between NIH (US) and Pasteur Institute (France) and among various lawyers. This greatly delayed lifesaving advances in HIV research and damaged reputations.

Once the financial settlement was reached between the US (NIH) and France (Pasteur Institute) and once the lawyers in NY representing the French government were paid, this horrifying period abruptly ended. What are the lessons from this for the future? I would say that a scientist should be more assertive in taking charge of matters that could affect his life. If I had been aggressive in promoting a sharing of royalties from the start, then all of this could have been avoided.

Progress in HIV/AIDS Research and Future

Some have called the years (1983–85) the fastest progress in the science of a new disease in medical history. This need not be reviewed here. We can simply summarize by stating that we know as much or more about HIV and AIDS as we know about any microbe and any disease in medicine, and I refer the reader to my book, Virus Hunting AIDS, Cancer & The Human Retrovirus: A Story of Scientific Discovery (Basic Books, publishers, New York, 1991).

The essence of HIV pathogenesis is the impairment and killing of many infected cells as well as uninfected cells of the immune system. The molecular bases for these events are multiple and still being clarified. The increased incidence of some cancers appears to be secondary to HIV enhancement of replication of viruses that can cause tumors. Many of the other abnormalities involve hyper-activation of the immune system, but the most important mechanism(s) for the hyper-activation is not clear.

Current chemical anti-HIV therapy has been a great advance, but it is too limited (targeting only RT or protease), and its lifelong need leads to toxicity, virus resistance, compliance problems, and difficulties in reaching the developing world. We think new, and whenever possible, more biological forms of therapy are needed, and I think that these will come from studies aimed at (1) blockage of HIV entry, (2) reduction in the hyperactivity of immune cells, (3) inhibiting extracellular "toxins" which we think include some viral proteins as well as some over-produced cytokines, and (4) learning the mechanisms of all the naturally occurring HIV suppressive factors and imitating their effects, e.g., our finding that certain chemokines block HIV infection in 1995 helped open a new avenue of therapy (*41*).

The major goal of much of HIV/AIDS research today is, of course, the development of a preventive vaccine. Our premises are: (1) that we must continue to strive for sterilizing immunity and (2) that sterilizing immunity will have to include neutralizing antibodies with broad effects on various HIV clades (a major scientific challenge). This should provide enough challenges for the young investigator entering this field.

Closure

I want to use this opportunity to congratulate the many Japanese scientists who contributed brilliantly to the opening "Chapter" of human retrovirology. I wish them continued great success, and I hope our scientific paths will cross many more times in the future.

REFERENCES

Selected Personal References

The discovery of reverse transcriptase

1. Baltimore, D. RNA-dependent DNA polymerase in virions of RNA tumor viruses. *Nature*, **226**, 1209-1211 (1970).
2. Temin, H. M. and Mizutani, S. RNA-dependent DNA polymerase in virions of Rous sarcoma virus. *Nature*, **226**, 1211–1213 (1970).
3. Bobrow, S. N., Smith, R. G., Reitz, M. S., and Gallo, R. C. Stimulated normal human lymphocytes contain a ribonuclease-sensitive DNA polymerase distinct from viral RNA-directed DNA polymerase. *Proc. Natl. Acad. Sci. USA*, **69**, 3228–3232 (1972).
4. Sarngadharan, M. G., Sarin, P. S., Reitz, M. S., and Gallo, R. C. Reverse transcriptase activity of human acute leukemic cells: purification of the enzyme, response to AMV 70S RNA, and characterization of the DNA product. *Nature New Biol.*, **240**, 67–72 (1972).
5. Smith, R. G. and Gallo, R. C. DNA-dependent DNA polymerase I and II from normal human-blood lymphocytes. *Proc. Natl. Acad. Sci. USA*, **69**, 2879–2884 (1972).
6. Abrell, J. W. and Gallo, R. C. Purification, characterization, and comparison of the DNA polymerases from two primate RNA tumor viruses, *J. Virol.*, **12**, 431–439 (1973).

Examples of the development of sensitive and specific assays for reverse transcriptase for applicability to finding a human retrovirus

7. Todaro, G. J. and Gallo, R. C. Immunological relationship of DNA polymerase from human acute leukaemia cells and primate and mouse leukaemia virus reverse transcriptase. *Nature*, **244**, 206–209 (1973).
8. Reitz, M. S., Jr., Smith, R. C., Roseberry, E. A., and Gallo, R. C. DNA-directed and RNA-primed DNA synthesis in microsomal and mitochondrial fractions of normal human lymphocytes. *Biochem. Biophys. Res. Commun.*, **57**, 934–948 (1974).
9. Smith, R. G., Abrell, J. W., Lewis, B. J., and Gallo, R. C. Serologic analysis of human

deoxyribonucleic acid polymerases. *J. Biol. Chem.*, **250**, 1702–1709 (1975).

10. Robert-Guroff, M. and Gallo, R. C. Serological analysis of cellular and viral DNA polymerases by an antiserum of DNA polymerase γ of human lymphoblasts. *Biochemistry,* **16**, 2874–2880 (1977).

The discovery of interleukin-2

11. Morgan, D. A., Ruscetti, F. W., and Gallo, R. C. Selective *in vitro* growth of T lymphocytes from normal human bone marrows. *Science,* **193**, 1007–1008 (1976).
12. Ruscetti, F. W., Morgan, D. A., and Gallo, R. C. Functional and morphologic characterization of human T cells continuously grown *in vitro. J. Immunol.*, **119**, 131–138 (1977).
13. Mier, J. W. and Gallo, R. C. Purification and some characteristics of human T-cell growth factor from phytohemagglutinin-stimulated lymphocyte-conditioned media. *Proc. Natl. Acad. Sci. USA,* **77**, 6134–6138 (1980).

The discovery and early characterization of HTLV-I in the US

14. Poiesz, B. J., Ruscetti, F. W., Gazdar, A. F., Bunn, P. A., Minna, J. D., and Gallo, R. C. Detection and isolation of type C retrovirus particles from fresh and cultured lymphocytes of a patient with cutaneous T-cell lymphoma. *Proc. Natl. Acad. Sci. USA,* **77**, 7415–7419 (1980).
15. Kalyanaraman, V. S., Sarngadharan, M. G., Bunn, P. A., Minna, J. D., and Gallo, R. C. Antibodies in human sera reactive against an internal structural protein of human T-cell lymphoma virus. *Nature,* **294**, 271–273 (1981).
16. Kalyanaraman, V. S., Sarngadharan, M. G., Poiesz, B. J., Ruscetti, F. W., and Gallo, R. C. Immunological properties of a type C retrovirus isolated from cultured human T-lymphoma cells and comparison to other mammalian retroviruses. *J. Virol.*, **38**, 906–915 (1981).
17. Reitz, M. S., Jr., Poiesz, B. J., Ruscetti, F. W., and Gallo, R. C. Characterization and distribution of nucleic acid sequences of a novel type C retrovirus isolated from neoplastic human T lymphocytes. *Proc. Natl. Acad. Sci. USA,* **78**, 1887–1891 (1981).
18. Rho, H. M., Poiesz, B., Ruscetti, F. W., and Gallo, R. C. Characterization of the reverse transcriptase from a new retrovirus (HTLV) produced by a human cutaneous T-cell lymphoma cell line. *Virology,* **112**, 355–360 (1981).
19. Robert-Guroff, M., Ruscetti, F. W., Posner, L. E., Poiesz, B. J., and Gallo, R. C. Detection of the human T cell lymphoma virus p19 in cells of some patients with cutaneous T-cell lymphoma and leukemia using a monoclonal antibody. *J. Exp. Med.*, **154**, 1957–1964 (1981).
20. Poiesz, B. J., Ruscetti, F. W., Reitz, M. S., Jr., Kalyanaraman, V. S., and Gallo, R. C. Isolation of a new type C retrovirus (HTLV) in primary uncultured cells of a patient with Sezary T-cell leukemia. *Nature,* **294**, 268–271 (1981).
21. Kalyanaraman, V. S., Sarngadharan, M. G., Nakao, Y., Ito, Y., Aoki, T., and Gallo, R. C. Natural antibodies to the structural core protein (p24) of the human T-cell leukemia (lymphoma) retrovirus found in sera of leukemia patients in Japan. *Proc. Natl. Acad. Sci. USA,* **79**, 1653–1657 (1982).

Linkage of HTLV-I to the cause of ATL

22. Blattner, W. A., Kalyanaraman, V. S., Robert-Guroff, M., Lister, T. A., Galton, D. A., Sarin, P. S., Crawford, M. H., Catovsky, D., Greaves, M., and Gallo, R. C. The human type-C retrovirus, HTLV, in Blacks from the Caribbean region, and relationship to

adult T-cell leukemia/lymphoma. *Int. J. Cancer*, **30**, 257–264 (1982).
(For more details on this report see: "Virus Hunting AIDS, Cancer & the Human Retrovirus: A Story of Scientific Discovery," Robert Gallo, Basic Books—A Subsidiary of Perseus Books Group, L. L. C., New York, 1991.)

Examples of linkage of HTLV-I to T-cell leukemia/lymphoma (ATL) as the causative agent US and US-British studies

23. Kalyanaraman, V. S., Sarngadharan, M. G., Bunn, P. A., Minna, J. D., and Gallo, R. C. Antibodies in human sera reactive against an internal structural protein of human T-cell lymphoma virus. *Nature*, **294**, 271–273 (1981).
24. Robert-Guroff, M., Ruscetti, F. W., Posner, L. E., Poiesz, B. J., and Gallo, R. C. Detection of the human T cell lymphoma virus p19 in cells of some patients with cutaneous T cell lymphoma and leukemia using a monoclonal antibody. *J. Exp. Med.*, **154**, 1957–1964 (1981).
25. Catovsky, D., Greaves, M. F., Rose, M., Galton, D. A. G., Goolden, A. W. G., McCluskey, D. R., White, J. M., Lampert, I., Bourikas, G., Ireland, R., Brownell, A. I., Bridges, J. W., Blattner, W. A., and Gallo, R. C. Adult T-cell lymphoma-leukaemia in blacks from the West Indies. *Lancet*, **i**, 639–643 (1982).
26. Kalyanaraman, V. S., Sarngadharan, M. G., Nakao, Y., Ito, Y., Aoki, T., and Gallo, R. C. Natural antibodies of the structural core protein (p24) of the human T-cell leukemia (lymphoma) retrovirus found in sera of leukemia patients in Japan. *Proc. Natl. Acad. Sci. USA*, **79**, 1653–1657 (1982).

Selected References of Major Early Contributions from Japan on ATL and HTLV-I

Definition of disease

27. Takatsuki, K., Uchiyama, T., Sagawa, K., and Yodoi, J. Adult T cell leukemia in Japan (Abstract). *In* "Topics in Hematology," No. 1825, ed. S. Seno, F. Takaku, and S. Irino, pp. 73–78 (1976). Excerpta Medica Amsterdam, Amsterdam.
28. Uchiyama, T., Yodoi, J., Sagawa, K., Takatsuki, K., and Uchino, H. Adult T-cell leukemia: clinical and hematologic features of 16 cases. *Blood*, **50**, 481–492 (1977).

First isolate of HTLV-I from Japan

29. Hinuma, Y., Nagata, K., Hanaoka, M., Nakai, M., Matsumoto, T., Kinoshita, K. I., Shirakawa, S., and Miyoshi, I. Adult T-cell leukemia: antigen in an ATL cell line and detection of antibodies of the antigen in human sera. *Proc. Natl. Acad. Sci. USA*, **78**, 6476–6480 (1981).

T-cells immortalized by HTLV-I

30. Miyoshi, I., Kubonishi, I., Yoshimoto, S., Akagi, T., Ohtsuki, Y., Shiraishi, Y., Nagata, K., and Hinuma, Y. Type C virus particles in a cord T-cell line derived by co-cultivating normal human cord leukocytes and human leukaemic T cells. *Nature*, **294**, 770–771 (1981).
31. Yoshida, M., Seiki, M., Yamaguchi, K., and Takatsuki, K. Monoclonal integration of human T-cell leukemia provirus in all primary tumors of adult T-cell leukemia suggests causative role of human T-cell leukemia virus in the disease. *Proc. Natl. Acad. Sci. USA*, **81**, 2534–2537 (1984).

The discovery of HTLV-II

32. Kalyanaraman, V. S., Sarngadharan, M. G., Robert-Guroff, M., Miyoshi, I., Blayney, D., Golde, D., and Gallo, R. C. A new subtype of human T-cell leukemia virus

(HTLV-II) associated with a T-cell variant of hairy cell leukemia. *Science*, **218**, 571–573 (1982).

HTLV-I and neurologic disease

33. Gessain, A., Barin, F., Vernant, J. C., Gout, O., Maurs, L., Calender, A., and de The, G. Antibodies to human T-lymphotropic virus type I in patients with tropical spastic paraparesis. *Lancet*, **ii**, 407–410 (1975).
34. Osame, M., Usuku, K., Izumo, S., Ijichi, N., Amitani, H., Igata, A., Matsumoto, M., and Tara, M. HTLV-I associated myelopathy, a new clinical entity. *Lancet*, **i**, 1031–1032 (1986).
35. Osame, M., Matsumoto, M., Usuku, K. *et al.* Chronic progressive myelopathy associated with elevated antibodies of human T-lymphotropic virus type I and adult T-cell leukemia like cells. *Ann. Neurol.*, **21**, 117–122 (1987).

The first isolation of HIV

36. Barre-Sinoussi, F., Chermann, J.-C., Rey, F., Mugeyre, M. T., Chamaret, S., Gruest, J., Daugret, C., Axler-Blin, C., Vezinet-Brun, F., Rouzioux, C., Rozenbaum, W., and Montagnier, L. Isolation of a T-lymphotropic retrovirus from a patient at risk for acquired immune deficiency syndrome AIDS. *Science*, **220**, 868–871 (1983).

The first reports linking HIV to AIDS, the next independent isolates of HIV, and the development of a working blood test for HIV

37. Gallo, R. C., Salahuddin, S. Z., Popovic, M., Shearer, G. M., Kaplan, M., Haynes, B. F., Palker, T. J., Redfield, R., Oleske, J., Safai, B., White, G., Foster, P., and Markham, P. D. Frequent detection and isolation of cytopathic retroviruses (HTLV-III) from patients with AIDS and at risk for AIDS. *Science*, **224**, 500–503 (1984).
38. Popovic, M., Sarngadharan, M. G., Read, E., and Gallo, R. C. Detection, isolation, and continuous production of cytopathic retroviruses (HTLV-III) from patients with AIDS and pre-AIDS. *Science*, **224**, 497–500 (1984).
39. Sarngadharan, M. G., Popovic, M., Bruch, L., Schupbach, J., and Gallo, R. C. Antibodies reactive with human T-lymphotropic retroviruses (HTLV-III) in the serum of patients with AIDS. *Science*, **244**, 506–508 (1984).
40. Schupbach, J., Popovic, M., Gilden, R. V., Gonda, M. A., Sarngadharan, M. G., and Gallo, R. C. Serological analysis of a subgroup of human T-lymphotropic retroviruses (HTLV-III) associated with AIDS. *Science*, **244**, 503–505 (1984).

The discovery of naturally occurring HIV suppressive factors

41. Cocchi, F., DeVico, A. L., Garzino-Demo, A., Arya, S. K., Gallo, R. C., and Lusso, P. Identification of RANTES, MIP-1α and MIP-1β as the major HIV-suppressive factors produced by $CD8^+$ T cells. *Science*, **270**, 1811–1815 (1995).

Gann Monograph on Cancer Research 50, 2003

HUMAN RECEPTORS FOR RETROVIRUSES

Robin A. WEISS

*Department of Immunology and Molecular Pathology, University College London**

Retroviruses use specific cell surface receptors in order to bind to and subsequently gain entry into their host cells. Some retroviruses such as human immunodeficiency virus (HIV) utilize one receptor for high affinity binding, and a separate co-receptor to mediate fusion of the viral envelope with the cell membrane. Although many retrovirus receptors have been identified as distinct membrane proteins, the receptor(s) for human T-cell leukemia virus type 1 (HTLV-1) remains to be characterized. The primate T-cell lymphotropic retroviruses recognize a common receptor that is broadly expressed *in vitro* on many cell types of many species. However, the cellular tropism of HTLV-1 *in vivo* is more restricted. Retroviruses can infect cells by membrane fusion at the cell surface and by receptor-mediated endocytosis. Knowledge of the receptors involved has led to practical developments such as inhibitory drugs for HIV and to the design of improved retroviral vectors for gene therapy.

As intracellular parasites, viruses have evolved a variety of mechanisms to gain entry into the host cell. The molecules on the cell membrane to which viruses attach and interact in order to mediate entry are called receptors. Such receptors have natural functions at the cell surface and have been 'hijacked' by the virus to serve as entry gates into the cell. The recognition epitopes on receptors may be carbohydrates borne on glycoproteins and glycolipids, like the sialic acid residues used by influenza viruses, or they may be an integral part of the conformation of cell surface proteins, as is the case for the known retrovirus receptors.

In the 20 years since adult T-cell leukemia (ATL) was linked etiologically with human T-cell leukemia virus type 1 (HTLV-1), much has been elucidated concerning retrovirus receptors. The first to be defined in 1984 was the CD4 antigen for human immunodeficiency virus type 1 (HIV-1) (*13, 24*). The receptor for the γ-retrovirus, ecotropic murine leukemia virus (MLV-E) was found through cDNA cloning in 1989 (*2*) and for gibbon ape leukemia viruses in 1990 (*38*). During the 1990s receptors for several more α-, β- and γ-retroviruses and the HIV co-receptors were identified (*5, 40, 52*). Curiously, despite much interest, the HTLV-1 receptor has not yet been definitively identified (*44*). This is regrettable since a knowledge of the HTLV-1 receptors might throw light upon the pathogenesis of ATL and HTLV-associated myelopathy.

* 46 Cleveland Street, London W1T 4JF, UK

HTLV-1 Adsorption to and Infection of Cells

It is well known that HTLV-1 infection is a cell-associated phenomenon. However, cell-free infection has been demonstrated *in vitro* (*7*), and both vehicles of infection require interaction with specific cell surface receptors. Neutralization of HTLV-1 by antibodies in the sera of infected individuals blocks interaction of the viral glycoproteins with cellular receptors (*41*). The various genotypes of HTLV-1 belong to a single neutralizing serotype, and weak cross-neutralization of HTLV-2 is also seen (*8, 19*).

Because cell-free infection is difficult to achieve and quantify, indirect techniques have been used to investigate HTLV-cell surface interactions. One of these assays is cell-to-cell fusion leading to the induction of syncytia. This occurs when chronically infected, HTLV-producing cells are co-cultivated with uninfected cells bearing appropriate receptors (*20, 35*) or with a reporter assay (*39*). Another useful assay is the generation of pseudotype viruses bearing HTLV envelope antigens. These can be formed on vesicular stomatitis virus (*8*) or on retroviral vectors derived from MLV or HIV (*26, 57, 62*). Syncytium induction and pseudotype infection can be blocked by neutralizing antisera. These functional assays help to determine receptor expression. Binding of recombinant soluble HTLV-1 SU protein, gp46, is also a useful probe for receptors (*22*).

It is a paradox that HTLV-1 can infect a broad range of cell types *in vitro* (*36, 61*), but appears to be much more restricted *in vivo* (*46*). The main target *in vivo* is the CD4-positive helper T-lymphocyte, which is the phenotype of ATL. Monocytes might also be susceptible to infection. The universal expression of HTLV-1 receptors in dividing cells in culture implies that a cell-cycle dependent molecule might be the receptor. However, it is not yet clear that the restriction of infection *in vivo* is related to receptors or to post-penetration regulation of cell permissivity for HTLV-1.

The *in vitro* species range for HTLV-1 infection is also widespread, extending to cultures derived from birds as well as many species of mammal (*36, 50, 54, 57, 61*). This broad host range is reflected in the capacity of HTLV-1 to infect non-human species *in vivo*, when experimentally inoculated into primates, lagomorphs (*31*) and rodents (see papers by Yoshiki and by Kannagi, this monograph). Whatever the cell surface receptor for HTLV-1 may be, it would appear to be well conserved between different species of host.

The pseudotype and syncytium assays for HTLV-receptor interaction have been useful in defining whether related retroviruses utilize the same receptors. This can be accomplished by cross-interference studies, in which cells already productively infected by HTLV-1 become insensitive to superinfection by pseudotypes or to syncytium formation by co-cultivation with virus-producing cells. In this way it was possible to demonstrate that all the primate δ-retroviruses, HTLV-1, HTLV-2 and various strains of simian T-cell leukemia viruses utilize the same receptor (*8, 53*).

In contrast, bovine leukosis virus (BLV) utilizes a different receptor, on human and non-human cells (*53*). A candidate receptor for BLV was reported (*4*) although the murine homolog is not a cell surface protein (*58*). As a well-understood δ-retrovirus infection affecting a domestic animal species, the BLV receptor merits further investigation.

Most murine cell lines are less sensitive to HTLV infection than human cells.

Murine-human somatic cell hybrids, therefore, have been exploited in attempts to map the host gene(s) responsible for determining the HTLV-1 receptors. A locus on the long arm of human chromosome 17 was reported (*16, 54*). However, this finding has not been followed up by fine-mapping of the chromosome region, and other investigators have queried whether this represents the major receptor gene (*23, 47, 57*). Attempts to identify the HTLV receptor gene by expression cloning have been frustrated by the absence of a genuinely receptor-negative cell line for use in transfection assays. Insect cells, however, may prove to be a suitable target (*23*).

A number of candidate receptor molecules for HTLV-1 have been reported based on inhibiting antibodies (*1, 15, 21*). A suggestion that the heat shock cognate protein, Hsc70 is the receptor (*47*), based on inhibition of syncytium by the gp46 peptide P-197 was not confirmed because the same peptide is inhibitory for HTLV-1 on Hsc70 negative cells (*6*). By analogy with HIV-1, it seems likely that several distinct receptors are required for syncytium formation and viral entry, whereas primary high affinity binding of gp46 may need only one type of receptor. This complexity may explain why gp46 binding data are not consistent with syncytium sensitivity, and why there is disagreement on the role of the genetic locus on chromosome 17q encoding a receptor (*23, 57*). Furthermore, cell adhesion molecules such as ICAM-1, ICAM-3 and VCAM-1 (*12, 18*) may be required for syncytium induction, not as specific receptors for HTLV-1, but rather by enabling the cells to adhere closely before virus-mediated cell fusion can occur.

HIV and Other Retrovirus Receptors

A number of specific cell membrane proteins are known to act as receptors for retroviruses (*39, 52*). Figure 1 depicts them in diagrammatical form. The CD4 receptor for HIV is a member of the immunoglobulin superfamily (as are the receptors for polio and common cold viruses).

Soon after the discovery that CD4 is the HIV receptor (*13, 24*) it was found that CD4 is necessary but not sufficient for HIV entry into cells. For example, mouse cells expressing human CD4 were permissive for high affinity binding of HIV-1, but not for fusion or entry (*27*). While all strains of HIV-1, HIV-2 and simian immunodeficiency virus (SIV) recognized CD4 (*48*), different isolates appeared to require distinct secondary receptors in order to undergo fusion and entry (*9*).

The second receptors or co-receptors were eventually identified as chemokine receptors, initially by Berger and colleagues (*3, 14*). Many groups soon showed that several kinds of chemokine receptor could serve as co-receptors to CD4 for HIV entry (*5*). However, *in vivo* only two co-receptors have thus far been shown to be utilized by HIV. Primary isolates of HIV infect macrophages and peripheral blood mononuclear cells (PBMC) of the T-helper lymphocyte lineage. These strains of HIV interact with the CCR5 receptor, which naturally binds the chemokines RANTES, MIP-1α and MIP-1β. Late in the course of HIV infection progressing to AIDS, HIV variants frequently appear that become readily adapted to propagate in immortalized CD4-positive T-cell lines (including HTLV-1 transformed cells such as MT2). These HIV substrains utilize the CXCR4 chemokine receptor as a co-receptor with CD4. The natural ligand for CXCR4 is stromal derived factor (SDF-1). Some primary isolates are dual-tropic and can recognize both CCR5 and CXCR4.

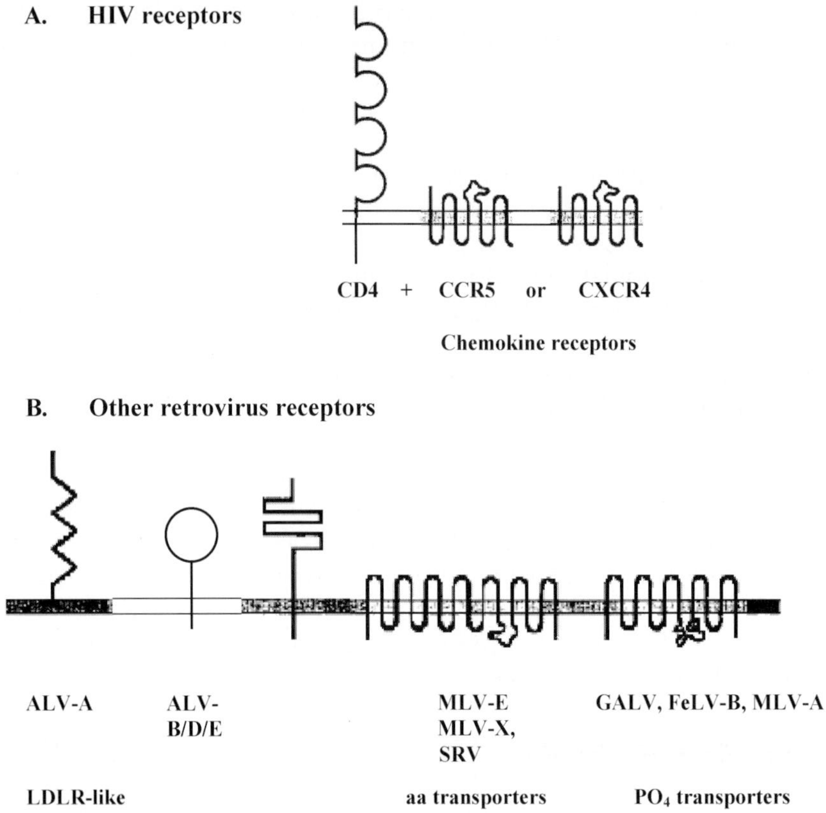

FIG. 1. Schematic diagram of cell surface molecules that serve as receptors for retroviruses
A. Receptors for HIV. B. Receptors for other retroviruses. ALV, avian leukosis virus; BLV, bovine leukosis virus; FeLV, feline leukemia virus; GALV, gibbon ape leukemia virus; MLV, murine leukemia virus; SRV, simian retrovirus (B-type).

As shown in Fig. 1, CD4 has a single membrane spanning domain. On the other hand, the chemokine co-receptors for HIV are seven-transmembrane spanning receptors that belong to the family of G-protein coupled receptors. The receptors for MLV strains and for the feline and gibbon ape leukemia viruses are also multiple spanning molecules that normally serve as transporters for amino acids and for inorganic phosphate. The receptor for avian leukosis virus subgroup A (ALV-A) is related to the low-density lipoprotein receptors. It does not span the cell membrane but is tethered to it by a glycosylphosphatidylinositol anchor. Some of these viruses may possibly use unknown primary or secondary receptors in addition to those already identified.

Retrovirus Attachment, Receptor Interaction and Entry

The first step in virus infection is adsorption to the cell surface. Even before interaction with the specific, high-affinity receptor occurs, adsorption involves extracellular molecules such as glyco-amino-glycans. Heparan has been shown to act as an initial adsorption for HIV onto adherent cell lines (*32*) although it is not strongly expressed on T-lymphocytes. Heparan adsorption may not require viral glycoproteins.

Using direct visualization of virus particles, similar levels of adsorption by MLV to adherent cell lines were observed for virions lacking envelope glycoproteins (*43*).

Binding to the specific virus receptor induces conformational changes in the SU glycoprotein to expose a hydrophobic domain in the TM protein which in turn effects membrane fusion. This series of events can be complex, as exemplified with HIV. The gp120-CD4 interaction induces conformational changes in both molecules allowing access to the co-receptors which in turn expose the fusion domain in gp41 (*33*). These interconnections occur in the context of trimeric gp120/gp41 envelope spikes interacting with more than one CD41 molecule and co-receptor (*5*). The epitopes on receptors that are recognized by the retrovirus envelope glycoproteins have been mapped in several cases. There does not appear to be a common recognition motif on different receptors, except that the binding epitopes fit into pockets on the virion surface. Fine structure data at the crystallographic level has been achieved thus far only for HIV gp120 with CD4 (*25*).

In apparent contrast to HTLV-1, the cellular tropism of HIV is largely determined by the cell surface receptors it uses for binding and entry. HIV infects and eventually destroys T-helper lymphocytes but not T-killer lymphocytes because the T-helper cells express CD4 whereas the cytotoxic T-cells express CD8. Macrophages express lower levels of CD4 and are also susceptible to HIV infection. Other receptors may play a role in HIV adsorption and infection in special circumstances. Certain kinds of dendritic cells (DC), including the Langerhans cells of the genital mucosa, express DC-SIGN, a C-type lectin recognized by HIV gp120. It is thought that this receptor may not permit membrane fusion, but can deliver virus to T-helper cells when taken up by DC cells that migrate to the lymph node (*17*). Another route to adsorption and entry is via Fc and complement receptors (CR) when HIV is coated with antibody. Since macrophages express FcR and CR, these receptors may play a role in antibody-enhancement of HIV infection (*56*). Antibody enhancement via FcR or CR has not been investigated for HTLV-1 infection, but it could play a role in milk transmission.

Thus retrovirus entry into susceptible cells involves a series of events leading the fusion of the viral envelope with a cellular membrane. This results in the delivery of the virus core or replication complex to an appropriate location in the cytoplasm to initiate reverse transcription and subsequent migration to the nucleus. There are two major pathways by which viruses penetrate cell membranes to gain access to the cytoplasm and to commence replication. One is direct fusion of the plasma membrane at the cell surface. This route is typical of paramyxoviruses such as measles virus or simian virus 5. The second route is via receptor-mediated endocytosis, typical of influenza virus and vesicular stomatitis virus. These viruses are dependent on the low pH of endocytic vesicles for conformational changes in their envelope glycoproteins that expose hydrophobic membrane fusion domains. Fusion can be inhibited by raising endosomal pH with agents such as monensin, NH_4Cl and chloroquine.

Studies with retroviruses indicate that they do not require acidic endosomes for infection, with the exception of ecotropic MLV (*28, 29*). For example, Fig. 2 shows that infection by pseudotypes of vesicular stomatitis virus (VSV) with retrovirus envelopes is pH-independent, in contrast to native VSV bearing its own envelope G protein. In general, cell-to-cell fusion occurs at neutral pH, and it is therefore not surprising to see a correlation between sensitivity to syncytium induction and pH-

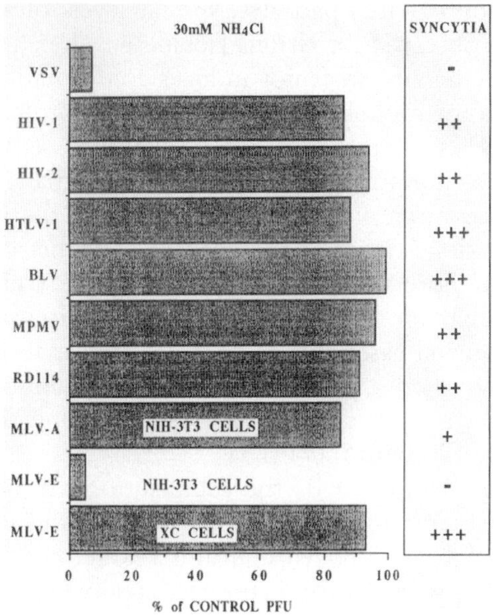

FIG. 2. Sensitivity of VSV pseudotypes with retroviral envelopes to raised endosomal pH
Cells were treated with 30 mM NH_4Cl for 1 h before and 3 h after challenge with virus. The efficiency of plating was calculated from the VSV plaque titer (PFU) compared to that on cells incubated in control medium. The cells were human sarcoma cells (HT1080) except where indicated as rat XC cells or murine NIH-3T3 cells. The sensitivity to cell fusion on co-cultivation with cells producing the retrovirus is also indicated as syncytia; +, 10% nuclei in syncytia; ++, 10–25%; +++ >25%. Whereas VSV in its own envelope was dependent on low endosomal pH, only MLV-E among the retroviral envelope pseudotypes were pH-dependent, for NIH-3T3 cells but not for XC cells. As with NIH-3T3 cells (shown), HT1080 human cells (not shown) were pH-independent for MLV-A entry. Data from McClure *et al.* (*29*).

independence. In the case of ecotropic MLV-E this correlation is consistent even when the cell types in which the assays are performed exhibit different routes of virus entry (NIH-3T3 versus XC, Fig. 2). The avian leukosis virus, ALV-A, undergoes pH-independent binding to its receptor, but this interaction converts the viral glycoprotein to a pH-sensitive state for infection via endosomes (*34*).

Although most retroviruses are pH-independent for infection, they may still be able to enter via endosomes. Figure 3 illustrates HIV-1 particles fusing at the cell surface and in the endosome, and either pathway can result in infection. The route of entry, however, will determine which region of the host cell's cytoplasm the virus reaches. This can influence restrictions on virus replication that occur after envelope fusion but before migration to the nucleus. For instance, a strain of HIV-2 that is restricted for infection of HeLa-CD4 cells could replicate efficiently after delivery to a different cellular compartment by pseudotyping with VSV G protein (*30*). HTLV-1 entry appears to be pH-independent (Fig. 2) although virions are also frequently associated with endocytic vesicles (*60*).

Receptors in Prevention and Therapy

The expression of retroviral receptors can have a profound effect on infection.

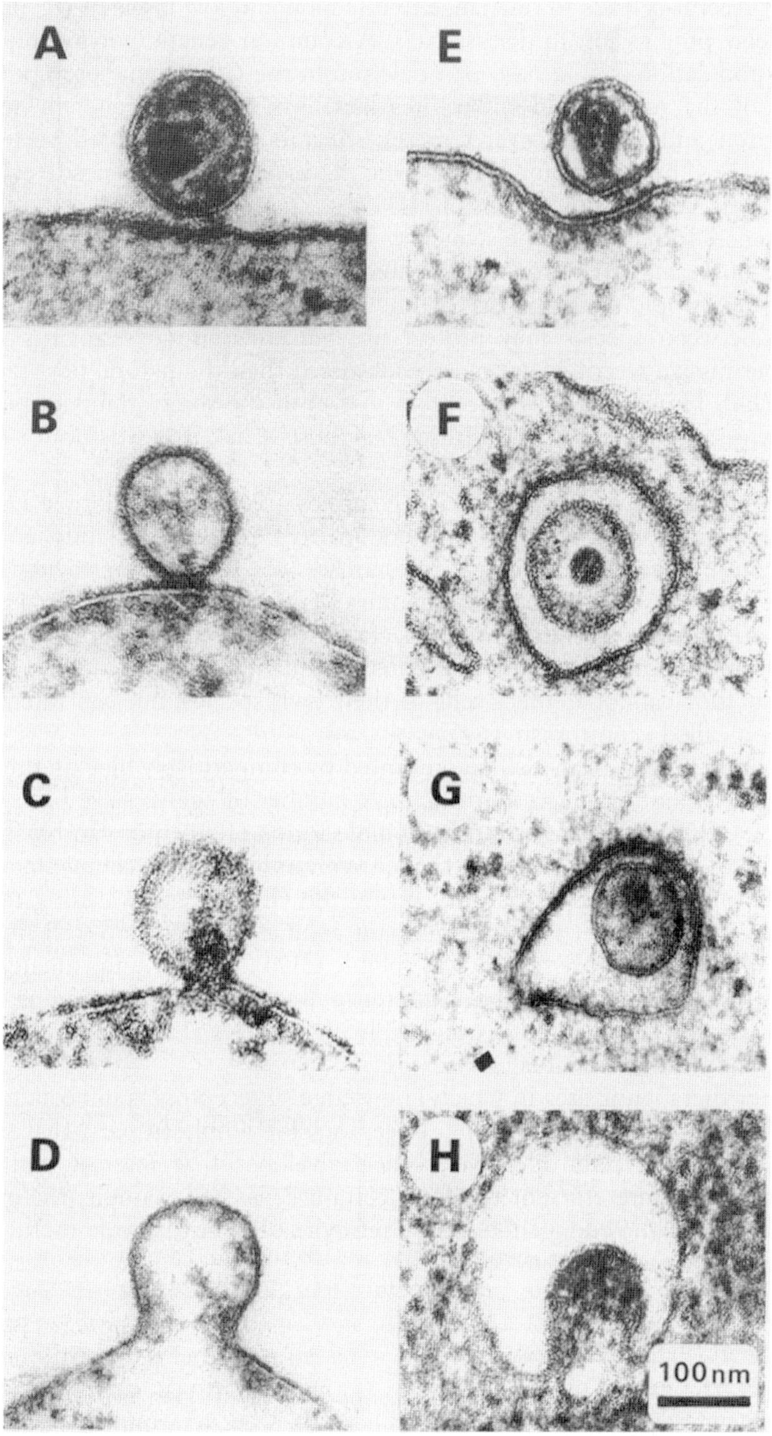

FIG. 3. Electron micrographs showing HIV-1 infection by fusion at the cell surface membrane (A–D) and by receptor mediated endocytosis (E–H)

Reproduced with permission from Weiss (*63*) with micrographs kindly provided by H. Gelderblom.

Absence of receptors leads to resistance to infection and low levels of receptor expression can delay progression to disease (*37*). A common genetic polymorphism in the Caucasian population is a 32 base-pair deletion in the *CCR5* gene preventing surface expression of the receptor. Homozygotes lacking CCR5 are generally resistant to HIV infection and they are over-represented among HIV-exposed yet uninfected people (*5, 37*). The few homozygotes who have become HIV-positive have been infected with CXCR4-using strains of HIV.

Differences in receptor gene sequences between different species helps to explain in part the host range of retroviruses. Thus human cells can be infected by amphotropic and xenotropic strains of MLV but not by ecotropic strains. Receptor polymorphism between species may protect us against zoonotic retrovirus infections, although permissive receptors are more widespread than the natural host range of the retrovirus (*64*). Humans express receptors to two of the three strains of pig endogenous retrovirus (PERV), so these viruses may pose a risk if pig tissues or organs are used in xenotransplantation (*42, 59*).

Endogenous retroviruses are proviral genomes present in the normal DNA of the host that are transmitted from one generation to the next as Mendelian elements. They are common among the simple retroviruses (α-, β- and γ-retroviruses) but are not seen in retroviruses with complex genomes (lenti- and foamy viruses and HTLV). If viremia arises from an endogenous infection, it could be pathogenic. Mutation of the host receptors can prevent viremia. Thus a frequent feature of endogenous retroviruses is that they cannot reinfect cells of their own species, but can infect others, a phenomenon called xenotropism (*64*).

Retrovirus infection can also be prevented by competitively blocking receptors so that the virus cannot gain access to them. A number of agents have been developed that block CCR5 or CXCR4 (*49, 55*). Soluble forms of receptor can block infection by neutralizing the virus either by competitively inhibiting attachment to the cell-bound receptors or by prematurely triggering the viral envelope leading to inactivation. This was first shown with recombinant soluble CD4 (*51*). T-cell line adapted strains of HIV-1 are highly sensitive to soluble CD4 (*10*), but unfortunately primary strains are semi-resistant (*11*). The difference was explained by finding that with CXCR4-using virus, soluble CD4 irreversibly uncouples gp120 from the TM protein, gp41 thus inactivating the virus (*34*).

Another therapeutic area in which knowledge of receptors is important is the use of retroviral vectors in gene therapy. Receptors for amphotropic MLV are widely expressed on human cells. However, other envelopes that interact with different receptors may be more efficient to target certain cells. For example, the gibbon ape leukemia virus and the endogenous feline retrovirus RD114 provide useful envelopes for infecting bone marrow stem cells (*45*).

Conclusions and Prospects

Much knowledge has been recently gained in identifying cell surface receptors for retroviruses. This information has helped our understanding of mechanisms of virus entry into cells and in selected cases, how to block infection. It could be relevant to vaccine design in the future. Where receptor expression correlates with the cellular tropism of the virus, it has helped to explain aspects of retroviral pathogenesis. Cell receptors for retroviruses can be complex, as in HIV infection, where the virus first docks onto CD4, and then infects via either the CCR5 or the CXCR4 co-receptor.

Functional receptors for HTLV-1 and HTLV-2 are expressed on many cell types from many host species. In contrast, the cell tropism for HTLV-1 appears largely to be limited to T-helper cells within the infected person. The molecular nature of the HTLV receptor(s) has yet to be conclusively defined. The renewed interest in identifying the HTLV receptor will be welcomed by those involved in studying HTLV infection and disease.

Acknowledgments

I am most grateful to Dr. Áine McKnight for discussion and comments on this manuscript. The author's research is funded by the Medical Research Council and by Cancer Research UK.

REFERENCES

1. Agadjanyan, M. G., Ugen, K. E., Wang, B., Williams, W. V., and Weiner, D. B. Identification of an 80-kilodalton membrane glycoprotein important for human T-cell leukemia virus type I and type II syncytium formation and infection. *J. Virol.*, **68**, 485–493 (1994).
2. Albritton, L. M., Tseng, L., Scadden, D., and Cunningham, J. M. A putative murine ecotropic retrovirus receptor gene encodes a multiple membrane-spanning protein and confers susceptibility to virus infection. *Cell*, **57**, 659–666 (1989).
3. Alkhatib, G., Combadiere, C., Broder, C. C., Feng, Y., Kennedy, P. E., Murphy, P. M., and Berger, E. A. CC CKR5: a RANTES, MIP-1alpha, MIP-1beta receptor as a fusion cofactor for macrophage-tropic HIV-1. *Science*, **272**, 1955-1958 (1996).
4. Ban, J., Portetelle, D., Altaner, C., Horion, B., Milan, D., Krchnak, V., Burny, A., and Kettmann, R. Isolation and characterization of a 2.3-kilobase-pair cDNA fragment encoding the binding domain of the bovine leukemia virus cell receptor. *J. Virol.*, **67**, 1050–1057 (1993).
5. Berger, E. A., Murphy, P. M., and Farber, J. M. Chemokine receptors as HIV-1 coreceptors: roles in viral entry, tropism, and disease. *Annu. Rev. Immunol.*, **17**, 657–700 (1999).
6. Brighty, D. W. and Jassal, S. R. The synthetic peptide P-197 inhibits human T-cell leukemia virus type 1 envelope-mediated syncytium formation by a mechanism that is independent of Hsc70. *J. Virol.*, **75**, 10472–10478 (2001).
7. Clapham, P., Nagy, K., Cheingsong-Popov, R., Exley, M., and Weiss, R. A. Productive infection and cell-free transmission of human T-cell leukemia virus in a nonlymphoid cell line. *Science*, **222**, 1125–1127 (1983).
8. Clapham, P., Nagy, K., and Weiss, R. A. Pseudotypes of human T-cell leukemia virus types 1 and 2: neutralization by patients' sera. *Proc. Natl. Acad. Sci. USA*, **81**, 2886–2889 (1984).
9. Clapham, P. R., Blanc, D., and Weiss, R. A. Specific cell surface requirements for the infection of CD4-positive cells by human immunodeficiency virus types 1 and 2 and by simian immunodeficiency virus. *Virology*, **181**, 703–715 (1991).
10. Clapham, P. R., Weber, J. N., Whitby, D., McIntosh, K., Dalgleish, A. G., Maddon, P.

J., Deen, K. C., Sweet, R. W., and Weiss, R. A. Soluble CD4 blocks the infectivity of diverse strains of HIV and SIV for T cells and monocytes but not for brain and muscle cells. *Nature*, **337**, 368–370 (1989).

11. Daar, E. S., Li, X. L., Moudgil, T., and Ho, D. D. High concentrations of recombinant soluble CD4 are required to neutralize primary human immunodeficiency virus type 1 isolates. *Proc. Natl. Acad. Sci. USA*, **87**, 6574–6578 (1990).
12. Daenke, S., McCracken, S. A., and Booth, S. Human T-cell leukaemia/lymphoma virus type 1 syncytium formation is regulated in a cell-specific manner by ICAM-1, ICAM-3 and VCAM-1 and can be inhibited by antibodies to integrin beta2 or beta7. *J. Gen. Virol.*, **80**, 1429–1436 (1999).
13. Dalgleish, A. G., Beverley, P. C., Clapham, P. R., Crawford, D. H., Greaves, M. F., and Weiss, R. A. The CD4 (T4) antigen is an essential component of the receptor for the AIDS retrovirus. *Nature*, **312**, 763–767 (1984).
14. Feng, Y., Broder, C. C., Kennedy, P. E., and Berger, E. A. HIV-1 entry cofactor: functional cDNA cloning of a seven-transmembrane, G protein-coupled receptor. *Science*, **272**, 872–877 (1996).
15. Gavalchin, J., Fan, N., Lane, M. J., Papsidero, L., and Poiesz, B. J. Identification of a putative cellular receptor for HTLV-I by a monoclonal antibody, Mab 34-23. *Virology*, **194**, 1–9 (1993).
16. Gavalchin, J., Fan, N., Waterbury, P. G., Corbett, E., Faldasz, B. D., Peshick, S. M., Poiesz, B. J., Papsidero, L., and Lane, M. J. Regional localization of the putative cell surface receptor for HTLV-I to human chromosome 17q23.2-17q25.3. *Virology*, **212**, 196–203 (1995).
17. Geijtenbeek, T. B., Kwon, D. S., Torensma, R., van Vliet, S. J., van Duijnhoven, G. C., Middel, J., Cornelissen, I. L., Nottet, H. S., KewalRamani, V. N., Littman, D. R., Figdor, C. G., and van Kooyk, Y. DC-SIGN, a dendritic cell-specific HIV-1-binding protein that enhances trans-infection of T cells. *Cell*, **100**, 587–597 (2000).
18. Hildreth, J. E., Subramanium, A., and Hampton, R. A. Human T-cell lymphotropic virus type 1 (HTLV-1)-induced syncytium formation mediated by vascular cell adhesion molecule-1: evidence for involvement of cell adhesion molecules in HTLV-1 biology. *J. Virol.*, **71**, 1173–1180 (1997).
19. Hoshino, H., Clapham, P. R., Weiss, R. A., Miyoshi, I., Yoshida, M., and Miwa, M. Human T-cell leukemia virus type I: pseudotype neutralization of Japanese and American isolates with human and rabbit sera. *Int. J. Cancer*, **36**, 671–675 (1985).
20. Hoshino, H., Shimoyama, M., Miwa, M., and Sugimura, T. Detection of lymphocytes producing a human retrovirus associated with adult T-cell leukemia by syncytia induction assay. *Proc. Natl. Acad. Sci. USA*, **80**, 7337–7341 (1983).
21. Imai, T., Fukudome, K., Takagi, S., Nagira, M., Furuse, M., Fukuhara, N., Nishimura, M., Hinuma, Y., and Yoshie, O. C33 antigen recognized by monoclonal antibodies inhibitory to human T cell leukemia virus type 1-induced syncytium formation is a member of a new family of transmembrane proteins including CD9, CD37, CD53, and CD63. *J. Immunol.*, **149**, 2879–2886 (1992).
22. Jassal, S. R., Lairmore, M. D., Leigh-Brown, A. J., and Brighty, D. W. Soluble recombinant HTLV-1 surface glycoprotein competitively inhibits syncytia formation and viral infection of cells. *Virus Res.*, **78**, 17–34 (2001).

23. Jassal, S. R., Pohler, R. G., and Brighty, D. W. Human T-cell leukemia virus type 1 receptor expression among syncytium-resistant cell lines revealed by a novel surface glycoprotein-immunoadhesin. *J. Virol.*, **75**, 8317–8328 (2001).
24. Klatzmann, D., Champagne, E., Chamaret, S., Gruest, J., Guetard, D., Hercend, T., Gluckman, J. C., and Montagnier, L. T-lymphocyte T4 molecule behaves as the receptor for human retrovirus LAV. *Nature*, **312**, 767–768 (1984).
25. Kwong, P. D., Wyatt, R., Robinson, J., Sweet, R. W., Sodroski, J., and Hendrickson, W. A. Structure of an HIV gp120 envelope glycoprotein in complex with the CD4 receptor and a neutralizing human antibody. *Nature*, **393**, 648–659 (1998).
26. Landau, N. R., Page, K. A., and Littman, D. R. Pseudotyping with human T-cell leukemia virus type I broadens the human immunodeficiency virus host range. *J. Virol.*, **65**, 162–169 (1991).
27. Maddon, P. J., Dalgleish, A. G., McDougal, J. S., Clapham, P. R., Weiss, R. A., and Axel, R. The T4 gene encodes the AIDS virus receptor and is expressed in the immune system and the brain. *Cell*, **47**, 333–348 (1986).
28. McClure, M. O., Marsh, M., and Weiss, R. A. Human immunodeficiency virus infection of CD4-bearing cells occurs by a pH-independent mechanism. *EMBO J.*, **7**, 513–518 (1988).
29. McClure, M. O., Sommerfelt, M. A., Marsh, M., and Weiss, R. A. The pH independence of mammalian retrovirus infection. *J. Gen. Virol.*, **71**, 767–773 (1990).
30. McKnight, A., Griffiths, D. J., Dittmar, M., Clapham, P., and Thomas, E. Characterization of a late entry event in the replication cycle of human immunodeficiency virus type 2. *J. Virol.*, **75**, 6914–6922 (2001).
31. Miyoshi, I., Yoshimoto, S., Kubonishi, I., Fujishita, M., Ohtsuki, Y., Yamashita, M., Yamato, K., Hirose, S., Taguchi, H., Niiya, K. *et al.* Infectious transmission of human T-cell leukemia virus to rabbits. *Int. J. Cancer*, **35**, 81–85 (1985).
32. Mondor, I., Ugolini, S., and Sattentau, Q. J. Human immunodeficiency virus type 1 attachment to HeLa CD4 cells is CD4 independent and gp120 dependent and requires cell surface heparans. *J. Virol.*, **72**, 3623–3634 (1998).
33. Moore, J. P., McKeating, J. A., Weiss, R. A., and Sattentau, Q. J. Dissociation of gp120 from HIV-1 virions induced by soluble CD4. *Science*, **250**, 1139–1142 (1990).
34. Mothes, W., Boerger, A. L., Narayan, S., Cunningham, J. M., and Young, J. A. T. Retroviral entry mediated by receptor priming and low pH triggering of an envelope glycoprotein. *Cell*, **103**, 679–689 (2000).
35. Nagy, K., Clapham, P., Cheingsong-Popov, R., and Weiss, R. A. Human T-cell leukemia virus type I: induction of syncytia and inhibition by patients' sera. *Int. J. Cancer*, **32**, 321–328 (1983).
36. Nagy, K., Weiss, R. A., Clapham, P. R., and Cheingsong-Popov, R. Biological properties of human T-cell leukemia virus envelope antigens. *In* "Human T-cell Leukemia/Lymphoma Viruses," ed. R. C. Gallo, pp.121–131 (1984). Cold Spring Harbor Laboratory Press, Cold Spring Harbor, New York.
37. O'Brien, S. J. and Moore, J. P. The effect of genetic variation in chemokines and their receptors on HIV transmission and progression to AIDS. *Immunol. Rev.*, **177**, 99–111 (2000).
38. O'Hara, B., Johann, S. V., Klinger, H. P., Blair, D. G., Rubinson, H., Dunn, K. J., Sass,

P., Vitek, S. M., and Robins, T. Characterization of a human gene conferring sensitivity to infection by gibbon ape leukemia virus. *Cell Growth Differ.*, **1**, 119–127 (1990).
39. Okuma, K., Nakamura, M., Nakano, S., Niho, Y., and Matsuura, Y. Host range of human T-cell leukemia virus type I analyzed by a cell fusion-dependent reporter gene activation assay. *Virology*, **254**, 235–244 (1999).
40. Overbaugh, J., Miller, A. D., and Eiden, M. V. Receptors and entry cofactors for retroviruses include single and multiple transmembrane-spanning proteins as well as newly described glycophosphatidylinositol-anchored and secreted proteins. *Microbiol. Mol. Biol. Rev.*, **65**, 371–389 (2001).
41. Palker, T. J., Riggs, E. R., Derry, E. P., Muir, A. J., Scearce, R. M., Randall, R. R., McAdams, M. W., McKnight, A., Clapham, P. R., Weiss, R. A., and Haynes, B. F. Mapping of homologous, amino-terminal neutralizing regions of human T-cell lymphotropic virus type I and II gp46 envelope glycoproteins. *J. Virol.*, **66**, 5879–5889 (1992).
42. Patience, C., Takeuchi, Y., and Weiss, R. A. Infection of human cells by an endogenous retrovirus of pigs. *Nature Med.*, **3**, 282–286 (1997).
43. Pizzato, M., Marlow, S. A., Blair, E. D., and Takeuchi, Y. Initial binding of murine leukemia virus particles to cells does not require specific Env-receptor interaction. *J. Virol.*, **73**, 8599–8611 (1999).
44. Poiesz, B. J. The HTLV cell surface receptor. *Cancer Invest.*, **20**, 154–155 (2002).
45. Porter, C. D., Collins, M. K., Tailor, C. S., Parkar, M. H., Cosset, F. L., Weiss, R. A., and Takeuchi, Y. Comparison of efficiency of infection of human gene therapy target cells via four different retroviral receptors. *Hum. Gene Ther.*, **7**, 913–919 (1996).
46. Richardson, J. H., Edwards, A. J., Cruickshank, J. K., Rudge, P., and Dalgleish, A. G. *In vivo* cellular tropism of human T-cell leukemia virus type 1. *J. Virol.*, **64**, 5682–5687 (1990).
47. Sagara, Y., Ishida, C., Inoue, Y., Shiraki, H., and Maeda, Y. 71-kilodalton heat shock cognate protein acts as a cellular receptor for syncytium formation induced by human T-cell lymphotropic virus type 1. *J. Virol.*, **72**, 535–541 (1998).
48. Sattentau, Q. J., Clapham, P. R., Weiss, R. A., Beverley, P. C., Montagnier, L., Alhalabi, M. F., Gluckmann, J. C., and Klatzmann, D. The human and simian immunodeficiency viruses HIV-1, HIV-2 and SIV interact with similar epitopes on their cellular receptor, the CD4 molecule. *AIDS*, **2**, 101–105 (1988).
49. Simmons, G., Clapham, P. R., Picard, L., Offord, R. E., Rosenkilde, M. M., Schwartz, T. W., Buser, R., Wells, T. N., and Proudfoot, A. E. Potent inhibition of HIV-1 infectivity in macrophages and lymphocytes by a novel CCR5 antagonist. *Science*, **276**, 276–279 (1997).
50. Sinangil, F., Harada, S., Purtilo, D. T., and Volsky, D. J. Host cell range of adult T-cell leukemia virus. I. Viral infectivity and binding to various cells as detected by flow cytometry. *Int. J. Cancer*, **36**, 191–198 (1985).
51. Smith, D. H., Byrn, R. A., Marsters, S. A., Gregory, T., Groopman, J. E., and Capon, D. J. Blocking of HIV-1 infectivity by a soluble, secreted form of the CD4 antigen. *Science*, **238**, 1704–1707 (1987).
52. Sommerfelt, M. A. Retrovirus receptors. *J. Gen. Virol.*, **80**, 3049–3064 (1999).
53. Sommerfelt, M. A. and Weiss, R. A. Receptor interference groups of 20 retroviruses plating on human cells. *Virology*, **176**, 58–69 (1990).

54. Sommerfelt, M. A., Williams, B. P., Clapham, P. R., Solomon, E., Goodfellow, P. N., and Weiss, R. A. Human T cell leukemia viruses use a receptor determined by human chromosome 17. *Science*, **242**, 1557–1559 (1988).
55. Strizki, J. M., Xu, S., Wagner, N. E., Wojcik, L., Liu, J., Hou, Y., Endres, M., Palani, A., Shapiro, S., Clader, J. W., Greenlee, W. J., Tagat, J. R., McCombie, S., Cox, K., Fawzi, A. B., Chou, C. C., Pugliese-Sivo, C., Davies, L., Moreno, M. E., Ho, D. D., Trkola, A., Stoddart, C. A., Moore, J. P., Reyes, G. R., and Baroudy, B. M. SCH-C (SCH 351125), an orally bioavailable, small molecule antagonist of the chemokine receptor CCR5, is a potent inhibitor of HIV-1 infection *in vitro* and *in vivo*. *Proc. Natl. Acad. Sci. USA*, **98**, 12718–12723 (2001).
56. Sullivan, N. J. Antibody-mediated enhancement of viral disease. *Curr. Top. Microbiol. Immunol.*, **260**, 145–169 (2001).
57. Sutton, R. E. and Littman, D. R. Broad host range of human T-cell leukemia virus type 1 demonstrated with an improved pseudotyping system. *J. Virol.*, **70**, 7322–7326 (1996).
58. Suzuki, T. and Ikeda, H. The mouse homolog of the bovine leukemia virus receptor is closely related to the delta subunit of adaptor-related protein complex AP-3, not associated with the cell surface. *J. Virol.*, **72**, 593–599 (1998).
59. Takeuchi, Y., Patience, C., Magre, S., Weiss, R. A., Banerjee, P. T., Le Tissier, P., and Stoye, J. P. Host range and interference studies of three classes of pig endogenous retrovirus. *J. Virol.*, **72**, 9986–9991 (1998).
60. Timar, J., Nagy, K., Robertson, D., Clapham, P., and Weiss, R. A. Ultrastructure and immunoelectron microscopy of human T-cell leukaemia virus type I-producing lymphoid and non-lymphoid human tumour cells. *J. Gen. Virol.*, **68**, 1011–1020 (1987).
61. Trejo, S. R. and Ratner, L. The HTLV receptor is a widely expressed protein. *Virology*, **268**, 41–48 (2000).
62. Vile, R. G., Schulz, T. F., Danos, O. F., Collins, M. K., and Weiss, R. A. A murine cell line producing HTLV-I pseudotype virions carrying a selectable marker gene. *Virology*, **180**, 420–424 (1991).
63. Weiss, R. A. Cellular receptors and viral glycoproteins involved in retrovirus entry. *In* "The Retroviridae," vol. 2, ed. J. A. Levy, pp. 1–108 (1993). Plenum Press, New York.
64. Weiss, R. A. and Tailor, C. S. Retrovirus receptors. *Cell*, **82**, 531–533 (1995).

MOLECULAR PATHOGENESIS

MOLECULAR MECHANISMS OF Tax-MEDIATED CELL GROWTH OF HUMAN T LYMPHOCYTES

Kiyoshi OHTANI and Masataka NAKAMURA

*Human Gene Sciences Center, Tokyo Medical and Dental University**

Human T-cell leukemia virus type I (HTLV-I) Tax can stimulate cell growth, an important process in the initial step of leukemogenesis by HTLV-I. Cell growth is controlled by cell cycle progression and prevention of apoptosis, and both processes could be modulated by Tax. Expression of Tax can induce cell cycle progression from G0/G1 to S-G2/M phases in interleukin (IL)-2-deprived resting human T cells. Tax-induced cell cycle progression is mediated by activation of cyclin-dependent kinases (CDKs), which phosphorylate the retinoblastoma tumor suppressor family, resulting in activation of the transcription factor E2F. Tax-induced activation of CDKs is mediated, at least in part, by *trans*-acting transcriptional activation of the cell cycle regulatory genes such as *cyclin D2*, *cdk2*, *cdk4* and *cdk6*. Tax-dependent reduction in CDK inhibitors, $p19^{INK4d}$ and $p27^{Kip1}$, also contributes to activation of CDKs. In addition, Tax prevents apoptosis of IL-2-deprived murine T cells, allowing the cells to grow continuously independent of IL-2. Tax-induced inhibition of apoptosis is mediated, at least in part, by *trans*-activation of the anti-apoptotic gene *bcl-xl.* Cell cycle progression and prevention of apoptosis by Tax form the basis of proliferation of host T cells. This allows expansion of the virus-infected T cells, which are also probably predisposed to secondary genetic changes due to genome instability induced by Tax, resulting in full transformation of T cells.

Human T-cell leukemia virus type I (HTLV-I) is the causative agent of adult T-cell leukemia (ATL) (*17, 40, 57*). The *trans*-acting transcriptional activator Tax encoded by HTLV-I is thought to play critical roles in leukemogenesis. Introduction of the *Tax* gene induces phenotypic transformation in fibroblast cell lines (*50*), neoplastic transformation of primary rat fibroblasts in co-operation with the *ras* oncogene (*41*), persistent interleukin (IL)-2-dependent growth of primary T cells *in vitro* (*1, 13*), and development of tumors and leukemia in mice *in vivo* (*14, 31*). These results indicate that Tax has growth stimulatory properties, which are important for the initial step in transformation of host T cells.

Cell growth is primarily controlled by the cell cycle, which consists of five phases: the first gap (G1) phase, the DNA synthetic (S) phase, the second gap (G2) phase, the mitotic (M) phase and the resting (G0) phase. Cell cycle progression in mammalian cells is dependent on mitogenic stimulation. The latter promotes cells in G0/G1 phases to proceed into the S phase. The commitment to replicate cellular DNA is the

* 1-5-45 Yushima, Bunkyo-ku, Tokyo 113-8510, Japan

most important step in regulating the cell cycle. This decision is made at a point in late G1 phase, which is called the restriction point. Once they pass the restriction point, cells are destined to undergo one round of the cell cycle without mitogenic stimulation. Periodic activation and inactivation of cyclin-dependent kinases (CDKs) are pivotally involved in the cell cycle. Progression through G0/G1 phases into S phase is controlled by G1 CDKs, which associate cyclin Ds and cyclin E. One role of these kinases in G1/S progression is phosphorylation of members of the retinoblastoma tumor suppressor protein (pRb) family, consequently activating the transcription factor E2F (*45*). E2F plays crucial roles in induction of the S phase by regulating the expression of genes that encode various molecules involved in DNA replication and cell cycle progression (Fig. 1) (*33, 34*).

Tax-induced growth stimulation has been attributed, at least in part, to *trans*-activation of growth-related cellular genes. Tax was initially identified as a *trans*-acting transcriptional activator of the HTLV-I promoter in the long terminal repeat (LTR) (*12, 44, 46*). It was subsequently demonstrated to *trans*-activate transcription of various cellular genes involved in cell growth promotion, such as genes for growth factors (IL-2 and IL-15), growth factor receptors (IL-2Rα, IL-2Rγ and IL-15Rα), cytoplasmic signal transducers (lyn), and nuclear transcription factors (c-fos, c-jun, jun D, fra-1 and c-egr) (*56*) (Fig. 2). *Trans*-activation of IL-2 and IL-2 receptors suggested the possible involvement of the autocrine loop in growth promotion of virus-infected cells (*28*). *Trans*-activation of these growth-related cellular genes by Tax is thought to render cellular environment advantageous for cell growth. At least three cellular transcription factors, cAMP responsive element binding factor (CREB),

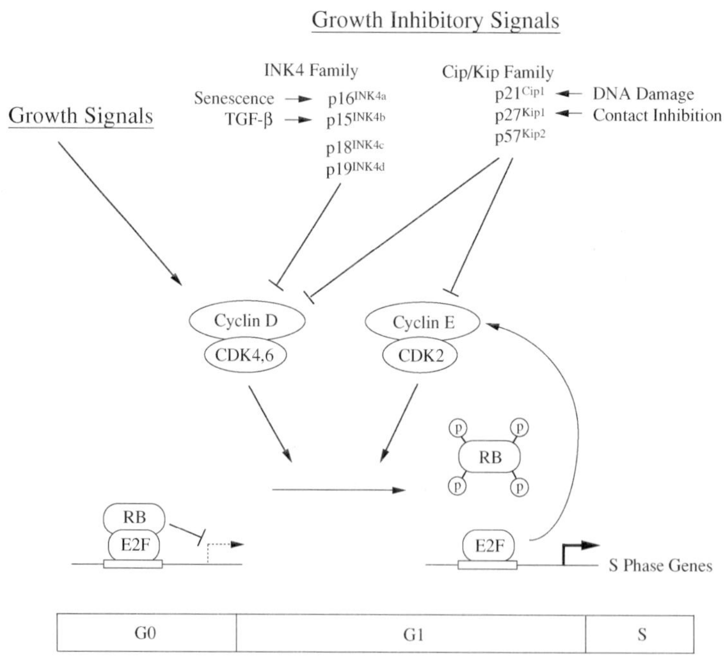

FIG. 1. Regulatory mechanisms of G0/G1 to S phase transition
See text for details.

serum responsive factor (SRF) and nuclear factor (NF)-κB are known to be modulated by Tax (*56*) (Fig. 3). Tax also interacts with co-activators CREB binding protein (CBP)/p300, which mediate transcriptional activation by CREB and NF-κB. In addition to *trans*-activating growth-related cellular genes, Tax is reported to *trans*-repress some cellular genes; down-regulation of DNA polymerase β, which is involved in host cell DNA repair, suggests involvement of Tax in induction of genome instability (*22*), and repression of the $p18^{INK4c}$ promoter is suggested to contribute to

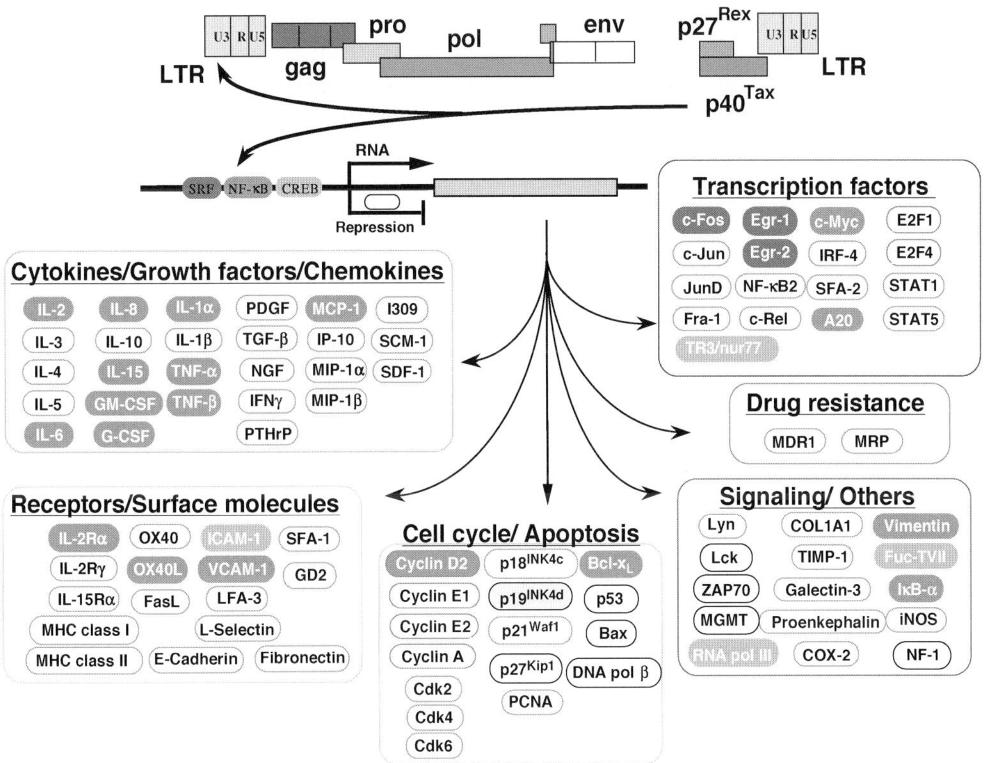

FIG. 2. HTLV-I genome structure and Tax-mediated *trans*-activation of cellular genes

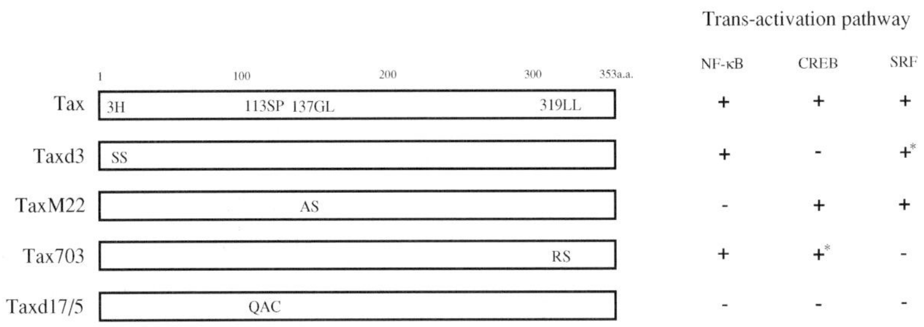

FIG. 3. Tax mutants used in these studies

Amino acid substitutions in each Tax mutant are indicated in left panel. *Trans*-activation abilities of the Tax mutants in Kit 225 cells are indicated in right panel. Asterisks indicate that *trans*-activation abilities may differ between cell lines.

cell cycle progression (*48*) (Fig. 2). Sequestration of co-activators CBP/p300 by binding of Tax to the proteins is proposed as a mechanism of *trans*-repression (*56*). In addition to transcriptionally modulating growth-related cellular genes, Tax is thought to be directly involved in cell cycle progression by protein-protein interaction. It has also been reported to bind and inactivate $p16^{INK4a}$ and $p15^{INK4b}$, inhibitors of D-type CDKs (*27, 47, 48*). The exact roles of Tax in cell cycle progression are yet to be elucidated.

Cell growth is a consequence of not only cell cycle progression but also of prevention of apoptosis (programmed cell death); mitogenic stimulation promotes cell cycle as well as protection of cells from apoptosis. Inappropriate promotion of the cell cycle such as deregulated activation of oncogenes is often accompanied by apoptosis. Apoptosis seems to be one of the important mechanisms preventing generation of transformed cells. In addition, virus-infected cells are often eliminated by apoptosis. It is thus important for the virus to protect cells from apoptosis. Several groups have reported the anti-apoptotic function of Tax (*4, 10, 25*), however, the exact role of Tax-mediated regulation of apoptosis during cell cycle progression remains uncertain.

We investigated the above issues related to the roles of Tax on the cell cycle and prevention of apoptosis using IL-2-dependent T-cell lines. Our results demonstrate that Tax can directly stimulate cell cycle progression and at the same time prevent cells from apoptosis. These Tax-related properties are thought to be important for proliferation of host T cells. This chapter focuses on recent advances in our understanding of the cell cycle promotion and prevention of apoptosis by Tax.

Effects of Tax on Cell Cycle Progression

It is important to analyze the effect of Tax on cell growth in T cells, because the growth regulatory mechanisms are quite different between T cells and fibroblasts. In addition, knowledge gained on the role of Tax in fibroblasts cannot simply be applied to Tax-induced leukemogenesis. Resting T cells are definitely required to examine the exact effect of Tax on the cell cycle. Using human T-cell lines and normal peripheral blood lymphocytes (PBLs), we and others have reported that Tax could induce cell cycle progression in human T cells (*20, 32, 35, 43*).

We utilized an IL-2-dependent human T-cell line, Kit 225 (*18*). Upon withdrawal of IL-2, Kit 225 cells were arrested at G0/G1 phase of the cell cycle without significant loss of live cells. Tax-expressing recombinant adenoviruses were used for efficient introduction of Tax into resting T cells. Kit 225 cells were made quiescent by deprivation of IL-2 for 48 h and Tax was expressed by infecting these cells with the Tax-recombinant viruses. Half of the cell population expressed Tax under these conditions (Fig. 4B). Expression of Tax in resting Kit 225 cells was found to induce cell cycle progression from G0/G1 to S-G2/M phases (Fig. 4A). This effect was a cell intrinsic event, because cell cycle progression was noted only in Tax-expressing cells (Fig. 4B). Tax has been reported to bind and inactivate $p16^{INK4a}$, and thus is implicated in cell cycle progression (*47*). Kit 225 cells are, however, negative for $p16^{INK4a}$ as examined by the reverse transcriptase-PCR method, indicating that Tax-induced cell cycle progression is independent of $p16^{INK4a}$. It should be noted that the ability of Tax to induce cell cycle progression was not specific to Kit 225 cells. Tax also induced cell cycle progression in phytohemagglutinin (PHA)-stimulated normal PBLs.

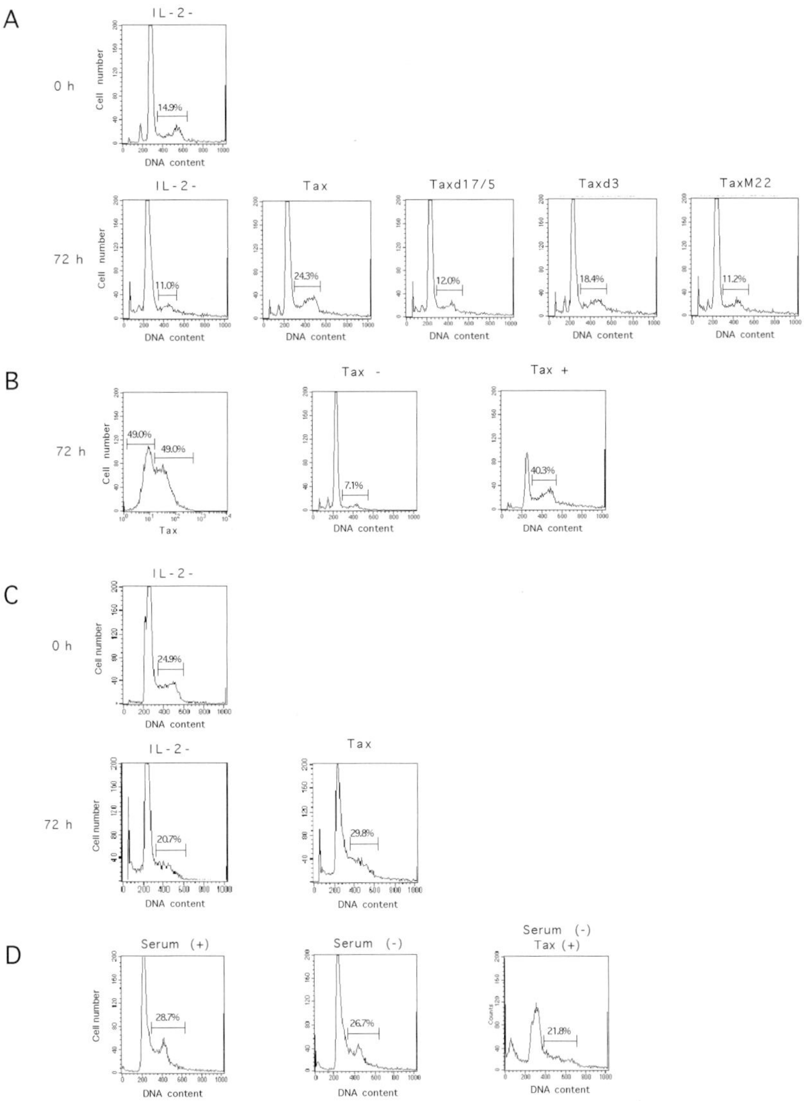

FIG. 4. Cell cycle progression induced by Tax

(A) Cell cycle progression of Kit 225 cells infected with Tax recombinant viruses. Kit 225 cells starved of IL-2 for 48 h (0 h in figure) were infected with indicated recombinant adenoviruses for Tax expression: AxCAIY-Tax, AxCAIY-Taxd17/5, AxCAIY-Taxd3 and AxCAIY-TaxdM22. Cells were further cultured in the absence of IL-2 for 72 h. As controls, mock-infected cells were cultured in the absence of IL-2 for 72 h. Cells were harvested for DNA staining with propidium iodide, followed by flow cytometric examination. Percentages of cells in S and G2/M phases are indicated. (B) Cell cycle progression in Tax-expressing cells. Kit 225 cells infected with AxCAIY-Tax as in (A) were stained for Tax and DNA with anti-Tax monoclonal antibody, Lt-4, and propidium iodide, respectively. Cells expressing Tax (Tax +) and not expressing Tax (Tax −) were gated and examined for DNA content with a flow cytometer. Percentages of cells positive and negative for Tax and cells in S and G2/M phases are indicated. (C) Cell cycle progression in PHA-PBLs induced by Tax. PBLs were stimulated by PHA for 3 days and after washing out PHA cultured in the absence of PHA and IL-2 for 24 h. Cells were infected with the Tax recombinant adenovirus, further cultured for 72 h in the absence of IL-2, and examined for DNA content. (D) Effect of Tax in Jurkat cells. Tax-inducible Jurkat cell line, JPX-9, was cultured in the presence or absence of serum for 20 h. Expression of Tax was induced [Tax (+)] by addition of $CdCl_2$ at 20 μM for 17 h, and cells were examined for DNA content.

Thus, we conclude that Tax alone could induce cell cycle progression in resting human T cells. Using Tax mutants, Taxd3 and TaxM22, defective in *trans*-activation pathways of CREB and NF-κB, respectively, we also showed that the NF-κB pathway (or a pathway parallel with the NF-κB pathway) is indispensable for Tax-mediated cell cycle progression (Fig. 4A).

We established JPX-9, a Tax-inducible human T-cell line Jurkat, which was used to examine the effect of Tax on the cell cycle by Neuveut *et al.* (*32*). Since JPX-9 cells grew independently of IL-2, the cells were deprived of serum to assess the effects of Tax expression induced. These experiments showed that induction of Tax could promote cell cycle progression under serum-starvation conditions. In a series of preliminary experiments, we also showed that JPX-9 cells were not arrested at the G0/G1 even under serum-starvation conditions (Fig. 4D). This could be due, at least in part, to the absence of functional pRb and p53 in Jurkat cells. In addition, induction of expression of Tax under serum-starvation conditions did not increase the number of cells in the S-G2/M phases (Fig. 4D). Instead, cells with DNA content of more than 4C increased and many cells died by apoptosis. This could be due to blocking of the cell cycle at the G2/M phase by over-expression of Tax (*23*). The reason for the discrepancy between the two groups is not known at present.

Cord blood T lymphocytes were transformed by infection with recombinant rhadinovirus inducibly expressing Tax, generating Tesi cells. Shut-down of Tax expression in the cells abrogated cell growth and re-expression of Tax re-initiated cell growth by suggesting that Tax expression induced cell cycle progression (*43*). The cells, however, seem to be cultured in the presence of IL-2, thus the interpretation would be that IL-2-dependent growth of the cells requires Tax. Consistent with this, expression of IL-2Rα on the immortalized cells correlated with that of Tax (*43*).

Activation of Transcription Factor E2F by Tax

Accumulating evidence indicates that the pRb/E2F pathway plays pivotal roles in progression of the cell cycle. E2F regulates the expression of various genes for cell cycle progression and DNA replication, including the *cyclin E* and *Cdc6* genes, and thus activation of E2F is essential for progression into S phase (*54*). The E2F family consists of at least six members, E2F1 through E2F6. Expression of *E2F1* to *E2F3* genes is cell cycle-dependent and is induced at the G1/S boundary by E2F itself. E2F4 and E2F5 are expressed constitutively throughout the cell cycle. Transcriptional activity of E2F is repressed by binding to the hypophosphorylated forms of pRb family members. E2F1 through E2F3 bind predominantly to pRb, and E2F4 and E2F5 are mainly associated with p130 and p107. Phosphorylation of pRb family members by G1 CDK releases free E2F from the repressor complex, and thus activates E2F-dependent transcription, allowing cells to proceed into the S phase (Fig. 1).

As a first step in analyzing the molecular mechanism of Tax-mediated cell cycle progression, we examined the effect of Tax on endogenous E2F activity (*35*). Expression of Tax in quiescent Kit 225 cells induced activation of endogenous E2F as measured by E2F-dependent reporter assay (Fig. 5A). Tax activated a variety of the S-phase gene promoters harboring the E2F sites such as DHFR, DNA polymerase α and Cdc6. Interestingly, Tax also *trans*-activated promoters of cell cycle regulatory genes with the E2F sites of c-myc, cyclin D2, cyclin E, cyclin A, E2F1 and E2F2

(Fig. 5B). The cyclin D3 promoter, which does not contain typical E2F sites, was not *trans*-activated by Tax.

The molecular mechanism of Tax-induced E2F activation was further examined by monitoring the expression of *E2F* genes and E2F binding activity (*20*). Tax increased the expression of E2F1 at both mRNA and protein levels consistent with the result of reporter assay with the E2F1 promoter (Table I). In contrast to *trans*-activation of the E2F2 promoter by Tax, expression of *E2F2* gene was relatively reduced by Tax. Expression of the *E2F3* gene was not significantly affected, and a slight increase (up to 1.5-fold) was observed with E2F4 mRNA by Tax. It should be noted that Tax-induced expression of the *E2F1* gene was seen earlier than that of cell cycle progression. This presumably indicates that Tax directly induces *E2F1* gene expression, which contributes to activation of E2F transcriptional activity. Tax increased free E2F4 binding activity as examined by gel mobility shift assay (Fig. 6A). Since Tax did not significantly increase *E2F4* gene expression, the profound increase in E2F4 binding activity may be due to the release of free E2F4 from complexes with the pRb family members. The induction of E2F1 expression and increase in E2F4 binding activity by Tax was dependent on the NF-κB pathway.

Similar results were obtained with an IL-2-independent human T-cell line

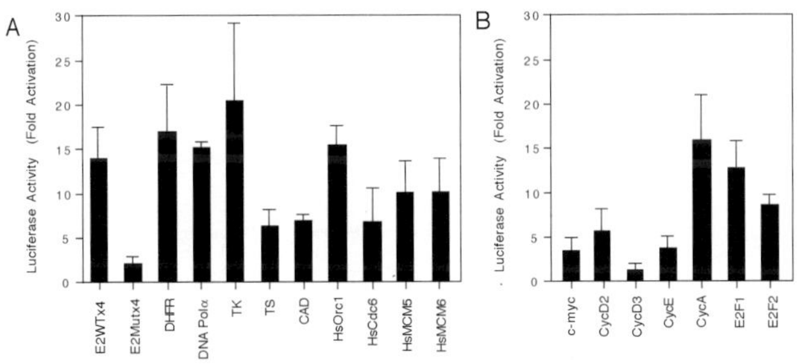

FIG. 5. (A) Activation of E2F binding sites by Tax. Kit 225 cells were transfected with the E2F luciferase reporter plasmid, pE2WTx4-Luc, or its mutant, pE2MTx4-Luc, along with Tax expression plasmid, pMT-2Tax. Luciferase reporter plasmids with upstream regulatory regions of indicated genes were also used. Cells were further cultured for 48 h without IL-2 and harvested for luciferase activity measurement. Values represent fold activation relative to luciferase activity of cells transfected with the backbone vector pH β APr-1-neo used for Tax expression. (B) Effect of Tax on cell cycle regulatory genes. Luciferase reporter plasmids with upstream regulatory regions of indicated genes were introduced along with pMT-2Tax or the backbone vector into Kit 225 cells and luciferase activities were determined as in (A). Data are expressed as mean±SD values of three experiments.

TABLE I. Tax-mediated Modulation of Molecules Associated with the Cell Cycle

E2F	mRNA	Protein	Binding activity
E2F1	↑↑	↑↑	ND
E2F2	↓	ND	ND
E2F3	→	ND	ND
E2F4	↗	ND	↑↑

ND: not determined.

Cyclin/CDK	mRNA	Protein
Cyclin D2	↑↑	↑↑
Cyclin D3	→	→
Cyclin E1	↑	→
Cyclin E2	↑	ND
CDK2	↑	↑
CDK4	↑	↑
CDK6	↑	↑

CDK inhibitor	mRNA	Protein
p15	→	ND
p16	(−)	(−)
p18	↑	→
p19	↓	↓↓
p21	↑↑	↑↑
p27	↓	↓

(−): not expressed.

CEM, which is also negative for p16^{INK4a}, and JPX-9 cells by Lemasson *et al.* (*26*). Their studies showed that Tax could induce exogenous E2F1 promoter activation and endogenous E2F1 gene expression in asynchronously growing CEM and JPX-9 cells, respectively. There is, however, a difference in the pathways required for Tax-dependent activation of the E2F1 promoter. In resting Kit 225 cells, Tax-mediated activation of the E2F1 promoter is dependent on activation of the NF-κB pathway. However, in growing CEM cells, the effect of Tax was mainly dependent on the CREB pathway. The difference in the cell lines used and conditions of the cell culture may explain the difference in the two studies: resting Kit 225 vs. growing CEM cells.

Increased binding activity of free E2F4 resulted from Tax-promoted phosphorylation of all members of the pRb family as shown by the slow migration pattern in western blot analyses (Fig. 6B). We also showed that the ability of Tax to induce phosphorylation of the pRb family members was also dependent on the NF-κB pathway. Considered together, these studies indicate that activation of E2F by Tax is mediated, at least in part, by induction of E2F1 expression by direct *trans*-activation and by an increase in free E2F4 binding by Tax-induced phosphorylation of the pRb family members.

Tax-induced Activation of G1 CDKs

Progression through G0/G1 phases into the S phase is controlled by cyclin D-

A

B

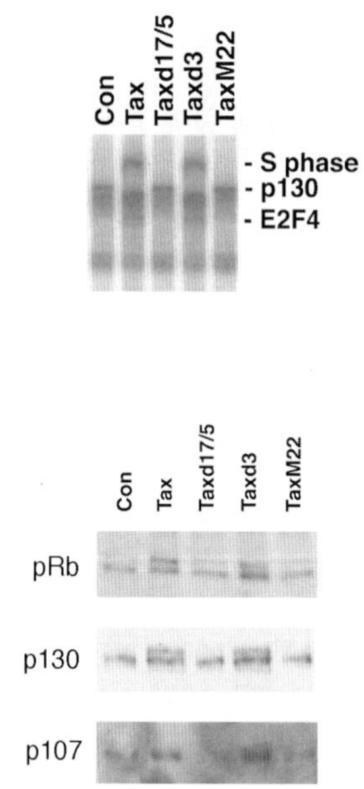

FIG. 6. (A) Effects of Tax on E2F binding activity. Whole cell lysates were prepared from Kit 225 cells infected with indicated Tax recombinant adenoviruses and cells left IL-2 starved (Con). E2F binding activity in the lysates was examined by gel mobility shift assay using the typical E2F site from the *DHFR* gene as a probe. S phase and p130 complexes, and free E2F4 are indicated. (B) Phosphorylation of pRb family proteins induced by Tax. Cell lysates were prepared from IL-2-starved Kit 225 cells infected with the indicated recombinant adenoviruses and further cultured in the absence of IL-2 for 72 h. As a control, mock-infected cells were also used (Con). Western blots were carried out with 40 μg of sample proteins and pRb, p130 and p107 were detected with corresponding antibodies.

and cyclin E-dependent kinases (G1 CDKs), which phosphorylate members of the pRb family and activate E2F. Cyclin Ds bind to and activate CDK4 and CDK6. Similarly, cyclin E is a partner of CDK2. The fact that all of the pRb family members are phosphorylated by expression of Tax prompted us to examine the effect of Tax on the CDK activity (*20*). CDK4 and CDK2 were immunoprecipitated using specific antibodies and immunoprecipitates were examined for their kinase activity by *in vitro* kinase assay. Both CDKs were profoundly activated upon expression of Tax (Fig. 7). This result indicates that expression of Tax in resting human T cells induces activation of CDK, which is central for cell cycle progression. In this regard, Neuveut *et al.* (*32*) reported that expression of Tax in serum-starved JPX-9 cells induced CDK4, CDK6 and cyclin D3 associated kinase activity. Furthermore, Schmitt *et al.* (*43*) reported that re-expression of Tax in Tesi cells induced CDK4 and CDK6 associated kinase activity. Regardless of the integrity of experimental systems, these results are consistent with ours in that the Tax expression induced CDK activities in human T cells.

The significance of CDK activation on cell cycle progression was examined by evaluating the effect of a CDK inhibitor on Tax-mediated activation of E2F. In these studies, we used $p19^{INK4d}$, which is specific for D-type CDK and does not bind to Tax, unlike $p16^{INK4a}$. Co-expression of $p19^{INK4d}$ with Tax significantly abolished Tax-mediated activation of E2F (*19*), suggesting that Tax induces D-type CDK activation, which is crucial for activation of E2F and consequently for progression of the cell cycle.

CDK activity is primarily regulated by expression of cyclin and CDK molecules. To gain insight into the molecular mechanism of Tax-mediated activation of CDKs, we first examined the effect of Tax on expression of cyclins and CDKs (*20*). T cells predominantly express cyclin D2 and cyclin D3 relative to cyclin D1. Expression of Tax in IL-2-starved Kit 225 cells induced expression of cyclin D2 at mRNA and protein levels (Table I); in contrast, cyclin D3 expression was not altered by Tax. This is consistent with our results of the reporter analyses in which Tax *trans*-activated the cyclin D2 promoter but not cyclin D3 promoter (Fig. 5B). *Cyclin E1* and *cyclin E2*

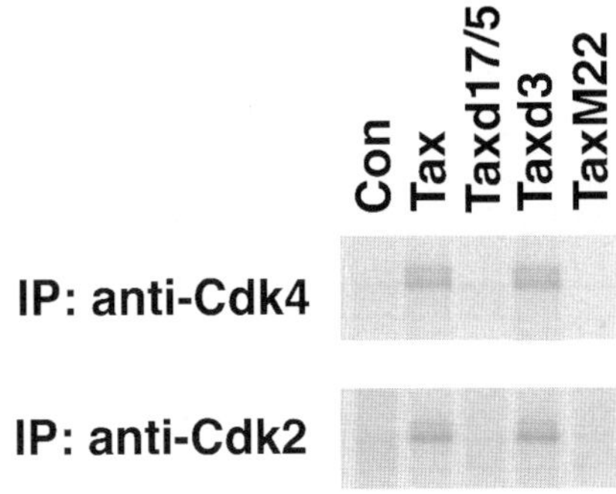

FIG. 7. Activation of CDK4 and CDK2 by Tax

Cell lysates were prepared from Kit 225 cells IL-2 starved for 48 h and infected with the indicated recombinant adenoviruses and further cultured for 72 h. As a control (Con), cells were mock-infected and further cultured in the absence of IL-2 for 72 h. Cell lysates (100 μg) were immunoprecipitated with anti-CDK4 and anti-CDK2 antibodies and kinase activities of the immunoprecipitates were measured by *in vitro* kinase assay using GST-Rb as a substrate.

genes were slightly induced but the amount of cyclin E1 protein was not altered by Tax. Tax induced the expression of CDK4, CDK6 and CDK2 at both mRNA and protein levels. It is likely that Tax-induced CDK activation in resting human T cells is based on activation of expression of *cyclin D2* and *CDK* genes.

Akagi and colleagues (*2*) indicated previously that HTLV-I-infected and Tax-immortalized T-cell lines predominantly expressed cyclin D2 relative to cyclin D3. Furthermore, induction of cyclin D2 and down-regulation of cyclin D3 in HTLV-I-infected T-cell lines have also been reported by other groups (*16, 42*). These findings are consistent with our results that Tax induces the expression of *cyclin D2* gene but not that of *cyclin D3*.

The cyclin D2 promoter possesses a typical E2F site, raising the possibility that activation of this promoter by Tax is a consequence of Tax-induced E2F activation. Since the induction of D-type CDK activity is crucial for activation of E2F and expression of *cyclin D2* gene is induced by Tax, we examined whether Tax directly induced the transcriptional activation of cyclin D2. Chloramphenicol acetyltransferase (CAT) reporter analyses with cloned cyclin D2 promoter indicated that Tax significantly activated cyclin D2 promoter (*19*). Mutational analyses of the cyclin D2 promoter revealed that Tax directly induced the expression of *cyclin D2* gene mainly through an NF-κB-like site with slight assistance of a CRE-like site. These results indicate that direct *trans*-activation of the cell cycle regulatory genes is implicated in Tax-mediated cell cycle progression.

The activity of cyclin/CDK complexes is negatively regulated by two classes of CDK inhibitors based on their structures and CDK targets (Fig. 1). One group is inhibitors of the CDK4 (INK4) family consisting of $p16^{INK4a}$, $p15^{INK4b}$, $p18^{INK4c}$ and $p19^{INK4d}$, which contain multiple ankyrin motifs and bind only to CDK4 and CDK6 either in complexes with D-type cyclins or in a catalytic subunit alone. The other group of inhibitors is the Cip/Kip family, which includes $p21^{Cip1}$, $p27^{Kip1}$ and $p57^{Kip2}$. The Cip/Kip family interferes with activities of not only cyclin D-dependent kinases (CDK4 and CDK6) but also cyclin E- and cyclin A-dependent kinases (CDK2). We thus examined the effect of Tax on the expression of these CDK inhibitors (Table I) (*20*). Expression of $p19^{INK4d}$ and $p27^{Kip1}$ was reduced at both mRNA and protein levels upon expression of Tax. Interestingly, reduction in the $p19^{INK4d}$ protein level was far greater than that in mRNA level, indicating that Tax also down-regulates $p19^{INK4d}$ expression at protein level. Tax slightly increased expression of the $p18^{INK4c}$ gene but did not affect $p18^{INK4c}$ at the protein level. No significant change in expression of the $p15^{INK4b}$ gene was noted. These results indicate that reduction in $p19^{INK4d}$ and $p27^{Kip1}$ proteins may contribute to activation of CDKs in resting Kit 225 cells upon expression of Tax. Intriguingly, Tax induced expression of $p21^{Cip1}$ at both mRNA and protein levels. Other groups have also reported induction of $p21^{Cip1}$ expression in HTLV-I-infected T-cell lines (*2, 7, 11*). The biological significance of the induction of $p21^{Cip1}$ expression remains to be elucidated.

Previous studies reported significant reduction of $p18^{INK4c}$ expression in HTLV-I-infected and Tax-immortalized T-cell lines (*2*) and Tax-mediated suppression of promoter activity of the $p18^{INK4c}$ gene (*48*), in contrast to our results with Kit 225 cells in which expression of the $p18^{INK4c}$ gene was only slightly induced by Tax. This may be due to differences in the cell lines used in the assays.

Tissue-specific E2F Activation by Tax

HTLV-I induces T-cell-specific malignancy *in vivo*, indicating that Tax seems to exert a tissue-specific effect on cell cycle progression. To address this issue, we examined the ability of Tax to activate endogenous E2F in fibroblasts. Rat embryonic fibroblasts, REF52, undergo growth arrest upon serum withdrawal. Tax expression in REF52 cells under serum starvation did not activate endogenous E2F activity although it activated the CREB and NF-κB pathways (*35*). These results emphasize the tissue-specific properties of Tax in activating E2F and possibly cell cycle progression.

In normal fibroblasts, cell cycle progression is inhibited when cells become confluent (cell to cell contact) even in the presence of serum which provides growth stimulation. Loss of contact inhibition causes phenotypic transformation of fibroblasts. Tax has been reported to induce phenotypic transformation of rat fibroblasts (*50*), indicating that it can overcome the growth inhibitory signal(s) generated by cell to cell contact. Our study clearly demonstrated that, in fibroblasts made quiescent by serum deprivation, Tax could not activate E2F, indicating that Tax by itself fails to promote cell cycle in resting fibroblasts. This is in contrast to Tax-induced promotion of cell cycle progression in resting human T cells. This tissue-specific ability of Tax to promote cell cycle progression could be important for T-cell-specific leukemogenesis by HTLV-I.

IL-2-independent Growth Conferred by Tax

HTLV-I can transform normal PBLs not only in an IL-2-dependent manner but also in an IL-2-independent manner, although less efficiently. Introduction of Tax alone has been shown to transform normal PBLs in an IL-2-dependent manner (*1, 13*). The molecular mechanism of conversion from IL-2-dependent growth to IL-2-independent growth remains to be elucidated. Tax induced conversion of IL-2-independent growth from IL-2-dependent growth was addressed using a murine IL-2-dependent T-cell line, CTLL-2. Growth of this T-cell line is completely dependent on IL-2, and upon deprivation of IL-2 the cell line ceases cell proliferation and dies by apoptosis. Stable introduction of Tax-expression vector into CTLL-2 cells rendered the cells capable of continuous growth independent of IL-2 (*21*). Tax mutant, Tax703, capable of activating the NF-κB pathway, but not TaxM22 which is not capable of activating this pathway, showed similar results, indicating that the pathway is important for Tax to render the cells capable of continuous growth independent of IL-2. Consistent with this conclusion, we noted that endogenous NF-κB binding activity was induced by Tax and Tax703 but not by TaxM22 in CTLL-2 cells. Thus, Tax can promote cell cycle progression and enables cells to grow independently of IL-2 without apoptosis even in the absence of IL-2 in CTLL-2 Tax transformants (*21*).

Prevention of Apoptosis by Tax

Cell proliferation could be the consequence of coordination of cell cycle progression and prevention of apoptosis. Apoptosis represents an intrinsic active process of programmed cell death, characterized by activation of caspases and DNA fragmentation. If cells die by apoptosis, cells cannot grow in number even when cells divide by cell cycle progression. For example, inappropriate promotion of the cell cycle such as by deregulated activation of oncogenes and pRb/E2F pathway, is often accompanied by apoptosis. This is thought to be an important mechanism that protects cells from uncontrolled cell growth. It is thus necessary to protect cells from apoptosis for cell growth.

In Tax transformant CTLL-2 cells, Tax prevented the cells from apoptosis in the absence of IL-2, enabling the cells to grow continuously independent of exogenous growth factors (*21*). The molecular mechanism of Tax-mediated protection from apoptosis was examined by measuring apoptosis-related genes of *bcl-xl*, *bcl-xs*, *bak*, *bcl-2* and *bad* (*51*). In control CTLL-2 cells, IL-2 deprivation resulted in a marked decrease of *bcl-xl* gene expression. Expression of the *bak* gene was slightly decreased by IL-2 deprivation, but none of the other members of the *bcl* family tested was significantly affected by this. In contrast, Tax- and Tax703-introduced CTLL-2 clones showed a constitutive expression of *bcl-xl* mRNA even in the absence of IL-2, whereas expression of the *bcl-xl* gene was dependent on IL-2 in TaxM22-introduced clones. Constitutive expression of Bcl-x_L protein was confirmed by western blot analysis in Tax- and Tax703-introduced CTLL-2 clones but not in TaxM22-introduced clones. The constitutive expression of Bcl-x_L by Tax in the absence of IL-2 contributes to Tax-induced resistance to apoptosis in the absence of IL-2, and the ability of Tax to induce Bcl-x_L expression is dependent on the NF-κB pathway.

To explore the mechanism of Bcl-x_L induction by Tax, mouse *bcl-xl* promoter was cloned and analyzed by transient transfection assay (*51*). The results revealed that a 3.2-kbp segment of 5′ flanking region of the mouse *bcl-xl* gene was responsible for transcriptional activation in response to Tax and Tax703 but not TaxM22 in Jurkat cells. An NF-κB-like site located at −848 relative to translation initiation site exerted the Tax-dependent activation. Consistent with this finding, nuclear extracts from the Tax transformants showed higher binding activity to the NF-κB-like site than control cells. A supershift experiment with specific antibodies to the NF-κB family proteins indicated that p65, p50 and RelB were bound to the NF-κB-like site. These results demonstrate that Tax induced *bcl-xl* gene expression by direct *trans*-activation of the promoter through the NF-κB pathway. Induction of Tax in JPX-9 cells induced endogenous *bcl-xl* gene expression (*29*). In addition, *bcl-xl* gene expression was higher in HTLV-I-infected cell lines than in HTLV-I-uninfected cell lines. Transient transfection assay using Jurkat cells showed that human *bcl-xl* gene was activated by Tax through the NF-κB and CREB pathways (*29*).

Gene expression profiles of HTLV-I-immortalized T cells were determined by Harhaj *et al.* (*16*) with identification of deregulated expression of genes involved in apoptosis regulation. *bax* gene expression was markedly inhibited in HTLV-I-immortalized T cells compared to PHA-stimulated PBLs. However, the expression levels of *bcl-2* and *bcl-xl* genes were not altered in these cells. Repression of *bax* gene expression has also been reported by Brauweiler *et al.* (*6*). Tax induced activation of *bcl-xl* and repression of *bax* imply that Tax inhibits apoptosis by increasing the Bcl-x_L/Bax ratio in HTLV-I-infected cells.

Several groups have reported enhanced resistance to apoptosis in HTLV-I-infected cells or in cells expressing Tax (*4, 6, 10, 25*). On the other hand, enhanced apoptosis by Tax has been observed (*8, 9, 15, 24, 55*). The difference may reflect variation of cellular circumstances. Deprivation of serum has been shown to induce apoptosis of Tax-transformed rat fibroblast Rat-1 cells (*55*). Tax has been reported to sensitize cells to apoptosis induced by DNA damaging agents (*24*). These effects could be due to inappropriate cell cycle promotion by Tax under conditions in which cell cycle has to be halted by serum starvation or DNA damage. Similarly, tumor cells derived from Tax transgenic mice showed enhanced level of apoptosis and oncogene expression (*15*). This could also be due to inappropriate cell cycle promotion by Tax. Moreover, Chen *et al.* (*8*) also reported induction of apoptosis in JPX-9 cells upon induction of Tax. This could be due to induction of Fas ligand by Tax where Fas is constitutively expressed. Assessed together, these studies indicate that enhanced apoptosis does not seem to be directly mediated by Tax, but rather that it represents an indirect effect of Tax depending on the cellular circumstances.

p53 is often involved in induction of apoptosis. Deregulated mitogenic signals such as activation of oncogenes and pRb/E2F pathway leads to the activation of p53, which provides an important mechanism to prevent abnormal cell proliferation that could lead to tumor development (*53*). Thus, inactivation of p53 is a critical process in tumorigenesis. Several groups have reported functional impairment and aberrant expression of p53 in HTLV-I-infected T-cell lines (*3, 7*). Tax could inactivate p53 through several mechanisms: it could suppress transcriptional activity of p53 by interfering with the N-terminal *trans*-activation domain (*36*), and induce phosphorylation of p53 at Ser15 and Ser392, which could consequently inactivate the protein (*37, 39*). Other studies indicated that Tax suppresses transcriptional activity of p53 by sequestering co-activator CBP (*5, 30, 49, 52*). Tax may utilize alternative pathways for inactivation of p53 depending on the cell lines (*38*). Thus, Tax-induced prevention of T-cell apoptosis is mediated, at least in part, by inactivation of p53.

Current Model for Tax-mediated Cell Growth in Human T Cells

Based on results from our laboratories and those of other investigators, a model can be drawn for the mechanisms of Tax-induced promotion of cell growth (Fig. 8). Broadly speaking, Tax can promote cell growth by stimulating cell cycle progression and by preventing cells from apoptosis. Tax-induced cell cycle progression is primarily mediated by activation of CDKs, the central players of cell cycle progression. Tax induces direct *trans*-activation of *cyclin D2* and *CDK4*, *CDK6* and *CDK2* genes. Reduction of $p27^{Kip1}$ and $p19^{INK4d}$ is closely linked to activation of CDKs. Such activation initiates a cascade of events associated with cell cycle progression followed by phosphorylation of pRb family proteins and activation of E2F, leading to progression into the S phase. Tax also induces expression of *bcl-xl* gene and repression of *bax* gene, which protect cells from apoptosis. Inactivation of p53 by Tax also contributes to prevention of cells from apoptosis. The prevention of apoptosis allows growth of host T cells which are promoted in the cell cycle by Tax, resulting in expansion of HTLV-I-infected T cells, some of which undergo secondary genetic changes due to genome instability induced by inactivation of p53 and repression of *DNA polymerase* β gene expression by Tax.

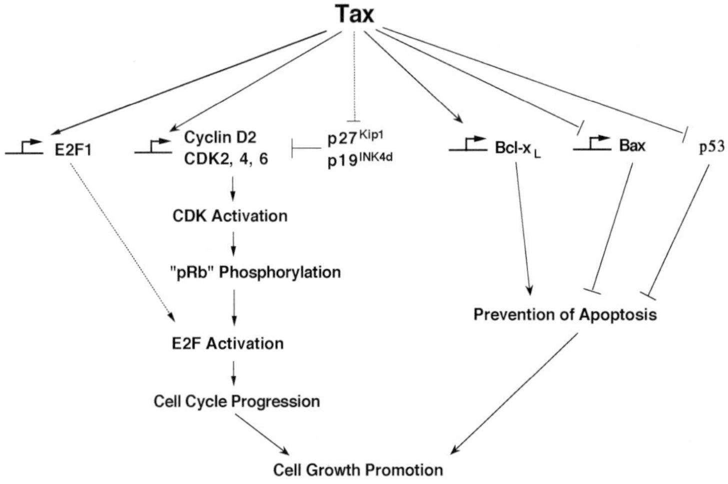

FIG. 8. A schematic model of Tax-mediated cell growth in human T cells See text for details. "pRb" indicates pRb family proteins.

Future Perspectives

Our results, together with those of other investigators, clearly demonstrate that Tax can induce cell cycle progression in resting human T cells by activating CDKs. Tax-induced activation of CDK is at least partially mediated by direct *trans*-activation of cell cycle regulatory genes such as *cyclin D2*. There seem to be multiple effects of Tax on induction of CDK activities, which remain to be elucidated. For example, our results show that expression levels of CDK inhibitors $p19^{INK4d}$ and $p27^{Kip1}$ are reduced upon expression of Tax. In particular, reduction of $p19^{INK4d}$ protein is far greater than that of $p19^{INK4d}$ mRNA, indicating that Tax alters $p19^{INK4d}$ at a protein level. The ubiquitin/proteasome-mediated protein degradation pathway is involved in regulation of CDK inhibitors, suggesting that Tax may affect this degradation pathway. Neuveut *et al.* (*32*) reported that Tax directly bound to cyclin D3 and induced phosphorylation of the protein, although its exact role in this binding is yet to be elucidated. Activation of CDKs by Tax seems to be tissue-specific, because Tax could not activate endogenous E2F in resting fibroblasts. The molecular mechanisms of tissue-specific activation of CDKs by Tax are totally unknown at present. Considering that HTLV-I induces T-cell-specific malignancy, tissue-specific activation of CDKs by Tax could be the underlying basis for the pathogenesis and is an important issue to be explored.

Tax alone is sufficient to promote cell growth. However, accumulating evidence indicates that Tax is not sufficient for full transformation of host T cells, indicating that additional genetic changes are required. Genome instability induced by Tax could contribute to the development of additional genetic changes. However, the exact nature of the additional changes is yet to be elucidated. Understanding the molecular mechanism of T-cell-specific growth promotion by Tax and of the exact nature and effects of additional genetic changes on full transformation may allow the development of new therapeutic agents that could suppress growth of the transformed T cells.

REFERENCES

1. Akagi, T., Ono, H., and Shimotohno, K. Characterization of T cells immortalized by Tax1 of human T-cell leukemia virus type 1. *Blood*, **86**, 4243–4249 (1995).
2. Akagi, T., Ono, H., and Shimotohno, K. Expression of cell-cycle regulatory genes in HTLV-I infected T-cell lines: possible involvement of Tax1 in the altered expression of cyclin D2, $p18^{Ink4}$ and $p21^{Waf1/Cip1/Sdi1}$. *Oncogene*, **12**, 1645–1652 (1996).
3. Akagi, T., Ono, H., Tsuchida, N., and Shimotohno, K. Aberrant expression and function of p53 in T-cells immortalized by HTLV-I Tax1. *FEBS Lett.*, **406**, 263–266 (1997).
4. Arai, M., Kannagi, M., Matsuoka, M., Sato, T., Yamamoto, N., and Fujii, M. Expression of FAP-1 (Fas-associated phosphatase) and resistance to Fas-mediated apoptosis in T cell lines derived from human T cell leukemia virus type 1-associated myelopathy/tropical spastic paraparesis patients. *AIDS Res. Hum. Retroviruses*, **14**, 261–267 (1998).
5. Ariumi, Y., Kaida, A., Lin, J. Y., Hirota, M., Masui, O., Yamaoka, S., Taya, Y., and Shimotohno, K. HTLV-1 tax oncoprotein represses the p53-mediated trans-activation function through coactivator CBP sequestration. *Oncogene*, **19**, 1491–1499 (2000).
6. Brauweiler, A., Garrus, J. E., Reed, J. C., and Nyborg, J. K. Repression of bax gene expression by the HTLV-1 Tax protein: implications for suppression of apoptosis in virally infected cells. *Virology*, **231**, 135–140 (1997).
7. Cereseto, A., Diella, F., Mulloy, J. C., Cara, A., Michieli, P., Grassmann, R., Franchini, G., and Klotman, M. E. p53 functional impairment and high $p21^{waf1/cip1}$ expression in human T-cell lymphotropic/leukemia virus type I-transformed T cells. *Blood*, **88**, 1551–1560 (1996).
8. Chen, X., Zachar, V., Zdravkovic, M., Guo, M., Ebbesen, P., and Liu, X. Role of the Fas/Fas ligand pathway in apoptotic cell death induced by the human T cell lymphotropic virus type I Tax transactivator. *J. Gen. Virol.*, **78**, 3277–3285 (1997).
9. Chlichlia, K., Moldenhauer, G., Daniel, P. T., Busslinger, M., Gazzolo, L., Schirrmacher, V., and Khazaie, K. Immediate effects of reversible HTLV-1 tax function: T-cell activation and apoptosis. *Oncogene*, **10**, 269–277 (1995).
10. Copeland, K. F., Haaksma, A. G., Goudsmit, J., Krammer, P. H., and Heeney, J. L. Inhibition of apoptosis in T cells expressing human T cell leukemia virus type I Tax. *AIDS Res. Hum. Retroviruses*, **10**, 1259–1268 (1994).
11. de La Fuente, C., Santiago, F., Chong, S. Y., Deng, L., Mayhood, T., Fu, P., Stein, D., Denny, T., Coffman, F., Azimi, N., Mahieux, R., and Kashanchi, F. Overexpression of $p21^{waf1}$ in human T-cell lymphotropic virus type 1-infected cells and its association with cyclin A/cdk2. *J. Virol.*, **74**, 7270–7283 (2000).
12. Felber, B. K., Paskalis, H., Kleinman-Ewing, C., Wong-Staal, F., and Pavlakis, G. N. The pX protein of HTLV-I is a transcriptional activator of its long terminal repeats. *Science*, **229**, 675–679 (1985).
13. Grassmann, R., Dengler, C., Muller-Fleckenstein, I., Fleckenstein, B., McGuire, K., Dokhelar, M. C., Sodroski, J. G., and Haseltine, W. A. Transformation to continuous growth of primary human T lymphocytes by human T-cell leukemia virus type I X-region genes transduced by a *Herpesvirus saimiri* vector. *Proc. Natl. Acad. Sci. USA*, **86**, 3351–3355 (1989).
14. Grossman, W. J., Kimata, J. T., Wong, F. H., Zutter, M., Ley, T. J., and Ratner, L. Development of leukemia in mice transgenic for the tax gene of human T-cell leukemia

virus type I. *Proc. Natl. Acad. Sci. USA*, **92**, 1057–1061 (1995).

15. Hall, A. P., Irvine, J., Blyth, K., Cameron, E. R., Onions, D. E., and Campbell, M. E. Tumours derived from HTLV-I tax transgenic mice are characterized by enhanced levels of apoptosis and oncogene expression. *J. Pathol.*, **186**, 209–214 (1998).
16. Harhaj, E. W., Good, L., Xiao, G., and Sun, S. C. Gene expression profiles in HTLV-I-immortalized T cells: deregulated expression of genes involved in apoptosis regulation. *Oncogene*, **18**, 1341–1349 (1999).
17. Hinuma, Y., Nagata, K., Hanaoka, M., Nakai, M., Matsumoto, T., Kinoshita, K. I., Shirakawa, S., and Miyoshi, I. Adult T-cell leukemia: antigen in an ATL cell line and detection of antibodies to the antigen in human sera. *Proc. Natl. Acad. Sci. USA*, **78**, 6476–6480 (1981).
18. Hori, T., Uchiyama, T., Tsudo, M., Umadome, H., Ohno, H., Fukuhara, S., Kita, K., and Uchino, H. Establishment of an interleukin 2-dependent human T cell line from a patient with T cell chronic lymphocytic leukemia who is not infected with human T cell leukemia/lymphoma virus. *Blood*, **70**, 1069–1072 (1987).
19. Huang, Y., Ohtani, K., Iwanaga, R., Matsumura, Y., and Nakamura, M. Direct transactivation of the human cyclin D2 gene by the oncogene product Tax of human T-cell leukemia virus type I. *Oncogene*, **20**, 1094–1102 (2001).
20. Iwanaga, R., Ohtani, K., Hayashi, T., and Nakamura, M. Molecular mechanism of cell cycle progression induced by the oncogene product Tax of human T-cell leukemia virus type I. *Oncogene*, **20**, 2055–2067 (2001).
21. Iwanaga, Y., Tsukahara, T., Ohashi, T., Tanaka, Y., Arai, M., Nakamura, M., Ohtani, K., Koya, Y., Kannagi, M., Yamamoto, N., and Fujii, M. Human T-cell leukemia virus type 1 tax protein abrogates interleukin-2 dependence in a mouse T-cell line. *J. Virol.*, **73**, 1271–1277 (1999).
22. Jeang, K. T., Widen, S. G., Semmes, O. J. T., and Wilson, S. H. HTLV-I trans-activator protein, tax, is a trans-repressor of the human β-polymerase gene. *Science*, **247**, 1082–1084 (1990).
23. Jin, D. Y., Spencer, F., and Jeang, K. T. Human T cell leukemia virus type 1 oncoprotein Tax targets the human mitotic checkpoint protein MAD1. *Cell*, **93**, 81– 91 (1998).
24. Kao, S. Y., Lemoine, F. J., and Mariott, S. J. HTLV-1 Tax protein sensitizes cells to apoptotic cell death induced by DNA damaging agents. *Oncogene*, **19**, 2240–2248 (2000).
25. Kishi, S., Saijyo, S., Arai, M., Karasawa, S., Ueda, S., Kannagi, M., Iwakura, Y., Fujii, M., and Yonehara, S. Resistance to fas-mediated apoptosis of peripheral T cells in human T lymphocyte virus type I (HTLV-I) transgenic mice with autoimmune arthropathy. *J. Exp. Med.*, **186**, 57–64 (1997).
26. Lemasson, I., Thebault, S., Sardet, C., Devaux, C., and Mesnard, J. M. Activation of E2F-mediated transcription by human T-cell leukemia virus type I Tax protein in a $p16^{INK4A}$-negative T-cell line. *J. Biol. Chem.*, **273**, 23598–23604 (1998).
27. Low, K. G., Dorner, L. F., Fernando, D. B., Grossman, J., Jeang, K. T., and Comb, M. J. Human T-cell leukemia virus type 1 Tax releases cell cycle arrest induced by $p16^{INK4a}$. *J. Virol.*, **71**, 1956–1962 (1997).
28. Maruyama, M., Shibuya, H., Harada, H., Hatakeyama, M., Seiki, M., Fujita, T., Inoue, J., Yoshida, M., and Taniguchi, T. Evidence for aberrant activation of the interleukin-2 autocrine loop by HTLV-1-encoded p40x and T3/Ti complex triggering. *Cell*,

48, 343–350 (1987).

29. Mori, N., Fujii, M., Cheng, G., Ikeda, S., Yamasaki, Y., Yamada, Y., Tomonaga, M., and Yamamoto, N. Human T-cell leukemia virus type I tax protein induces the expression of anti-apoptotic gene Bcl-xL in human T-cells through nuclear factor-κB and c-AMP responsive element binding protein pathways. *Virus Genes*, **22**, 279–287 (2001).
30. Mulloy, J. C., Kislyakova, T., Cereseto, A., Casareto, L., LoMonico, A., Fullen, J., Lorenzi, M. V., Cara, A., Nicot, C., Giam, C., and Franchini, G. Human T-cell lymphotropic/leukemia virus type 1 Tax abrogates p53-induced cell cycle arrest and apoptosis through its CREB/ATF functional domain. *J. Virol.*, **72**, 8852–8860 (1998).
31. Nerenberg, M., Hinrichs, S. H., Reynolds, R. K., Khoury, G., and Jay, G. The tat gene of human T-lymphotropic virus type 1 induces mesenchymal tumors in transgenic mice. *Science*, **237**, 1324–1329 (1987).
32. Neuveut, C., Low, K. G., Maldarelli, F., Schmitt, I., Majone, F., Grassmann, R., and Jeang, K. T. Human T-cell leukemia virus type 1 Tax and cell cycle progression: role of cyclin D-cdk and p110Rb. *Mol. Cell. Biol.*, **18**, 3620–3632 (1998).
33. Nevins, J. R., Leone, G., DeGregori, J., and Jakoi, L. Role of the Rb/E2F pathway in cell growth control. *J. Cell. Physiol.*, **173**, 233–236 (1997).
34. Ohtani, K. Implication of transcription factor E2F in regulation of DNA replication. *Front. Biosci.*, **4**, D793–D804 (1999).
35. Ohtani, K., Iwanaga, R., Arai, M., Huang, Y., Matsumura, Y., and Nakamura, M. Cell type-specific E2F activation and cell cycle progression induced by the oncogene product Tax of human T-cell leukemia virus type I. *J. Biol. Chem.*, **275**, 11154–11163 (2000).
36. Pise-Masison, C. A., Choi, K. S., Radonovich, M., Dittmer, J., Kim, S. J., and Brady, J. N. Inhibition of p53 transactivation function by the human T-cell lymphotropic virus type 1 Tax protein. *J. Virol.*, **72**, 1165–1170 (1998).
37. Pise-Masison, C. A., Mahieux, R., Jiang, H., Ashcroft, M., Radonovich, M., Duvall, J., Guillerm, C., and Brady, J. N. Inactivation of p53 by human T-cell lymphotropic virus type 1 Tax requires activation of the NF-κB pathway and is dependent on p53 phosphorylation. *Mol. Cell. Biol.*, **20**, 3377–3386 (2000).
38. Pise-Masison, C. A., Mahieux, R., Radonovich, M., Jiang, H., and Brady, J. N. Human T-lymphotropic virus type I Tax protein utilizes distinct pathways for p53 inhibition that are cell type-dependent. *J. Biol. Chem.*, **276**, 200–205 (2001).
39. Pise-Masison, C. A., Radonovich, M., Sakaguchi, K., Appella, E., and Brady, J. N. Phosphorylation of p53: a novel pathway for p53 inactivation in human T-cell lymphotropic virus type 1-transformed cells. *J. Virol.*, **72**, 6348–6355 (1998).
40. Poiesz, B. J., Ruscetti, F. W., Gazdar, A. F., Bunn, P. A., Minna, J. D., and Gallo, R. C. Detection and isolation of type C retrovirus particles from fresh and cultured lymphocytes of a patient with cutaneous T-cell lymphoma. *Proc. Natl. Acad. Sci. USA*, **77**, 7415–7419 (1980).
41. Pozzatti, R., Vogel, J., and Jay, G. The human T-lymphotropic virus type I tax gene can cooperate with the ras oncogene to induce neoplastic transformation of cells. *Mol. Cell. Biol.*, **10**, 413–417 (1990).
42. Santiago, F., Clark, E., Chong, S., Molina, C., Mozafari, F., Mahieux, R., Fujii, M., Azimi, N., and Kashanchi, F. Transcriptional up-regulation of the cyclin D2 gene and acquisition of new cyclin-dependent kinase partners in human T-cell leukemia virus

type 1-infected cells. *J. Virol.*, **73**, 9917–9927 (1999).
43. Schmitt, I., Rosin, O., Rohwer, P., Gossen, M., and Grassmann, R. Stimulation of cyclin-dependent kinase activity and G1- to S-phase transition in human lymphocytes by the human T-cell leukemia/lymphotropic virus type 1 Tax protein. *J. Virol.*, **72**, 633–640 (1998).
44. Seiki, M., Inoue, J., Takeda, T., and Yoshida, M. Direct evidence that p40x of human T-cell leukemia virus type I is a trans-acting transcriptional activator. *EMBO J.*, **5**, 561–565 (1986).
45. Sherr, C. J. G1 phase progression: cycling on cue. *Cell*, **79**, 551–555 (1994).
46. Sodroski, J., Rosen, C., Goh, W. C., and Haseltine, W. A transcriptional activator protein encoded by the x-lor region of the human T-cell leukemia virus. *Science*, **228**, 1430–1434 (1985).
47. Suzuki, T., Kitao, S., Matsushime, H., and Yoshida, M. HTLV-1 Tax protein interacts with cyclin-dependent kinase inhibitor $p16^{INK4A}$ and counteracts its inhibitory activity towards CDK4. *EMBO J.*, **15**, 1607–1614 (1996).
48. Suzuki, T., Narita, T., Uchida-Toita, M., and Yoshida, M. Down-regulation of the INK4 family of cyclin-dependent kinase inhibitors by tax protein of HTLV-1 through two distinct mechanisms. *Virology*, **259**, 384–391 (1999).
49. Suzuki, T., Uchida-Toita, M., and Yoshida, M. Tax protein of HTLV-1 inhibits CBP/p300-mediated transcription by interfering with recruitment of CBP/p300 onto DNA element of E-box or p53 binding site. *Oncogene*, **18**, 4137–4143 (1999).
50. Tanaka, A., Takahashi, C., Yamaoka, S., Nosaka, T., Maki, M., and Hatanaka, M. Oncogenic transformation by the tax gene of human T-cell leukemia virus type I *in vitro*. *Proc. Natl. Acad. Sci. USA*, **87**, 1071–1075 (1990).
51. Tsukahara, T., Kannagi, M., Ohashi, T., Kato, H., Arai, M., Nunez, G., Iwanaga, Y., Yamamoto, N., Ohtani, K., Nakamura, M., and Fujii, M. Induction of $Bcl\text{-}x_L$ expression by human T-cell leukemia virus type 1 Tax through NF-κB in apoptosis-resistant T-cell transfectants with Tax. *J. Virol.*, **73**, 7981–7987 (1999).
52. Van Orden, K., Giebler, H. A., Lemasson, I., Gonzales, M., and Nyborg, J. K. Binding of p53 to the KIX domain of CREB binding protein: a potential link to human T-cell leukemia virus type I-associated leukemogenesis. *J. Biol. Chem.*, **274**, 26321–26328 (1999).
53. Vousden, K. H. p53: death star. *Cell*, **103**, 691–694 (2000).
54. Wu, L., Timmers, C., Maiti, B., Saavedra, H., Sang, I. L., Chong, G. T., Nuckolls, F., Giangrande, P., Wright, F. A., Field, S. J., Greenberg, M. E., Orkin, S., Nevins, J. R., Robinson, M. L., and Leone, G. The E2F1–3 transcription factors are essential for cellular proliferation. *Nature*, **414**, 457–462 (2001).
55. Yamada, T., Yamaoka, S., Goto, T., Nakai, M., Tsujimoto, Y., and Hatanaka, M. The human T-cell leukemia virus type I Tax protein induces apoptosis which is blocked by the Bcl-2 protein. *J. Virol.*, **68**, 3374–3379 (1994).
56. Yoshida, M. Multiple viral strategies of HTLV-1 for dysregulation of cell growth control. *Annu. Rev. Immunol.*, **19**, 475–496 (2001).
57. Yoshida, M., Miyoshi, I., and Hinuma, Y. Isolation and characterization of retrovirus from cell lines of human adult T-cell leukemia and its implication in the disease. *Proc. Natl. Acad. Sci. USA*, **79**, 2031–2035 (1982).

MULTIPLE ROLES OF CELLULAR EXPORT MACHINERY IN HTLV-1 Rex FUNCTIONING

Hisatoshi SHIDA and Yoshiyuki HAKATA
*Institute for Genetic Medicine, Hokkaido University**

Rex protein is a post-transcriptional regulator that is required for the expression of unspliced and incompletely spliced viral mRNAs that encode Gag and Env proteins. Rex binds directly Rex response element (RxRE) in human T cell leukemia virus type 1 (HTLV-1) RNAs and then undergoes multimerization on the RNA, which is necessary to export viral RNAs. Human CRM1, which is cellular machinery to export proteins from the nucleus to the cytoplasm, has been found to bind to Rex directly. Therefore, Rex executes its function by bridging viral RNAs and the cellular export machinery.

In the first half of this review article we describe additional roles of CRM1 in Rex functioning, such as involvement in Rex multimerization on RNA. In the latter half, we report that rat CRM1 export little viral RNA because it has poor ability to induce the Rex multimerization. Our results suggest that the poor ability of rat CRM1 as a cofactor of Rex may be responsible for the poor replication of HTLV-1 in rats. Such results provide a clue to means of improving presently incomplete rat models of HTLV-1-related diseases.

Human T cell leukemia virus type 1 (HTLV-1) is a causative agent of both adult T cell leukemia (ATL), which is an aggressive malignancy of T cells (*16, 22*), and the chronic neurodegenerative disorder, HTLV-1-associated myelopathy/tropical spastic paraparesis (HAM/TSP) (*21, 33*). Detailed molecular biological studies at cellular levels revealed the structure of HTLV-1 genomes and their functions in relation to viral replication and cellular transformation. Despite these fruitful studies the pathogenesis and oncogenesis of HTLV-1 still remain to be fully understood. To link the recent advances in molecular biology and immunology on HTLV-1 to full understanding of HTLV-1 pathology, which will lead to the development of therapeutic and preventative methods, convenient animal models of HTLV-1 infection are required. In this review article we will describe our recent progress on the molecular biology of HTLV-1, and then propose a way to develop a novel small animal model.

Molecular Mode of Rex Action

HTLV-1 belongs to a group of complex retroviruses that encode transactivator proteins for their gene expression. These transactivator proteins include Tax protein,

* Hokkaido 060-0815, Japan

which activates its own viral and cellular transcription (*45*), and Rex protein, a post-transcriptional regulator that is required for the expression of unspliced and incompletely spliced viral mRNAs that encode Gag and Env proteins, respectively (*6, 15, 19*). These proteins are therefore essential for viral replication.

Rex is a 27-kDa phosphoprotein that shuttles between the nucleus and the cytoplasm (*3, 24*). In the nucleus, Rex binds directly and specifically to the highly structured *cis*-acting Rex response element (RxRE) which is encoded by sequences within the 3′ long terminal repeat (LTR) of HTLV-1 (*1, 37*). The arginine-rich basic domain comprised by amino acids (a.a.) 1 to 19 of the 189 a.a. Rex protein has been shown to mediate its binding to the RxRE (*1, 4, 9, 12*). This region also serves as a nuclear/nucleolar localization signal (NLS) and binds to importin β directly (*5, 32, 35, 36, 39*) (Fig. 1).

Another important Rex domain is a leucine-rich region spanning a.a. 81 to 94 (*3, 24*) that functions as a nuclear export signal (NES) (Fig. 1). This domain binds directly to human CRM1 (human CRM1, exportin 1, XPO1) (*11*), a member of the importin β family, in cooperation with a GTP bound form of small G-protein, Ran (Ran-GTP). It is well known that Ran-GTP is located in the nucleus, while a GDP bound form of Ran is in the cytoplasm. This gradient of Ran across the nuclear envelope is a key in the decision of directionality of a transport machine. Human CRM1 has been shown to be a component of the cellular machinery involved in the protein export of various leucine-rich NES-bearing proteins (*7, 8, 26, 31, 34, 40*). The interaction of Rex, human CRM1, Ran-GTP, and Rex target viral mRNA generates the export complex which is then transported to the cytoplasm. In the cytoplasm Ran-GTP is converted to Ran-GDP with the aid of RanBP1 and Ran-GAP, leading to dissociation of the export complex. The liberated Rex protein binds to importin β to

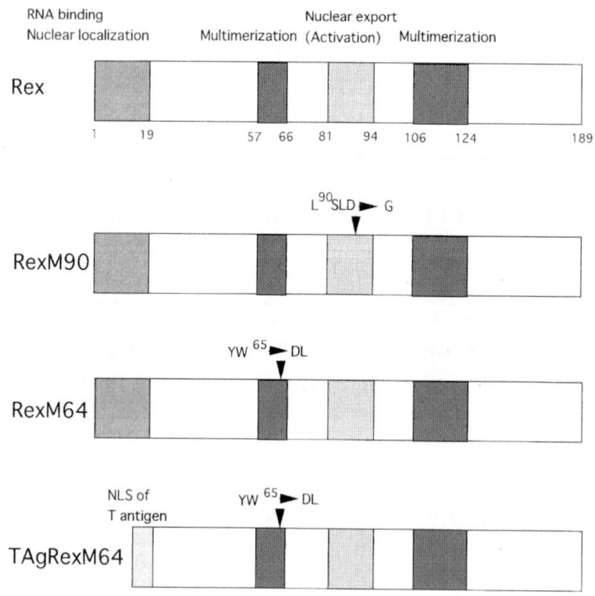

FIG. 1. Schematic presentation of domain structures of Rex protein

return to the nucleus. In the nucleus Ran-GTP associates with importin β, resulting in release of Rex (Fig. 2).

Although the view described above appears to fully elucidate the molecular mode of Rex action, it does not include the necessity and mechanism of multimerization of Rex proteins on the viral RNA. It has been generally agreed that multimerization of Rex on the target viral mRNA is critical for Rex-mediated mRNA transport (*1, 4, 9, 12*), although Rex protein itself does not require multimerization to exit the nucleus (*14*). A third Rex domain that maps to a.a. 57 to 66 and 106 to 124 has been reported to mediate Rex multimerization (*2, 14, 44*) (Fig. 1). The ability of Rex protein to form homodimer is prerequisite for Rex multimerization on its cognate RNA (*2*).

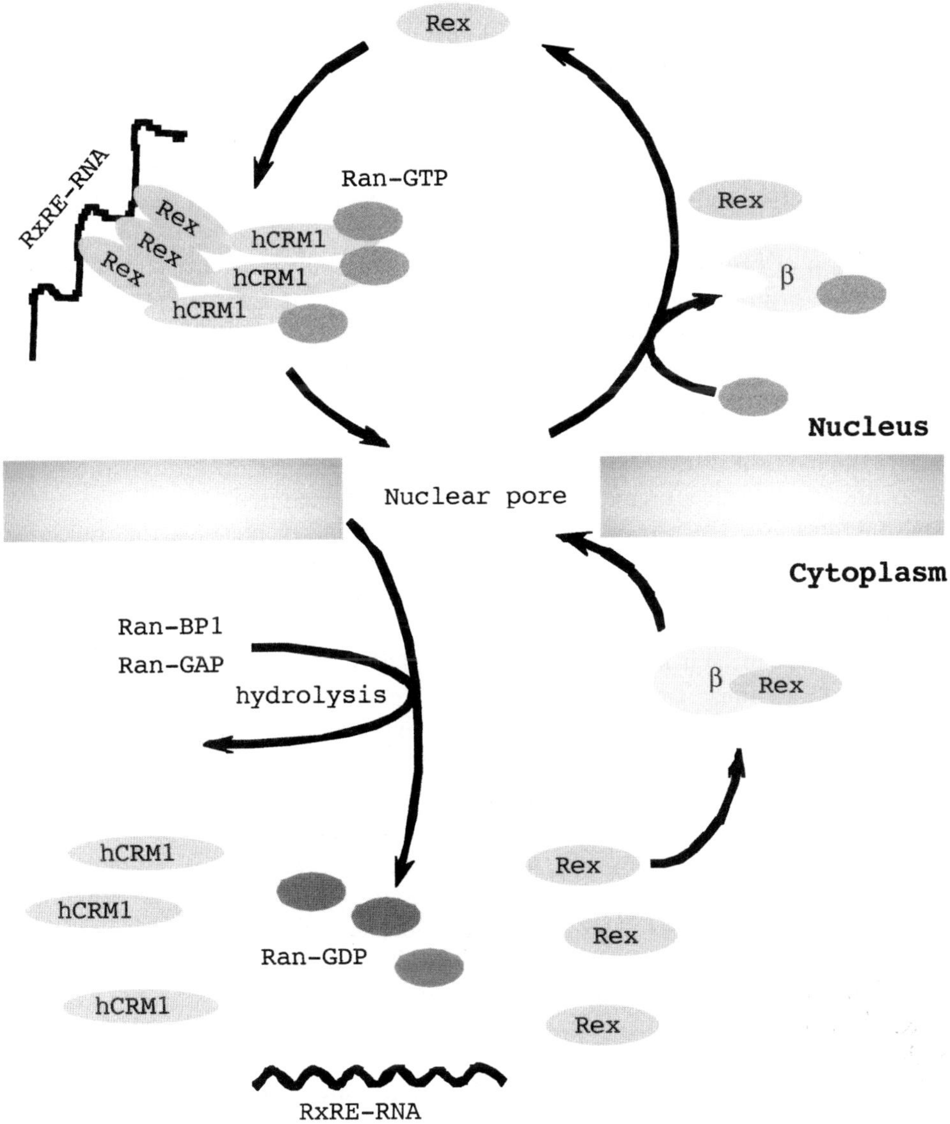

FIG. 2. Overview of Rex action

Involvement of CRM1 in Homodimer Formation of Rex Protein

We conducted an examination of whether a cellular cofactor takes part in the oligomerization of Rex as a first step to form the process of the export complex formation.

At first we constructed a Rex mutant, RexM90, which has a mutation in the activation (NES) domain, and examined its affinity for human CRM1 by mammalian two-hybrid assay. Although RexM90 had a greatly reduced affinity for human CRM1, unexpectedly it still had the residual ability (approximately 20% of wild-type Rex) to interact with human CRM1. Thus, we were interested whether RexM90 still retains the residual activity to produce Env protein. RexM90 showed very low activity (approximately 3% of wild-type Rex at most). The low activity of RexM90 appeared to be inconsistent with its ability to interact with human CRM1. To explore the cause of this low activity, we analyzed its capacity to form homodimer by mammalian two-hybrid analysis. The combination of GAL-RexM90 with RexM90-VP did not result in the formation of homo-oligomers. Thus, the multimerization defect of RexM90 may account for its very weak capacity to induce Env production (*11*).

The above results raise the possibility that a cofactor may be involved in the multimerization of Rex protein. If so, the activities of RexM64, which has a mutation in the multimerization domain and retains an intact NES, might be complemented by overexpression of human CRM1. We tested this possibility by cotransfecting CRM1 expression plasmid along with the wild-type or mutant Rex expressing plasmid. Overexpression of human CRM1, in fact, restored both Env producing activity and multimerization of RexM64 up to one third of the level of wild-type Rex in a dose-dependent manner, whereas it had a little effect on RexM90. The simultaneous restoration of RexM64 activity and multimerization by overexpression of human CRM1 supports the hypothesis that human CRM1 is involved in the multimerization of Rex protein, and that this is a prerequisite for its biological activity (*11*).

If human CRM1 is involved in the multimerization of not only the mutant Rex but also wild-type Rex protein, it is conceivable that reduction of intracellular concentration of free human CRM1 may abrogate multimerization of wild-type Rex proteins. To test this possibility, we used TAgRexM64, which has an intact NES and the NLS of SV40 T antigen in place of the normal RNA binding/NLS domain of the RexM64 protein, since TAgRexM64 inhibits Rex function by sequestering human CRM1 (*23*). We cotransfected a TAgRexM64 expression plasmid in the two hybrid assay. The dimer formation of Rex was severely inhibited by coexpression of TAgRexM64. Next, we examined whether or not overexpression of human CRM1 suppresses the inhibitory effect of TAgRexM64 on multimerization. The extent of multimerization of Rex gradually increased with the amount of human CRM1 expression plasmid cotransfected into cells. These results further support a hypothesis on the involvement of human CRM1 in the multimerization of Rex proteins (*11*).

Taken together, the above results conclusively indicate that human CRM1 participates both in export of Rex proteins and multimerization of Rex proteins on target viral RNA. These two activities of human CRM1 are necessary to export viral mRNA complex.

With respect to the formation of the transport complex composed of viral RNA, Rex, human CRM1 and possibly other cellular factors, we propose the following

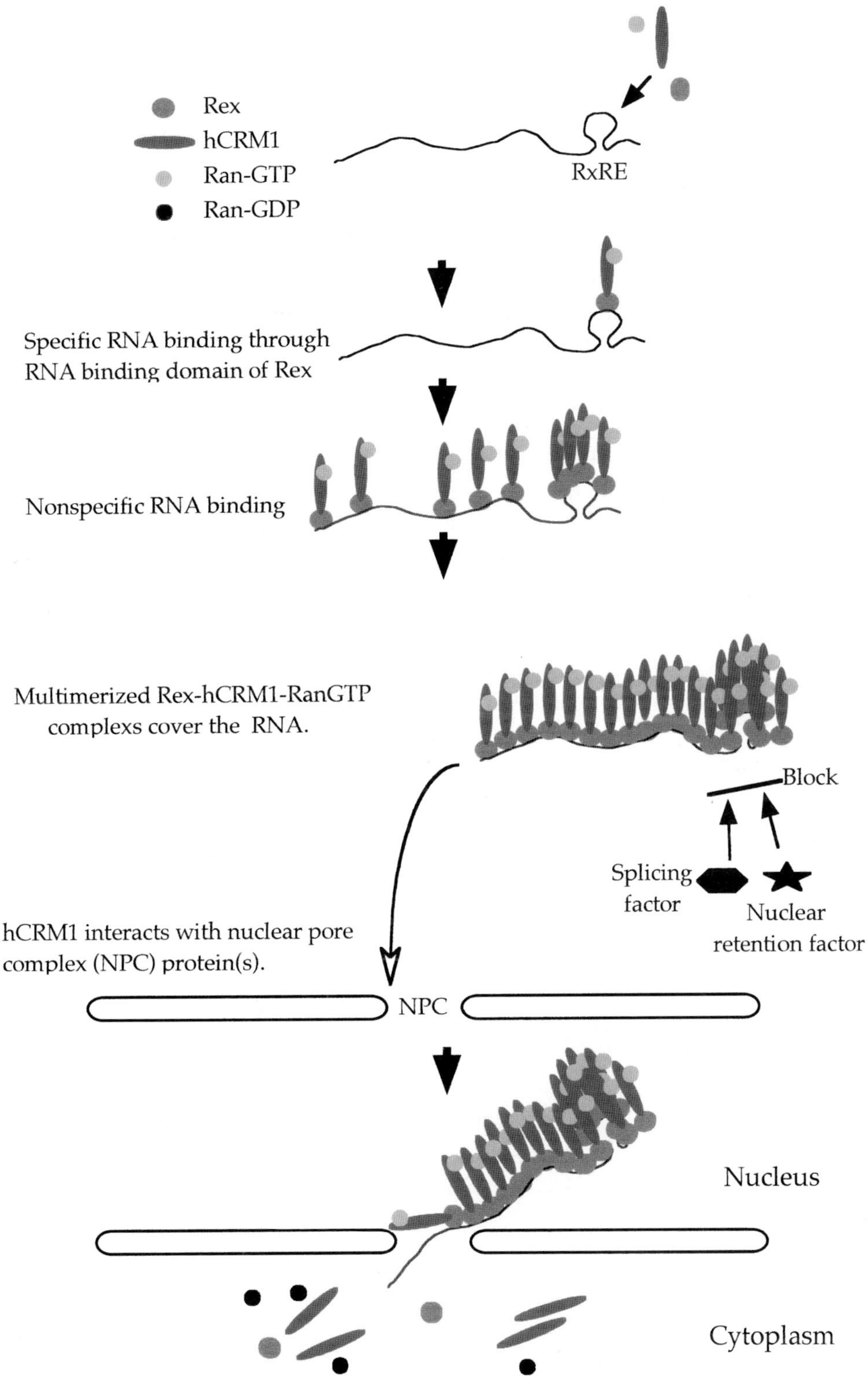

FIG. 3. Effect of multimerization of Rex on viral RNA export

schemes. First, a Rex-human CRM1-Ran-GTP complex is formed and mobilized to the high affinity site in RxRE. Additional complexes then bind to the viral RNA through its nonspecific RNA-binding activity of Rex (*13*), eventually provoking Rex polymerization along the viral RNA transcript via its multimerization domain with the aid of human CRM1. Alternatively, Rex-human CRM1-Ran-GTP complexes may be formed on the RNA. In any case this process would allow a number of Rex proteins to cover the RNA, thus preventing attachment of the factors, which are involved in splicing and nuclear retention of RNA, and increasing the number of human CRM1 molecules that associate with the complex. The association of multiple CRM1 molecules may be important to overcome the factors that retain RNAs in the nucleus (*30*), and would allow the RNAs to pass efficiently through nuclear pores. This overview is illustrated in Fig. 3.

Animal Models of HTLV-1 Infection

HTLV-1 is known to be able to infect experimental animals such as monkeys, rabbits, and rats. Although monkey and rabbit models have been used in vaccine development and to study mother-to-child transmission caused by breast-feeding, and in some cases ATL (*18, 38, 42, 43*), the utility of these models is limited, because of difficulty in dealing with a number of these animals and the lack of inbred strains. Consequently, it would be more convenient to use small animal models, in particular mouse and rat models because their inbred strains are well characterized and can be genetically manipulated. With regard to mouse models, although attempts to infect mice with HTLV-1 have been reported, the infection was limited and could be detected only with PCR that measured the integrated genomes of HTLV-1 (*6*). Rat models, on the other hand, have been extensively developed to study HTLV-1-associated diseases, such as HAM/TSP-like and lymphoproliferative diseases to analyze their pathogenic mechanisms (*20, 28*). However, these models do not completely reproduce human diseases. To better understand the mechanism by which ATL or HAM/TSP develops, a better rat model is required. The poor replication of HTLV-1 in rats may be the most important reason for the differences between human and rat HTLV-1-associated diseases. Presently available animal models of HTLV-1 infection are summarized in Table I.

HTLV-1 has been reported capable of infecting a number of rat cells, which indicates that these cells possess the receptors for viral attachment and penetration

TABLE I. Animal Models of HTLV-1 Infection

Animal	Advantages	Disadvantages
Rabbits	Medium infection efficiency Established mother-to-child infection Use for prophylactic vaccine	Variable viral loads Difficult to deal with many animals
Monkeys	Medium infection efficiency Use for prophylactic vaccine	Difficult to deal with many animals
Rats	Ease of dealing with many animals Onset of HAM-like disease Feasibility of genetic manipulation	Low infection efficiency
Mice	Ease of dealing with many animals Establishment of genetic manipulation	Very low infection efficiency

into the cells (*29, 41*). Thus, the blocking of viral propagation must occur at subsequent steps within the rat cells. Identification of the blocking step and the responsible host factor(s) could lead to the construction of transgenic rats which express the critical human factor(s) and which are highly susceptible to HTLV-1 infection. Such transgenic rats might be expected to develop ATL and HAM/TSP-like diseases that more closely resemble the human diseases.

Rat CRM1 Is a Poor Cofactor of Rex

The first hint of a blocking step was a report that showed that the viral mRNAs encoding Gag and Env proteins are produced at low levels in rat cells despite the fact that the mRNA for Tax/Rex protein is abundant (*25*). On the basis of these observations, we hypothesized that Rex function may be impaired in rat cells. To test Rex function in rat and human cells, we performed a reporter assay using the plasmid pDM128RxRE which allows Rex activity to be quantitated in a transient expression system (*11, 17*), because this plasmid expresses chloramphenicol acetyltransferase (CAT) protein in a manner that is dependent on Rex function. Human HeLa and rat REF52 cells were transfected with varying amounts of the Rex expression plasmid, and the level of CAT and Rex proteins in both cell types were evaluated by CAT ELISA and western blotting, respectively. Introduction of Rex into the human cells enhanced CAT production strikingly, whereas in the rat cells Rex had almost no effect on CAT production in spite of an adequate expression level of Rex protein (*10*). These results indicate that Rex functions very poorly in the rat cells.

To test whether the poor activity of Rex in rat cells may be due to the weaker ability of rat CRM1 to support Rex functions relative to human CRM1, we cloned a cDNA encoding rat CRM1. Sequence analysis revealed that the nucleotide sequence of the rat CRM1 cDNA was 89% identical to that of human CRM1, while the predicted a.a. sequences were 98% identical, with only 24 a.a. being different (*10*). To investigate the capacity of rat CRM1 to support Rex function, we introduced rat or human CRM1 into REF52 cells and measured Rex activity. Human CRM1 augmented Rex activity in a dose-dependent manner, whereas overexpression of rat CRM1 had little effect (*10*). These results indicate that rat CRM1 can only weakly support Rex function, suggesting that this may be the primary reason why Rex has limited activity in rat cells. It is also possible, however, that there may be another limiting factor that hampers the cofactor activity of rat CRM1 but not human CRM1 in rat cells. To exclude this possibility, we tested the ability of rat and human CRM1 to support Rex function in human cells. We expressed human or rat CRM1 exogenously in HeLa cells in the presence of leptomycin B (LMB) and measured Rex activity. LMB specifically inactivates CRM1 (*27*) and thus inhibits Rex activity. Overexpression of rat CRM1 should overcome the inhibitory effect of LMB if it works well as a cofactor of Rex in HeLa cells. We found, however, that overexpression of only human CRM1, but not rat CRM1, restored Rex activity (*10*). Taken together, these results indicate that rat CRM1, unlike human CRM1, has only very weak activity as a cofactor of Rex.

As the data described above indicate that rat CRM1 does not help Rex export RxRE-bearing mRNA, we wondered whether rat CRM1 was even able to export Rex protein itself. Subcellular localization studies showed that Rex protein was predominantly localized in the nucleus and nucleolus in HeLa and REF52 cells. However, when either human CRM1 or rat CRM1 was overexpressed exogenously, the Rex protein was mainly detected in the cytoplasm, probably due to the enhanced transport to the cytoplasm. This notion was supported by strong accumulation of Rex protein in the nuclei in the presence of LMB even under the condition of overexpression of human CRM1 or rat CRM1 (*10*). Thus, our data indicate that rat CRM1 as well as human CRM1 are able to support Rex protein export.

Next, we examined the affinities of Rex protein for the rat and human CRM1s in rat cells by the two-hybrid assay. Here, rat and human CRM1 proteins were expressed as GAL-fusion proteins in the presence of the Rex-VP fusion protein. Rat CRM1 interacts with Rex as efficiently as human CRM1, although the rat and human GAL-CRM1 fusion proteins were expressed at equivalent levels. The pull-down assay employing recombinant Rex protein was done to ascertain the binding of Rex to each human CRM1. Rex bound to human CRM1 and rat CRM1 with similar efficiency in the presence of GTP-charged RanQ69L protein, a GTPase-deficient mutant of Ran (*10*). Together with the subcellular localization data described above, our results suggest that rat CRM1 can bind to and export Rex protein as efficiently as human CRM1 in rat cells.

We then evaluated whether rat CRM1 can similarly support Rex-Rex interaction by a two-hybrid assay in REF52 cells. In the absence of exogenously expressed CRM1 protein, the extent of Rex-Rex interaction was very weak because it was similar to that of the negative control. Overexpression of human CRM1 efficiently enhanced Rex-Rex interaction. However, when rat CRM1 was overexpressed, the enhancement of Rex-Rex interactions could hardly be detected.

We also tested the effect of LMB on the Rex-Rex interaction in HeLa cells; this interaction was reduced, which confirms that Rex dimerization requires human CRM1. We were interested in whether Rex-Rex interactions that had been inhibited by LMB could be restored by overexpression of rat CRM1 or human CRM1, and found that human CRM1 could almost completely restore Rex-Rex interactions but that rat CRM1 could not (*10*). Thus, it appears that rat CRM1 is unable to support Rex-Rex interaction.

To confirm the above results, interaction between differently-tagged Rexes (GAL-Rex and Rex-VP) was examined by co-immunoprecipitation method. In the presence of GTP-charged RanQ69L, GAL-Rex was coprecipitated with Rex-VP in HeLa cell lysate but hardly in REF52 cell lysate. However, addition of recombinant human CRM1 protein to REF52 cell lysate enhanced co-immunoprecipitation of GAL-Rex (*10*). These results indicated that the ability of rat CRM1 to support Rex-Rex interaction is considerably less than that of human CRM1, although both CRM1s can bind efficiently to Rex protein. This property of rat CRM1 may be responsible for the inability of the Rex protein in rat cells to transport gag/env mRNAs into the cytoplasm.

Concluding Remarks

We have described dual roles of human CRM1 in export and multimerization of Rex. The limited ability of rat CRM1 to support multimerization of Rex may be responsible for the poor replication of HTLV-1 in rats. Our results suggest that better rat models of HTLV-1 infection could be developed by constructing transgenic rats expressing human CRM1. This notion is partly supported by our observation that overexpression of human CRM1 was able to restore Rex activity in rat cells. Formal proof of this concept will be the observation that HTLV-1 actually does replicate better in rat cells that express human CRM1. However, a possible problem with such transgenic rat models may be that the presence of rat CRM1 in rat cells could have a dominant negative effect on human CRM1 because rat CRM1 binds to Rex as efficiently as human CRM1 does. This dominant negative effect on the Rex activity was detected when rat CRM1 was overexpressed in HeLa cells, although a relatively large amount of rat CRM1 was required. Thus, more investigation is necessary to optimize the conditions under which Rex will fully function in rat cells.

REFERENCES

1. Ballaun, C., Farrington, G. K., Dobrovnik, M., Rusche, J., Hauber, J., and Bohnlein, E. Functional analysis of human T-cell leukemia virus type I rex-response element: direct RNA binding of Rex protein correlates with *in vivo* activity. *J. Virol.*, **65**, 4408–4413 (1991).
2. Bogerd, H. and Greene, W. C. Dominant negative mutants of human T-cell leukemia virus type I Rex and human immunodeficiency virus type 1 Rev fail to multimerize *in vivo*. *J. Virol.*, **67**, 2496–2502 (1993).
3. Bogerd, H. P., Fridell, R. A., Benson, R. E., Hua, J., and Cullen, B. R. Protein sequence requirements for function of the human T-cell leukemia virus type 1 Rex nuclear export signal delineated by a novel *in vivo* randomization-selection assay. *Mol. Cell. Biol.*, **16**, 4207–4214 (1996).
4. Bogerd, H. P., Huckaby, G. L., Ahmed, Y. F., Hanly, S. M., and Greene, W. C. The type I human T-cell leukemia virus (HTLV-I) Rex trans-activator binds directly to the HTLV-I Rex and the type 1 human immunodeficiency virus Rev RNA response elements. *Proc. Natl. Acad. Sci. USA*, **88**, 5704–5708 (1991).
5. Bohnlein, E., Berger, J., and Hauber, J. Functional mapping of the human immunodeficiency virus type 1 Rev RNA binding domain: new insights into the domain structure of Rev and Rex. *J. Virol.*, **65**, 7051–7055 (1991).
6. Fang, J., Kushida, S., Feng, R., Tanaka, M., Kawamura, T., Abe, H., Maeda, N., Onobori, M., Hori, M., Uchida, K., and Miwa, M. Transmission of human T-cell leukemia virus type 1 to mice. *J. Virol.*, **72**, 3952–3957 (1998).
7. Fornerod, M., Ohno, M., Yoshida, M., and Mattaj, I. W. CRM1 is an export receptor for leucine-rich nuclear export signals. *Cell*, **90**, 1051–1060 (1997).
8. Fukuda, M., Asano, S., Nakamura, T., Adachi, M., Yoshida, M., Yanagida, M., and Nishida, E. CRM1 is responsible for intracellular transport mediated by the nuclear export signal. *Nature*, **390**, 308–311 (1997).
9. Grassmann, R., Berchtold, S., Aepinus, C., Ballaun, C., Boehnlein, E., and Fleckenstein, B. *In vitro* binding of human T-cell leukemia virus rex proteins to the rex-response element of viral transcripts. *J. Virol.*, **65**, 3721–3727 (1991).
10. Hakata, Y., Yamada, M., and Shida, H. Rat CRM1 is responsible for the poor activity of human T-cell leukemia virus type 1 Rex protein in rat cells. *J. Virol.*, **75**, 11515–11525 (2001).

11. Hakata, Y., Umemoto, T., Matsushita, S., and Shida, H. Involvement of human CRM1 (exportin 1) in the export and multimerization of the Rex protein of human T-cell leukemia virus type 1. *J. Virol.*, **72**, 6602–6607 (1998).
12. Hanly, S. M., Rimsky, L. T., Malim, M. H., Kim, J. H., Hauber, J., Duc Dodon, M., Le, S. Y., Maizel, J. V., Cullen, B. R., and Greene, W. C. Comparative analysis of the HTLV-I Rex and HIV-1 Rev trans-regulatory proteins and their RNA response elements. *Genes Dev.*, **3**, 1534–1544 (1989).
13. Heaphy, S., Dingwall, C., Ernberg, I., Gait, M. J., Green, S. M., Karn, J., Lowe, A. D., Singh, M., and Skinner, M. A. HIV-1 regulator of virion expression (Rev) protein binds to an RNA stem-loop structure located within the Rev response element region. *Cell*, **60**, 685–693 (1990).
14. Heger, P., Rosorius, O., Koch, C., Casari, G., Grassmann, R., and Hauber, J. Multimer formation is not essential for nuclear export of human T-cell leukemia virus type 1 Rex trans-activator protein. *J. Virol.*, **72**, 8659–8668 (1998).
15. Hidaka, M., Inoue, J., Yoshida, M., and Seiki, M. Post-transcriptional regulator (rex) of HTLV-1 initiates expression of viral structural proteins but suppresses expression of regulatory proteins. *EMBO J.*, **7**, 519–523 (1988).
16. Hinuma, Y., Nagata, K., Hanaoka, M., Nakai, M., Matsumoto, T., Kinoshita, K. I., Shirakawa, S., and Miyoshi, I. Adult T-cell leukemia: antigen in an ATL cell line and detection of antibodies to the antigen in human sera. *Proc. Natl. Acad. Sci. USA*, **78**, 6476–6480 (1981).
17. Hope, T. J., Huang, X. J., McDonald, D., and Parslow, T. G. Steroid-receptor fusion of the human immunodeficiency virus type 1 Rev transactivator: mapping cryptic functions of the arginine-rich motif. *Proc. Natl. Acad. Sci. USA*, **87**, 7787–7791 (1990).
18. Ibuki, K., Funahashi, S. I., Yamamoto, H., Nakamura, M., Igarashi, T., Miura, T., Ido, E., Hayami, M., and Shida, H. Long-term persistence of protective immunity in cynomolgus monkeys immunized with a recombinant vaccinia virus expressing the human T cell leukemia virus type I envelope gene. *J. Gen. Virol.*, **78**, 147–152 (1997).
19. Inoue, J., Yoshida, M., and Seiki, M. Transcriptional (p40x) and post-transcriptional (p27x-III) regulators are required for the expression and replication of human T-cell leukemia virus type I genes. *Proc. Natl. Acad. Sci. USA*, **84**, 3653–3657 (1987).
20. Ishiguro, N., Abe, M., Seto, K., Sakurai, H., Ikeda, H., Wakisaka, A., Togashi, T., Tateno, M., and Yoshiki, T. A rat model of human T lymphocyte virus type I (HTLV-I) infection. 1. Humoral antibody response, provirus integration, and HTLV-I-associated myelopathy/tropical spastic paraparesis-like myelopathy in seronegative HTLV-I carrier rats. *J. Exp. Med.*, **176**, 981–989 (1992).
21. Jacobson, S., Raine, C. S., Mingioli, E. S., and McFarlin, D. E. Isolation of an HTLV-1-like retrovirus from patients with tropical spastic paraparesis. *Nature*, **331**, 540–543 (1988).
22. Kalyanaraman, V. S., Sarngadharan, M. G., Nakao, Y., Ito, Y., Aoki, T., and Gallo, R. C. Natural antibodies to the structural core protein (p24) of the human T-cell leukemia (lymphoma) retrovirus found in sera of leukemia patients in Japan. *Proc. Natl. Acad. Sci. USA*, **79**, 1653–1657 (1982).
23. Katahira, J., Ishizaki, T., Sakai, H., Adachi, A., Yamamoto, K., and Shida, H. Effects of translation initiation factor eIF-5A on the functioning of human T-cell leukemia

virus type 1 Rex and human immunodeficiency virus Rev inhibited trans dominantly by a Rex mutant deficient in RNA binding. *J. Virol.*, **69**, 3125–3133 (1995).

24. Kim, F. J., Beeche, A. A., Hunter, J. J., Chin, D. J., and Hope, T. J. Characterization of the nuclear export signal of human T-cell lymphotropic virus type 1 Rex reveals that nuclear export is mediated by position-variable hydrophobic interactions. *Mol. Cell. Biol.*, **16**, 5147–5155 (1996).
25. Koya, Y., Ohashi, T., Kato, H., Hanabuchi, S., Tsukahara, T., Takemura, F., Etoh, K., Matsuoka, M., Fujii, M., and Kannagi, M. Establishment of a seronegative human T-cell leukemia virus type 1 (HTLV-1) carrier state in rats inoculated with a syngeneic HTLV-1-immortalized T-cell line preferentially expressing Tax. *J. Virol.*, **73**, 6436–6443 (1999).
26. Kudo, N., Khochbin, S., Nishi, K., Kitano, K., Yanagida, M., Yoshida, M., and Horinouchi, S. Molecular cloning and cell cycle-dependent expression of mammalian CRM1, a protein involved in nuclear export of proteins. *J. Biol. Chem.*, **272**, 29742–29751 (1997).
27. Kudo, N., Matsumori, N., Taoka, H., Fujiwara, D., Schreiner, E. P., Wolff, B., Yoshida, M., and Horinouchi, S. Leptomycin B inactivates CRM1/exportin 1 by covalent modification at a cysteine residue in the central conserved region. *Proc. Natl. Acad. Sci. USA*, **96**, 9112–9117 (1999).
28. Kushida, S., Mizusawa, H., Matsumura, M., Tanaka, H., Ami, Y., Hori, M., Yagami, K., Kameyama, T., Tanaka, Y., Yoshida, A., Nyunoya, H., Shimotono, K., Iwasaki, Y., Uchida, K., and Miwa, M. High incidence of HAM/TSP-like symptoms in WKA rats after administration of human T-cell leukemia virus type 1-producing cells. *J. Virol.*, **68**, 7221–7226 (1994).
29. Li, Q. X., Camerini, D., Xie, Y., Greenwald, M., Kuritzkes, D. R., and Chen, I. S. Syncytium formation by recombinant HTLV-II envelope glycoprotein. *Virology*, **218**, 279–284 (1996).
30. Nakielny, S. and Dreyfuss, G. The hnRNP C proteins contain a nuclear retention sequence that can override nuclear export signals. *J. Cell Biol.*, **134**, 1365–1373 (1996).
31. Neville, M., Stutz, F., Lee, L., Davis, L. I., and Rosbash, M. The importin-beta family member Crm1p bridges the interaction between Rev and the nuclear pore complex during nuclear export. *Curr. Biol.*, **7**, 767–775 (1997).
32. Nosaka, T., Siomi, H., Adachi, Y., Ishibashi, M., Kubota, S., Maki, M., and Hatanaka, M. Nucleolar targeting signal of human T-cell leukemia virus type I rex-encoded protein is essential for cytoplasmic accumulation of unspliced viral mRNA. *Proc. Natl. Acad. Sci. USA*, **86**, 9798–9802 (1989).
33. Osame, M., Usuku, K., Izumo, S., Ijichi, N., Amitani, H., Igata, A., Matsumoto, M., and Tara, M. HTLV-I associated myelopathy, a new clinical entity. *Lancet*, **i**, 1031–1032 (1986).
34. Ossareh-Nazari, B., Bachelerie, F., and Dargemont, C. Evidence for a role of CRM1 in signal-mediated nuclear protein export. *Science*, **278**, 141–144 (1997).
35. Palmeri, D. and Malim, M. Importin β can mediate the nuclear import of an arginine-rich nuclear localization signal in the absence of importin α. *Mol. Cell. Biol.*, **19**, 1218–1225 (1999).
36. Rimsky, L., Dodon, M. D., Dixon, E. P., and Greene, W. C. Trans-dominant inactiva-

tion of HTLV-I and HIV-1 gene expression by mutation of the HTLV-I Rex transactivator. *Nature*, **341**, 453–456 (1989).
37. Seiki, M., Inoue, J., Hidaka, M., and Yoshida, M. Two cis-acting elements responsible for posttranscriptional trans-regulation of gene expression of human T-cell leukemia virus type I. *Proc. Natl. Acad. Sci. USA*, **85**, 7124–7128 (1988).
38. Shida, H., Tochikura, T., Sato, T., Konno, T., Hirayoshi, K., Seki, M., Ito, Y., Hatanaka, M., Hinuma, Y., Sugimoto, M., T-Nishmaki, F., Maruyama, T., Miki, K., Suzuki, K., Morita, M., Sashiyama, H., Yoshimura, N., and Hayami, M. Effect of the recombinant vaccinia viruses that express HTLV-I envelope gene on HTLV-I infection. *EMBO J.*, **6**, 3379–3384 (1987).
39. Siomi, H., Shida, H., Nam, S. H., Nosaka, T., Maki, M., and Hatanaka, M. Sequence requirements for nucleolar localization of human T cell leukemia virus type I pX protein, which regulates viral RNA processing. *Cell*, **55**, 197–209 (1988).
40. Stade, K., Ford, C. S., Guthrie, C., and Weis, K. Exportin 1 (Crm1p) is an essential nuclear export factor. *Cell*, **90**, 1041–1050 (1997).
41. Takebe, Y., Seiki, M., Fujisawa, J., Hoy, P., Yokota, K., Arai, K., Yoshida, M., and Arai, N. SR alpha promoter: an efficient and versatile mammalian cDNA expression system composed of the simian virus 40 early promoter and the R-U5 segment of human T-cell leukemia virus type 1 long terminal repeat. *Mol. Cell. Biol.*, **8**, 466–472 (1988).
42. Tanaka, Y., Tanaka, R., Terada, E., Koyanagi, Y., Miyano-K, N., Yamamoto, N., Baba, E., Nakamura, M., and Shida, H. Induction of antibody responses that neutralize human T-cell leukemia virus type I infection *in vitro* and *in vivo* by peptide immunization. *J. Virol.*, **68**, 6323–6331 (1994).
43. Uemura, Y., Kotani, S., Yoshimoto, S., Fujishita, M., Yano, S., Ohtsuki, Y., and Miyoshi, I. Oral transmission of human T-cell leukemia virus type I in the rabbit. *Jpn. J. Cancer Res.*, **77**, 970–973 (1986).
44. Weichselbraun, I., Berger, J., Dobrovnik, M., Bogerd, H., Grassmann, R., Greene, W. C., Hauber, J., and Bohnlein, E. Dominant-negative mutants are clustered in a domain of the human T-cell leukemia virus type I Rex protein: implications for trans dominance. *J. Virol.*, **66**, 4540–4545 (1992).
45. Yoshida, M., Suzuki, T., Fujisawa, J., and Hirai, H. HTLV-1 oncoprotein tax and cellular transcription factors. *Curr. Top. Microbiol. Immunol.*, **193**, 79–89 (1995).

ABERRANT ACTIVATION OF NF-κB/Rel AND AP-1 IN ADULT T-CELL LEUKEMIA

Naoki MORI[*1] and Masahiro FUJII[*2]

*Department of Virology, Faculty of Medicine, University of the Ryukyus[*1] and Division of Virology, Niigata University Graduate School of Medical and Dental Sciences[*2]*

Adult T-cell leukemia (ATL) is a fatal T-lymphoproliferative disorder, and its development is associated with infection by human T-cell leukemia virus type I (HTLV-I). Several studies have suggested that aberrant transcription of cellular genes in HTLV-I infected cells plays a key role in the development of the disease. For instance, two transcription factors, nuclear factor kappa B (NF-κB)/Rel and activating protein-1 (AP-1), are transiently activated in normal T cells by growth-signals, whereas they are constitutively activated in HTLV-I infected T cells *in vitro* as well as *in vivo*. The HTLV-I oncoprotein Tax is responsible for the activation of NF-κB/Rel and AP-1 in HTLV-I transformed T-cell lines, and it induces a variety of cellular genes through NF-κB/Rel and AP-1. Unlike HTLV-I transformed T-cell lines, leukemic cells of ATL patients express extremely low levels of Tax protein, but NF-κB/Rel and AP-1 in leukemic cells are constitutively activated. Thus, the aberrant activation of NF-κB/Rel and AP-1 may be a key factor underlying the prolonged survival and proliferation of HTLV-I infected T cells both at the initial and late stages of leukemogenesis, although the mechanisms are different. In this review, we discuss the involvement of NF-κB/Rel and AP-1 in Tax-dependent and independent steps of leukemogenesis, and their role as potential therapeutic targets for treatment of ATL.

Adult T-cell leukemia (ATL) is an aggressive and often untreatable malignancy of CD4+ T cells, and its development is closely associated with human T-cell leukemia virus type I (HTLV-I) infection (*29, 84, 116*). ATL develops in a small proportion of HTLV-I infected subjects (3–5%) generally 40 to 60 years after the infection. This long latency suggests that a multistep oncogenic process including the accumulation of genetic mutations in infected T cells, in addition to HTLV-I infection, may be required for the induction of ATL. HTLV-I transforms primary human CD4+ T cells in interleukin (IL)-2-dependent and independent manners *in vitro*. Accumulating evidence has indicated that the viral protein Tax plays a crucial role in the transformation of human T cells as well as leukemogenesis. For instance, like HTLV-I, Tax can immortalize primary CD4+T cells, although the immortalized cells remain IL-2-dependent (*2, 21*). Tax can induce colony formation of rat fibroblast cell lines in soft agar (*92*), and immortalize primary rat embryo fibroblasts (*85*). Tax in coopera-

[*1] 207 Uehara, Nishihara, Okinawa 903-0215, Japan
[*2] 1-757 Asahimachi-dori, Niigata 951-8510, Japan

tion with the *ras* oncogene, transforms primary rat embryo fibroblasts such that they are tumorigenic in nude mice (*85, 96*). Furthermore, *tax* transgenic mice develop several malignancies including leukemia, mesenchymal tumors, and neurofibromas (*22, 28, 74*).

Tax was originally identified as a transcriptional activator of viral gene through the 21-bp repeats in the HTLV-I long terminal repeat (*93*). Since then, Tax has been shown to upregulate the expression of a number of cellular genes through several transcription factors, such as nuclear factor kappa B (NF-κB)/Rel and activating protein-1 (AP-1) (*6, 17*). The activation of cellular genes by Tax through these two transcription factors is thought to contribute to deregulated phenotypes of HTLV-I infected T cells, thereby the leukemogenesis, as discussed below.

Although Tax is considered crucial for the transformation of infected T cells at the initial and probably middle stages of leukemogenesis *in vivo*, leukemic cells from ATL patients express extremely low level of Tax protein (*45*). Moreover, leukemic cells from several ATL patients possess mutation and truncation of the Tax coding region, which inactivate its functions (*18, 81*). These results indicate that Tax is dispensable for the late stage of ATL leukemogenesis. Interestingly, NF-κB/Rel and AP-1 are constitutively activated in leukemic cells of ATL patients even in cells lacking Tax expression. Thus, it is questionable that Tax alone explains the activation of NF-κB/Rel and AP-1 in leukemic cells *in vivo* (*64, 65*), and the activation of NF-κB/Rel and AP-1 is probably due to genetic mutations in leukemic cells. This review will discuss recent advances in our understanding of how HTLV-I modulates cellular signaling machinery to activate NF-κB/Rel and AP-1, and what these transcription factors do in the transformation and the leukemogenesis, and, finally, discuss the role of NF-κB/Rel as a promising target for treatment of ATL.

NF-κB/Rel Pathway

1. Regulation of NF-κB/Rel

NF-κB/Rel regulates well over 150 target genes, whose products play important roles in various immune and inflammatory responses as well as proliferation of lymphocytes. The NF-κB/Rel family consists of five structurally related proteins c-Rel, RelA (p65), RelB, NF-κB1 (p50/p105), and NF-κB2 (p52/p100). These NF-κB proteins form homodimers or heterodimers with other family members. These proteins share an approximately 300-amino acid N-terminal Rel homology domain that is essential for dimerization, DNA binding, and nuclear transport. c-Rel, RelA, and RelB possess C-terminal activation domains that are important for transcriptional activation. The p50 and p52 proteins are the proteolytically processed forms of p105 and p100, respectively, and contain the N-terminal Rel homology domain. The C-terminus of p105 and p100 contains the so-called ankyrin repeat, which is found in IκB family members. The NF-κB proteins are normally sequestered in the cytoplasm through physical interaction with ankyrin repeat-containing inhibitors such as IκBα, IκBβ, IκBε, p105, p100, and Bcl-3. Interaction of NF-κB with IκB masks the nuclear localization signal and retains NF-κB in the cytoplasm. Activation of the NF-κB-signaling cascade results in the degradation of IκB or the processing of p105 and p100, allowing the release of NF-κB proteins. Free NF-κB then translocates to the nucleus, where it stimulates the transcription of genes containing the consensus sequence 5′-GGGRNNYYCC-3′. All IκB proteins contain two conserved serine residues within their N-terminal domain. Phosphorylation of these conserved serine residues in response to activation signals leads to the immediate polyubiquitination of IκB proteins by the SCF-β-TrCP complex. This modification targets IκB proteins for rapid degradation by the 26S proteasome. A high-molecular weight complex, IκB kinase (IKK) complex, phosphorylates IκBα or IκBβ. The IKK complex is composed of two catalytic subunits, IKKα and IKKβ, and a regulatory subunit, IKKγ/NEMO (*43*). IKKγ is not a kinase *per se* but is absolutely essential for NF-κB activation (*112*). IKKα and IKKβ share 50% sequence homology and form both homodimers and heterodimers through their central leucine zipper domains. Both proteins have C-terminal helix-loop-helix domains that are involved in the regulation of their kinase activity. The N-termini of IKKα and IKKβ contain a kinase domain. Although both IKKα and IKKβ exhibit IκB kinase activity, IKKβ appears to be a more potent kinase. For instance, dominant negative mutants of IKKβ have pronounced inhibitory effects on the nuclear translocation of NF-κB and NF-κB-dependent transcriptional activities, while dominant negative mutants of IKKα have little effect (*60, 107*). In IKKβ knockout fibroblasts, stimulation by tumor necrosis factor (TNF)-α and IL-1 induces little or no activation of NF-κB, while a normal level of activation is seen in IKKα knockout fibroblasts (*33, 51, 95, 97*). Recent studies indicate that members of the mitogen-activated protein kinase kinase kinase (MAP3K) protein kinase family mediate physiologic activation of IKK via phosphorylation of the MAP2K activation loops present in the IKKα and IKKβ catalytic subunits (*118*). These kinases include mitogen-activated protein kinase/extracellular signal-regulated kinase kinase 1 (MEKK1) (*50*), MEKK2 (*119*), MEKK3 (*119*), transforming growth factor (TGF)-β activating kinase (*78*), NF-κB-inducing kinase (NIK) (*107*), NF-κB-activating kinase (*98*), and mixed-lineage kinase 3 (MLK3) (*26*).

2. *Tax-mediated NF-κB/Rel activation*

Numerous stimuli including cytokines induce transient NF-κB activation in various types of cells, whereas Tax induces persistent activation of NF-κB. Thus, NF-κB is constitutively present in the nuclei of Tax-expressing cells and HTLV-I infected cells. Through NF-κB, Tax constitutively activates the expression of several cytokines (IL-1α, IL-2, IL-4, IL-6, IL-15, TNF-α, and TNF-β) (*3, 5, 52, 68, 71, 91, 99*), chemokines (IL-8, RANTES, and monocyte chemoattractant protein-1 [MCP-1]) (*66, 72*), growth factors (granulocyte-macrophage colony-stimulating factor [GM-CSF] and granulocyte colony-stimulating factor [G-CSF]) (*27*), cytokine receptors (IL-2 and IL-15 receptor α chains) (*6, 56*), cell adhesion molecules (VCAM-1, OX40, OX40 ligand, and fibronectin) (*80, 82, 105, 114*), transcription factors (c-*myc*, c-Rel, NF-κB2, and IκBα) (*14, 94*), apoptosis regulatory genes (*A-20*, *Bcl-*x_L, and *TRAIL*) (*48, 62, 76, 86, 101*), cell-cycle regulatory genes (*cyclin D1*, *cyclin D2*, *cyclin E*, *CDK2*, *CDK4*, *CDK6*, and *E2F1*) (*34, 39, 63*), matrix metalloproteinase (MMP)-9 (*69*), and vimentin (*53*) (Fig. 1).

Tax activates NF-κB through various mechanisms and has been reported to physically interact with some NF-κB/Rel family members (*31, 117*). In addition, Tax in part has been shown to colocalize with NF-κB in subnuclear regions that contain specific RNA transcripts from a promoter containing NF-κB binding sites (*7, 8, 90*).

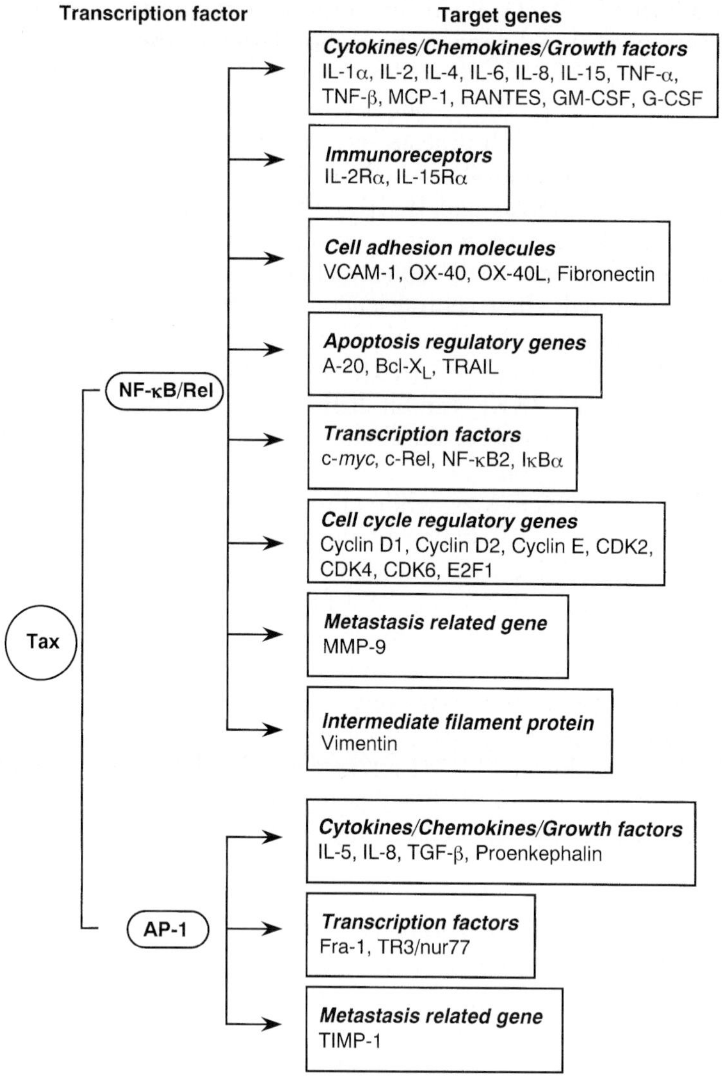

FIG. 1. Tax targets and potential transcription factors NF-κB/Rel and AP-1

These findings suggest that Tax functions as a transcriptional cofactor acting cooperatively with NF-κB in the nucleus.

Other reports showed that Tax-mediated NF-κB activation occurs primarily at the level of NF-κB nuclear translocation (*20, 42, 47, 94*). For example, a Tax mutant defective in nuclear localization can still activate NF-κB-dependent transcription (*77*). A series of elegant studies have shown that Tax induces the phosphorylation and degradation of IκBα and IκBβ (*9, 20, 42, 47, 59, 94*). Furthermore, recent studies showed that Tax physically associates with IKK (*11*) and stimulates the catalytic activity of this cellular kinase (*19, 104, 115*) (Fig. 2). The activation of IKK by Tax is dependent on IKKγ (*25, 112*), which serves as an adaptor for recruiting Tax to the IKK catalytic subunits (*12, 24, 41, 109*). The assembly of Tax-IKK complexes leads to constitutive phosphorylation of IKKβ and IKKγ (*10*). Tax induces the phosphorylation of IKKβ at Ser-177 and Ser-181. Since kinase-dead mutants of IKKβ are defective for Tax-induced phosphorylation of IKKβ, IKKβ may be autophosphorylated in a Tax-dependent manner. IKKγ is phosphorylated by IKKβ within this complex (*10*), although it has not been defined whether the phosphorylation status of IKKγ affects the IKK activity. Thus, Tax, through interaction with IKKγ, may induce conformational changes of IKKβ, triggering its catalytic activity. The Tax-associated IKK may be further activated via the autophosphorylation of IKKβ (Fig. 2A). The Tax-IKKγ interaction requires two homologous leucine zippers present in IKKγ and a leucine-rich region present in Tax (*109*). Mutation of these residues inactivates Tax function toward NF-κB.

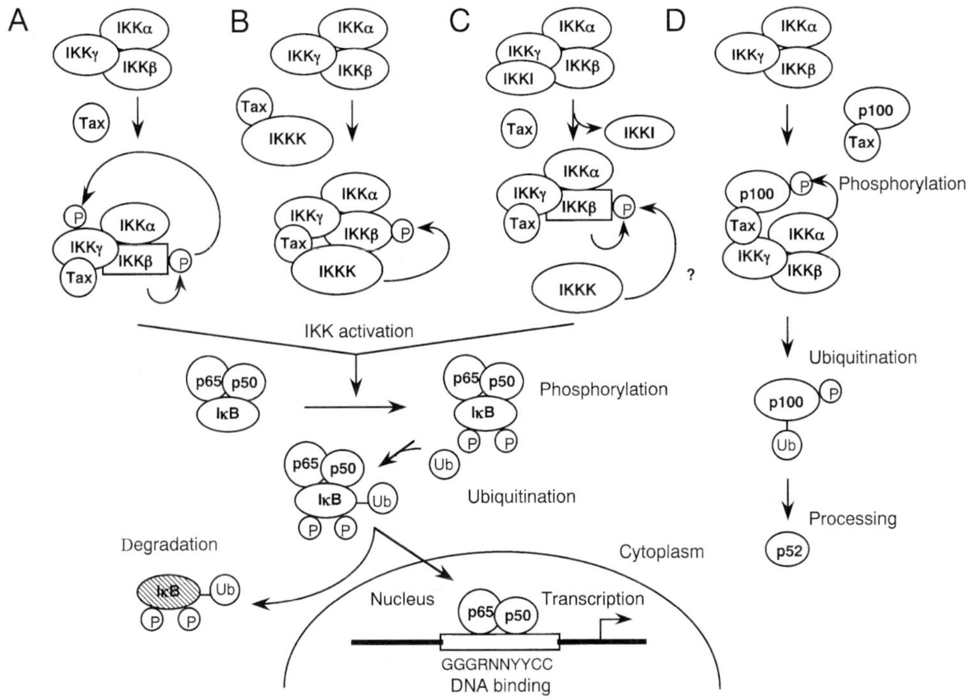

FIG. 2. Tax not only stimulates the catalytic activity of IKKβ (A–C) but also induces processing of NF-κB2/p100 by recruiting IKKα to p100 (D). Tax physically interacts with IKKγ via specific domains present in both IKKγ and Tax. Through this molecular interaction, Tax is recruited to the IKKβ catalytic subunit, where Tax may stimulate the activity of IKKβ by three potential mechanisms. (A) Tax may stimulate IKKβ by causing its conformational changes; the stimulated IKKβ may be further activated via its autophosphorylation. IKKγ is phosphorylated by IKKβ within the same complex. However, whether the phosphorylation status of IKKγ affects the regulation of IKK activity remains to be determined. (B) Tax may recruit an upstream activator (IKKK) to the IKK complex, triggering IKK activation by phosphorylation. (C) Tax may displace a cellular IKK inhibitor (IKKI) from the IKK holoenzyme, which is sufficient to induce IKK catalytic activity via enhanced autophosphorylation. Alternatively, removal of IKKI may render the IKKβ catalytic subunit more susceptible to activation by an upstream Tax-responsive IKKK. (D) Tax acts by physically recruiting IKKα to p100, triggering phosphorylation-dependent ubiquitination and processing of p100, resulting in active production of p52. Irrespective of the model, it is assumed that Tax-IKKγ interaction is required to maintain IKK activation.

Tax may recruit the IKK complex to an upstream activator, IKK kinase (IKKK) (Fig. 2B). Tax binds to and activates MEKK1 that phosphorylates IKK *in vitro* (*115*). However, Tax can activate NF-κB in cells derived from MEKK1 knockout mice (T. Yujiri, personal communication). Thus, MEKK1 is not essential for Tax-induced NF-κB activation, and MEKK1 is not a main Tax-responsive IKKK. NIK and MLK3 are candidate Tax-responsive IKKK (*19, 75, 104*), since a dominant-negative mutant of NIK and MLK3 blocks Tax induction of NF-κB. However, the mechanism by which NIK and MLK3 are involved in Tax activation remains to be determined.

In HTLV-I infected T-cell lines, IKKα is involved in the processing of the NF-κB2 precursor protein p100, resulting in the production of the cleaved product p52 (*108*). Tax interacts specifically with p100, and recruits IKKα to p100, triggering phosphorylation-dependent ubiquitination and processing of p100 (Fig. 2D). In fact, high levels of p52 are present in HTLV-I transformed T-cell lines (*49*).

Potential negative regulatory factors, IKK inhibitor (IKKI) in the large IKK complex have been suggested by the finding that the phosphatase MKP-1 associates with the IKK complex after TNF treatment (*60*). Tax may displace an IKKI from the IKK complex, yielding a more favorable conformation for constitutive activation via IKKβ autophosphorylation or the action of the Tax-responsive IKKK (Fig. 2C). In all models, it is assumed that Tax-IKK interactions are primarily required to maintain a constitutive activation of IKK by Tax.

3. Role of NF-κB/Rel in oncogenic function of Tax

Several studies have suggested that activation of NF-κB/Rel by Tax is essential for the transforming activity (*1, 57, 113*). Previous studies demonstrated that inhibition of NF-κB activity by antisense oligonucleotides corresponding to RelA (p65) interfered with the growth of an HTLV-I transformed T-cell line *in vitro* and Tax-transformed fibroblast cell lines *in vitro* and *in vivo* (*46*). Furthermore, studies using an infectious HTLV-I molecular clone indicated that Tax-induced activation of NF-κB pathway was critical for the immortalizing activity of primary human T cells (*87*). Among the genes activated by Tax through NF-κB, activation of IL-2/IL-2 receptor α and IL-15/IL-15 receptor α seems to be a critical event in polyclonal proliferation of HTLV-I infected T cells in the early stage of leukemogenesis. However, most HTLV-I infected T-cell lines exhibit negligible IL-2 activity. Other Tax-inducible genes through NF-κB also contribute to the control of cell growth. The ability of Tax to induce cell-cycle progression always parallels the activation of NF-κB pathway (*39*). Through NF-κB, Tax induces the expression of cell-cycle control genes such as *cyclin D1*, *cyclin D2*, *cyclin E*, *CDK2*, *CDK4*, *CDK6*, and *E2F1* (*34, 39, 63*). Among them, the expression of cyclin D1 and cyclin D2 is associated with the conversion of a mouse T-cell line (CTLL-2) from IL-2-dependent growth into an IL-2-independent one (*63*). In addition, activation of anti-apoptotic Bcl-x_L by Tax through NF-κB is associated with the resistance of CTLL-2 against apoptosis induced by IL-2 withdrawal (*40, 101*). We have recently demonstrated that chemical inhibition of NF-κB activity resulted in a marked increase of apoptosis in HTLV-I infected T-cell lines (*73*). Apoptosis was specifically associated with inhibition of NF-κB activity, but not that of AP-1. On the other hand, inhibition of NF-κB down-regulated the expression of Bcl-x_L in HTLV-I transformed T-cell lines. Hence, Bcl-x_L may be the factor responsible for Tax-induced inhibition of apoptosis through NF-κB.

Another mechanism by which Tax may contribute to T-cell transformation involves inactivation of the p53 tumor suppressor protein. Activation of NF-κB by Tax was associated with inactivation of p53 function in T-cell lines (*83*). For instance, Tax-induced p53 inactivation was abrogated by coexpression of an IκB mutant. Tax-induced p53 inactivation was not observed in RelA (p65)-deficient cells (*83*). This p53 inactivation correlates with the phosphorylation of p53 at Ser-15 and Ser-392. Since Tax is known to enhance the frequency of mutations of the cellular genome (*61*), inhibition of p53 by Tax may be a critical factor in genomic instability and progression to cellular transformation.

4. *Status of NF-κB/Rel in leukemic cells of ATL patients*

Leukemic cells from ATL patients express extremely low levels of viral genes including *tax* (*45*), suggesting that the expression of viral products is dispensable for leukemic proliferation at the late stage of the disease. However, several genes that can be transactivated by Tax through NF-κB are already constitutively expressed in leukemic cells of ATL patients, probably by a Tax-independent mechanism. For instance, leukemic cells *in vivo* express cytokine receptors (IL-2 and IL-15 receptor α chains), cytokines (IL-1α, IL-6, and TNF-β), chemokines (IL-8, MCP-1, and RANTES),

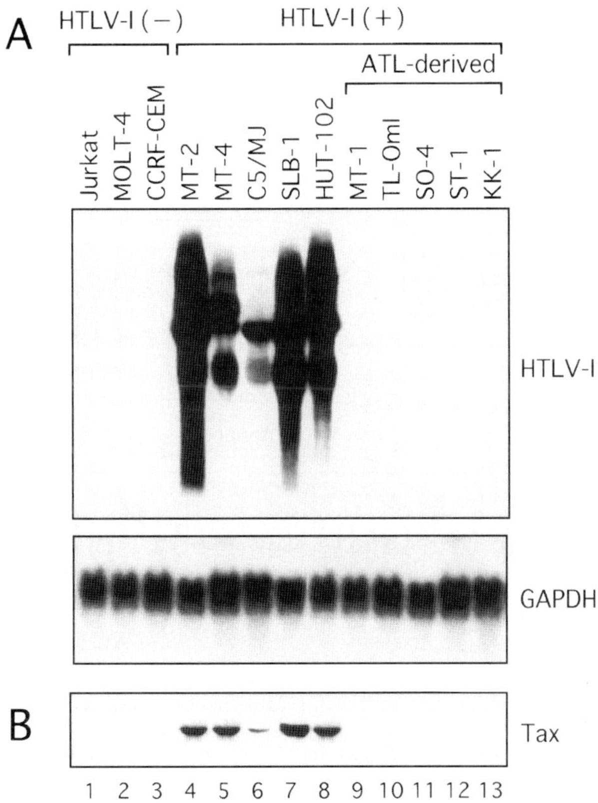

FIG. 3. Expression of HTLV-I mRNA determined by northern blot analysis (A) and expression of Tax protein determined by western blot analysis (B) in HTLV-I infected T-cell lines GAPDH expression was used as a control.

growth factors (GM-CSF and G-CSF), cell adhesion molecule (OX40), anti-apoptotic gene (*Bcl-x_L*), cell-cycle regulatory genes (*cyclin D1* and *cyclin D2*), and metastasis related gene (*MMP-9*) (*36, 37, 56, 58, 62, 63, 67, 69–71, 76, 79, 89, 103, 110*). Many HTLV-I infected T-cell lines have been established by culturing peripheral blood mononuclear cells (PBMC) from ATL patients in the presence or absence of IL-2 *in vitro*. Most of the cell lines established, however, were established from nonleukemic cell clones as determined by HTLV-I provirus integration sites and the gene rearrangement pattern of T-cell receptor β chains (*55*). The remaining few cell lines were established from leukemic cells. Such nonleukemic HTLV-I infected T-cell lines as well as HTLV-I transformed T-cell lines established from normal PBMC by *in vitro* coculture with virus-producing cell lines expressed high amounts of *HTLV-I* genes. In contrast, ATL-derived T-cell lines expressed undetectable or very low amounts of *HTLV-I* genes including *tax* like primary ATL cells (*35, 64, 65*) (Fig. 3). We found that both leukemic cells from ATL patients and ATL-derived T-cell lines displayed a constitutive NF-κB binding activity in nucleus and increased degradation of IκBα (*64*) (Fig. 4, A and C). Thus, it is likely that constitutive activation of NF-κB/Rel in ATL cells *in vivo* occurs in a Tax-independent mechanism. It is noteworthy that the components of the activated NF-κB in ATL-derived T-cell lines and primary ATL cells are different from those in Tax-expressing HTLV-I transformed T-cell lines. Gel-shift analysis demonstrated that NF-κB complexes in both ATL-

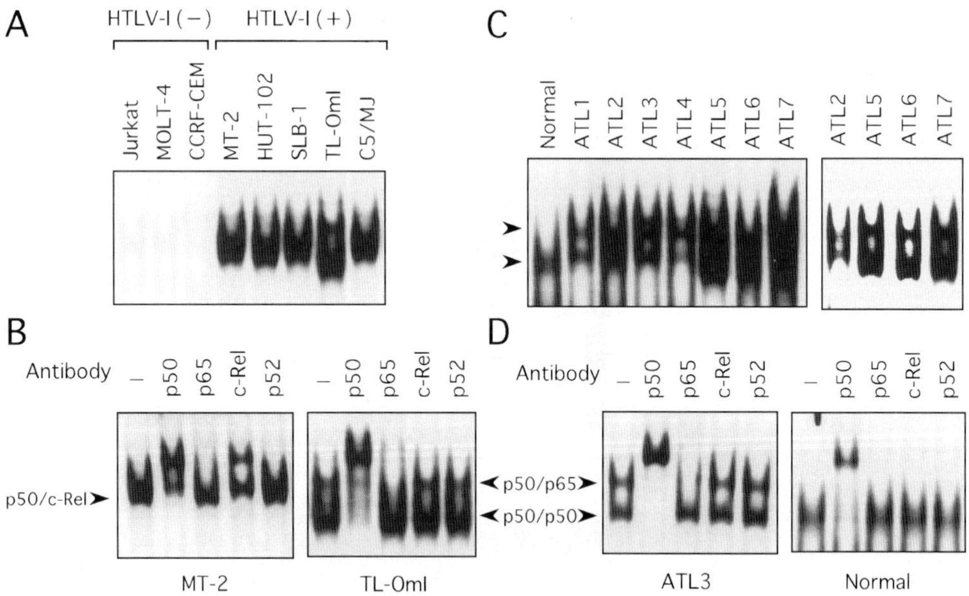

FIG. 4. NF-κB binding activity in T-cell lines and primary ATL cells

Nuclear extracts of T-cell lines (A) and PBMC (C) were incubated with a radiolabeled oligonucleotide that corresponds to the NF-κB sequence of the *IL-2R α chain* gene and analyzed by gel-shift assay. In super-shift assays, antibody specific for each NF-κB subunit (indicated above the lanes) was incubated with MT-2 or TL-OmI extracts (B) and primary ATL cells or normal PBMC extracts (D) before probe addition. MT-2, HUT-102, SLB-1, and C5/MJ are HTLV-I infected T-cell lines derived from nonleukemic cell clones. TL-OmI is an ATL-derived cell line. Expression of Tax is undetectable in TL-OmI, while other HTLV-I infected T-cell lines express high amounts of Tax (see Fig. 3).

derived T-cell lines and primary ATL cells are composed of p50/p50 and p50/RelA (p65), whereas the complexes in T-cell lines expressing Tax consist of p50/c-Rel (*64*) (Fig. 4, B and D). Thus, Tax-dependent transcriptional activation of c-Rel could account for the NF-κB complex containing c-Rel in HTLV-I transformed T-cell lines (*94*).

5. Inhibition of NF-κB/Rel pathway induces apoptosis of HTLV-I infected T-cell lines and primary ATL cells

ATL is associated with a poor prognosis mainly because of its resistance to conventional as well as high-dose chemotherapy. Therefore, there is an urgent need to establish new strategies for the treatment of ATL. NF-κB may be a suitable target for the design of new effective drugs against ATL, since its activation is involved in inhibition of apoptosis of HTLV-I infected T cells. Indeed, we recently demonstrated that chemical inhibition of NF-κB activity resulted in a marked increase of apoptosis not only of HTLV-I infected T-cell lines but also primary ATL cells (*73*). These results suggest that NF-κB is involved in inhibition of apoptosis of ATL cells, and could be considered as a therapeutic target for ATL.

AP-1 Pathway

1. Regulation of AP-1

AP-1 was initially identified as a transcription factor that contributes to phorbol 12-*O*-tetradecanoate-13-acetate (TPA)-inducible gene expression. However, many other stimuli, most notably serum, growth factors, and oncoproteins such as v-Src or Ha-Ras, are also potent inducers of AP-1 activity. The AP-1 proteins are homodimers or heterodimers composed of Jun (c-Jun, JunB, and JunD) and Fos (c-Fos, FosB, Fra-1, and Fra-2) family members. Jun and Fos proteins also dimerize with other transcription factors such as Jun dimerization partners (JDP1 and JDP2) and activating transcription factor (ATF) (ATF2, LRF1/ATF3, and B-ATF) family members. These proteins have a related basic region and a leucine zipper structure. Jun proteins can form homodimers that bind AP-1 DNA recognition elements (5′-TGAG/CTCA-3′), also known as TRE (TPA response element), based on their ability to mediate transcriptional induction in response to TPA. Fos proteins do not form stable homodimers but can bind DNA by forming heterodimers with Jun proteins that have higher affinity to TRE than Jun homodimers. Among AP-1 family members, c-Jun and c-Fos were originally isolated as cellular counterparts of viral Jun (v-Jun) and v-Fos, respectively. Thus, AP-1 plays a key role in growth control, and its aberration causes transformation.

Regulation of AP-1 activity occurs both at transcriptional and posttranscriptional levels. Expression of c-Fos, which is almost absent in normal quiescent cells, is controlled at a transcriptional level. External growth signals to quiescent cells activate three MAPK, extracellular signal-regulated protein kinase (ERK) 1/2, stress-activated protein kinase/Jun N-terminal kinase (SAPK/JNK), and p38, and they then phosphorylate the transcription factor Elk-1. Elk-1 binds the serum response element (SRE) in the c-*fos* promoter by forming a ternary complex with serum response factor (SRF), thereby inducing c-*fos* transcription (*30*). c-Fos proteins then heterodimerize with Jun proteins, and the heterodimers can lead to increased c-*jun* transcription through AP-1 sites in the c-*jun* promoter (*4*). ERK activation may also contribute to c-*jun* induction through MEF2 proteins, another group of transcription factors that bind to the c-*jun* promoter (*23*). Posttranscriptionally, c-Jun activity is potentiated through phosphorylation of N-terminal transcriptional activator domain by SAPK/JNK, resulting in increased transcriptional activity (*13*).

2. Tax-mediated AP-1 activation

T-cell lines transformed by HTLV-I express high levels of mRNA encoding c-Fos, Fra-1, c-Jun, JunB, and JunD (*17, 32, 65*), and exhibit high levels of AP-1 activity (*17, 65*) (Fig. 5A). Tax can activate the expression of c-Fos, Fra-1, c-Jun, and JunD mRNAs (*17*) in a human T-cell line. Tax-induced activation of AP-1 is involved in the transcriptional activation of several cellular genes such as cytokine (*IL-5*) (*111*), chemokine (*IL-8*) (*66*), growth factor (*TGF-β*) (*44*), transcription factors (*fra-1* and *TR3/nur77*) (*54, 100*), opioid peptide (*proenkephalin*) (*15*), and tissue inhibitor of MMP-1 (*TIMP-1*) (*102*) (Fig. 1). HTLV-I transformed T-cell lines and Tax expressing cells showing accumulation of AP-1 binding complex containing JunD, whereas other known Fos family members (c-Fos, FosB, Fra-1, and Fra-2) were undetectable in gel-shift assays (*54, 65*) (Fig. 5B). JunD homodimer forms unstable complexes with DNA (*88*). Thus, a factor(s) different from the known characterized AP-1 may be a component of the AP-1-binding complex in HTLV-I infected cells.

There are two well-known Tax mutants, M22 and M47/703. TaxM22 activates

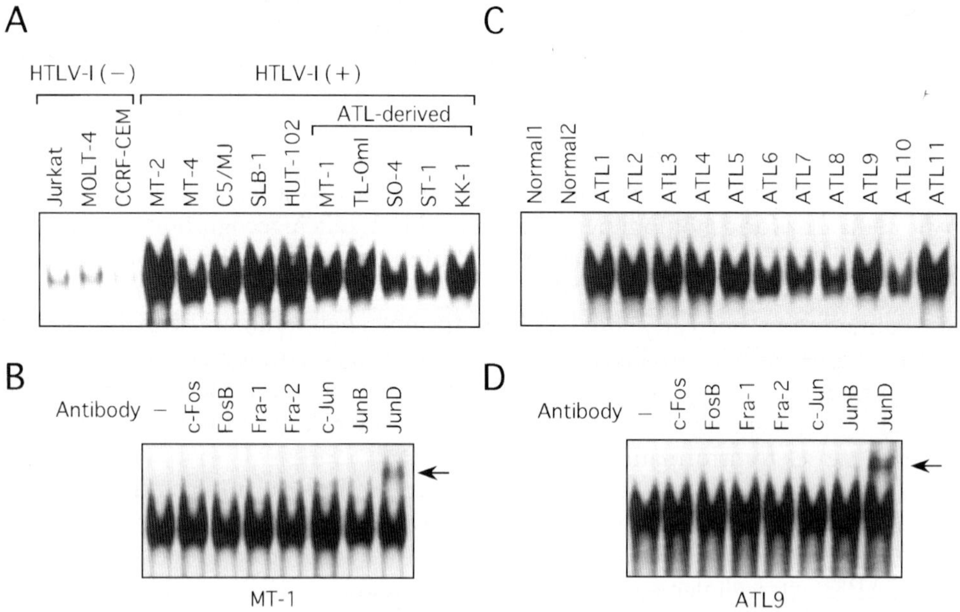

FIG. 5. AP-1-binding activity in T-cell lines and primary ATL cells

Nuclear extracts from T-cell lines (A) and PBMC (C) were incubated with a radiolabeled oligonucleotide that corresponds to the AP-1 sequence of the *IL-8* gene and analyzed by gel-shift assay. The antisera indicated above lanes were incubated with nuclear extracts from MT-1 cells (B) and leukemic cells from a representative patient with ATL (D) and subjected to gel-shift assay. A super-shift band was observed following the addition of antisera against JunD (arrows).

CREB/ATF (cAMP response element-binding protein/activating transcription factor) and SRF but not NF-κB, whereas TaxM47/703 preferentially activates NF-κB but not CREB/ATF and SRF. The properties of these mutants indicate that two pathways activated by Tax, NF-κB together with either CREB/ATF or SRF, are required for the activation of AP-1-dependent transcription (*38*). Since c-*jun* promoter has NF-κB-binding site (*16*), while c-*fos* promoter has SRF-binding site, activation of various family members of AP-1 by Tax may mediate the activation of AP-1-dependent transcription.

3. Status of AP-1 in leukemic cells from ATL

Both ATL-derived cell lines and primary ATL cells exhibit increased AP-1 activity containing JunD, although they lacked Tax (Fig. 5) (*65*). Consistently, ATL-derived T-cell lines and primary leukemic cells expressed high levels of JunD mRNA (*65*). The study of $JunD^{-/-}$ primary embryonic fibroblasts showed that these cells undergo premature replicative growth arrest, also termed senescence, and displayed increased sensitivity to death-promoting agents, including ultraviolet irradiation or TNF-α (*106*). JunD also interacts with the p53 pathway (*106*). These findings suggest that JunD protects normal cells against senescence or from apoptosis induced by stress stimuli by modulating a p53-dependent pathway (*106*). Thus, constitutive activation of AP-1 containing JunD may inhibit apoptosis of ATL cells and could substantially contribute to their growth.

Conclusion

Only a small proportion of individuals infected with HTLV-I develop ATL after a long latent period. This suggests that multiple events, including accumulation of genetic mutations in HTLV-I infected cells, take place during this period. During the late stage of leukemogenesis, mutation and/or truncation of *tax* gene in HTLV-I infected cells seems to be required for the leukemic outgrowth, since Tax is the main target of the host immune system. Leukemic cells that do not express Tax can grow freely in infected individuals, if the other oncogenic functions of Tax in infected cells, such as inhibition of apoptosis, are replaced by genetic mutations. Activation of NF-κB and AP-1 in leukemic cells is likely to be such an event required for their outgrowth during the late stage of leukemogenesis. Thus, inhibition of NF-κB and AP-1 is a promising therapeutic strategy for ATL. Indeed, we have already demonstrated that an inhibitor of NF-κB induced apoptosis of primary ATL cells. We believe that a better understanding of the mechanism(s) by which Tax modulates the cellular signaling machinery should be useful for the design of novel therapeutics for ATL.

REFERENCES

1. Akagi, T., Ono, H., Nyunoya, H., and Shimotohno, K. Characterization of peripheral blood T-lymphocytes transduced with HTLV-I Tax mutants with different *trans*-activating phenotypes. *Oncogene*, **14**, 2071–2078 (1997).
2. Akagi, T., Ono, H., and Shimotohno, K. Characterization of T cells immortalized by Tax1 of human T-cell leukemia virus type 1. *Blood*, **86**, 4243–4249 (1995).
3. Albrecht, H., Shakhov, A. N., and Jongeneel, C. V. *Trans* activation of the tumor

necrosis factor alpha promoter by the human T-cell leukemia virus type I tax_1 protein. *J. Virol.*, **66**, 6191–6193 (1992).

4. Angel, P., Hattori, K., Smeal, T., and Karin, M. The *jun* proto-oncogene is positively autoregulated by its product, Jun/AP-1. *Cell*, **55**, 875–885 (1988).
5. Azimi, N., Brown, K., Bamford, R. N., Tagaya, Y., Siebenlist, U., and Waldmann, T. A. Human T cell lymphotropic virus type I Tax protein trans-activates interleukin 15 gene transcription through an NF-κB site. *Proc. Natl. Acad. Sci. USA*, **95**, 2452–2457 (1998).
6. Ballard, D. W., Bohnlein, E., Lowenthal, J. W., Wano, Y., Franza, B. R., and Greene, W. C. HTLV-I tax induces cellular proteins that activate the κB element in the IL-2 receptor α gene. *Science*, **241**, 1652–1655 (1988).
7. Bex, F., McDowall, A., Burny, A., and Gaynor, R. The human T-cell leukemia virus type 1 transactivator protein Tax colocalizes in unique nuclear structures with NF-κB proteins. *J. Virol.*, **71**, 3483–3497 (1997).
8. Bex, F., Yin, M.-J., Burny, A., and Gaynor, R. B. Differential transcriptional activation by human T-cell leukemia virus type 1 Tax mutants is mediated by distinct interactions with CREB binding protein and p300. *Mol. Cell. Biol.*, **18**, 2392–2405 (1998).
9. Brockman, J. A., Scherer, D. C., McKinsey, T. A., Hall, S. M., Qi, X., Lee, W. Y., and Ballard, D. W. Coupling of a signal response domain in IκBα to multiple pathways for NF-κB activation. *Mol. Cell. Biol.*, **15**, 2809–2818 (1995).
10. Carter, R. S., Geyer, B. C., Xie, M., Acevedo-Suarez, C. A., and Ballard, D. W. Persistent activation of NF-κB by the Tax transforming protein involves chronic phosphorylation of IκB kinase subunits IKKβ and IKKγ. *J. Biol. Chem.*, **276**, 24445–24448 (2001).
11. Chu, Z.-L., DiDonato, J. A., Hawiger, J., and Ballard, D. W. The Tax oncoprotein of human T-cell leukemia virus type 1 associates with and persistently activates IκB kinases containing IKKα and IKKβ. *J. Biol. Chem.*, **273**, 15891–15894 (1998).
12. Chu, Z.-L., Shin, Y.-A., Yang, J.-M., DiDonato, J. A., and Ballard, D. W. IKKγ mediates the interaction of cellular IκB kinases with the Tax transforming protein of human T cell leukemia virus type 1. *J. Biol. Chem.*, **274**, 15297–15300 (1999).
13. Derijard, B., Hibi, M., Wu, I.-H., Barrett, T., Su, B., Deng, T., Karin, M., and Davis, R. J. JNK1: a protein kinase stimulated by UV light and Ha-Ras that binds and phosphorylates the c-Jun activation domain. *Cell*, **76**, 1025–1037 (1994).
14. Duyao, M. P., Kessler, D. J., Spicer, D. B., Bartholomew, C., Cleveland, J. L., Siekevitz, M., and Sonenshein, G. E. Transactivation of the c-*myc* promoter by human T cell leukemia virus type 1 *tax* is mediated by NFκB. *J. Biol. Chem.*, **267**, 16288–16291 (1992).
15. Fu, W., Shah, S. R., Jiang, H., Hilt, D. C., Dave, H. P. G., and Joshi, J. B. Transactivation of proenkephalin gene by HTLV-1 tax_1 protein in glial cells: involvement of Fos/Jun complex at an AP-1 element in the proenkephalin gene promoter. *J. Neurovirol.*, **3**, 16–27 (1997).
16. Fujii, M., Minamino, T., Nomura, M., Miyamoto, K., Tanaka, J., and Seiki, M. Selective activation of the proto-oncogene c-jun promoter by the transforming protein of v-Rel. *Oncogene*, **12**, 2193–2202 (1996).
17. Fujii, M., Niki, T., Mori, T., Matsuda, T., Matsui, M., Nomura, N., and Seiki, M.

HTLV-1 Tax induces expression of various immediate early serum responsive genes. *Oncogene*, **6**, 1023–1029 (1991).

18. Furukawa, Y., Kubota, R., Tara, M., Izumo, S., and Osame, M. Existence of escape mutant in HTLV-I tax during the development of adult T-cell leukemia. *Blood*, **97**, 987–993 (2001).
19. Geleziunas, R., Ferrell, S., Lin, X., Mu, Y., Cunningham, E. T., Jr., Grant, M., Connelly, M. A., Hambor, J. E., Marcu, K. B., and Greene, W. C. Human T-cell leukemia virus type 1 Tax induction of NF-κB involves activation of the IκB kinase α (IKKα) and IKKβ cellular kinases. *Mol. Cell. Biol.*, **18**, 5157–5165 (1998).
20. Good, L. and Sun, S.-C. Persistent activation of NF-κB/Rel by human T-cell leukemia virus type 1 Tax involves degradation of IκBβ. *J. Virol.*, **70**, 2730–2735 (1996).
21. Grassmann, R., Berchtold, S., Radant, I., Alt, M., Fleckenstein, B., Sodroski, J. G., Haseltine, W. A., and Ramstedt, U. Role of human T-cell leukemia virus type 1 X region proteins in immortalization of primary human lymphocytes in culture. *J. Virol.*, **66**, 4570–4575 (1992).
22. Grossman, W. J., Kimata, J. T., Wong, F.-H., Zutter, M., Ley, T. J., and Ratner, L. Development of leukemia in mice transgenic for the *tax* gene of human T-cell leukemia virus type I. *Proc. Natl. Acad. Sci. USA*, **92**, 1057–1061 (1995).
23. Han, T. H. and Prywes, R. Regulatory role of MEF2D in serum induction of the c-jun promoter. *Mol. Cell. Biol.*, **15**, 2907–2915 (1995).
24. Harhaj, E. and Sun, S.-C. IKKγ serves as a docking subunit of the IκB kinase (IKK) and mediates interaction of IKK with the human T-cell leukemia virus Tax protein. *J. Biol. Chem.*, **274**, 22911–22914 (1999).
25. Harhaj, E. W., Good, L., Xiao, G., Uhlik, M., Cvijic, M. E., Rivera-Walsh, I., and Sun, S.-C. Somatic mutagenesis studies of NF-κB signaling in human T cells: evidence for an essential role of IKKγ in NF-κB activation by T-cell costimulatory signals and HTLV-I Tax protein. *Oncogene*, **19**, 1448–1456 (2000).
26. Hehner, S. P., Hofmann, T. G., Ushmorov, A., Dienz, O., Wing-Lan Leung, I., Lassam, N., Scheidereit, C., Droge, W., and Schmitz, M. L. Mixed-lineage kinase 3 delivers CD3/CD28-derived signals into the IκB kinase complex. *Mol. Cell. Biol.*, **20**, 2556–2568 (2000).
27. Himes, S. R., Coles, L. S., Katsikeros, R., Lang, R. K., and Shannon, M. F. HTLV-1 tax activation of the GM-CSF and G-CSF promoters requires the interaction of NF-κB with other transcription factor families. *Oncogene*, **8**, 3189–3197 (1993).
28. Hinrichs, S. H., Nerenberg, M., Reynolds, R. K., Khoury, G., and Jay, G. A transgenic mouse model for human neurofibromatosis. *Science*, **237**, 1340–1343 (1987).
29. Hinuma, Y., Nagata, K., Hanaoka, M., Nakai, M., Matsumoto, T., Kinoshita, K., Shirakawa, S., and Miyoshi, I. Adult T-cell leukemia: antigen in an ATL cell line and detection of antibodies to the antigen in human sera. *Proc. Natl. Acad. Sci. USA*, **78**, 6476–6480 (1981).
30. Hipskind, R. A., Rao, V. N., Mueller, C. G., Reddy, E. S. P., and Nordheim, A. Ets-related protein Elk-1 is homologous to the c-fos regulatory factor p62TCF. *Nature*, **354**, 531–534 (1991).
31. Hiscott, J., Petropoulos, L., and Lacoste, J. Molecular interactions between HTLV-1 Tax protein and the NF-κB/IκB transcription complex. *Virology*, **214**, 3–11 (1995).

32. Hooper, W. C., Rudolph, D. L., Lairmore, M. D., and Lal, R. B. Constitutive expression of *c-jun* and *jun-B* in cell lines infected with human T-lymphotropic virus types I and II. *Biochem. Biophys. Res. Commun.*, **181**, 976–980 (1991).
33. Hu, Y., Baud, V., Delhase, M., Zhang, P., Deerinck, T., Ellisman, M., Johnson, R., and Karin, M. Abnormal morphogenesis but intact IKK activation in mice lacking the IKKα subunit of IκB kinase. *Science*, **284**, 316–320 (1999).
34. Huang, Y., Ohtani, K., Iwanaga, R., Matsumura, Y., and Nakamura, M. Direct trans-activation of the human cyclin D2 gene by the oncogene product Tax of human T-cell leukemia virus type I. *Oncogene*, **20**, 1094–1102 (2001).
35. Imada, K., Takaori-Kondo, A., Akagi, T., Shimotohno, K., Sugamura, K., Hattori, T., Yamabe, H., Okuma, M., and Uchiyama, T. Tumorigenicity of human T-cell leukemia virus type I-infected cell lines in severe combined immunodeficient mice and characterization of the cells proliferating *in vivo*. *Blood*, **86**, 2350–2357 (1995).
36. Imura, A., Hori, T., Imada, K., Kawamata, S., Tanaka, Y., Imamura, S., and Uchiyama, T. OX40 expressed on fresh leukemic cells from adult T-cell leukemia patients mediates cell adhesion to vascular endothelial cells: implication for the possible involvement of OX40 in leukemic cell infiltration. *Blood*, **89**, 2951–2958 (1997).
37. Ishibashi, K., Ishitsuka, K., Chuman, Y., Otsuka, M., Kuwazuru, Y., Iwahashi, M., Utsunomiya, A., Hanada, S., Sakurami, T., and Arima, T. Tumor necrosis factor-β in the serum of adult T-cell leukemia with hypercalcemia. *Blood*, **77**, 2451–2455 (1991).
38. Iwai, K., Mori, N., Oie, M., Yamamoto, N., and Fujii, M. Human T-cell leukemia virus type 1 tax protein activates transcription through AP-1 site by inducing DNA binding activity in T cells. *Virology*, **279**, 38–46 (2001).
39. Iwanaga, R., Ohtani, K., Hayashi, T., and Nakamura, M. Molecular mechanism of cell cycle progression induced by the oncogene product Tax of human T-cell leukemia virus type I. *Oncogene*, **20**, 2055–2067 (2001).
40. Iwanaga, Y., Tsukahara, T., Ohashi, T., Tanaka, Y., Arai, M., Nakamura, M., Ohtani, K., Koya, Y., Kannagi, M., Yamamoto, N., and Fujii, M. Human T-cell leukemia virus type 1 Tax protein abrogates interleukin-2 dependence in a mouse T-cell line. *J. Virol.*, **73**, 1271–1277 (1999).
41. Jin, D.-Y., Giordano, V., Kibler, K. V., Nakano, H., and Jeang, K.-T. Role of adapter function in oncoprotein-mediated activation of NF-κB: human T-cell leukemia virus type I Tax interacts directly with IκB kinase γ. *J. Biol. Chem.*, **274**, 17402–17405 (1999).
42. Kanno, T., Brown, K., Franzoso, G., and Siebenlist, U. Kinetic analysis of human T-cell leukemia virus type I Tax-mediated activation of NF-κB. *Mol. Cell. Biol.*, **14**, 6443–6451 (1994).
43. Karin, M. and Ben-Neriah, Y. Phosphorylation meets ubiquitination: the control of NF-κB activity. *Annu. Rev. Immunol.*, **18**, 621–663 (2000).
44. Kim, S.-J., Kehrl, J. H., Burton, J., Tendler, C. L., Jeang, K.-T., Danielpour, D., Thevenin, C., Kim, K. Y., Sporn, M. B., and Roberts, A. B. Transactivation of the transforming growth factor β1 (TGF-β1) gene by human T lymphotropic virus type 1 Tax: a potential mechanism for the increased production of TGF-β1 in adult T cell leukemia. *J. Exp. Med.*, **172**, 121–129 (1990).
45. Kinoshita, T., Shimoyama, M., Tobinai, K., Ito, M., Ito, S., Ikeda, S., Tajima, K.,

Shimotohno, K., and Sugimura, T. Detection of mRNA for the *tax1/rex1* gene of human T-cell leukemia virus type I in fresh peripheral blood mononuclear cells of adult T-cell leukemia patients and viral carriers by using the polymerase chain reaction. *Proc. Natl. Acad. Sci. USA*, **86**, 5620–5624 (1989).

46. Kitajima, I., Shinohara, T., Bilakovics, J., Brown, D. A., Xu, X., and Nerenberg, M. Ablation of transplanted HTLV-1 Tax-transformed tumors in mice by antisense inhibition of NF-κB. *Science*, **258**, 1792–1795 (1992).
47. Lacoste, J., Petropoulos, L., Pepin, N., and Hiscott, J. Constitutive phosphorylation and turnover of IκBα in human T-cell leukemia virus type I-infected and Tax-expressing T cells. *J. Virol.*, **69**, 564–569 (1995).
48. Laherty, C. D., Perkins, N. D., and Dixit, V. M. Human T cell leukemia virus type I Tax and phorbol 12-myristate 13-acetate induce expression of the A20 zinc finger protein by distinct mechanisms involving nuclear factor κB. *J. Biol. Chem.*, **268**, 5032–5039 (1993).
49. Lanoix, J., Lacoste, J., Pepin, N., Rice, N., and Hiscott, J. Overproduction of NFKB2 (*lyt-10*) and c-Rel: a mechanism for HTLV-I Tax-mediated trans-activation via the NF-κB signalling pathway. *Oncogene*, **9**, 841–852 (1994).
50. Lee, F. S., Peters, R. T., Dang, L. C., and Maniatis, T. MEKK1 activates both IκB kinase α and IκB kinase β. *Proc. Natl. Acad. Sci. USA*, **95**, 9319–9324 (1998).
51. Li, Q., Van Antwerp, D., Mercurio, F., Lee, K. F., and Verma, I. M. Severe liver degeneration in mice lacking the IκB kinase 2 gene. *Science*, **284**, 321–325 (1999).
52. Li-Weber, M., Giaisi, M., Chlichlia, K., Khazaie, K., and Krammer, P. H. Human T cell leukemia virus type I Tax enhances IL-4 gene expression in T cells. *Eur. J. Immunol.*, **31**, 2623–2632 (2001).
53. Lilienbaum, A. and Paulin, D. Activation of the human vimentin gene by the Tax human T-cell leukemia virus I. Mechanisms of regulation by the NF-κB transcription factor. *J. Biol. Chem.*, **268**, 2180–2188 (1993).
54. Liu, X., Chen, X., Zachar, V., Chang, C., and Ebbesen, P. Transcriptional activation of human *TR3/nur77* gene expression by human T-lymphotropic virus type I Tax protein through two AP-1-like elements. *J. Gen. Virol.*, **80**, 3073–3081 (1999).
55. Maeda, M., Shimizu, A., Ikuta, K., Okamoto, H., Kashihara, M., Uchiyama, T., Honjo, T., and Yodoi, J. Origin of human T-lymphotrophic virus I-positive T cell lines in adult T cell leukemia. Analysis of T cell receptor gene rearrangement. *J. Exp. Med.*, **162**, 2169–2174 (1985).
56. Mariner, J. M., Lantz, V., Waldmann, T. A., and Azimi, N. Human T cell lymphotropic virus type I Tax activates IL-15Rα gene expression through an NF-κB site. *J. Immunol.*, **166**, 2602–2609 (2001).
57. Matsumoto, K., Shibata, H., Fujisawa, J., Inoue, H., Hakura, A., Tsukahara, T., and Fujii, M. Human T-cell leukemia virus type 1 Tax protein transforms rat fibroblasts via two distinct pathways. *J. Virol.*, **71**, 4445–4451 (1997).
58. Matsushita, K., Arima, N., Yamaguchi, K., Matsumoto, T., Ohtsubo, H., Hidaka, S., Fujiwara, H., Arimura, K., Kukita, T., Tokito, Y., Ozaki, A., and Tei, C. Granulocyte colony-stimulating factor production by adult T-cell leukaemia cells. *Br. J. Haematol.*, **111**, 208–215 (2000).
59. McKinsey, T. A., Brockman, J. A., Scherer, D. C., Al-Murrani, S. W., Green, P. L.,

and Ballard, D. W. Inactivation of IκBβ by the Tax protein of human T-cell leukemia virus type 1: a potential mechanism for constitutive induction of NF-κB. *Mol. Cell. Biol.*, **16**, 2083–2090 (1996).

60. Mercurio, F., Zhu, H., Murray, B. W., Shevchenko, A., Bennett, B. L., Li, J., Young, D. B., Barbosa, M., Mann, M., Manning, A., and Rao, A. IKK-1 and IKK-2: cytokine-activated IκB kinases essential for NF-κB activation. *Science*, **278**, 860–866 (1997).
61. Miyake, H., Suzuki, T., Hirai, H., and Yoshida, M. Trans-activator Tax of human T-cell leukemia virus type 1 enhances mutation frequency of the cellular genome. *Virology*, **253**, 155–161 (1999).
62. Mori, N., Fujii, M., Cheng, G., Ikeda, S., Yamasaki, Y., Yamada, Y., Tomonaga, M., and Yamamoto, N. Human T-cell leukemia virus type I Tax protein induces the expression of anti-apoptotic gene Bcl-x_L in human T-cells through nuclear factor-κB and c-AMP responsive element binding protein pathways. *Virus Genes*, **22**, 279–287 (2001).
63. Mori, N., Fujii, M., Hinz, M., Nakayama, K., Yamada, Y., Ikeda, S., Yamasaki, Y., Kashanchi, F., Tanaka, Y., Tomonaga, M., and Yamamoto, N. Activation of cyclin D1 and D2 promoters by human T-cell leukemia virus type I Tax protein is associated with IL-2-independent growth of T cells. *Int. J. Cancer*, **99**, 378–385 (2002).
64. Mori, N., Fujii, M., Ikeda, S., Yamada, Y., Tomonaga, M., Ballard, D. W., and Yamamoto, N. Constitutive activation of NF-κB in primary adult T-cell leukemia cells. *Blood*, **93**, 2360–2368 (1999).
65. Mori, N., Fujii, M., Iwai, K., Ikeda, S., Yamasaki, Y., Hata, T., Yamada, Y., Tanaka, Y., Tomonaga, M., and Yamamoto, N. Constitutive activation of transcription factor AP-1 in primary adult T-cell leukemia cells. *Blood*, **95**, 3915–3921 (2000).
66. Mori, N., Mukaida, N., Ballard, D. W., Matsushima, K., and Yamamoto, N. Human T-cell leukemia virus type I Tax transactivates human *interleukin 8* gene through acting concurrently on AP-1 and nuclear factor-κB-like sites. *Cancer Res.*, **58**, 3993–4000 (1998).
67. Mori, N., Murakami, S., Oda, S., Prager, D., and Eto, S. Production of interleukin 8 in adult T-cell leukemia cells: possible transactivation of the *interleukin 8* gene by human T-cell leukemia virus type I *tax*. *Cancer Res.*, **55**, 3592–3597 (1995).
68. Mori, N. and Prager, D. Transactivation of the interleukin-1α promoter by human T-cell leukemia virus type I and type II Tax proteins. *Blood*, **87**, 3410–3417 (1996).
69. Mori, N., Sato, H., Hayashibara, T., Senba, M., Hayashi, T., Yamada, Y., Kamihira, S., Ikeda, S., Yamasaki, Y., Morikawa, S., Tomonaga, M., Geleziunas, R., and Yamamoto, N. Human T-cell leukemia virus type I Tax transactivates the matrix metalloproteinase-9 gene: potential role in mediating adult T-cell leukemia invasiveness. *Blood*, **99**, 1341–1349 (2002).
70. Mori, N., Shirakawa, F., Murakami, S., Oda, S., and Eto, S. Interleukin-4 inhibits the production of interleukin-1 by adult T-cell leukemia cells. *Eur. J. Haematol.*, **55**, 121–125 (1995).
71. Mori, N., Shirakawa, F., Shimizu, H., Murakami, S., Oda, S., Yamamoto, K., and Eto, S. Transcriptional regulation of the human interleukin-6 gene promoter in human T-cell leukemia virus type I-infected T cell lines: evidence for the involvement of NF-κB. *Blood*, **84**, 2904–2911 (1994).

72. Mori, N., Ueda, A., Ikeda, S., Yamasaki, Y., Yamada, Y., Tomonaga, M., Morikawa, S., Geleziunas, R., Yoshimura, T., and Yamamoto, N. Human T-cell leukemia virus type I Tax activates transcription of the human monocyte chemoattractant protein-1 gene through two nuclear factor-κB sites. *Cancer Res.*, **60**, 4939–4945 (2000).
73. Mori, N., Yamada, Y., Ikeda, S., Yamasaki, Y., Tsukasaki, K., Tanaka, Y., Tomonaga, M., Yamamoto, N., and Fujii, M. Bay 11-7082 inhibits transcription factor NF-κB and induces apoptosis of HTLV-I-infected T-cell lines and primary adult T-cell leukemia cells. *Blood*, **100**, 1828–1834 (2002).
74. Nerenberg, M., Hinrichs, S. H., Reynolds, R. K., Khoury, G., and Jay, G. The tat gene of human T-lymphotropic virus type 1 induces mesenchymal tumors in transgenic mice. *Science*, **237**, 1324–1329 (1987).
75. Ng, P. W. P., Iha, H., Iwanaga, Y., Bittner, M., Chen, Y., Jiang, Y., Gooden, G., Trent, J. M., Meltzer, P., Jeang, K.-T., and Zeichner, S. L. Genome-wide expression changes induced by HTLV-1 Tax: evidence for MLK-3 mixed lineage kinase involvement in Tax-mediated NF-κB activation. *Oncogene*, **20**, 4484–4496 (2001).
76. Nicot, C., Mahieux, R., Takemoto, S., and Franchini, G. Bcl-X_L is up-regulated by HTLV-I and HTLV-II *in vitro* and in *ex vivo* ATLL samples. *Blood*, **96**, 275–281 (2000).
77. Nicot, C., Tie, F., and Giam, C.-Z. Cytoplasmic forms of human T-cell leukemia virus type 1 Tax induce NF-κB activation. *J. Virol.*, **72**, 6777–6784 (1998).
78. Ninomiya-Tsuji, J., Kishimoto, K., Hiyama, A., Inoue, J., Cao, Z., and Matsumoto, K. The kinase TAK1 can activate the NIK-IκB as well as the MAP kinase cascade in the IL-1 signalling pathway. *Nature*, **398**, 252–256 (1999).
79. Ogata, M., Ogata, Y., Kohno, K., Uno, N., Ohno, E., Ohtsuka, E., Saburi, Y., Kamberi, P., Nasu, M., and Kikuchi, H. Eosinophilia associated with adult T-cell leukemia: role of interleukin 5 and granulocyte-macrophage colony-stimulating factor. *Am. J. Hematol.*, **59**, 242–245 (1998).
80. Ohtani, K., Tsujimoto, A., Tsukahara, T., Numata, N., Miura, S., Sugamura, K., and Nakamura, M. Molecular mechanisms of promoter regulation of the gp34 gene that is trans-activated by an oncoprotein Tax of human T cell leukemia virus type I. *J. Biol. Chem.*, **273**, 14119–14129 (1998).
81. Okazaki, S., Moriuchi, R., Yoshizuka, N., Sugahara, K., Maeda, T., Jinnai, I., Tomonaga, M., Kamihira, S., and Katamine, S. HTLV-1 proviruses encoding non-functional TAX in adult T-cell leukemia. *Virus Genes*, **23**, 123–135 (2001).
82. Pankow, R., Durkop, H., Latza, U., Krause, H., Kunzendorf, U., Pohl, T., and Bulfone-Paus, S. The HTLV-I Tax protein transcriptionally modulates OX40 antigen expression. *J. Immunol.*, **165**, 263–270 (2000).
83. Pise-Masison, C. A., Mahieux, R., Jiang, H., Ashcroft, M., Radonovich, M., Duvall, J., Guillerm, C., and Brady, J. N. Inactivation of p53 by human T-cell lymphotropic virus type 1 Tax requires activation of the NF-κB pathway and is dependent on p53 phosphorylation. *Mol. Cell. Biol.*, **20**, 3377–3386 (2000).
84. Poiesz, B. J., Ruscetti, F. W., Gazdar, A. F., Bunn, P. A., Minna, J. D., and Gallo, R. C. Detection and isolation of type C retrovirus particles from fresh and cultured lymphocytes of a patient with cutaneous T-cell lymphoma. *Proc. Natl. Acad. Sci. USA*, **77**, 7415–7419 (1980).

85. Pozzatti, R., Vogel, J., and Jay, G. The human T-lymphotropic virus type I tax gene can cooperate with the ras oncogene to induce neoplastic transformation of cells. *Mol. Cell. Biol.*, **10**, 413–417 (1990).
86. Rivera-Walsh, I., Waterfield, M., Xiao, G., Fong, A., and Sun, S.-C. NF-κB signaling pathway governs TRAIL gene expression and human T-cell leukemia virus-I Tax-induced T-cell death. *J. Biol. Chem.*, **276**, 40385–40388 (2001).
87. Robek, M. D. and Ratner, L. Immortalization of $CD4^+$ and $CD8^+$ T lymphocytes by human T-cell leukemia virus type 1 Tax mutants expressed in a functional molecular clone. *J. Virol.*, **73**, 4856–4865 (1999).
88. Ryder, K., Lanahan, A., Perez-Albuerne, E., and Nathans, D. jun-D: a third member of the jun gene family. *Proc. Natl. Acad. Sci. USA*, **86**, 1500–1503 (1989).
89. Santiago, F., Clark, E., Chong, S., Molina, C., Mozafari, F., Mahieux, R., Fujii, M., Azimi, N., and Kashanchi, F. Transcriptional up-regulation of the cyclin D2 gene and acquisition of new cyclin-dependent kinase partners in human T-cell leukemia virus type 1-infected cells. *J. Virol.*, **73**, 9917–9927 (1999).
90. Semmes, O. J. and Jeang, K.-T. Localization of human T-cell leukemia virus type 1 Tax to subnuclear compartments that overlap with interchromatin speckles. *J. Virol.*, **70**, 6347–6357 (1996).
91. Shibuya, H. and Taniguchi, T. Identification of multiple cis-elements and trans-acting factors involved in the induced expression of human IL-2 gene. *Nucleic Acids Res.*, **17**, 9173–9184 (1989).
92. Smith, M. R. and Greene, W. C. Type I human T cell leukemia virus tax protein transforms rat fibroblasts through the cyclic adenosine monophosphate response element binding protein/activating transcription factor pathway. *J. Clin. Invest.*, **88**, 1038–1042 (1991).
93. Sodroski, J., Rosen, C., Goh, W. C., and Haseltine, W. A transcriptional activator protein encoded by the x-lor region of the human T-cell leukemia virus. *Science*, **228**, 1430–1434 (1985).
94. Sun, S.-C., Elwood, J., Beraud, C., and Greene, W. C. Human T-cell leukemia virus type I Tax activation of NF-κB/Rel involves phosphorylation and degradation of IκBα and RelA (p65)-mediated induction of the c-*rel* gene. *Mol. Cell. Biol.*, **14**, 7377–7384 (1994).
95. Takeda, K., Takeuchi, O., Tsujimura, T., Itami, S., Adachi, O., Kawai, T., Sanjo, H., Yoshikawa, K., Terada, N., and Akira, S. Limb and skin abnormalities in mice lacking IKKα. *Science*, **284**, 313–316 (1999).
96. Tanaka, A., Takahashi, C., Yamaoka, S., Nosaka, T., Maki, M., and Hatanaka, M. Oncogenic transformation by the *tax* gene of human T-cell leukemia virus type I *in vitro*. *Proc. Natl. Acad. Sci. USA*, **87**, 1071–1075 (1990).
97. Tanaka, M., Fuentes, M. E., Yamaguchi, K., Durnin, M. H., Dalrymple, S. A., Hardy, K. L., and Goeddel, D. V. Embryonic lethality, liver degeneration, and impaired NF-κB activation in IKK-β-deficient mice. *Immunity*, **10**, 421–429 (1999).
98. Tojima, Y., Fujimoto, A., Delhase, M., Chen, Y., Hatakeyama, S., Nakayama, K., Kaneko, Y., Nimura, Y., Motoyama, N., Ikeda, K., Karin, M., and Nakanishi, M. NAK is an IκB kinase-activating kinase. *Nature*, **404**, 778–782 (2000).
99. Tschachler, E., Bohnlein, E., Felzmann, S., and Reitz, M. S., Jr. Human T-lympho-

tropic virus type I *tax* regulates the expression of the human lymphotoxin gene. *Blood*, **81**, 95–100 (1993).

100. Tsuchiya, H., Fujii, M., Niki, T., Tokuhara, M., Matsui, M., and Seiki, M. Human T-cell leukemia virus type 1 Tax activates transcription of the human *fra-1* gene through multiple *cis* elements responsive to transmembrane signals. *J. Virol.*, **67**, 7001–7007 (1993).

101. Tsukahara, T., Kannagi, M., Ohashi, T., Kato, H., Arai, M., Nunez, G., Iwanaga, Y., Yamamoto, N., Ohtani, K., Nakamura, M., and Fujii, M. Induction of Bcl-x_L expression by human T-cell leukemia virus type 1 Tax through NF-κB in apoptosis-resistant T-cell transfectants with Tax. *J. Virol.*, **73**, 7981–7987 (1999).

102. Uchijima, M., Sato, H., Fujii, M., and Seiki, M. Tax proteins of human T-cell leukemia virus type 1 and 2 induce expression of the gene encoding erythroid-potentiating activity (tissue inhibitor of metalloproteinases-1, TIMP-1). *J. Biol. Chem.*, **269**, 14946–14950 (1994).

103. Uchiyama, T., Hori, T., Tsudo, M., Wano, Y., Umadome, H., Tamori, S., Yodoi, J., Maeda, M., Sawami, H., and Uchino, H. Interleukin-2 receptor (Tac antigen) expressed on adult T cell leukemia cells. *J. Clin. Invest.*, **76**, 446–453 (1985).

104. Uhlik, M., Good, L., Xiao, G., Harhaj, E. W., Zandi, E., Karin, M., and Sun, S.-C. NF-κB-inducing kinase and IκB kinase participate in human T-cell leukemia virus I Tax-mediated NF-κB activation. *J. Biol. Chem.*, **273**, 21132–21136 (1998).

105. Valentin, H., Lemasson, I., Hamaia, S., Casse, H., Konig, S., Devaux, C., and Gazzolo, L. Transcriptional activation of the vascular cell adhesion molecule-1 gene in T lymphocytes expressing human T-cell leukemia virus type 1 Tax protein. *J. Virol.*, **71**, 8522–8530 (1997).

106. Weitzman, J. B., Fiette, L., Matsuo, K., and Yaniv, M. JunD protects cells from p53-dependent senescence and apoptosis. *Mol. Cell*, **6**, 1109–1119 (2000).

107. Woronicz, J. D., Gao, X., Cao, Z., Rothe, M., and Goeddel, D. V. IκB kinase-β: NF-κB activation and complex formation with IκB kinase-α and NIK. *Science*, **278**, 866–869 (1997).

108. Xiao, G., Cvijic, M. E., Fong, A., Harhaj, E. W., Uhlik, M. T., Waterfield, M., and Sun, S.-C. Retroviral oncoprotein Tax induces processing of NF-κB2/p100 in T cells: evidence for the involvement of IKKα. *EMBO J.*, **20**, 6805–6815 (2001).

109. Xiao, G., Harhaj, E. W., and Sun, S.-C., Domain-specific interaction with the IκB kinase (IKK) regulatory subunit IKKγ is an essential step in Tax-mediated activation of IKK. *J. Biol. Chem.*, **275**, 34060–34067 (2000).

110. Yamada, Y., Sugawara, K., Hata, T., Tsuruta, K., Moriuchi, R., Maeda, T., Atogami, S., Murata, K., Fujimoto, K., Kohno, T., Tsukasaki, K., Tomonaga, M., Hirakata, Y., and Kamihira, S. Interleukin-15 (IL-15) can replace the IL-2 signal in IL-2-dependent adult T-cell leukemia (ATL) cell lines: expression of IL-15 receptor α on ATL cells. *Blood*, **91**, 4265–4272 (1998).

111. Yamagata, T., Mitani, K., Ueno, H., Kanda, Y., Yazaki, Y., and Hirai, H. Triple synergism of human T-lymphotropic virus type 1-encoded Tax, GATA-binding protein, and AP-1 is required for constitutive expression of the interleukin-5 gene in adult T-cell leukemia cells. *Mol. Cell. Biol.*, **17**, 4272–4281 (1997).

112. Yamaoka, S., Courtois, G., Bassia, C., Whiteside, S. T., Weil, R., Agou, F., Kirk, H.

E., Kay, R. J., and Israel, A. Complementation cloning of NEMO, a component of the IκB kinase complex essential for NF-κB activation. *Cell*, **93**, 1231–1240 (1998).

113. Yamaoka, S., Inoue, H., Sakurai, M., Sugiyama, T., Hazama, M., Yamada, T., and Hatanaka, M. Constitutive activation of NF-κB is essential for transformation of rat fibroblasts by the human T-cell leukemia virus type I Tax protein. *EMBO J.*, **15**, 873–887 (1996).

114. Yi, T., Lee, B.-H., Park, R.-W., and Kim, I.-S. Transactivation of fibronectin promoter by HTLV-I Tax through NF-κB pathway. *Biochem. Biophys. Res. Commun.*, **276**, 579–586 (2000).

115. Yin, M.-J., Christerson, L. B., Yamamoto, Y., Kwak, Y.-T., Xu, S., Mercurio, F., Barbosa, M., Cobb, M. H., and Gaynor, R. B. HTLV-I Tax protein binds to MEKK1 to stimulate IκB kinase activity and NF-κB activation. *Cell*, **93**, 875–884 (1998).

116. Yoshida, M., Miyoshi, I., and Hinuma, Y. Isolation and characterization of retrovirus from cell lines of human adult T-cell leukemia and its implication in the disease. *Proc. Natl. Acad. Sci. USA*, **79**, 2031–2035 (1982).

117. Yoshida, M., Suzuki, T., Fujisawa, J., and Hirai, H. HTLV-1 oncoprotein Tax and cellular transcription factors. *Curr. Top. Microbiol. Immunol.*, **193**, 79–89 (1995).

118. Zandi, E. and Karin, M. Bridging the gap: composition, regulation, and physiological function of the IκB kinase complex. *Mol. Cell. Biol.*, **19**, 4547–4551 (1999).

119. Zhao, Q. and Lee, F. S. Mitogen-activated protein kinase/ERK kinase kinases 2 and 3 activate nuclear factor-κB through IκB kinase-α and IκB kinase-β. *J. Biol. Chem.*, **274**, 8355–8358 (1999).

CLASTOGENIC AND ANEUGENIC DNA DAMAGE IN HTLV-I Tax-EXPRESSING CELLS

Kerstin HALLER and Kuan-Teh JEANG

*Molecular Virology Section, Laboratory of Molecular Microbiology, National Institute of Allergy and Infectious Diseases, National Institutes of Health**

The probability of developing cancer increases with advancing age. Presentation of cancer is the net outcome between processes which generate abnormal cells and those which eliminate them. It has been estimated that the number of sporadic genetic changes needed to convert a normal to a malignant cell are so numerous that one might not mathematically expect cancers to emerge during a human lifespan. However, the fact that cancers are commonly seen suggests that cells are not simply subject to ambient mistakes. Indeed, it is now clear that in many cases cells show an accelerated proclivity for genetic mutations after the acquisition of a "mutator" phenotype. Such mutator phenotype could either arise spontaneously or be induced after infection with an oncogenic virus such as human T cell leukemia virus type I (HTLV-I). In this chapter, we explore the evidence supporting the notion that HTLV-I Tax functions early in ATL-transformation by conferring a mutator phenotype to cells. We review how Tax affects cellular genetic stability by dysregulation of the mitotic spindle assembly checkpoint and other repair pathways to produce aneuploidy and clastogenic DNA damage.

The transformation of a normal cell to malignancy is marked by dysregulated cell cycle control (*56*) and the presentation of genomic instability (*32*). Tumorigenesis is a multistep process. For tumors to successfully take hold *in vivo*, such a collection of cells must acquire the capacities for limitless replication, sustained angiogenesis, evasion of apoptosis, self sufficiency in growth, insensitivity to anti-growth signals, and tissue invasion and metastasis (*15*). It is likely that these numerous phenotypic changes accumulate progressively over time resulting from discrete step-wise genetic alterations in somatic cells (*29*). The empirical observation that older humans develop cancer at a higher rate than their younger counterparts provides evidence linking the time required to accumulate multiple genetic lesions and ultimate malignant conversion (*1*).

Because man has evolved many sophisticated cellular as well as bodily surveillance mechanisms to check against malignancies and because numerous mutations are required to defeat these checks, it has been suggested that an entire lifetime would not provide sufficient time to acquire all the genetic changes needed for cancer. Accordingly, it has been argued that in the early stages of malignant transformation there is a selection for cells with a "mutator" phenotype (*34*). This mutator phenotype

* Bethesda, Maryland 20892-0460, USA

can be triggered by exogenous insult with a viral oncogene and/or by mutation in one or more endogenous proto-oncogenes and/or tumor suppressor genes. Acquisition of a mutator phenotype makes the affected cell more susceptible to accumulation of further mutations leading to underlying genetic instability. In principle, two different sets of mechanisms can produce increased mutation rates: 1) sequence instability (SIN) effects caused by defects in the DNA repair machinery, and 2) chromosomal instability (CIN) effects that are characterized by rearrangements, losses or gains of whole chromosomes leading to aneuploidy which is associated with the majority of human cancers (*32*).

Human T cell leukemia virus type I (HTLV-I) is the etiological agent of adult T cell leukemia (ATL). The virus immortalizes and transforms infected T lymphocytes (*62*). However, only a minor population of HTLV-I-infected individuals (about 5%) develop ATL after a long period of latency (more than 20 years) (*44*). This disease course suggests that genetic events, in addition to infection of cells by a virus, are required for the development of ATL. Although HTLV-I is thought to integrate randomly into the host genome, patients with ATL harbor only a small number of transformed T cell clones (*6*) pointing to rare transformational events. Cytogenetic analysis of ATL revealed prevalent abnormal karyotypes (*51, 65*) with pleomorphic multinuclei (*60*) but no consistently identifiable chromosomal abnormalities. In this regard, one could consider the HTLV-I/ATL system as a general virus/cell model for studying the complex processes linking genetic instability to the development of full-fledged malignancy.

Tax and Aneuploidy

The transforming agent of HTLV-I is the Tax protein. Tax can immortalize primary human T lymphocytes (*11*) and induces leukemia in transgenic mice (*12*). The 40 kDa phosphoprotein is a very pleiotropic factor. Although the exact events of Tax-induced transformation are not fully understood, some consistent principles are emerging. One such principle is that Tax interferes with critical cellular processes through direct protein-protein interactions.

Tax can activate or repress gene expression through at least four distinct transcriptional pathways (*23*). Interestingly, it can also bind to several regulators of cellular proliferation and cell cycle checkpoints (*42*). For purposes of understanding how repair of ambient DNA-damage might be subverted by Tax, it is important to note that this oncoprotein, in a growth signal independent manner, promotes rapid G1- to S-phase progression (*53*) by binding to the CDK inhibitor p16 (*35, 58*) and the cyclin D/CDK complex (*13, 14, 41*) (Fig. 1). G1 normally provides a time period for the cell to correct mistakes that might have occurred during mitosis (M). By abbreviating G1, Tax reduces the opportunity for correction and increases the likelihood that genetic errors will become imprinted through replication of the damaged DNA in S.

Recent findings that Tax engenders DNA errors during the M-phase of the cell cycle makes the occurrence of a shortened G1 more than a purely theoretical concern. During mitosis the newly replicated chromosomes in one mother cell assemble on the mitotic spindle to be distributed equally to two daughter cells. Eucaryotic cells have evolved the mitotic spindle assembly checkpoint (MSC) (*55*) in order to monitor the fidelity of chromosomal segregation and to prevent aneuploidy (*2*). At least 8 proteins

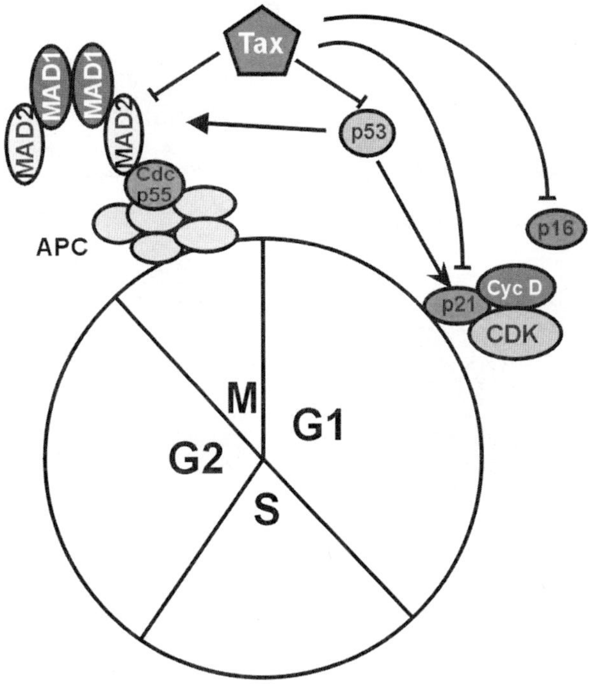

FIG. 1. Deregulation of different cell cycle checkpoints by Tax
Tax stimulates G1/S-phase transition (*53*) by interacting with the CDK4/6 inhibitor p16INK4A and blocking its association with kinase (*35, 57*). Additionally, Tax can bind to the holo-cyclin D/CDK complex and directly stimulate its activity (*13, 14, 41*). Beyond G1/S, Tax interferes with M-checkpoint by binding the MAD1 protein (*24*).

have been identified as playing roles in the spindle checkpoint. These are the MAD1, 2, 3 (mitotic arrest deficiency; *33*) proteins as well as budding uninhibited by benzimidazole factors 1–3 (BUB) (*19*), the monopolar spindle kinase 1 (MPS1) (*16, 64*) and cell division cycle factor 55 (CDC55/Cdc20) (*63*). In the course of a yeast two-hybrid survey, Jin *et al.* (*24*) identified a binding partner of Tax which turned out to be the human homologue of the yeast mitotic arrest deficiency factor 1 (HsMAD1).

HsMAD1 binds hsMAD2 in human cells (*5, 24*). The current thinking is that MAD1 directs MAD2 to associate in a dynamic way with microtubule-unattached kinetochores (Fig. 2A) (*17, 18*). Failure of microtubules to attach to a kinetochore is an error which results in chromosomal mis-segregation. Once bound to unattached kinetochores MAD2 is "activated" to associate with CDC55. Such association inactivates the anaphase promoting complex/cyclosome (APC/C, a multi-subunit E3 ubiquitin ligase associated with proteasome-mediated proteolysis; *66*) resulting in mitotic arrest which affords the temporal opportunity for correcting chromosomal mis-segregation. By binding to hsMAD1, Jin *et al.* (*24*) showed that Tax interfered with normal hsMAD1-hsMAD2 interaction leading to a loss in MSC function (Fig. 2B). Cells are thus permitted to exit mitosis despite incorrect segregation of sister chromatids, resulting in aneuploidy which is commonly observed for Tax-expressing cells (*37, 54*).

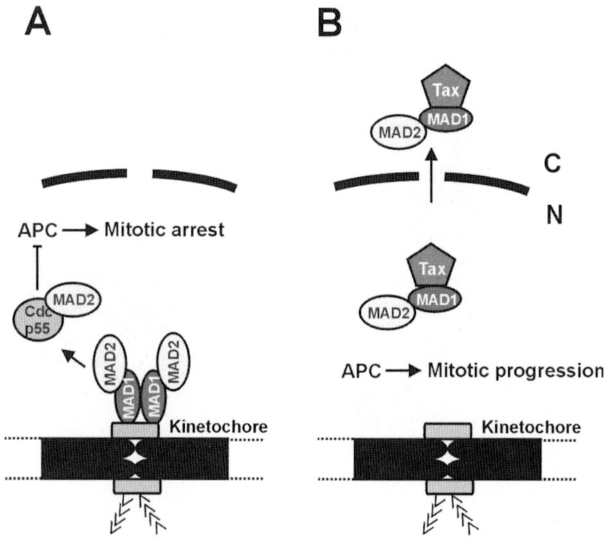

FIG. 2. Tax directly interferes with the mitotic spindle assembly checkpoint
In this schematic Tax and the main participants of MSC are depicted. A) In a normal cell, at the transition of metaphase to anaphase unattached kinetochores are recognized by MAD1/MAD2 complexes. By the following recruitment and inhibition of the Cdcp55 kinase MAD2 can deactivate the anaphase promoting complex (cyclosome) which leads to mitotic arrest (*55*). B) In HTLV-I-infected cells, Tax attaches to MAD1 (*24*). Presumably using its shuttling function (*3*), Tax relocates MAD1 and the associated MAD2 from the nucleus into the cytoplasm (*24*). By depleting the nucleus of active MAD2 Tax prevents an efficient function of the mitotic spindle assembly checkpoint.

A direct role of dysregulated mitotic checkpoint in cancer development is supported by studies which have identified loss-of-function mutations in genes controlling the MSC (i.e., hsMAD1, hsBUB1, and hsBUBR1; *4, 20, 43*). Because the spindle checkpoint serves dual roles in preventing metaphase-to-anaphase transition and inhibiting the exit from mitosis (*10*), binding of hsMAD1 by Tax could be envisioned to propagate cell cycling in a manner wholly insensitive to DNA-mistakes accumulated during M. Aneuploidy is a common finding in sporadic as well as virus-induced cancers. Hence, transforming agents such as SV40, EBV (*31*) and HTLV-I have evolved the ability to utilize common cellular mistakes to benefit the virus.

Tax and Clastogenic Damage

Apart from its effect on the MSC, Tax also subverts mechanisms which check against clastogenic damage (Fig. 3) (*25, 37–39, 48, 54*). First, Tax inactivates the "genome-guardian" function of p53 through mechanism(s) which is incompletely understood (Fig. 3C) (*8, 36, 40, 49, 50, 58*). Loss of p53-function in Tax-expressing cells is one explanation for a heightened mutational rate. This is of added interest because the *hsmad1* gene was found to be upregulated by p53 (*21, 52*) thereby linking MSC control to DNA damage surveillance. Second, Tax is a repressor of DNA polymerase β (*22*), an enzyme with an important role in base excision DNA repair (Fig. 3A). This pathway of DNA repair was demonstrated to be defective in Tax-expressing cells (*48*). Third, other studies have also revealed a Tax-induced impairment in

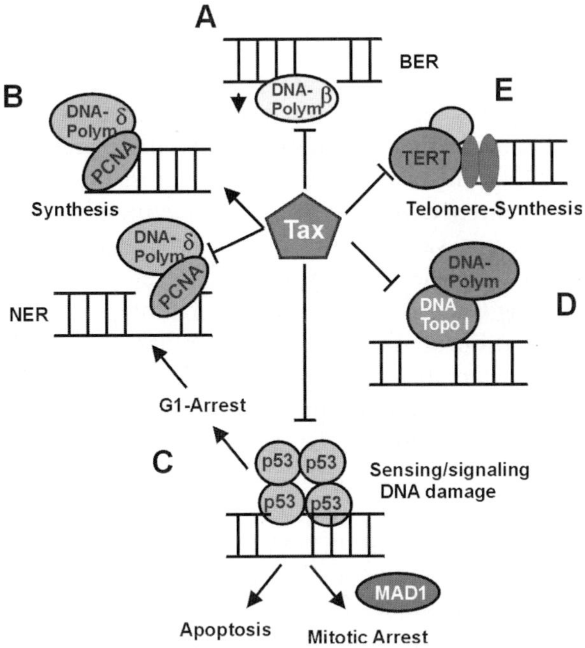

FIG. 3. Tax allows clastogenic damage by interfering with a variety of DNA damage-repair pathways

Tax has been reported to influence no fewer than 5 different pathways of DNA repair. A) Through repression of DNA polymerase β expression (*22*) Tax inhibits base-excision repair in HTLV-infected cells (*48*). B) Kao and Marriott (*25*) found that Tax by simultaneously up-regulating PCNA and inactivating p53 disrupts nucleotide excision repair. C) p53 signaling is defective in HTLV-positive cells (*8, 36, 40, 49, 50, 58*). D) Tax associates with topoisomerase I and down-modulates enzymatic activity (*59*). E) Tax is shown to have a repressive effect on telomerase function (*38*).

the nucleotide excision repair pathway through the upregulation of proliferating cell nuclear antigen (PCNA) (*25*) (Fig. 3B). This seems to be paradoxical since PCNA is proposed to be involved in DNA repair. However, the PCNA-repair function can only take place if DNA damage is in a setting of wild-type p53 function (*47*). Because p53 does not function properly in Tax-expressing cells the overall effect of up-regulation of PCNA is to increase DNA synthesis without PCNA-dependent DNA repair (*26*). Fourth, Tax can also interfere with DNA repair through topoisomerase I (Fig. 3D). Tax was reported to interact with topoisomerase I, reducing its enzymatic activity (*59*). The extent contributed by this interaction to genome stability remains to be quantified. Finally, Tax may also reduce telomerase activity in HTLV-I-infected cells (Fig. 3E) (*38*). While a reduction in such activity is seemingly contradictory to transformation since many have reported increased telomerase activity in human tumors (*27*), recent findings have re-evaluated this line of thinking (*7*). A current perspective is that reduced telomerase activity in the setting of other cellular changes promotes a crisis from which surviving cells emerge with extensive chromosomal rearrangements which benefit malignant transformation (*6*). Post-crisis, telomerase activity is reactivated in order to stabilize the transformed cell. Potentially, the kinetics of Tax-expression in T cells (i.e., the absence of functional Tax in many primary ATL cells; *9, 28, 45*) reflects the temporal profiles expected of telomerase activity during transformation.

Concluding Remarks

HTLV-I Tax by suppressing DNA repair mechanisms and neutralizing cell cycle checkpoints does not directly create mutations but rather prevents the correction of ambiently generated DNA mistakes in cells. While most mistakes may be detrimental, a subset confers heightened viability to cells. *In vivo*, one can imagine that some HTLV-I-infected Tax-expressing CD4+ cells survive multiple genetic changes to ultimately develop over time the phenotype of ATL cells. In this scenario, Tax plays an early role by endowing the cell with a mutator phenotype. Tax-expression is, therefore, critical to the initiation of ATL. However a mutator phenotype may not be of major advantage during late stages of transformation. Hence, how important Tax might be to the maintenance and the progression of ATL is yet to be fully determined. Indeed, several studies have indicated that Tax is barely detectable in most freshly isolated ATLs and that many patients harbor HTLV-I proviruses coding for non-functional Tax protein (*45*). On the other hand, a continuous role of Tax in leukemogenesis is suggested by the finding that most defective HTLV-I proviruses bear mutations outside of the pX region (*30, 46, 61*). Furthermore, in studies using a highly sensitive RT-PCR technique, Tax transcripts from peripheral blood mononuclear cells (PBMCs) of ATL patients, albeit at a level lower than that from asymptomatic carriers and HAM/TSP patients (*9*), were consistently detected. Regardless of the final answer, the emerging picture of havoc wreaked by Tax on cellular control mechanisms offers a humbling and challenging opportunity to learn more about cell life and cell death.

REFERENCES

1. ACS. Cancer facts and figures 2000. *Am. Cancer Soc.*, 1–7 (2000).
2. Amon, A. The spindle checkpoint. *Curr. Opin. Genet. Dev.*, **9**, 69–75 (1999).
3. Burton, M., Upadhyaya, C. D., Maier, B., Hope, T. J., and Semmes, O. J. Human T-cell leukaemia virus type 1 Tax shuttles between functionally discrete subcellular targets. *J. Virol.*, **74**, 2351–2364 (2000).
4. Cahill, D. P., Lengauer, Yu. J., Riggins, G. J., Willson, J. K., Markowitz, S. D., Kinzler, K. W., and Vogelstein, B. Mutations of mitotic checkpoint genes in human cancers. *Nature*, **392**, 300–303 (1998).
5. Campbell, M. S., Chan, G. K. T., and Yen, T. J. Mitotic checkpoint proteins HsMAD1 and HsMAD2 are associated with nuclear core complexes in interphase. *J. Cell Sci.*, **114**, 953–963 (2001).
6. Cavrois, M., Leclercq, I., Gout, O., Gessain, A., Wain-Hobson, S., and Wattel, E. Persistent oligoclonal expansion of human T-cell leukaemia virus type 1-infected circulating cells in patients with tropical spastic paraparesis/HTLV-1 associated myelopathy. *Oncogene*, **17**, 77–82 (1998).
7. Chang, S., Khoo, C., and DePinho, R. A. Modeling chromosomal instability and epithelial carcinogenesis in the telomerase-deficient mouse. *Semin. Cancer Biol.*, **11**, 227–238 (2001).
8. Cereseto, A., Diella, F., Mulloy, J. C., Cara, A., Michieli, P., Grassmann, R., Franchini,

G., and Klotman, M. E. p53 functional impairment and high p21 waf/cip1 expression in human T-cell lymphotropic/leukaemia virus type I-transformed T cells. *Blood*, **88**, 1551–1560 (1996).

9. Furukawa, Y., Osame, M., Kubota, R., Tara, M., and Yoshida, M. Human T-cell leukaemia virus type-1 (HTLV-1) Tax is expressed at the same level in infected cells of HTLV-1 associated myelopathy or tropical spastic paraparesis patients as in asymptomatic carriers but at a lower level in adult T-cell leukaemia cells. *Blood*, **85**, 1865–1870 (1995).
10. Gardner, R. D. and Burke, D. J. The spindle checkpoint: two transitions, two pathways. *Trends Cell Biol.*, **10**, 154–158 (2000).
11. Grassmann, R., Berchthold, S., Radant, I., Alt, M., Fleckenstein, B., Sodorski, J. G., Haseltine, W. A., and Ramstedt, U. Role of human T-cell leukaemia virus type 1 X region proteins in immortalization of primary human lymphocytes in culture. *J. Virol.*, **66**, 4570–4575 (1992).
12. Grossman, W. J., Kimata, J. T., Wong, F. H., Zutter, M., Ley, T. J., and Ratner, L. Development of leukaemia in mice transgenic for tax gene of human T-cell leukaemia virus type I. *Proc. Natl. Acad. Sci. USA*, **92**, 1057–1061 (1995).
13. Haller, K., Ruckes, T., Schmitt, I., Saul, D., Derow, E., and Grassmann, R. Tax-dependent stimulation of G1 phase-specific cyclin-dependent kinases and increased expression of signal transduction genes characterize HTLV type 1-transformed T cells. *AIDS Res. Hum. Retroviruses*, **16**, 1683–1688 (2000).
14. Haller, K., Wu, Y., Derow, E., Schmitt, I., Jeang, K. T., and Grassmann, R. Physical interaction of HTLV-1 Tax with cyclin dependent kinase 4 stimulates the phosphorylation of Rb protein. *Mol. Cell. Biol.*, **22**, 3327–3338 (2002).
15. Hanahan, D. and Weinberg, R. A. The hallmarks of cancer. *Cell*, **100**, 57–70 (2000).
16. Hardwick, K. G., Weiss, E., Luca, F. C., Winey, M., and Murray, A. W. Activation of the budding yeast spindle assembly checkpoint without mitotic spindle disruption. *Science*, **273**, 953–956 (1996).
17. Hardwick, K. G., Johnston, R. C., Smith, D. L., and Murray, A. W. MAD3 encodes a novel component of the spindle checkpoint, which interacts with Bub3p, Cdc20p, and Mad2p. *J. Cell Biol.*, **148**, 871–882 (2000).
18. Howell, B. J., Hoffman, D. B., Fang, J., Murray, A. W., and Salmon, E. D. Visualization of Mad2 dynamics at kinetochores, along spindle fibers, and at spindle poles in living cells. *J. Cell Biol.*, **150**, 1233–1250 (2000).
19. Hoyt, M. A., Totis, L., and Roberts, B. T. *S. cerevisiae* genes required for cell cycle arrest in response to loss of microtubule function. *Cell*, **66**, 507–517 (1991).
20. Imai, Y., Shiratori, Y., Kato, N., Inoue, T., and Omata, M. Mutational inactivation of mitotic checkpoint genes, *hsMAD2* and *hBUB1*, is rare in sporadic digestive tract cancers. *Jpn. J. Cancer Res.*, **90**, 837–840 (1999).
21. Iwanaga, Y. and Jeang, K. T. Expression of mitotic spindle checkpoint protein hsMAD1 correlates with cellular proliferation and is activated by a gain-of-function p53 mutant. *Cancer Res.*, **62**, 2618–2624 (2002).
22. Jeang, K. T., Widen, S. G., Semmes, O. J., and Wilson, S. H. HTLV-I trans-activator protein, Tax is a trans-repressor of the human β-polymerase gene. *Science*, **247**, 1028–1084 (1990).

23. Jeang, K. T. Functional activities of the human T-cell leukaemia virus type I Tax oncoprotein: cellular signalling through NF-kappa B. *Cytokine Growth Factor Rev.*, **12**, 207–217 (2001).
24. Jin, D. Y., Spencer, F., and Jeang, K. T. Human T cell leukemia virus type 1 oncoprotein Tax targets the human mitotic checkpoint protein MAD1. *Cell*, **93**, 81–91 (1998).
25. Kao, S. Y. and Marriott, S. J. Disruption of nucleotide excision repair by the human T-cell leukemia virus type 1 Tax protein. *J. Virol.*, **73**, 4299–4304 (1999).
26. Kao, S. Y., Lemoine, F. J., and Marriott, S. J. Suppression of DNA repair by human T cell leukemia virus type 1 Tax is rescued by a functional p53 signaling pathway. *J. Biol. Chem.*, **275**, 35926–35931 (2000).
27. Kim, N. W., Piatyszek, M. A., Prowse, K. R., Harley, C. B., West, M. D., Ho, P. L., Coviello, G. M., Wright, W. E., Weinrich, S. L., and Shay, J. W. Specific association of human telomerase activity with immortal cells and cancer. *Science*, **266**, 2011–2015 (1994).
28. Kinoshita, T., Shimoyama, M., Tobinai, K., Ito, M., Ito, S., Ikeda, S., Tajima, K., and Shimotohno, K. Detection of mRNA for the tax1/rex1 gene of human T-cell leukaemia virus type I in fresh peripheral blood mononuclear cells of adult T-cell leukaemia patients and viral carriers by using the polymerase chain reaction. *Proc. Natl. Acad. Sci. USA*, **86**, 5620–5624 (1989).
29. Kinzler, K. W. and Vogelstein, B. Lessons from hereditary colorectal cancer. *Cell*, **87**, 159–170 (1996).
30. Korber, B., Okayama, A., Donnelly, R., Tachibana, N., and Essex, M. Polymerase chain reaction analysis of defective human T-cell leukaemia virus type 1 proviral genomes in leukemic cells of patients with adult T-cell leukaemia. *J. Virol.*, **65**, 5471–5476 (1991).
31. Laird, A. D. and Shalloway, D. Oncoprotein signaling and mitosis. *Cell. Signal.*, **9**, 249–255 (1997).
32. Lengauer, C., Kinzler, K. W., and Vogelstein, B. Genetic instabilities in human cancers. *Nature*, **396**, 643–649 (1998).
33. Li, R. and Murray, A. W. Feedback control of mitosis in budding yeast. *Cell*, **66**, 519–531 (1991).
34. Loeb, L. A. A mutator phenotype in cancer. *Cancer Res.*, **61**, 3230–3239 (2001).
35. Low, K. G., Dorner, L. F., Fernando, D. B., Grossman, J., Jeang, K. T., and Comb, M. J. Human T-cell leukaemia virus type 1 Tax releases cell cycle induced by p16INK4a. *J. Virol.*, **71**, 1956–1962 (1997).
36. Mahieux, R., Pise-Masison, C. A., Lambert, P. F., Nicot, C., DeMarchis, L., Gessain, A., Green, P., Hall, W., and Brady, J. N. Differences in the ability of human T-cell lymphotropic virus type 1 (HTLV-1) and HTLV-2 tax to inhibit p53 function. *J. Virol.*, **74**, 6866–6874 (2000).
37. Majone, F., Semmes, O. J., and Jeang, K. T. Induction of micronuclei by HTLV-I Tax: a cellular assay for function. *Virology*, **193**, 456–459 (1993).
38. Majone, F. and Jeang, K. T. Clastogenic effect of the human T-cell leukemia virus type I tax oncoprotein correlates with unstabilized DNA breaks. *J. Biol. Chem.*, **275**, 32906–32910 (2000).
39. Miyake, H., Suzuki, T., Hirai, H., and Yoshida, M. Trans-activator Tax of human T-

cell leukaemia virus type 1 enhances mutation frequency of the cellular genome. *Virology*, **253**, 155–161 (1999).

40. Mulloy, J. C., Kislyakova, T., Cereseto, A., Casareto, A., LoMonico, A., Fullen, J., Lorenzi, M. V., Cara, A., Nicot, C., Giam, C., and Franchini, G. Human T-cell lymphotropic/leukaemia virus type 1 Tax abrogates p53-induced cell cycle arrest and apoptosis through its CREB/ATF functional domain. *J. Virol.*, **72**, 8852–8860 (1998).
41. Neuveut, C., Low, K. G., Maldarelli, F., Schmitt, I., Majone, F., Grassmann, R., and Jeang, K. T. Human T-cell leukemia virus type 1 Tax and cell cycle progression: role of cyclin D-cdk and p110Rb. *Mol. Cell. Biol.*, **18**, 3620–3632 (1998).
42. Neuveut, C. and Jeang, K. T. Cell cycle dysregulation by HTLV-I: role of the tax oncoprotein. *Front. Biosci.*, **7**, D157–D163 (2002).
43. Nomoto, S., Haruki, N., Takahashi, T., Masuda, A., Koshikawa, T., Takahashi, T., Fujii, Y., Osada, H., and Takahashi, T. Search for *in vivo* somatic mutations in the mitotic checkpoint gene hMAD1, in human lung cancers. *Oncogene*, **18**, 7180–7183 (1999).
44. Okamoto, T., Ohno, Y., Tsugane, S., Watanabe, S., Shimoyama, M., Tajima, K., Miwa, M., and Shimotohno, K. Multi-step carcinogenesis model for adult T-cell leukemia. *Jpn. J. Cancer Res.*, **80**, 191–195 (1989).
45. Okazaki, S., Moriuchi, R., Yosizuka, N., Sugahara, K., Maeda, T., Jinnai, I., Tomonaga, M., Kamihira, S., and Katamine, S. HTLV-1 proviruses encoding non-functional Tax in adult T-cell leukemia. *Virus Genes*, **23**, 123–135 (2001).
46. Ohshima, K., Kikuchi, M., Masuda, Y., Kobari, S., Sumiyoshi, Y., Eguchi, F., Mohtai, H., and Yoshida, M. Defective provirus form of human T-cell leukaemia/lymphoma: Clinicopathological features. *Cancer Res.*, **51**, 4639–4642 (1991).
47. Paunesku, T., Mittal, S., Protic, M., Oryhon, J., Korolev, S. V., Joachimiak, A., and Woloschak, G. E. Proliferating cell nuclear antigen (PCNA): ringmaster of the genome. *Int. J. Radiat. Biol.*, **77**, 1007–1021 (2001).
48. Philpott, S. M. and Buehring, G. C. Defective DNA repair in cells with human T-cell leukemia/bovine leukemia viruses: role of tax gene. *J. Natl. Cancer Inst.*, **91**, 933–942 (1999).
49. Pise-Masison, C. A., Choi, K. S., Radonovich, M., Dittmer, J., Kim, S. J., and Brady, J. N. Inhibition of p53 transactivation function by the human T-cell lymphotropic virus type 1 Tax protein. *J. Virol.*, **72**, 1165–1170 (1998).
50. Pise-Masison, C. A., Mahieux, R., Jiang, H., Ashcroft, M., Radonovich, M., Duvall, J., Guillerm, C., and Brady, J. N. Inactivation of p53 by human T-cell lymphotropic virus type 1 Tax requires activation of the NF-kappaB pathway and is dependent on p53 phosphorylation. *Mol. Cell. Biol.*, **20**, 3377–3386 (2000).
51. Poiesz, B. J., Ruscetti, F. W., Gadzar, A. F., Bunn, P. A., Minna, J. D., and Gallo, R. C. Detection and isolation of type C retrovirus particles from fresh and cultured lymphocytes of a patient with cutaneous T-cell lymphoma. *Proc. Natl. Acad. Sci. USA,* **78**, 7415–7419 (1980).
52. Polyak, K., Xia, Y., Zweier, J. L., Kinzler, K. W., and Vogelstein, B. A model for p53-induced apoptosis. *Nature*, **389**, 300–305 (1997).
53. Schmitt, I., Rosin, O., Rohwer, P., Gossen, M., and Grassmann, R. Stimulation of cyclin-dependent kinase activity and G1- to S-phase transition in human lymphocytes

by the human T-cell leukaemia/lymphotropic virus type 1 Tax protein. *J. Virol.*, **72**, 633–640 (1998).

54. Semmes, O. J., Majone, F., Cantemir, C., Turchetto, L., Hjelle, B., and Jeang, K. T. HTLV-I and HTLV-II Tax: differences in induction of micronuclei in cells and transcriptional activation of viral LTRs. *Virology*, **217**, 373–379 (1996).
55. Shah, J. V. and Cleveland, D. W. Waiting for anaphase: Mad2 and the spindle assembly checkpoint. *Cell*, **103**, 997–1000 (2000).
56. Sherr, C. J. Cancer cell cycles. *Science*, **274**, 1672–1677 (1996).
57. Suzuki, T., Kitao, S., Matsushime, H., and Yoshida, M. HTLV-1 Tax protein interacts with cyclin-dependent kinase inhibitor p16INK4A and counteracts its inhibitory activity towards CDK4. *EMBO J.*, **15**, 1607–1614 (1996).
58. Suzuki, T., Uchida-Toita, M., and Yoshida, M. Tax protein of HTLV-1 inhibits CDP/p300-mediated transcription by interfering with recruitment of CBP/p300 onto DNA element of E-box or p53 binding site. *Oncogene*, **18**, 4137–4143 (1999).
59. Suzuki, T., Uchida-Toita, M., Andoh, T., and Yoshida, M. HTLV-1 Tax oncoprotein binds to DNA topoisomerase I and inhibits its catalytic activity. *Virology*, **270**, 291–298 (2000).
60. Taguchi, H., Sonobe, H., Yamato, K., Takeuchi, T., Ookawa, K., Kodama, H., Othsuki, Y., and Miyoshi, I. Central nervous system involvement of adult T-cell leukaemia-lymphoma with multinucleated giant cells in the brain, skin and kidney. *Cancer*, **71**, 133–137 (1993).
61. Tamiya, S., Matsuoka, M., Etoh, K., Watanabe, T., Kamihira, S., Yamaguchi, K., and Takatsuki, K. Two types of defective human T-lymphotropic virus type 1 provirus in adult T-cell leukaemia. *Blood*, **88**, 3065–3073 (1996).
62. Uchiyama, T. Human T cell leukaemia virus type I (HTLV-I) and human diseases. *Annu. Rev. Immunol.*, **15**, 15–37 (1997).
63. Wang, Y. and Burke, D. J. Cdc55p, the B-type regulatory subunit of protein phosphatase 2A, has multiple function in mitosis and is required for the kinetochore/spindle checkpoint in *Saccharomyces cerevisiae. Mol. Cell. Biol.*, **17**, 620–626 (1997).
64. Weiss, E. and Winey, M. The *Saccharomyces cerevisiae* spindle pole body duplication gene MSP1 is part of a mitotic checkpoint. *J. Cell Biol.*, **132**, 111–123 (1996).
65. Yoshida, M., Miyoshi, I., and Hinuma, Y. Isolation and characterization of retrovirus from cell lines of human adult T-cell leukaemia cells and its implication in disease. *Proc. Natl. Acad. Sci. USA*, **79**, 2031–2035 (1982).
66. Zachariae, W. and Nasmyth, K. Whose end is destruction: cell division and anaphase-promoting complex. *Genes Dev.*, **13**, 2039–2058 (1999).

FUNCTIONAL ROLE OF THE HTLV-1 p12^{I} PROTEIN

Julie M. JOHNSON, Risaku FUKUMOTO, and Genoveffa FRANCHINI

*Basic Research Laboratory, National Cancer Institute**

Human T-cell leukemia/lymphotropic virus type 1 (HTLV-1) is the etiologic agent of adult T-cell leukemia/lymphoma (ATLL) (*23, 26, 59, 73*), as well as a variety of immune-mediated disorders including HTLV-1 associated myelopathy/tropical spastic paraparesis (HAM/TSP) (*24, 54, 61*). In addition to the structural and enzymatic genes *gag*, *pol*, and *env*, HTLV-1 carries a unique region at the 3′ end of the genome that was designated pX (*64*) because the coding potential of the open reading frame (ORF) was unknown at the time (Fig. 1). Later, however, it was appreciated that the four ORF within the pX region encode the regulatory proteins Rex and Tax (ORF III and IV, respectively) (*10, 25, 28, 65*). ORF I and II encode additional proteins (p12^{I}, p30^{II}, and p13^{II}) whose function is under investigation (*6, 8, 11, 12, 16, 27, 35, 36, 38, 39, 74, 75*).

ORF I encodes a 12-kDa protein (p12^{I}), which is a hydrophobic protein that resides in the endoplasmic reticulum (ER) and golgi (Fig. 2) (*20, 33, 35, 36*), forms dimers (*69*), exhibits weak oncogenic activity and shares amino acid similarities with the bovine papillomavirus (BPV) type 1 E5 oncoprotein (*21*). Two natural variants of the p12^{I} protein have been identified: p12^{I}K88 has a lysine at position 88 and is more commonly found in HTLV-1 strains from TSP/HAM patients; p12^{I}R88 has an arginine at position 88 and is mainly found in HTLV-1 strains from all ATLL patients and healthy carriers studied (*69*). The p12^{I}K88 protein is much less stable than p12^{I}R88, as it is ubiquitinated and rapidly degraded by the proteasome (*69*), suggesting that this sequence variation might have a functional relevance. p12^{I} expression has been difficult to demonstrate in HTLV-1-infected cells; however, indirect evidence suggests its synthesis *in vivo*. Transcripts encoding p12^{I} have been detected in *in vitro* and *ex vivo* HTLV-1-infected T-cells and macrophages (*36, 37*). Serum antibodies from rabbits experimentally infected with HTLV-1 and sera from humans infected with HTLV-1 have been shown to recognize the ORF I product (*18*). In addition, a cytotoxic T-lymphocyte (CTL) response to ORF I products can be detected in HTLV-1-infected individuals (*56*). While p12^{I} does not appear necessary for HTLV-1 replication *in vitro* (*19, 60*), the ablation of the acceptor splice site for the p12^{I} mRNA results in impairment of viral infectivity *in vivo* (*15*).

The HTLV-1 p12^{I} and Nef proteins of human immunodeficiency virus/simian immunodeficiency virus (HIV/SIV) appear to have functional similarities. As with p12^{I} (*15*), Nef is dispensable *in vitro*, but is required for *in vivo* replication and pathogenicity (*34*). *In vitro*, p12^{I} has been shown to bind the 16-kDa subunit of the vacuolar ATPase (*21*) and Nef binds to the catalytic subunit of the same enzyme (*43*).

HTLV-1 infects and immortalizes primary human T-cells *in vitro*. These infect-

* 41/D804, Bethesda, Maryland, USA

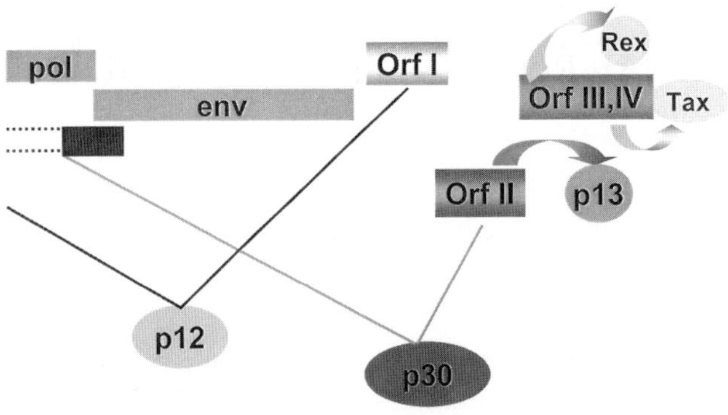

FIG. 1. Schematic representation of the HTLV-1 pX region

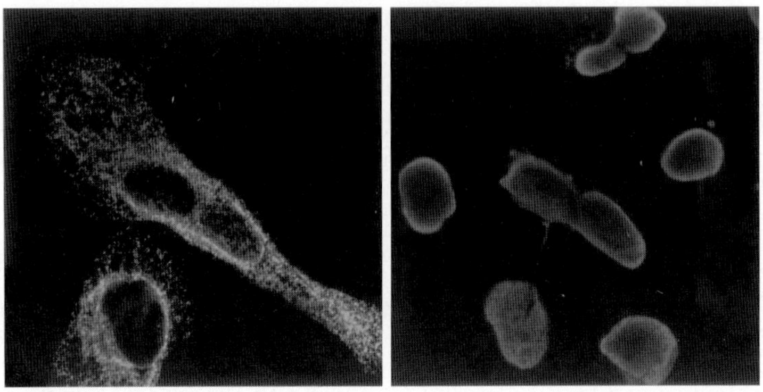

FIG. 2. Intracellular localization of p12^{I}
HeLa cells were transfected with p12^{I} plasmid tagged with HA1 epitope. p12^{I} is detected as a green signal (left); blue is DAPI stain (right).

ed cells acquire the ability to grow in the absence of interleukin 2 (IL-2) after long-term culture. At this time, the T-cells express high levels of the IL-2 receptor (IL-2R) α chain, which is characteristic of ATLL cells, as well as HTLV-1 transformed cells *in vitro.* In most cases, IL-2 independence correlates with the constitutive activation of Jak/Stat pathways (*46, 49, 72*), as well as a decreased expression of the src homology 2-containing tyrosine phosphatase 1 protein, which is important for the regulation of signaling from several hematopoietic surface receptors (*41, 45*). The proliferation of *ex vivo* derived ATLL cells was found to be associated with the constitutive activation of Jak/Stat proteins in some patients, and this observation has led to the hypothesis that activation of this pathway may contribute to neoplastic growth of ATLL cells (*67*). The constitutive activation of Jak and/or Stat proteins has been correlated with cell transformation in other viral models: Abelson murine leukemia virus (*17*), Epstein-Barr virus (*71*), and spleen focus-forming virus (*53*). p12^{I} binds to the IL-2-receptor (IL-2R) β and γ_c chains and affects their trafficking to the cell surface (*47*). This finding has led to the hypothesis that p12^{I} may play a role in modulating the intracellular signaling pathways mediated by these receptors in HTLV-1-infected cells and contribute to T-cell proliferation. The binding of IL-2 to the IL-2Rβ and γ_c chains results in the activation and recruitment of Jak 1 and Jak 3 and the subsequent phosphorylation and nuclear translocation of STAT5a and STAT5b (Fig. 3). Indeed, ectopic expression in HeLa-Tat cells or overexpression of p12^I in primary peripheral blood mononuclear cells is associated with enhancement of STAT5 activation and decreased IL-2 requirement for proliferation (*52*). The enhancement of STAT5 activation is, however, dependent on the previous activation of T-cells (*52*). This effect is not observed *in vitro* in HTLV-1 immortalized T-cells since the culture conditions usually include excess amounts of IL-2 that, in turn, may compensate for ORF I expression (*14*). However, the deletion/mutation of p12^I results in decreased viral infectivity in primary lymphocytes *in vitro* (*3*) and decreased viral burden *in vivo* (*15*). These data suggest that p12^I plays an important role in the establishment and maintenance of infection. The *in vitro* data coupled with observations in animal models suggest the hypothesis that, in the presence of suboptimal antigen stimulation, the ability of p12 to increase T-cell responsiveness to IL-2 (*52*) combined with the ability of Tax to induce cell-cycle progression (*31*) may contribute to T-cell proliferation. This may result in an increased rate of infected-cell duplication leading to the efficient establishment of chronic infection. In support of this hypothesis, patients with concomitant *Stronglyloides stercoralis* infection have a higher frequency of ATLL (*1, 4, 5*).

p12^I may also play a role in HTLV-1 evasion of immune detection through its interaction with the major histocompatibility complex class I heavy chain (MHC I-Hc) (*33*). Many viruses that induce long-standing infection encode proteins that target or modulate the host's immune response (*58, 68*). The E3/19K adenoviral protein binds to MHC I in the ER and prevents its transport to the cell surface (*68*). In addi-

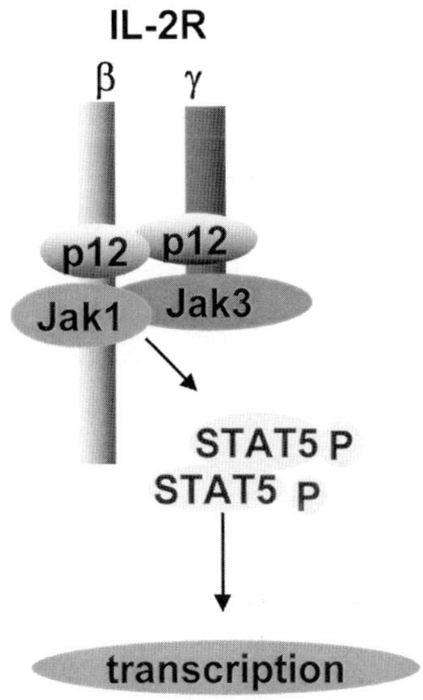

FIG. 3. Diagram of p12^I bound to the IL-2Rβ and γ_c chains, which results in the recruitment of Jak1 and Jak3 and leads to the subsequent STAT5 activation.

tion, E3/19K binds the transporter associated with antigen processing (TAP) and prevents TAP/class I association, thereby interfering with peptide loading (*7*). Human cytomegalovirus (HCMV) encodes multiple proteins that use different mechanisms to target MHC I synthesis, peptide loading, and transport (*58, 68*). Murine cytomegalovirus glycoprotein, gp34, also interacts with the Hc/β_2-microglobulin complex in the ER and has been recently shown to target MHC I for degradation in the lysosomes (*68*). Herpes simplex virus encodes a protein, ICP47, that inhibits the TAP transporter (*58, 68*). K3 and K5 proteins encoded by Kaposi's sarcoma-associated herpesvirus (KSHV) down-regulate MHC I from the cell surface (*29*). The HIV/SIV Nef protein was shown to down-regulate both CD4 and MHC I expression at the cell surface by accelerated endocytosis through its interaction with the intracellular sorting machinery of the cell (*2, 13, 44, 55, 63*). Nef binds to the vacuolar ATPase, which results in the internalization and degradation of CD4 (*43*). In addition, Nef's interaction with the μ subunit of adaptor protein complexes results in misrouting of MHC I complexes to the clathrin-coated vesicles (*40*).

p12^I has been shown to bind to the newly synthesized free MHC I heavy chain (MHC I-Hc) before its association with β_2-microglobulin and target it for degradation by the proteasome (*33*). This represents a novel mechanism of viral immune evasion, whereby the p12^I protein specifically targets the free MHC I-Hc for proteasomal mediated destruction. To accomplish this, p12^I may take advantage of a pathway termed ERAD (ER-associated degradation pathway), whose purpose is to

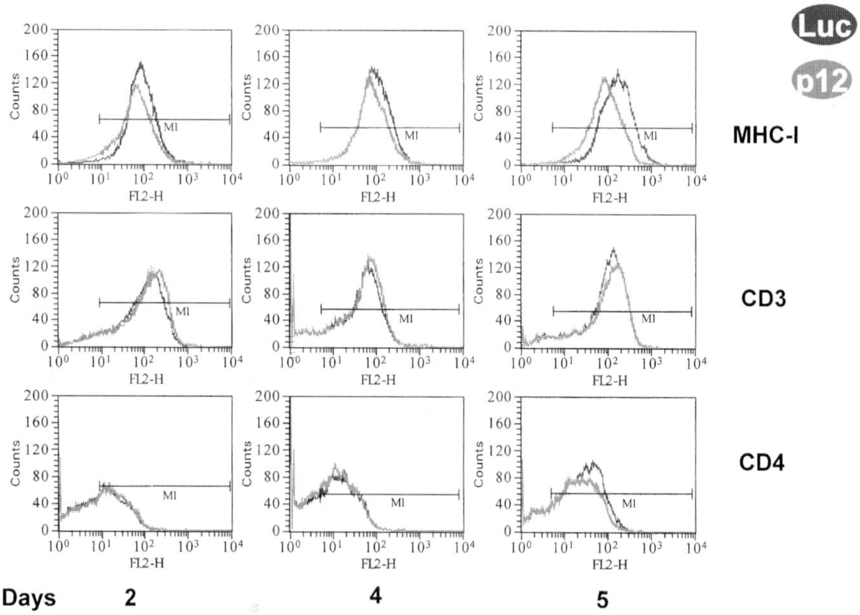

FIG. 4. p12^I alters the trafficking of MHC I to the plasma membrane

The surface expression of MHC I, CD3, and CD4 were examined on Jurkat T-cells expressing p12^I. There was an approximate 50% reduction in the level of MHC I at the cell surface in cell expressing p12^I (light gray lines). The levels of CD3 and CD4 remained constant throughout. There was no change in MHC I, CD3, or CD4 in control cells (charcoal gray lines). This figure is reprinted from *Journal of Virology* (*33*).

remove misfolded, inappropriately glycosylated or improperly assembled proteins from the ER (*9*). The Hc/p12^I interaction results in disruption of intracellular trafficking of the MHC I complex and a significant decrease in MHC I at the surface of human T cells (Fig. 4) (*33*). Whether this event may cause a sufficient decrease of MHC I/viral peptide complexes on the cell surface to confer protection of HTLV-1-infected cells from immune surveillance remains to be ascertained. p12^I has been detected in the lipid rafts of human T-cells (R. Fukumoto *et al.*, unpublished data), which raises the possibility that p12^I trafficking to the cell surface may also interfere with the MHC I turnover at the cell surface.

Future studies are necessary to determine whether the decrease in MHC I levels at the cell surface induced by p12^I may be sufficient to alter recognition by CTL, as well as determine the exact mechanism(s) involved. Further, the two natural alleles of p12^I may differ in their ability to affect antigen presentation and modulate the host-specific immune response. The preliminary finding that the ubiquitinated form of p12^I (p12^{I}K88) is found more frequently in TSP/HAM (*69*), an immune mediated disease (*30, 54*), suggests that studies of the two natural alleles of p12^I may further our understanding of HTLV-1 pathogenesis. p12^I has been suggested to play an important role early during HTLV-1 infection where it might enable the virus to establish infection in the host. p12^I may decrease the density of MHC I complexes presenting viral peptides on the cell surface by targeting the MHC I-Hc for degradation, protecting infected cells from lysis by cytotoxic T-cells and at the same time providing proliferation signals. Tax, the viral transactivator, targets numerous cell-cycle regulators and plays a role in the proliferation/survival of HTLV-1-infected cells as well (*32, 42, 48, 50, 51, 57, 62, 66*). However, Tax is highly immunogenic (*31*). Therefore, the down-regulation of MHC I complexes containing Tax peptides may be key for the survival of Tax-expressing cells *in vivo*. Further, the augmentation of the IL-2R signaling pathway by p12^I may contribute to the continual low level expansion of infected cells in the presence of antigen stimulation, contributing to a persistent viral infection and the eventual progression to ATLL. Thus, p12^I may contribute to the expansion and maintenance of infected T-cell clones (*70*). Over time, one of these cell clones may acquire mutations sufficient to expand in an uncontrolled manner and cause leukemia.

In conclusion, we believe that the ability of HTLV-1 to cause proliferation of T-cells and ultimately leukemia does not rely on a single viral product, as often has been implied by studies on Tax. Tax alone is not sufficient to transform T-cells and its expression may not be necessary in ATLL (*22*). Several other *HTLV-1* genes, including p12^I, in concert with Tax are likely necessary to confer a proliferative advantage to HTLV-1-infected cells. The study of the protein products of the pX region may be important to fully understand the complexity of virus host interaction, which leads over time to disease development.

REFERENCES

1. Agape, P., Copin, M. C., Cavrois, M., Panelatti, G., Plumelle, Y., Ossondo-Landeau, M., Quist, D., Grossat, N., Gosselin, B., Fenaux, P., and Wattel, E. Implication of HTLV-I infection, strongyloidiasis, and P53 overexpression in the development, response to treatment, and evolution of non-Hodgkin's lymphomas in an endemic area

(Martinique, French West Indies). *J. Acquir. Immune Defic. Syndr. Hum. Retrovirol.*, **20**, 394–402 (1999).

2. Aiken, C., Konner, J., Landau, N. R., Lenburg, M. E., and Trono, D. Nef induces CD4 endocytosis: requirement for a critical dileucine motif in the membrane-proximal CD4 cytoplasmic domain. *Cell*, **76**, 853–864 (1994).
3. Albrecht, B., Collins, N. D., Burniston, M. T., Nisbet, J. W., Ratner, L., Green, P. L., and Lairmore, M. D. Human T-lymphotropic virus type 1 open reading frame I p12(I) is required for efficient viral infectivity in primary lymphocytes. *J. Virol.*, **74**, 9828–9835 (2000).
4. Arakaki, T., Asato, R., Ikeshiro, T., Sakiyama, K., and Iwanaga, M. Is the prevalence of HTLV-1 infection higher in Strongyloides carriers than in non-carriers? *Trop. Med. Parasitol.*, **43**, 199–200 (1992).
5. Atkins, N. S., Lindo, J. F., Lee, M. G., Hanchard, B., Robinson, R. D., and Bundy, D. A. Immunomodulatory effects of concurrent HTLV-I infection in strongyloidiasis. *J. Acquir. Immune Defic. Syndr. Hum. Retrovirol.*, **18**, 188–190 (1998).
6. Bartoe, J. T., Albrecht, B., Collins, N. D., Robek, M. D., Ratner, L., Green, P. L., and Lairmore, M. D. Functional role of pX open reading frame II of human T-lymphotropic virus type 1 in maintenance of viral loads *in vivo*. *J. Virol.*, **74**, 1094–1100 (2000).
7. Bennett, E. M., Bennink, J. R., Yewdell, J. W., and Brodsky, F. M. Cutting edge: adenovirus E19 has two mechanisms for affecting class I MHC expression. *J. Immunol.*, **162**, 5049–5052 (1999).
8. Berneman, Z. N., Gartenhaus, R. B., Reitz, M. S., Jr., Blattner, W. A., Manns, A., Hanchard, B., Ikehara, O., Gallo, R. C., and Klotman, M. K. Expression of alternatively spliced human T-lymphotropic virus type I (HTLV-I) pX mRNA in infected cell lines and in primary uncultured cells from patients with adult T-cell leukemia/lymphoma and healthy carriers. *Proc. Natl. Acad. Sci. USA*, **89**, 3005–3009 (1992).
9. Bonifacino, J. S. and Klausner, R. D. Degradation of proteins retained in the endoplasmic reticulum. *In* "Cellular Proteolytic Systems," ed. A. Ciechanover and A. L. Schwartz, pp. 137 (1994). Wiley-Liss, New York.
10. Chen, I. S., McLaughlin, J., Gasson, J. C., Clark, S. C., and Golde, D. W. Molecular characterization of genome of a novel human T-cell leukaemia virus. *Nature*, **305**, 502–505 (1983).
11. Ciminale, V., Pavlakis, G. N., Derse, D., Cunningham, C. P., and Felber, B. K. Complex splicing in the human T-cell leukemia virus (HTLV) family of retroviruses: novel mRNAs and proteins produced by HTLV-I. *J. Virol.*, **66**, 1737–1745 (1992).
12. Ciminale, V., Zotti, L., D'Agostino, D. M., Ferro, T., Casareto, L., Franchini, G., Bernardi, P., and Chieco-Bianchi, L. Mitochondrial targeting of the p13II protein coded by the x-II ORF of human T-cell leukemia/lymphotropic virus type I (HTLV-I). *Oncogene*, **18**, 4505–4514 (1999).
13. Cohen, G. B., Gandhi, R. T., Davis, D. M., Mandelboim, O., Chen, B. K., Strominger, J. L., and Baltimore, D. The selective downregulation of class I major histocompatibility complex proteins by HIV-1 protects HIV-infected cells from NK cells. *Immunity*, **10**, 661–671 (1999).
14. Collins, N. D., D'Souza, C., Albrecht, B., Robek, M. D., Ratner, L., Ding, W., Green, P. L., and Lairmore, M. D. Proliferation response to interleukin-2 and Jak/Stat activa-

tion of T cells immortalized by human T-cell lymphotropic virus type 1 is independent of open reading frame I expression. *J. Virol.*, **73**, 9642–9649 (1999).

15. Collins, N. D., Newbound, G. C., Albrecht, B., Beard, J. L., Ratner, L., and Lairmore, M. D. Selective ablation of human T-cell lymphotropic virus type 1 p12^{I} reduces viral infectivity *in vivo. Blood,* **91**, 4701–4707 (1998).
16. D'Agostino, D. M., Zotti, L., Ferro, T., Franchini, G., Chieco-Bianchi, L., and Ciminale, V. The p13II protein of HTLV type 1: comparison with mitochondrial proteins coded by other human viruses. *AIDS Res. Hum. Retroviruses*, **16**, 1765–1770 (2000).
17. Danial, N. N., Pernis, A., and Rothman, P. B. Jak-STAT signaling induced by the v-abl oncogene. *Science*, **269**, 1875–1877 (1995).
18. Dekaban, G. A., Peters, A. A., Mulloy, J. C., Johnson, J. M., Trovato, R., Rivadeneira, E., and Franchini, G. The HTLV-I orf I protein is recognized by serum antibodies from naturally infected humans and experimentally infected rabbits. *Virology*, **274**, 86–93 (2000).
19. Derse, D., Mikovits, J., and Ruscetti, F. X-I and X-II open reading frames of HTLV-I are not required for virus replication or for immortalization of primary T-cells *in vitro. Virology*, **237**, 123–128 (1997).
20. Ding, W., Albrecht, B., Luo, R., Zhang, W., Stanley, J. R., Newbound, G. C., and Lairmore, M. D. Endoplasmic reticulum and cis-Golgi localization of human T-lymphotropic virus type 1 p12(I): association with calreticulin and calnexin. *J. Virol.*, **75**, 7672–7682 (2001).
21. Franchini, G., Mulloy, J. C., Koralnik, I. J., Lo Monico, A., Sparkowski, J. J., Andresson, T., Goldstein, D. J., and Schlegel, R. The human T-cell leukemia/lymphotropic virus type I p12^{I} protein cooperates with the E5 oncoprotein of bovine papillomavirus in cell transformation and binds the 16-kilodalton subunit of the vacuolar H+ ATPase. *J. Virol.*, **67**, 7701–7704 (1993).
22. Franchini, G., Wong-Staal, F., and Gallo, R. C. Human T-cell leukemia virus (HTLV-I) transcripts in fresh and cultured cells of patients with adult T-cell leukemia. *Proc. Natl. Acad. Sci. USA*, **81**, 6207–6211 (1984).
23. Gallo, R. C. The first human retrovirus. *Sci. Am.*, **255**, 88–98 (1986).
24. Gessain, A., Barin, F., Vernant, J.-C., Gout, O., Maurs, L., Calendar, A., and de The, G. Antibodies to human T-lymphotropic virus type I in patients with tropical spastic paraparesis. *Lancet,* **ii**, 407–410 (1985).
25. Hidaka, M., Inoue, J., Yoshida, M., and Seiki, M. Post-transcriptional regulator (rex) of HTLV-1 initiates expression of viral structural proteins but suppresses expression of regulatory proteins. *EMBO J.*, **7**, 519–523 (1988).
26. Hinuma, Y., Nagata, K., Hanaoka, M., Nakai, M., Matsumoto, T., Kinoshita, K. I., Shirakawa, S., and Miyoshi, I. Adult T-cell leukemia: antigen in an ATL cell line and detection of antibodies to the antigen in human sera. *Proc. Natl. Acad. Sci. USA,* **78**, 6476–6480 (1981).
27. Hou, X., Foley, S., Cueto, M., and Robinson, M. A. The human T-cell leukemia virus type I (HTLV-I) X region encoded protein p13(II) interacts with cellular proteins. *Virology*, **277**, 127–135 (2000).
28. Inoue, J., Seiki, M., and Yoshida, M. The second pX product p27 chi-III of HTLV-1 is required for gag gene expression. *FEBS Lett.*, **209**, 187–190 (1986).

29. Ishido, S., Wang, C., Lee, B. S., Cohen, G. B., and Jung, J. U. Downregulation of major histocompatibility complex class I molecules by Kaposi's sarcoma-associated herpesvirus K3 and K5 proteins. *J. Virol.*, **74**, 5300–5309 (2000).
30. Jacobson, S., Raine, C. S., Mingioli, E. S., and McFarlin, D. E. Isolation of an HTLV-1-like retrovirus from patients with tropical spastic paraparesis. *Nature*, **331**, 540–543 (1988).
31. Jacobson, S., Shida, H., McFarlin, D. E., Fauci, A. S., and Koenig, S. Circulating CD8+ cytotoxic T lymphocytes specific for HTLV-I pX in patients with HTLV-I associated neurological disease. *Nature*, **348**, 245–248 (1990).
32. Jin, D.-Y., Spencer, F., and Jeang, K.-T. Human T-cell leukemia virus type 1 oncoprotein Tax targets the human mitotic checkpoint protein MAD1. *Cell*, **93**, 1–20 (1998).
33. Johnson, J. M., Nicot, C., Fullen, J., Ciminale, V., Casareto, L., Mulloy, J. C., Jacobson, S., and Franchini, G. Free major histocompatibility complex class I heavy chain is preferentially targeted for degradation by human T-cell leukemia/lymphotropic virus type 1 p12(I) protein. *J. Virol.*, **75**, 6086–6094 (2001).
34. Kestler, H. W., III, Ringleer, D. J., Mori, K., Panicali, D. L., Sehgal, P. K., Daniel, M. D., and Desrosiers, R. C. Importance of the nef gene for maintenance of high virus loads and for development of AIDS. *Cell*, **65**, 651–662 (1991).
35. Koralnik, I. J., Fullen, J., and Franchini, G. The $p12^{I}$, $p13^{II}$, and $p30^{II}$ proteins encoded by human T-cell leukemia/lymphotropic virus type I open reading frames I and II are localized in three different cellular compartments. *J. Virol.*, **67**, 2360–2366 (1993).
36. Koralnik, I., Gessain, A., Klotman, M. E., Lo Monico, A., Berneman, Z. N., and Franchini, G. Protein isoforms encoded by the pX region of the human T-cell leukemia/lymphotropic virus type I. *Proc. Natl. Acad. Sci. USA*, **89**, 8813–8817 (1992).
37. Koralnik, I., Lemp, J. F., Jr., Gallo, R. C., and Franchini, G. *In vitro* infection of human macrophages by human T-cell leukemia/lymphotropic virus type I (HTLV-I). *AIDS Res. Hum. Retroviruses*, **8**, 1845–1849 (1992).
38. Lairmore, M. D., Albrecht, B., D'Souza, C., Nisbet, J. W., Ding, W., Bartoe, J. T., Green, P. L., and Zhang, W. *In vitro* and *in vivo* functional analysis of human T cell lymphotropic virus type 1 pX open reading frames I and II. *AIDS Res. Hum. Retroviruses*, **16**, 1757–1764 (2000).
39. Lefebvre, L., Vanderplasschen, A., Ciminale, V., Heremans, H., Dangoisse, O., Jauniaux, J. C., Toussaint, J. F., Zelnik, V., Burny, A., Kettmann, R., and Willems, L. Oncoviral bovine leukemia virus G4 and human T-cell leukemia virus type 1 p13(II) accessory proteins interact with farnesyl pyrophosphate synthetase. *J. Virol.*, **76**, 1400–1414 (2002).
40. Le Gall, S., Erdtmann, L., Benichou, S., Berlioz-Torrent, C., Liu, L., Benarous, R., Heard, J., and Schwartz, O. Nef interacts with the μ subunit of clathrin adaptor complexes and reveals a cryptic sorting signal in MHC I molecules. *Immunity,* **8**, 483–495 (1998).
41. Leonard, W. J. and O'Shea, J. J. JAKs and STATs: biological implications. *Annu. Rev. Immunol.,* **16**, 293–322 (1998).
42. Low, K. G., Dorner, L. F., Fernando, D. B., Grossman, J., Jeang, K.-T., and Comb, M. J. Human T-cell leukemia virus type 1 tax releases cell cycle arrest induced by p16INK4a. *J. Virol.*, **71**, 1956–1962 (1997).

43. Lu, X., Yu, H., Liu, S., Brodsky, F., and Peterlin, B. Interactions between HIV-1 Nef and vacuolar ATPase facilitate the internalization of CD4. *Immunity*, **8**, 647–656 (1998).
44. Mangasarian, A., Foti, M., Aiken, C., Chin, D., Carpentier, J., and Trono, D. The HIV-1 Nef protein acts as a connector with sorting pathways in the Golgi and at the plasma membrane. *Immunity*, **6**, 67–77 (1997).
45. Migone, T. S., Cacalano, N. A., Taylor, N., Yi, T., Waldmann, T. A., and Johnston, J. A. Recruitment of SH2-containing protein tyrosine phosphatase SHP-1 to the interleukin 2 receptor; loss of SHP-1 expression in human T-lymphotropic virus type I-transformed T cells. *Proc. Natl. Acad. Sci. USA*, **95**, 3845–3850 (1998).
46. Migone, T. S., Lin, J. X., Cereseto, A., Mulloy, J. C., O'Shea, J. J., Franchini, G., and Leonard, W. J. Constitutively activated JAK-STAT pathway in T-cells transformed with HTLV-I. *Science*, **269**, 79–81 (1995).
47. Mulloy, J. C., Crowley, R. W., Fullen, J., Leonard, W. J., and Franchini, G. The human T-cell leukemia/lymphotropic virus type I p12^I protein binds the interleukin-2 receptor β and γ_c chains and affects their expression on the cell surface. *J. Virol.*, **70**, 3599–3605 (1996).
48. Mulloy, J. C., Kislyakova, T., Cereseto, A., Casareto, L., Lo Monico, A., Fullen, J., Lorenzi, M. V., Cara, A., Nicot, C., Giam, C.-Z., and Franchini, G. Human T-cell lymphotropic/leukemia virus type 1 Tax abrogates p53-induced cell cycle arrest and apoptosis through its CREB/ATF functional domain. *J. Virol.*, **72**, 8852–8860 (1998).
49. Mulloy, J. C., Migone, T.-S., Ross, T. M., Ton, N., Green, P. L., Leonard, W. J., and Franchini, G. Human and simian T-cell leukemia viruses type 2 (HTLV-2 and STLV-2pan-p) transform T-cells independently of Jak/STAT activation. *J. Virol.*, **72**, 4408–4412 (1998).
50. Neuveut, C., Low, K. G., Maldarelli, F., Schmitt, I., Majone, F., Grassmann, R., and Jeang, K. T. Human T-cell leukemia virus type 1 Tax and cell cycle progression: role of cyclin D-cdk and p110Rb. *Mol. Cell. Biol.*, **18**, 3620–3632 (1998).
51. Nicot, C., Mahieux, R., Takemoto, S., and Franchini, G. Bcl-XL is upregulated by HTLV type I and type II *in vitro* and in *ex vivo* ATLL samples. *Blood*, **96**, 275–281 (2000).
52. Nicot, C., Mulloy, J. C., Ferrari, M. G., Johnson, J. M., Fu, K., Fukumoto, R., Trovato, R., Fullen, J., Leonard, W. J., and Franchini, G. HTLV-1 p12(I) protein enhances STAT5 activation and decreases the interleukin-2 requirement for proliferation of primary human peripheral blood mononuclear cells. *Blood*, **98**, 823–829 (2001).
53. Ohashi, T., Masuda, M., and Ruscetti, S. K. Induction of sequence-specific DNA-binding factors by erythropoietin and the spleen focus-forming virus. *Blood*, **85**, 1454–1462 (1995).
54. Osame, M., Usuku, K., Izumo, S., Ijichi, N., Amitani, H., and Igata, A. HTLV-I associated myelopathy, a new clinical entity. *Lancet*, **i**, 1031–1032 (1986).
55. Piguet, V., Wan, L., Borel, C., Mangasarian, A., Demaurex, N., Thomas, G., and Trono, D. HIV-1 Nef protein binds to the cellular protein PACS-1 to downregulate class I major histocompatibility complexes. *Nat. Cell Biol.*, **2**, 163–167 (2000).
56. Pique, C., Ureta-Vidal, A., Gessain, A., Chancerel, B., Gout, O., Tamouza, R., Agis, F., and Dokhelar, M. C. Evidence for the chronic *in vivo* production of human T cell

leukemia virus type I Rof and Tof proteins from cytotoxic T lymphocytes directed against viral peptides. *J. Exp. Med.*, **191**, 567–572 (2000).
57. Pise-Masison, C. A., Choi, K.-S., Radonovich, M., Dittmer, J., Kim, S.-J., and Brady, J. N. Inhibition of p53 transactivation function by the human T-cell lymphotropic virus type 1 Tax protein. *J. Virol.*, **72**, 1165–1170 (1998).
58. Ploegh, H. L. Viral strategies of immune evasion. *Science*, **280**, 248–253 (1998).
59. Poiesz, B. J., Ruscetti, F. W., Gazdar, A. F., Bunn, P. A., Minna, J. D., and Gallo, R. C. Detection and isolation of type C retrovirus particles from fresh and cultured lymphocytes of a patient with cutaneous T-cell lymphoma. *Proc. Natl. Acad. Sci. USA,* **77**, 7415–7419 (1980).
60. Robek, M. D., Wong, F.-H., and Ratner, L. Human T-cell leukemia virus type 1 pX-I and pX-II open reading frames are dispensable for the immortalization of primary lymphocytes. *J. Virol.*, **72**, 4458–4462 (1998).
61. Rodgers-Johnson, P., Gajdusek, D. C., Morgan, O. S., Zaninavic, V., Sarin, P. S., and Graham, D. S. HTLV-I and HTLV-III antibodies and tropical spastic paraparesis. *Lancet*, **ii**, 1247–1249 (1985).
62. Schmitt, I., Rosin, O., Rohwer, P., Gossen, M., and Grassmann, R. Stimulation of cyclin-dependent kinase activity and G1- to S-phase transition in human lymphocytes by the human T-cell leukemia/lymphotropic virus type 1 Tax protein. *J. Virol.*, **72**, 633–640 (1998).
63. Schwartz, O., Marechal, V., Le Gall, S., Lemonnier, F., and Heard, J. Endocytosis of major histocompatibility complex class I molecules is induced by the HIV-1 nef protein. *Nat. Med.*, **2**, 338–342 (1996).
64. Seiki, M., Hattori, S., Hirayama, Y., and Yoshida, M. Human adult T-cell leukemia virus: complete nucleotide sequence of the provirus genome integrated in leukemia cell DNA. *Proc. Natl. Acad. Sci. USA*, **80**, 3618–3622 (1983).
65. Sodroski, J., Rosen, C., Goh, W. C., and Haseltine, W. A transcriptional activator protein encoded by the x-lor region of the human T-cell leukemia virus. *Science*, **228**, 1430–1434 (1985).
66. Suzuki, T., Kitao, S., Matsushime, H., and Yoshida, M. HTLV-I Tax protein interacts with cyclin-dependent kinase inhibitor $p16^{ink4a}$ and counteracts its inhibitory activity towards CDK4. *EMBO J.*, **15**, 1607–1614 (1996).
67. Takemoto, S., Mulloy, J. C., Cereseto, A., Migone, T. S., Patel, B. K. R., Matsuoka, M., Yamaguchi, K., Takatsuki, K., Kamihira, S., White, J. D., Leonard, W. J., Waldmann, T., and Franchini, G. Proliferation of adult T-cell leukemia/lymphoma cells is associated with the constitutive activation of JAK/STAT proteins. *Proc. Natl. Acad. Sci. USA*, **94**, 13897–13902 (1997).
68. Tortorella, D., Gewurz, B., Schust, D., Furman, M., and Ploegh, H. Down-regulation of MHC class I antigen presentation by HCMV; lessons for tumor immunology. *Immunol. Invest.*, **29**, 97–100 (2000).
69. Trovato, R., Mulloy, J. C., Johnson, J. M., Takemoto, S., de Oliveira, M. P., and Franchini, G. A lysine-to-arginine change found in natural alleles of the HTLV-I $p12^{I}$ protein greatly influences its stability. *J. Virol.*, **73**, 6460–6467 (1999).
70. Wattel, E., Vartanian, J. P., Pannetier, C., and Wain-Hobson, S. Clonal expansion of human T-cell leukemia virus type I-infected cells in asymptomatic and symptomatic

carriers without malignancy. *J. Virol.*, **69**, 2863–2868 (1995).

71. Weber-Nordt, R. M., Egen, C., Wehinger, J., Ludwig, W., Gouilleux-Gruart, V., Mertelsmann, R., and Finke, J. Constitutive activation of STAT proteins in primary lymphoid and myeloid leukemia cells and in Epstein-Barr virus (EBV)-related lymphoma cell lines. *Blood*, **88**, 809–816 (1996).
72. Xu, X., Kang, S. H., Heidenrich, O., Okerholm, M., O'Shea, J. J., and Nerenberg, M. I. Constitutive activation of different Jak tyrosine kinases in human T-cell leukemia virus type 1 (HTLV-1) tax protein or virus-transformed cells. *J. Clin. Invest.*, **96**, 1548–1555 (1995).
73. Yoshida, M., Miyoshi, I., and Hinuma, Y. Isolation and characterization of retrovirus from cell lines of human adult T-cell leukemia and its implication in the disease. *Proc. Natl. Acad. Sci. USA*, **79**, 2031–2035 (1982).
74. Zhang, W., Nisbet, J. W., Albrecht, B., Ding, W., Kashanchi, F., Bartoe, J. T., and Lairmore, M. D. Human T-lymphotropic virus type 1 p30(II) regulates gene transcription by binding CREB binding protein/p300. *J. Virol.*, **75**, 9885–9895 (2001).
75. Zhang, W., Nisbet, J. W., Bartoe, J. T., Ding, W., and Lairmore, M. D. Human T-lymphotropic virus type 1 p30(II) functions as a transcription factor and differentially modulates CREB-responsive promoters. *J. Virol.*, **74**, 11270–11277 (2000).

MECHANISMS OF DISEASE DEVELOPMENT

SOMATIC ALTERATIONS IN ADULT T-CELL LEUKEMIA CELLS

Masao MATSUOKA

*Institute for Virus Research, Kyoto University**

Viral genes encoded by human T-cell leukemia virus type I (HTLV-I), especially *tax* gene, play an important role in immortalization of T-lymphocytes *in vitro*, and clonal proliferation *in vivo*. However, the enigma about Tax is that regardless of its significance in the leukemogenesis, adult T-cell leukemia (ATL) cells often have defect(s) in provirus that cannot produce Tax protein. The defects include loss of 5′-long terminal repeat (LTR), which is the enhancer/promoter of viral genes, deletion, and non-sense mutation of *tax* gene. These changes of *tax* gene were predominantly observed in the aggressive forms of ATL. During its long latent period, HTLV-I infected cells are thought to evolve into leukemic cells by several somatic alterations. Mutation of *p53* gene, and deletion of *CDKN2A* (*p16*) gene were reported in ATL cells although their frequencies are not so high. We reported the deletion of *Fas* gene, which conferred resistance from death signals. In addition to deletion, DNA methylation of *CDKN2A* gene was increased according to disease progression, implicating epigenetic change in suppression of CDKN2A and the progression of ATL. From these data, it is speculated that Tax is necessary to maintain the leukemic state during early stage of ATL, such as smoldering and chronic forms. Since Tax is a major target of cytotoxic T-cells, the growth of leukemic cells was controlled. However, at the late stage of leukemogensis, such as acute or lymphoma-type ATL, somatic alterations of the host genome including mutation, deletion, chromosomal changes and epigenetic changes accumulate, which can replace the function of Tax, enabling ATL cells to proliferate without Tax. Analyses of somatic alterations in ATL cells shed light on multi-step leukemogenesis.

It was estimated that about 2–6% of human T-cell leukemia virus type I (HTLV-I) carriers would have adult T-cell leukemia (ATL) during their lifetime (*2*). Since the average age of patients with ATL was around 60 years old in Japan and most carriers were infected with HTLV-I by mother-to-child fashion, there is a long latent period to onset of ATL after infection of HTLV-I (*23*). Tax has been identified as an oncogenic protein in viral proteins encoded by HTLV-I, by its pleiotropic actions influencing cellular functions (*8*, *29*). Transduced *tax* gene is enough to immortalize CD4 positive T-lymphocyte *in vitro* (*1*). However, the expression of Tax was unclear in leukemic cells, some of which had the defective HTLV-I provirus not to produce the intact Tax protein. Although HTLV-I is the etiologic agent of ATL, the onset rate of

* Kyoto 606-8507, Japan

ATL among carriers and long latency suggested that not only HTLV-I, but also the changes of the host genome contribute to the oncogenesis.

Clonal Proliferation of ATL Cells

Among viral proteins encoded by HTLV-I, Tax plays a central role in the leukemogenesis according to the following observations: 1) the expression of *tax* is sufficient to immortalize T-lymphocytes *in vitro* (*1*); 2) the fresh leukemic cells often have defective provirus, however, they still retain *tax* gene; 3) Tax shows the pleiotropic actions, which included activation of NFκB, SRF and CREB pathway, the functional inhibition of CDKN2A and p53, and transcriptional repression of *lck*, DNA polymerase β and *p18* genes. Such actions of Tax function to promote the proliferation, inhibit the apoptosis, and impair the repair system of infected cells, resulting in clonal proliferation of HTLV-I infected cells that increases proviral load, and finally malignant transformation, ATL.

Variation of sequences in HTLV-I provirus was more rare than that in the other human retrovirus, human immunodeficiency virus type 1 (HIV-1). In HIV-1, virus vigorously replicates in infected cells, which produces a tremendous quantity of viruses (about 10^{10} viruses/day) *in vivo*. Since reverse transcriptase of retrovirus does not have proof-reading activity, which was the reason for being the error prone enzyme (*5*), such replication generates numerous variations in the sequence of HIV-1, which also lead to acquisition of drug resistance and escape from the host immune system. The low frequency of variation in the HTLV-I provirus suggests that the replication rate of HTLV-I itself is low *in vivo*, and HTLV-I increases its provirus not by replication of virus, but by proliferation of HTLV-I infected cells *in vivo*. The fidelity of host DNA polymerase is much higher than viral reverse transcriptase, which can account for the rare variation in the sequence of HTLV-I provirus. Since infection of HTLV-I needs cell-to-cell interaction, increased HTLV-I infected cells are speculated to enhance the incidence of vertical or horizontal transmission.

The presence of HTLV-I provirus in HTLV-I infected cells and ATL cells enables us to analyze its state in each step of leukemogenesis, like the carrier state, smoldering, chronic and acute ATL. To prove the clonal proliferation of HTLV-I infected cells *in vivo*, we studied clonality of these cells using inverse PCR in asymptomatic carriers and in patients with adult T-cell leukemia (*6*). Combining inverse PCR with long PCR, we amplified the genomic DNA regions flanking the integration sites of the HTLV-I provirus in order to detect clones of infected cells (Fig. 1). Since integration sites of HTLV-I proviruses were random, genomic sequences adjacent to provirus are unique to each HTLV-I infected cell. Genomic DNAs from HTLV-I infected individuals were digested with *Eco*RI, which did not cut inside the HTLV-I provirus. Then, the digested DNAs were circularized by ligation with T4 DNA ligase. To prevent the amplification of HTLV-I itself, ligated mixtures were digested with *Mlu*I that cut HTLV-I provirus at the pX region. Then, long PCR was performed with primers in long terminal repeats (LTR) to amplify the adjacent genomic sequences. The band detected by this method represents HTLV-I infected clone. To ensure that the detected band reflected clonal expansion, cycles of long PCR were limited.

The HTLV-I provirus load differs more than 100-fold in the infected individuals

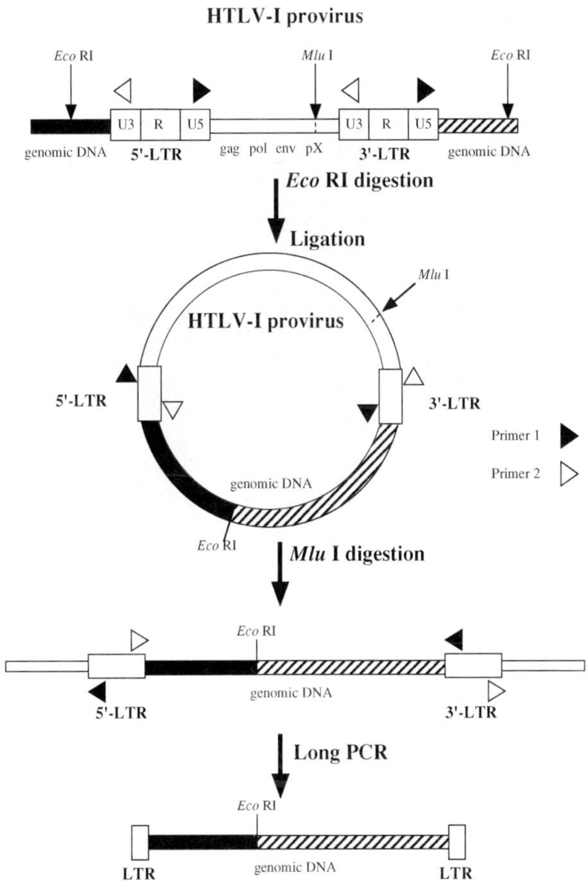

FIG. 1. Schematic diagram of inverse long PCR

Genomic DNA was digested with *Eco*RI, and then ligated with T4 DNA ligase to form circular DNA. This circular DNA contained flanking genomic DNA and HTLV-I provirus. Arrowheads indicate the PCR primers used in IL PCR. Since these primers can amplify both flanking genomic DNAs and the provirus itself, circular DNA was digested with *Mlu*I to avoid amplification of the provirus.

(*7*). Inverse long (IL) PCR revealed that increased provirus load was associated with an increase in both the number of cells in each clone and the number of clones. Clonal proliferations were found in both CD4 positive and CD8 positive cells, which was consistent with the finding that although most of provirus was detected in a population of CD4 positive T-cells, about 10% of HTLV-I provirus was found in CD8 positive T-lymphocytes (*28*). Most importantly, these HTLV-I infected clones persisted over several years in the same carriers (Fig. 2), and most of the persistent clones were CD4 positive in a HTLV-I carrier.

Expression of the *tax* gene in HTLV-I carriers was detected by reverse transcriptase mediated PCR, and anti-Tax antibodies were detected in carriers (*10, 15*), showing that *tax* gene was expressed *in vivo* in at least a small number of cells. The immortalization of T-lymphocytes, especially CD4 positive T-lymphocytes, is thought to be important in the pathogenesis of HTLV-I associated diseases and the

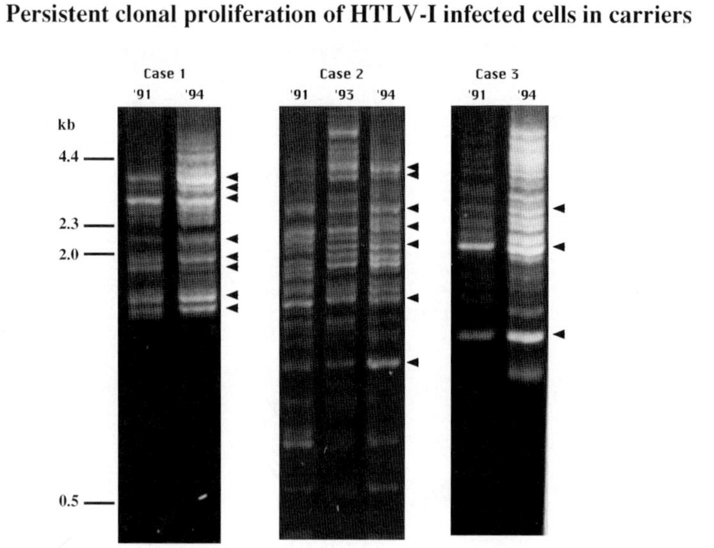

FIG. 2. Sequential analysis by IL PCR
Genomic DNAs from 3 different carriers at different time points were analyzed with IL PCR (33 cycles). The time when the samples were collected is shown at the top of each lane. Arrows show the persistent clones identified by IL PCR.

leukemogenesis of ATL. It also suggests that long surviving T-lymphocytes accumulate somatic changes of oncogenes or tumor-suppressor genes, leading to the onset of leukemia. This hypothesis would also explain the long latency of ATL after infection.

Virus load differs more than 100-fold among carriers, but did not change in separate individuals during follow-up over several years, indicating that long-surviving clones may be the molecular basis of a stable virus load (*7*). What mechanism controls this clonal proliferation and determines the virus load in HTLV-I carriers? The immune response, including cellular immunity and humoral immunity, may be responsible for controlling the number of HTLV-I infected cells. It has been reported that different human leukocyte antigen (HLA) haplotypes were associated with HTLV-I associated myelopathy/tropical spastic paraparesis (HAM/TSP) and ATL (*27*). It is possible that carriers with high cytotoxic T-lymphocyte (CTL) activity against cells presenting viral peptides suppress the number of HTLV-I infected cells, and that carriers with low CTL activity allow the proliferation of infected cells.

HTLV-I Provirus in ATL Cells

Since Tax is thought to play a central role in leukemogenesis of ATL, the analysis of HTLV-I provirus and *tax* gene should shed light on its detailed mechanism. Molecular analysis of HTLV-I provirus in the fresh leukemic cells from 72 cases disclosed three subtypes of HTLV-I provirus: 1) complete, 2) type 1 defective, and 3) type 2 defective provirus (Fig. 3) (*25*). In the complete provirus, there are no deletions detected by Southern blot analysis and long PCR (Fig. 3A). Defective provirus was found in 40 of 72 cases (56%). Type 1 defective provirus retained 5′- and 3′-LTR, but lacked the internal sequences that were predominantly 5′ regions such as *gag* and *pol*

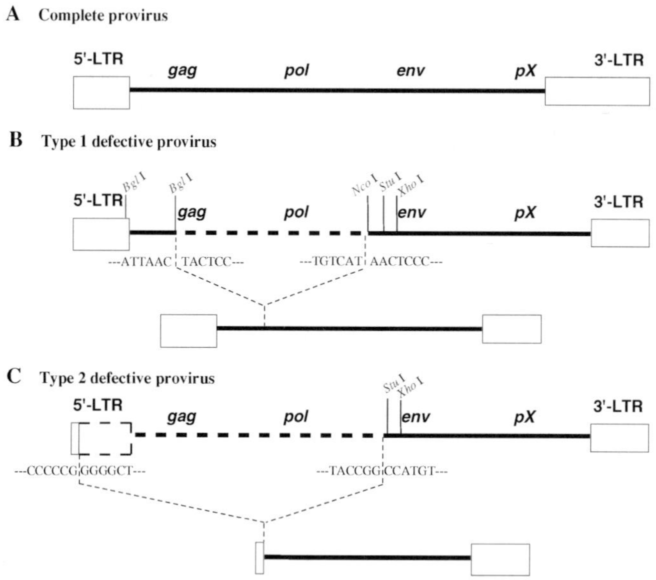

FIG. 3. Three types of HTLV-I provirus in ATL cells
(A) Complete provirus, (B) type 1 defective provirus; both LTRs retained, but internal sequence was deleted. (C) Type 2 defective provirus; 5′-LTR and internal sequence were deleted.

(Fig. 3B). Type 2 defective provirus lacked 5′-LTR in addition to internal sequences, as observed in 22 of 72 cases (31%) (Fig. 3C). Although the frequency of defective provirus (type 1+type 2) was not significantly different between chronic and acute or lymphoma-type ATL, type 2 defective provirus was more frequent in acute or lymphoma-type ATL (21/54 cases) than in the chronic ATL (1/18 cases), suggesting the relation between type 2 defective provirus and disease progression. Detailed analysis of type 2 defective provirus in a case of acute ATL showed 30 bp of the U3 region in 5′-LTR remained in provirus, suggesting that this deletion occurred after integration into the host genome.

5′-LTR was the promoter/enhancer for transcription of viral genes, indicating that loss of 5′-LTR impaired the expression of these genes. It is speculated that the *tax* gene is not transcribed in ATL cells with type 2 defective provirus. In addition to loss of 5′-LTR, deletion or non-sense mutation of *tax* gene were reported in fresh ATL cells that could not produce Tax protein (*9*). Such provirus with *tax* gene mutation was also frequently observed in acute ATL cases, suggesting the relation with disease progression. Tax is thought to play a central role in the survival and proliferation of HTLV-I infected cells whereas it is the major target of CTL *in vivo.* Although Tax increased the number of HTLV-I infected cells by its pleiotropic actions, CTLs, which recognize Tax peptide presented by infected cells, decreased the HTLV-I infected cells. Therefore, Tax is necessary in the carrier state or early stage of leukemogenesis for survival and proliferation of HTLV-I infected cells. However, when somatic alternations that can replace the pleiotropic actions of Tax accumulate during the progression, such ATL cells do not need the expression of Tax, and mutation of *tax* gene or deletion of 5′-LTR enables ATL cells to abandon the Tax, which is a disadvantage of fully transformed ATL cells.

Prospective follow-up of HTLV-I infected individuals in a Miyazaki cohort study showed that five carriers developed ATL during the follow-up. Titer of anti-Tax antibody was low in such carriers when compared with other carriers, suggesting that impaired immune response against Tax might be implicated in leukemogenesis (*15*). This observation also showed the close linkage between leukemogenesis of ATL and host immune system against Tax.

Genetic and Epigenetic Changes in ATL Cells

Genetic instability of cancer is implicated in the oncogenesis by mutating the tumor suppressor genes and oncogenes. Although somatic mutation of *p53* gene in ATL cells is reported to be about 20% in all cases analyzed (*4, 18, 21*), the majority of such cases were acute or lymphoma-type ATL, while the remaining cases showed no mutation of *p53* gene. In addition, deletion of *CDKN2A* gene, which encodes the inhibitor of cyclin dependent kinase 4 and 6, was reported to be present in about 20% of ATL cases. Again, most cases with deletion of *CDKN2A* gene were acute or lymphoma-type ATL (*12, 26*).

CD95 antigen (also known as Fas or Apo-1) and Fas ligand play key roles in apoptosis of cells of the immune system, function as effector molecules of cytotoxic T-lymphocytes, and function in the elimination of activated lymphocytes during the down-regulation of the immune response (*17*). The critical roles of the Fas-Fas ligand system in apoptosis suggest that its inactivation may be involved in malignant transformation. We analyzed the expression of Fas antigen on ATL cells by flow cytometry and found that the expression was absent in one case of ATL and markedly decreased in another case among 47 cases examined (*24*). Apoptosis could not be induced in the Fas-negative ATL cells by antibody against Fas antigen. Sequencing of RT-PCR products of the *Fas* genes in the Fas negative cells revealed two types of aberrant transcripts: one had a 5 bp deletion and a 1 bp insertion in exon 2, and the other transcript lacked exon 4 (Fig. 4). These mutations caused the premature termination of both alleles, resulting in the loss of expression of surface Fas antigen. These aberrant transcripts were not detected in a non-leukemic B cell line from the same patient. An RNase protection assay of the *Fas* gene revealed mutations in two additional cases with Fas-positive ATL cells out of 35 cases examined: one case lacked exon 4 and the other was a silent mutation. In the Fas antigen negative case, leukemic cells were resistant to anti-cancer drugs *in vivo*, indicating that the loss of expression of Fas antigen may be associated with a poor response to anti-cancer drugs. Indeed, Fas negative ATL cells were resistant to adriamycin-induced apoptosis *in vitro*, which is consistent with the finding that ATL in this case was resistant to chemotherapy. These findings indicate that mutation of the *Fas* gene may be associated with the progression of ATL and with resistance to anti-cancer drugs. Another ATL case with mutation of *Fas* gene was reported, in which deletion of the cytoplasmic region conferred a dominant negative effect to mutated *Fas* gene, resulting in resistance against Fas signals (*16*).

Somatic change of *CDKN2A* gene, which was deletion in most cases, was reported to be about 20% among ATL cases, it was predominantly observed in aggressive forms of ATL such as acute and lymphoma-type. However, the remaining cases did not have genetic changes. Epigenetic change like DNA methylation has been shown

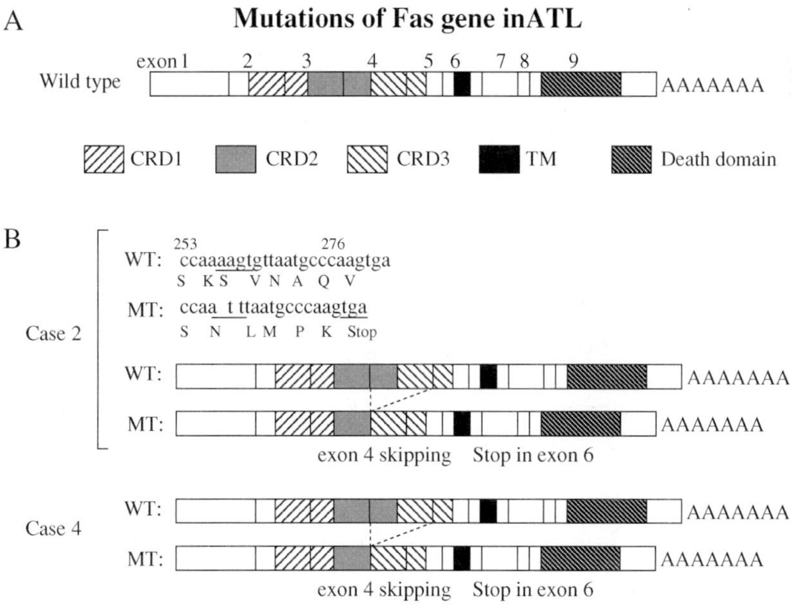

FIG. 4. Abnormalities of the *Fas* gene in ATL cells
(A) Structure of the human *Fas* gene. The three CRDs (cysteine-rich domains), TM (transmembrane domain) and death domain are shown. (B) Mutations of the *Fas* genes. Detected *Fas* gene mutations are shown.

to be implicated in silencing of tumor-suppressor genes including *CDKN2A* (*14, 20*), *E-cadherin* (*30*) and the *VHL*-tumor-suppressor gene (*13*). These reports suggest that methylation is a common mechanism of inhibiting tumor suppressor genes.

Therefore, we examined the methylation status of the *CDKN2A* gene in patients with different forms of ATL using Southern blot analysis, methylation-specific PCR (MSPCR), and nucleotide sequencing (*19*). We found that the *CDKN2A* gene was more frequently methylated in fresh tumor cells isolated from patients with acute ATL (47%) or lymphoma-type ATL (73%) than in those with less malignant chronic (17%) and smoldering ATL (17%). In addition, deletions of the *CDKN2A* gene were found in 24% of acute ATL patients; thus abnormalities of this gene were in 71% of acute ATL patients. In contrast, no *CDKN2A* gene methylation was found in asymptomatic carriers or uninfected individuals. Methylation of the *p15* gene was not found in any samples from 36 ATL patients. Direct sequencing of the *CDKN2A* gene after sodium bisulfite treatment of genomic DNA revealed that the methylation of CpG sites had occurred in 24 of 32 ATL cases (75%) including chronic and smoldering ATL, even when MS PCR and the Southern blot had failed to detect *CDKN2A* gene methylation. Among fresh ATL samples with methylation, methylation was detected in the promoter region and exon in 17 out of 24 cases, and methylation in the exon without upstream was detected in 7 cases of the 24 cases, showing an increasing methylation as the disease progressed. Increasing methylation is also associated with progressive suppression of *CDKN2A* expression, as shown by analysis of *CDKN2A* transcripts. Examination of ATL cells at each clinical stage showed that methylation first occurred in the exon region, and progressed to the promoter region with clinical progression of ATL. Since it has been reported that methylation in the exon region has no effect on the expression of the *CDKN2A* gene (*3, 11*), the progressive DNA methylation in the promoter region is associated with suppression of *CDKN2A* gene transcripts. Analysis of DNA methylation in the promoter region and exon 1 of *CDKN2A* gene is shown in Fig. 5. Each bar represents the extent of methylation in each CpG site. PCR product was subcloned into plasmid DNA, and then analyzed by sequencing. This result showed that DNA methylation tended to accumulate in the same allele, suggesting methylation in the exon 1 progressed to the promoter region.

In one case, the pattern of methylation proved to be different between peripheral blood cells and lymph node cells, suggesting the presence of multiple subclones with regard to methylation patterns despite the same HTLV-I integration site. Cells with different methylation pattern of the *CDKN2A* gene in the same patient may have different tissue affinities, perhaps reflecting different patterns of expression of adhesion molecules.

Of various proteins associated with Tax, CDKN2A has been shown to be functionally inactivated by Tax, resulting in cell cycle perturbation (*22*). However, at the

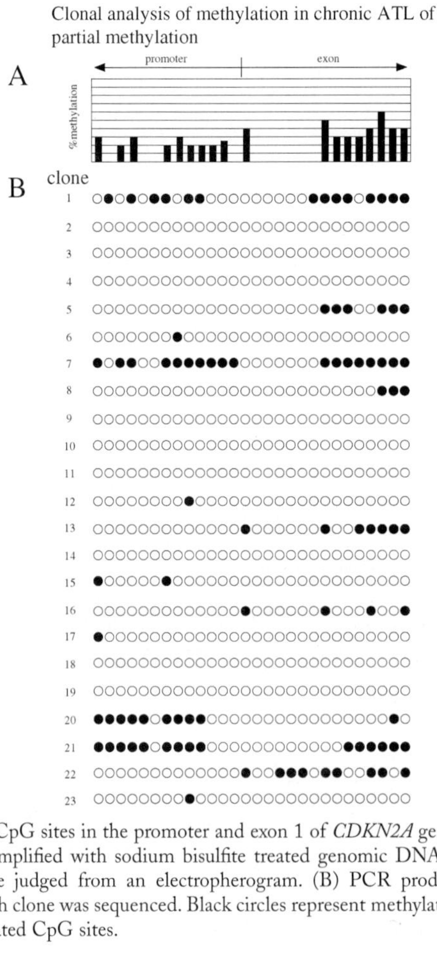

FIG. 5. Methylated CpG sites in the promoter and exon 1 of *CDKN2A* gene
(A) PCR product amplified with sodium bisulfite treated genomic DNA was sequenced. Methylated CpG sites were judged from an electropherogram. (B) PCR product was subcloned into plasmid DNA, and each clone was sequenced. Black circles represent methylated CpG sites, and white circles show unmethylated CpG sites.

late stage of leukemogenesis, such as acute or lymphoma-type ATL, the expression of *tax* is significantly lower than in asymptomatic carriers (*10*). Fresh leukemic cells in some acute ATL patients do not produce Tax *in vitro* because of deletion of the LTR and other mechanisms (*9, 25*). Thus, whether Tax expression is required in ATL cells remains unclear. On the other hand, such deletions in the HTLV-I provirus are rare in low-grade malignant ATL cases (smoldering or chronic ATL), suggesting that Tax may play an important role at such stages, but may not be essential for late stages of leukemia. The increasing methylation of the *CDKN2A* gene suggests that at an early stage of leukemogenesis, Tax is responsible for inactivating the CDKN2A protein, while at a late stage, methylation or even deletion of the *CDKN2A* gene abolishes the expression of CDKN2A. Methylation might be a way to transform HTLV-I infected T-cells in the absence of Tax expression. ATL cells express activation antigens like CD25, and secrete various cytokines; their phenotype resembles that of cells expressing Tax *in vitro*. Modification of methylation in ATL cells might fix a phenotype initially induced by Tax, and replace the functions of Tax.

Perspectives on Leukemogenesis of ATL

In the carrier state or early stage of leukemogenesis, Tax plays a critical role in the proliferation and survival of HTLV-I infected cells. However, at the same time, it renders HTLV-I infected cells susceptible to attack from CTLs *in vivo* since its peptide is the major target of CTL. During the long latent period, somatic alterations in the host genome increase and can replace the function of Tax. Such cells do not need Tax, rather, ATL cells without Tax are selected *in vivo*. Further analyses of somatic alterations will clarify the molecular mechanism of leukemogenesis, and help us to identify the molecular target to develop a novel treatment against this fatal disease. Intensification of host immune response against Tax might lead to prevention or immunotherapy against ATL.

Acknowledgments

These studies were performed by Sadahiro Tamiya, Ken-ichiro Etoh, Kisato Nosaka, and Tasunori Sakai. The authors also collaborated with Michiyuki Maeda, Kiyoshi Takatsuki, and Hiroaki Mitsuya.

REFERENCES

1. Akagi, T., Ono, H., and Shimotohno, K. Characterization of T cells immortalized by Tax1 of human T-cell leukemia virus type 1. *Blood,* **86**, 4243–4249 (1995).
2. Arisawa, K., Soda, M., Endo, S., Kurokawa, K., Katamine, S., Shimokawa, I., Koba, T., Takahashi, T., Saito, H., Doi, H., and Shirahama, S. Evaluation of adult T-cell leukemia/lymphoma incidence and its impact on non-Hodgkin lymphoma incidence in southwestern Japan. *Int. J. Cancer,* **85**, 319–324 (2000).
3. Bender, C. M., Pao, M. M., and Jones, P. A. Inhibition of DNA methylation by 5-aza-2′-deoxycytidine suppresses the growth of human tumor cell lines. *Cancer Res.*, **58**, 95–101 (1998).
4. Cesarman, E., Chadburn, A., Inghirami, G., Gaidano, G., and Knowles, D. M. Structural and functional analysis of oncogenes and tumor suppressor genes in adult T-cell

leukemia/lymphoma shows frequent p53 mutations. *Blood*, **80**, 3205–3216 (1992).

5. Coffin, J. M. Genetic diversity and evolution of retroviruses. *Curr. Top. Microbiol. Immunol.*, **176**, 143–164 (1992).
6. Etoh, K., Tamiya, S., Yamaguchi, K., Okayama, A., Tsubouchi, H., Ideta, T., Mueller, N., Takatsuki, K., and Matsuoka, M. Persistent clonal proliferation of human T-lymphotropic virus type I-infected cells *in vivo*. *Cancer Res.*, **57**, 4862–4867 (1997).
7. Etoh, K., Yamaguchi, K., Tokudome, S., Watanabe, T., Okayama, A., Stuver, S., Mueller, N., Takatsuki, K., and Matsuoka, M. Rapid quantification of HTLV-I provirus load: detection of monoclonal proliferation of HTLV-I-infected cells among blood donors. *Int. J. Cancer*, **81**, 859–864 (1999).
8. Franchini, G. Molecular mechanisms of human T-cell leukemia/lymphotropic virus type I infection. *Blood*, **86**, 3619–3639 (1995).
9. Furukawa, Y., Kubota, R., Tara, M., Izumo, S., and Osame, M. Existence of escape mutant in HTLV-I tax during the development of adult T-cell leukemia. *Blood*, **97**, 987–993 (2001).
10. Furukawa, Y., Osame, M., Kubota, R., Tara, M., and Yoshida, M. Human T-cell leukemia virus type-1 (HTLV-1) Tax is expressed at the same level in infected cells of HTLV-1-associated myelopathy or tropical spastic paraparesis patients as in asymptomatic carriers but at a lower level in adult T-cell leukemia cells. *Blood*, **85**, 1865–1870 (1995).
11. Gonzalgo, M. L., Hayashida, T., Bender, C. M., Pao, M. M., Tsai, Y. C., Gonzales, F. A., Nguyen, H. D., Nguyen, T. T., and Jones, P. A. The role of DNA methylation in expression of the p19/p16 locus in human bladder cancer cell lines. *Cancer Res.*, **58**, 1245–1252 (1998).
12. Hatta, Y., Hirama, T., Miller, C. W., Yamada, Y., Tomonaga, M., and Koeffler, H. P. Homozygous deletions of the p15 (MTS2) and p16 (CDKN2/MTS1) genes in adult T-cell leukemia. *Blood*, **85**, 2699–2704 (1995).
13. Herman, J. G., Latif, F., Weng, Y., Lerman, M. I., Zbar, B., Liu, S., Samid, D., Duan, D. S., Gnarra, J. R., Linehan, W. M., and Baylin, S. B. Silencing of the VHL tumor-suppressor gene by DNA methylation in renal carcinoma. *Proc. Natl. Acad. Sci. USA*, **91**, 9700–9704 (1994).
14. Herman, J. G., Merlo, A., Mao, L., Lapidus, R. G., Issa, J. P., Davidson, N. E., Sidransky, D., and Baylin, S. B. Inactivation of the CDKN2/p16/MTS1 gene is frequently associated with aberrant DNA methylation in all common human cancers. *Cancer Res.*, **55**, 4525–4530 (1995).
15. Hisada, M., Okayama, A., Shioiri, S., Spiegelman, D. L., Stuver, S. O., and Mueller, N. E. Risk factors for adult T-cell leukemia among carriers of human T-lymphotropic virus type I. *Blood*, **92**, 3557–3561 (1998).
16. Maeda, T., Yamada, Y., Moriuchi, R., Sugahara, K., Tsuruda, K., Joh, T., Atogami, S., Tsukasaki, K., Tomonaga, M., and Kamihira, S. Fas gene mutation in the progression of adult T cell leukemia. *J. Exp. Med.*, **189**, 1063–1071 (1999).
17. Nagata, S. and Golstein, P. The Fas death factor. *Science*, **267**, 1449–1456 (1995).
18. Nishimura, S., Asou, N., Suzushima, H., Okubo, T., Fujimoto, T., Osato, M., Yamasaki, H., Lisha, L., and Takatsuki, K. p53 gene mutation and loss of heterozygosity are associated with increased risk of disease progression in adult T cell leukemia. *Leukemia*,

9, 598–604 (1995).

19. Nosaka, K., Maeda, M., Tamiya, S., Sakai, T., Mitsuya, H., and Matsuoka, M. Increasing methylation of the CDKN2A gene is associated with the progression of adult T-cell leukemia. *Cancer Res.*, **60**, 1043–1048 (2000).
20. Reed, A. L., Califano, J., Cairns, P., Westra, W. H., Jones, R. M., Koch, W., Ahrendt, S., Eby, Y., Sewell, D., Nawroz, H., Bartek, J., and Sidransky, D. High frequency of p16 (CDKN2/MTS-1/INK4A) inactivation in head and neck squamous cell carcinoma. *Cancer Res.*, **56**, 3630–3633 (1996).
21. Sakashita, A., Hattori, T., Miller, C. W., Suzushima, H., Asou, N., Takatsuki, K., and Koeffler, H. P. Mutations of the p53 gene in adult T-cell leukemia. *Blood,* **79**, 477–480 (1992).
22. Suzuki, T., Kitao, S., Matsushime, H., and Yoshida, M. HTLV-1 Tax protein interacts with cyclin-dependent kinase inhibitor p16INK4A and counteracts its inhibitory activity towards CDK4. *EMBO J.*, **15**, 1607–1614 (1996).
23. Takatsuki, K., Yamaguchi, K., and Matsuoka, M. ATL and HTLV-I-related diseases. *In* "Adult T-cell Leukemia," ed. K. Takatsuki, pp. 1–27 (1994). Oxford University Press, New York.
24. Tamiya, S., Etoh, K., Suzushima, H., Takatsuki, K., and Matsuoka, M. Mutation of CD95 (Fas/Apo-1) gene in adult T-cell leukemia cells. *Blood,* **91**, 3935–3942 (1998).
25. Tamiya, S., Matsuoka, M., Etoh, K., Watanabe, T., Kamihira, S., Yamaguchi, K., and Takatsuki, K. Two types of defective human T-lymphotropic virus type I provirus in adult T-cell leukemia. *Blood,* **88**, 3065–3073 (1996).
26. Uchida, T., Kinoshita, T., Watanabe, T., Nagai, H., Murate, T., Saito, H., and Hotta, T. The CDKN2 gene alterations in various types of adult T-cell leukaemia. *Br. J. Haematol.*, **94**, 665–670 (1996).
27. Yashiki, S., Fujiyoshi, T., Arima, N., Osame, M., Yoshinaga, M., Nagata, Y., Tara, M., Nomura, K., Utsunomiya, A., Hanada, S., Tajima, K., and Sonoda, S. HLA-A*26, HLA-B*4002, HLA-B*4006, and HLA-B*4801 alleles predispose to adult T cell leukemia: the limited recognition of HTLV type 1 tax peptide anchor motifs and epitopes to generate anti-HTLV type 1 tax CD8(+) cytotoxic T lymphocytes. *AIDS Res. Hum. Retroviruses*, **17**, 1047–1061 (2001).
28. Yasunaga, J., Sakai, T., Nosaka, K., Etoh, K., Tamiya, S., Koga, S., Mita, S., Uchino, M., Mitsuya, H., and Matsuoka, M. Impaired production of naive T lymphocytes in human T-cell leukemia virus type I-infected individuals: its implications in the immunodeficient state. *Blood,* **97**, 3177–3183 (2001).
29. Yoshida, M. Multiple viral strategies of htlv-1 for dysregulation of cell growth control. *Annu. Rev. Immunol.*, **19**, 475–496 (2001).
30. Yoshiura, K., Kanai, Y., Ochiai, A., Shimoyama, Y., Sugimura, T., and Hirohashi, S. Silencing of the E-cadherin invasion-suppressor gene by CpG methylation in human carcinomas. *Proc. Natl. Acad. Sci. USA*, **92**, 7416–7419 (1995).

Gann Monograph on Cancer Research 50, 2003

LONG-TERM FOLLOW-UP OF HTLV-1 CARRIERS

Akihiko OKAYAMA*[1] and Sherri O. STUVER*[2]

Department of Internal Medicine II, Miyazaki Medical College[1] *and Department of Epidemiology, Harvard School of Public Health*[2]

To gain an understanding of the natural history of human T-lymphotropic virus type 1 (HTLV-1) infection, the Miyazaki Cohort Study (MCS) was performed in a highly endemic population in southwestern Japan beginning in 1984. As of the end of 2000, 2,046 subjects (1,179 women, 867 men) were enrolled in the cohort, with a baseline seroprevalence of 26%. The major findings were as follows: 1) Biomarkers of carriers. The HTLV-1 proviral DNA load among carriers was distributed over a wide range of values, from less than 50 to more than 5,000 copies per 10^5 peripheral blood mononuclear cells. Proviral load was found to be positively associated with anti-HTLV-1 titer, the serum concentration of soluble interleukin 2 receptor, and the presence of abnormal lymphocytes on blood smear. Proviral load levels did not change in individual subjects over time. 2) Transmission between married partners. The HTLV-1 seroconverters within the MCS were genetically confirmed to be the result of infection by the subjects' married partners. The proviral load and antibody titers were stabilized within a year of infection. We concluded that host factors are probably more important than viral factors in determining proviral load. 3) Outcome of HTLV-1 infection. Six adult T-cell leukemia (ATL) cases developed among asymptomatic carriers. The overall mortality rates due to ATL were 190.5 and 51.7 per 10^5 person-years for male and female carriers, respectively. No obvious case with HTLV-1-associated neurological disorder was found in this cohort. We hypothesize that carriers with high proviral load, oligoclonal expansion of HTLV-1-infected cells, and low anti-Tax immune response are at high risk to develop ATL; however, further studies, including the analysis of genetic abnormalities of infected cells of these carriers, are necessary.

Human T-lymphotropic virus type 1 (HTLV-1) is the causative agent of adult T-cell leukemia (ATL), HTLV-1-associated myelopathy/tropical spastic paraparesis (HAM/TSP), and HTLV-1-associated uveitis. In addition, chronic inflammatory diseases of the lung, arthritis, and Sjögren syndrome have been suggested to be related to this infection. However, it is difficult to know the whole picture of clinical outcomes among HTLV-1-infected subjects because of the relatively low incidence of diseases among HTLV-1 carriers as well as the long latent period from infection to disease onset. Population-based cohort studies represent a useful means to study the natural history of HTLV-1 infection (*2, 21, 26, 32*).

*[1] 5200 Kihara, Kiyotake, Miyazaki 889-1601, Japan
*[2] 677 Huntington Ave., Boston MA 02115, USA

The Miyazaki Cohort Study (MCS) was established in 1984 to evaluate the natural history of HTLV-1 (*19*). After follow-up of this cohort from 1984 to 2000, we have developed a database of clinical, virological, and epidemiological parameters, as well as a repository of serum specimens and peripheral blood mononuclear cell (PBMC) samples. The major questions we have tried to address in this study are as follows: 1) What are the viral or host factors that determine whether a person is at high risk for ATL? 2) What are the dynamics of HTLV-1 transmission between married partners? 3) What is the overall impact of HTLV-1 infection on the health of carriers?

Study Population and Data Collection

The subjects enrolled in the MCS included adult residents of two small villages in southwestern Japan, most of whom are 40 years of age or older. The cohort, representing more than 70% of the target population, was followed for 16 years with clinical, serological, demographic, health, and behavioral data collected annually, after obtaining informed consent. At the end of 2000, 2,046 subjects (1,179 women, 867 men) were enrolled in the cohort, with a baseline seroprevalence of 26% (Table I). The average attendance of the subjects was once every two years (about six screens per subject), representing a total of 15,355 person-years of follow-up.

Analysis of Biomarkers in Asymptomatic Carriers

Different biomarkers related to the virus itself, cellular markers, and host immune responses were measured. The HTLV-1 proviral DNA load in peripheral blood, which reflects the burden of infected cells in carriers, likely plays the central role with regard to the development of ATL or HAM/TSP. One hundred thirty asymptomatic anti-HTLV-1-positive subjects were tested for proviral load using a polymerase chain reaction (PCR)-based quantitative method, AmpliSensor assay (Fig. 1). The proviral DNA load among carriers was distributed over a wide range of values, from less than 50 to more than 5,000 copies per 10^5 PBMC. The proviral loads were not affected by the age of the subjects, although most of the subjects tested in this study were older than 40 years. Nor was any association reported between proviral load and age in a blood donor population in Japan which included younger subjects (7). No difference in the mean values of the proviral loads was found between males and females in the MCS; however, the carriers with high proviral loads (more than 5,000 copies per 10^5 PBMC) were more likely to be male than female. This

TABLE I. HTLV-1 Seroprevalence in the Miyazaki Cohort Study at Baseline

Age	Male	(%)	Female	(%)	Total	(%)
–40	13/148	(8.8)	17/187	(9.1)	30/335	(9.0)
41–50	25/147	(17.0)	46/242	(19.0)	71/389	(18.3)
51–60	66/273	(24.2)	104/369	(28.2)	170/642	(26.5)
61–70	53/193	(27.5)	107/252	(42.5)	160/445	(36.0)
71–80	28/83	(33.7)	53/101	(52.5)	81/184	(44.0)
81–	9/23	(39.1)	11/28	(39.3)	20/51	(39.2)
Total	194/867	(22.4)	338/1,179	(28.7)	532/2,046	(26.0)

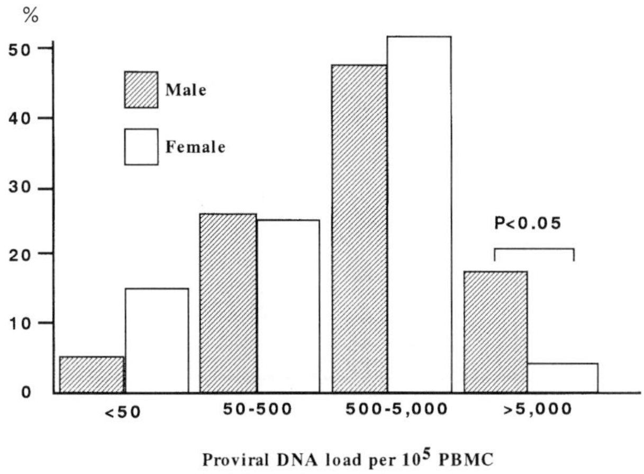

FIG. 1. Distribution of 130 carriers according to their HTLV-1 proviral DNA loads determined by AmpliSensor assay (Biotronics)

The percentage is the number of carriers belonging to each group, by gender. The carriers with high proviral loads (more than 5,000 copies per 10^5 PBMC) were more likely to be male than female (χ^2 test, P=0.02).

observation may be important when one considers that the incidence of ATL in males is higher than that in females (*10*).

HTLV-1 antibody titers are maintained over a long period among asymptomatic carriers and the proviral loads are found to be positively associated with HTLV-1 antibody titer (Fig. 2A). This result is consistent with our previous finding using the semiquantitative method (*14*). Proviral loads are also positively associated with the level of soluble interleukin 2 receptor (sIL-2R), which is generally the marker of activated T-cells (Fig. 2B). This finding is consistent with the observation in blood donors (*7*). The mean level of sIL-2R in the carriers was significantly higher than that in the seronegatives. HTLV-1 *tax/rex* messenger RNA was shown to be detectable more frequently in the carriers with a high proviral load than in those carriers with a lower proviral DNA load. Subjects with detectable *tax/rex* messenger RNA had a higher number of CD25 (interleukin 2 receptor α)-positive T-cells (*25*). A positive association was also found between the proviral load and the number of morphologically abnormal lymphocytes (Ably) on peripheral blood smear in the carriers (*11, 31*). In addition, levels of CD4- and CD25-positive cells were higher among asymptomatic carriers, with the highest levels seen in the subjects with Ably (*35*). The male carriers with high proviral load, high anti-HTLV-1 antibody, and low anti-Tax antibody were shown to have the highest Ably level (*11*). These data are informative, because both CD25 positivity and the morphological abnormality are features of leukemic cells of ATL. However, a small number of Ably can be seen on the peripheral blood smear in non-carriers, and Ably can disappear spontaneously. Therefore, it is still uncertain whether cells with these characteristics are the direct precursors of ATL cells.

The proviral load measured in individual subjects over a 5-year interval remains fairly stable (*7*). However, only very low levels of messenger RNA of HTLV-1, representing active transcription of the viral genome, can be detected in PBMC from

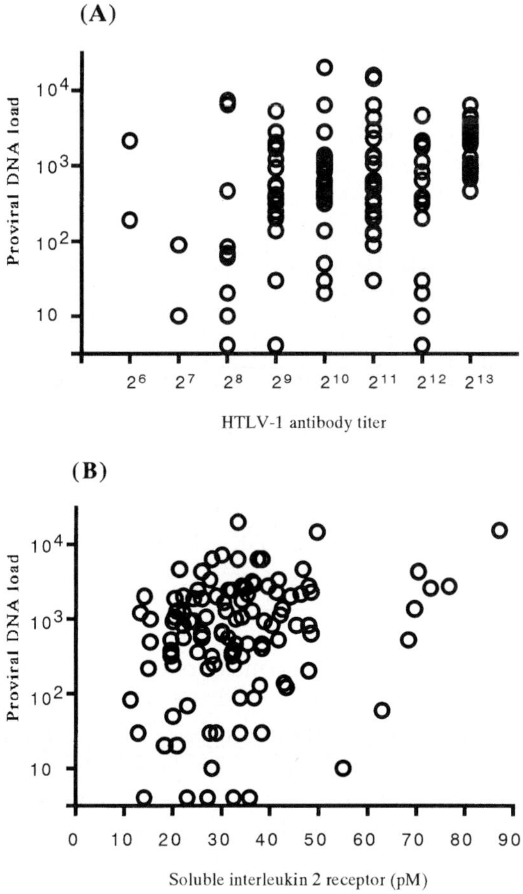

FIG. 2. The correlation of the proviral DNA loads of 130 carriers with their anti-HTLV-1 antibody titers (A) and serum levels of soluble interleukin 2 receptor (IL-2R) (B)

The antibody titer was determined by particle agglutination assay (Serodia HTLV, FujiRebio). The sIL-2R was measured by enzyme-linked immunoassay (sIL-2R, Immunotech). The proviral DNA load was positively associated with anti-HTLV-1 titer (Spearman's rank correlation, $P<0.0001$) and weakly associated with serum level of sIL-2R (Pearson's correlation coefficient, $r=0.243$, $P=0.009$).

asymptomatic carriers (*25*). Moreover, the HTLV-1 genome is highly conserved, even among different isolates, and the mutation rate of the virus is considered to be extremely low, suggesting that virus replication via error-prone reverse transcription is quite limited (*8, 34*). Therefore, the clonal proliferation of infected cells is an important mechanism for maintaining the number of HTLV-1-infected cells over time in each carrier. Clones of HTLV-1-infected cells are believed to persist for more than several years (*4, 6*). HTLV-1-infected T-cells may have a growth advantage, especially as a result of the transacting protein, Tax, which can promote cell proliferation through several different mechanisms (*33, 37*). However, at the same time, Tax is considered to be the major target of the host cytotoxic T-lymphocytes (CTL) (*15, 36*). The balance between these factors is probably important for the maintenance of a stable level of HTLV-1-infected cells in asymptomatic carriers.

Transmission of HTLV-1 within Married Couples

About 1,500 HTLV-1 seronegative subjects in the MCS were followed for anti-HTLV-1 seroconversion, and 23 seroconverters (four males and 19 females) were identified through 1996 (*24*). The average age at seroconversion was 58 years (48–68 years old). The incidence of seroconversion was 2.8 per 1,000 person-years (1.2 for males and 4.0 for females). Because 90% of the spouses of seroconverters were anti-HTLV-1 positive, sexual transmission from the HTLV-1-positive partners was the most likely route of infection. Indeed, the incidence of seroconversion was 25 per 1,000 person-years (10.5 for males and 54.0 for females), when the married partner was anti-HTLV-1 positive. Although the epidemiological findings strongly support the transmission of HTLV-1 between married partners, especially from male to female (*29, 32*), no direct evidence of such transmission existed. To address this issue, we determined sequence variations of the gp46 coding region of HTLV-1 in 13 seroconverters and their married partners (*13*). The gp46 coding sequences within each married couple were completely identical (Table II). This finding clearly demonstrates the transmission of HTLV-1 between married partners, both male-to-female and female-to-male, on a molecular level. The presence of anti-Tax in a carrier partner was thought to be predictive of transmission within married couples (*5, 27*).

By analyzing biomarkers in the seroconverters within the MCS, we were able to characterize the evolution of newly acquired HTLV-1 infection in adults. HTLV-1 proviral DNA loads in the initial years of seroconversion varied over a broad range from less than 10 to more than 1,000 copies per 10^5 PBMC (*24*). However, there was no obvious change in proviral load over several years within individual subjects (Table III). Anti-HTLV-1 antibody titers were moderately high after infection in most cases, and the average titer increased slightly, though not significantly, during the observation period (Table III). Similar findings have been reported for transfusion-acquired infection among HTLV-1 seroconverters in Jamaica (*17*). Therefore, both proviral load and humoral response in adult HTLV-1 infection appear to stabilize within a few years after initial infection. Of note, one subject tested positive for HTLV-1 provirus before seroconversion, suggesting the existence of a window period for community-acquired infection.

In general, the major factors affecting the viral load in retrovirus infection are the magnitude of replication and the host immune response. It has not been clear which of these two factors plays a more important role in determining HTLV-1 proviral load in asymptomatic carriers. Because a seroconverter and his/her HTLV-1-positive married partner share infection with ostensibly the same virus, we compared proviral loads between spouses within each married couple to see whether they were similar (*13*). The level of HTLV-1 proviral load in a seroconverter was often quite different from that of his/her partner (Table II), which suggests that virus-specific factors are not as important as host-related factors with respect to the determination of HTLV-1 proviral load in adult infection. In addition, the seropositive married partners, who should have had a longer duration of HTLV-1 infection than their newly infected spouses, did not always have a higher proviral load than the seroconverters. Thus, a longer infection period or an infection prior to adolescence may not always lead to a high proviral load in carriers.

TABLE II. Representative Data for the Comparison of HTLV-1 gp46 DNA Sequences and Proviral Loads between Seroconverters and Their HTLV-1 Carrier Spouses[a]

Nucleotide number		55	61	120	164	165	207	233	244	250	266	270	309	345	408	411	465	702	738	741	759	831	912	Proviral DNA load[b]
	ATK-1	T	G	C	T	G	C	C	C	C	C	G	T	G	A	C	A	C	T	T	C	T	C	
Couple 1	P[c]	—	A	T	—	—	—	—	—	—	T	A	—	—	G	—	—	—	—	—	—	—	—	1,490
	S[d]	—	A	T	—	—	—	—	—	—	T	A	—	—	G	—	—	—	—	—	—	—	—	740
Couple 2	P	—	—	T	—	—	—	—	—	—	T	A	—	—	G	—	—	—	—	—	—	—	—	60
	S	—	—	T	—	—	—	—	—	—	T	A	—	—	G	—	—	—	—	—	—	—	—	730
	FL[e]	—	—	T	—	—	—	—	—	—	T	A	—	—	G	—	—	—	—	—	—	—	—	50
	ML[f]	—	—	T	—	—	—	—	—	—	T	A	—	—	G	—	—	—	—	—	—	—	—	470
Couple 3	P	—	—	T	—	—	—	A	—	—	T	A	—	—	G	—	—	—	—	—	—	—	—	490
	S	—	—	T	—	—	—	A	—	—	T	A	—	—	G	—	—	—	—	—	—	—	—	60
Couple 4	P	—	—	—	—	A	—	—	—	—	T	A	C	—	G	—	—	—	—	—	—	—	—	220
	S	—	—	—	—	A	—	—	—	—	T	A	C	—	G	—	—	—	—	—	—	—	—	630
Couple 5	P	—	—	—	—	—	—	—	—	—	T	A	—	—	G	—	—	—	—	C	—	—	—	760
	S	—	—	—	—	—	—	—	—	—	T	A	—	—	G	—	—	—	—	C	—	—	—	80
Couple 6	P	—	—	—	—	—	—	—	A	—	T	A	—	—	G	—	—	—	—	—	—	—	—	1,710
	S	—	—	—	—	—	—	—	A	—	T	A	—	—	G	—	—	—	—	—	—	—	—	<10
Couple 7	P	C	—	—	C	—	T	A	—	—	T	A	—	—	G	A	G	A	—	—	A	—	—	800
	S	C	—	—	C	—	T	A	—	—	T	A	—	—	G	A	G	A	—	—	A	—	—	2,310

[a] Ref. *13*.
[b] Proviral DNA load, copies per 10^5 PBMC.
[c] P, partner of seroconverter. Dashes (—) indicate sequence identity with ATK-1.
[d] S, seroconverter. Female seroconverters (couple 1–5). Male seroconverters (couple 6, 7).
[e] FL, father-in-low of seroconverter.
[f] ML, mother-in-low of seroconverter.

TABLE III. Representative Data for HTLV-1 Proviral DNA Loads and Antibody Titers in Seroconverters[a]

Case[b]		Y–2[c]	Y–1	Y0	Y1	Y2	Y3	Y4	Y5	Y6	Y7	Y8	Y9	Y10
1	Proviral load[d]						740		850					
	Antibody titer[e]						1:2,048	1:2,048	1:2,048	1:2,048	1:2,048	1:2,048	1:2,048	1:2,048
2	Proviral load		N[f]								730		460	
	Antibody titer				1:512	1:512					1:1,024	1:512	1:1,024	1:1,024
3	Proviral load		N	N	60	140	130							
	Antibody titer				1:512	1:256	1:512							
4	Proviral load			N	630	180	140							
	Antibody titer				1:32	1:256	1:256							
5	Proviral load		N			80		190						
	Antibody titer					1:4,096		1:4,096	1:2,048					
6	Proviral load	N	N	780	<10			10						
	Antibody titer				1:512		1:512	1:512						
7	Proviral load	N	N		2,310	1,802	2,700	4,850		6,330				
	Antibody titer				1:512	1:512	1:1,024	1:1,024						

[a] Ref. *24*.
[b] The number of the case is consistent with the number of the couple in Table II.
[c] Years of the screens before and after seroconversion, with the year of the last seronegative sample as the reference year (Y0).
[d] Proviral load, proviral DNA load (copies per 10^5 PBMC).
[e] Antibody titer, anti-HTLV-1 antibody titer measured by particle agglutination assay.
[f] N, negative by the nested PCR assay.
Blank space indicates missing information due to nonattendance at annual health examination or because of insufficient volume of sample testing.

Of course, it will also be important to clarify the evolution of infection when HTLV-1 is acquired perinatally from mother to child. According to the results of a cohort study in Jamaica, the proviral loads among HTLV-1-positive children who were infected by their mothers varied over a wide range, as we have observed in adult infection. In most children, the proviral loads appear to stabilize within a year of initial infection, with no further increase during an observation period of several years. Among children who eventually developed infective dermatitis and/or seborrheic dermatitis, however, provirus loads had a tendency to increase until disease onset (Hisada, M. and Maloney, E., personal communications).

Impact of HTLV-1 Infection on the Health of Carriers

Six ATL cases (one prevalent case and five incident cases) were identified among the HTLV-1-positive subjects in the MCS through 1996 (*9*). Four were men and two were women, and the mean age at death of these six cases was 73 years. The overall mortality rates of ATL in the MCS were 190.5 and 51.7 per 10^5 person-years for male and female carriers, respectively (*10*). The five incident cases did not show any clinical symptoms or signs such as skin lesions or lymphadenopathy indicating smoldering ATL during the follow-up period prior to disease occurrence. The leukocyte count remained within normal limits in most of the cases. Only one case had a detectable level of Ably (more than 0.6% of leukocytes on the peripheral blood smear). When the prediagnostic biomarkers of the five incident cases were compared with age- and sex-matched HTLV-1-positive controls, anti-Tax antibody was found to be low for the cases up to 10 years preceding their diagnosis (*9*). Mutations in the coding region of Tax may lead to the loss of Tax expression of infected cells, allowing such cells to escape cell lysis by Tax-specific CTL. Such a scenario could explain the ATL cases with low anti-Tax reactivities. A comparison of the Tax coding sequence before and after the onset of ATL must be made to address this question. Another important issue concerning risk factors for ATL is whether carriers with a high proviral load develop ATL more frequently than do carriers with a lower proviral level. A study is ongoing in the MCS to compare the prediagnostic HTLV-1 proviral DNA loads of incident cases and asymptomatic carriers.

We did not find an increased number of carrier subjects with HAM/TSP-like symptoms in the MCS (*1*), which may simply reflect the lower incidence of this disease in Japan (*16*). Concerning uveitis, arthropathy, and other chronic inflammatory diseases reported to be related with HTLV-1 infection, we also have found no positive association between those diseases and HTLV-1 infection in the MCS. However, the study was not specifically designed to detect the symptoms or signs of these diseases, especially at an early stage. We have observed subclinical suppression of cellular immunity, as indexed by decreased delayed-type hypersensitivity to purified protein derivative (PPD), in HTLV-1 carrier subjects (*20*). This PPD-specific anergy in HTLV-1-infected individuals appears to be due in part to the inability to respond to IL-12 (*30*). In addition, HTLV-1 positivity was (nonsignificantly) associated with the increased prevalence of reported histories of anemia, renal diseases, asthma, and cardiac disease as well as with an abnormal electrocardiogram, although the exact reason for these associations is not clear (*28*). Of interest, a negative association of HTLV-1 positivity with a history of reported peptic ulcer was also observed. Peptic ulcers are believed to be related to the host immune response to *Helicobacter pylori* infection, which could be impaired in HTLV-1 carriers. We are currently performing additional studies to determine the relationship between HTLV-1, *H. pylori* infection and ulcer disease.

Recently, the prevalence of hepatitis C virus (HCV) infection was found to be relatively high in one of the study villages, and a prospective study was performed to see if HTLV-1 co-infection affects the course of HCV-associated disease (*3*). Dual infection with HCV and HTLV-1 seemed to have a synergistic effect on the incidence of liver diseases and death from liver cancer. It was also shown in the MCS that co-infected subjects were more likely to have PPD anergy compared with those infected with HTLV-1 alone (*12*). These findings indicate that progressive immune suppression among co-infected subjects may result in the acceleration of HCV-associated liver disease progression.

Natural History of HTLV-1 Infection Based on the Findings of the MCS

Our understanding of the model of the natural history of HTLV-1 infection in Japan, based on the analysis of data gathered as part of the MCS, is summarized in Fig. 3. The major natural routes of infection of HTLV-1 are from mother to infant and between married partners. The level of proviral load is likely determined within a year of infection. The development of a humoral and, probably, cellular immune response to virus occurs during this period. The nature of the host immune response may affect the burden of HTLV-1, which varies widely among carriers. The proviral load appears to remain stable, once a set point has been achieved. The maintenance of the level of proviral load seems largely due to the relationship between the clonal expansion of HTLV-1-infected cells and the CTL response to them. The carriers with high proviral load, low anti-Tax immune response, and oligoclonal expansion of

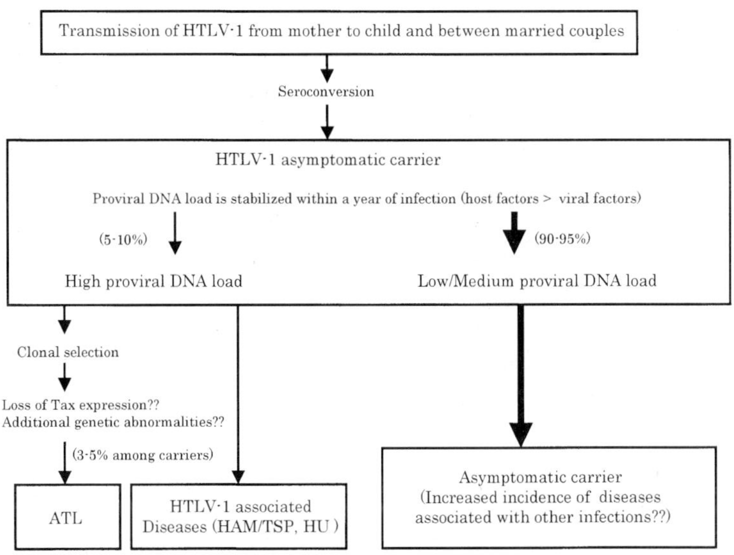

FIG. 3. A model of the natural history of HTLV-1 infection in Japan based on the findings of the MCS

HTLV-1-infected cells can be hypothesized to be at high risk to develop ATL. Of interest, carriers with the first two characteristics are more likely to be men. Genetic abnormalities have been found in ATL cells and likely are important for the malignant transformation of HTLV-1-infected cells (*18, 22, 23, 33*). Therefore, to predict the risk of ATL in each carrier, it is important to find a way to identify specific genetic abnormalities in the HTLV-1-infected cells of those carriers who have characteristic HTLV-1 markers.

Acknowledgments

This is a collaborative study between the Department of Epidemiology, Harvard School of Public Health (Dr. Nancy E. Muller as the former principal investigator), and the Department of Internal Medicine II, Miyazaki Medical College (Dr. Nobuyoshi Tachibana as the former principal investigator).

The authors are indebted to their many colleagues who have made substantial contributions to the work summarized in this review. These include Mr. Ali Abtahi, Ms. Brenda Birmann, Drs. Cynthia Boschi-Pinto, Michie Hisada, Ms. Karen Pawlish, Dr. Seth Welles (Department of Epidemiology, Harvard School of Public Health); Drs. Yi-ming A. Chen, Tun-Hou Lee, Max Essex (Department of Cancer Biology, Harvard School of Public Health); Dr. Renu Lal (Department of Human Retrovirology, Center for Diseases Control); Dr. Ken-ichiro Etoh (Second Department of Medicine, Kumamoto University); Dr. Masao Matsuoka (Institute for Viral Research, Kyoto University); Drs. Mutsunori Iga, Shiro Ishihara, Junzo Ishizaki, Koichi Murai, Shigemasa Shioiri, Muneo Suzuki, Hirohito Tsubouchi, Tsutomu Yokota (Department of Internal Medicine II, Miyazaki Medical College).

REFERENCES

1. Abtahi, A., Mueller, N. E., Okayama, A., and Stuver, S. O. Lack of evidence for a role of HTLV-I infection in the occurrence of subclinical HAM/TSP in the Miyazaki cohort study. *J. Acquir. Immune Defic. Syndr. Hum. Retrovirol.*, **24**, 86–87 (2000).
2. Arisawa, K., Soda, M., Endo, S., Kurokawa, K., Katamine, S., Shimokawa, I., Koba, T., Takahashi, T., Saito, H., Doi, H., and Shirahama, S. Evaluation of adult T-cell leukemia/lymphoma incidence and its impact on non-Hodgkin lymphoma incidence in southwestern Japan. *Int. J. Cancer*, **85**, 319–324 (2000).
3. Boschi-Pinto, C., Stuver, S., Okayama, A., Trichopoulos, D., Oray, E. J., Tsubouchi, H., and Mueller, N. A follow-up study of morbidity and mortality associated with hepatitis C virus infection and its interaction with human T lymphotropic virus type I in Miyazaki, Japan. *J. Infect. Dis.*, **181**, 35–41 (2000).
4. Cavrois, M., Leclercq, I., Gout, O., Gessain, A., Wain-Hobson, S., and Wattel, E. Persistent oligoclonal expansion of human T-cell leukemia virus type 1-infected circulating cells in patients with tropical spastic paraparesis/HTLV-1 associated myelopathy. *Oncogene,* **17**, 77-82 (1998).
5. Chen, Y. M., Okayama, A., Lee, T.-H., Tachibana, N., Mueller, N. E., and Essex, M. Sexual transmission of human T-cell leukemia virus type 1 associated with the presence of anti-Tax antibody. *Proc. Natl. Acad. Sci. USA*, **88**, 1182–1186 (1991).
6. Etoh, K., Tamiya, S., Yamaguchi, K., Okayama, A., Tsubouchi, H., Ideta, T., Mueller,

N., Takatsuki, K., and Matsuoka, M. Persistent clonal proliferation of human T-lymphotropic virus type I-infected cells *in vivo*. *Cancer Res.*, **57**, 4862–4867 (1997).

7. Etoh, K., Yamaguchi, K., Tokudome, S., Watanabe, T., Okayama, A., Stuver, S., Mueller, N., Takatsuki, K., and Matsuoka, M. Rapid quantification of HTLV-I provirus load: detection of monoclonal proliferation of HTLV-I-infected cells among blood donors. *Int. J. Cancer*, **81**, 859–864 (1999).
8. Gessain, A., Gallo, R. C., and Franchini, G. Low degree of human T-cell leukemia/lymphoma virus type I genetic drift *in vivo* as a means of monitoring viral transmission and movement of ancient human populations. *J. Virol.*, **66**, 2288–2295 (1992).
9. Hisada, M., Okayama, A., Shioiri, S., Spiegelman, D. L., Stuver, S. O., and Mueller, N. E. Risk factors for adult T-cell leukemia among carriers of human T-lymphotropic virus type I. *Blood*, **92**, 3557–3561 (1998).
10. Hisada, M., Okayama, A., Spiegelman, D., Mueller, N. E., and Stuver, S. O. Sex-specific mortality from adult T-cell leukemia among carriers of human lymphotropic virus type I. *Int. J. Cancer*, **91**, 497–499 (2001).
11. Hisada, M., Okayama, A., Tachibana, N., Stuver, S. O., Spiegelman, D. L., Tsubouchi, H., and Mueller, N. E. Predictors of level of circulating abnormal lymphocytes among human T-lymphotropic virus type I carriers in Japan. *Int. J. Cancer*, **77**, 188–192 (1998).
12. Hisada, M., Shima, T., Okayama, A., Mueller, N., Tsubouchi, H., and Stuver, S. Suppression of skin reactivity to purified protein derivative by hepatitis C virus among HTLV-I carriers in Japan. *J. Acquir. Immune Defic. Syndr. Hum. Retrovirol.*, **19**, 421–425 (1998).
13. Iga, M., Okayama, A., Stuver, S. O., Matsuoka, M., Mueller, N. E., Aoki, M., Mitsuya, H., Tachibana, N., and Tsubouchi, H. Genetic evidence of transmission of human T-lymphotropic virus type-I between spouses. *J. Infect. Dis.*, **185**, 691–695 (2002).
14. Ishihara, S., Okayama, A., Horinouchi, H., Shioiri, S., Murai, K., Kubota, T., Yamashita, R., Tachibana, N., Tsubouchi, H., Stuver, S., and Mueller, N. Association of HTLV-I antibody profile of asymptomatic carriers with proviral DNA levels of peripheral blood mononuclear cells. *J. Acquir. Immune Defic. Syndr.*, **7**, 199–203 (1994).
15. Kannagi, M., Matsushita, S., and Harada, S. Expression of the target antigen for cytotoxic lymphocytes on adult T-cell-leukemia cells. *Int. J. Cancer*, **54**, 582–588 (1993).
16. Kaplan, J. E., Osame, M., Kubota, H., Igata, A., Nishitani, H., Maeda, Y., Khabbaz, R. F., and Janssen, R. S. The risk of development of HTLV-I-associated myelopathy/tropical spastic paraparesis among persons infected with HTLV-I. *J. Acquir. Immune Defic. Syndr.*, **3**, 1096–1101 (1990).
17. Manns, A., Miley, W. J., Wilks, R. J., Morgan, O. S., Hanchard, B., Wharfe, G., Cranston, B., Maloney, E., Welles, S. L., Blattner, W. A., and Waters, D. Quantitative proviral DNA and antibody levels in the natural history of HTLV-I infection. *J. Infect. Dis.*, **180**, 1487–1493 (1999).
18. Mortreux, F., Leclercq, I., Gabet, A. S., Leroy, A., Westhof, E., Gessain, A., and Wain-Hobson, S., and Wattel, E. Somatic mutation in human T-cell leukemia virus type 1 provirus and flanking cellular sequences during clonal expansion *in vivo*. *J. Natl. Cancer Inst.*, **93**, 367–377 (2001).
19. Mueller, N., Okayama, A., Stuver, S., and Tachibana, N. Findings from the Miyazaki

Cohort Study. *J. Acquir. Immune Defic. Syndr. Hum. Retrovirol.*, **13-1**, S2–S7 (1996).

20. Murai, K., Tachibana, N., Shioiri, S., Shishime, E., Okayama, A., Ishizaki, J., Tsuda, K., and Mueller, N. Suppression of delayed-type hypersensitivity to PPD and PHA in elderly HTLV-I carriers. *J. Acquir. Immune Defic. Syndr.*, **3**, 1006–1009 (1990).
21. Murphy, E. L., Wilks, R., Morgan, O. S., Hanchard, B., Cranston, B., Figueroa, J. P., Gibbs, W. N., Murphy, J., and Blattner, W. A. Health effects of human T-lymphotropic virus type I (HTLV-I) in a Jamaican cohort. *Int. J. Epidemiol.*, **25**, 1090–1097 (1996).
22. Nosaka, K., Maeda, M., Tamiya, S., Sakai, T., Mitsuya, H., and Matsuoka, M. Increasing methylation of the CDKN2A gene is associated with the progression of adult T-cell leukemia. *Cancer Res.*, **6**, 1043–1048 (2000).
23. Okamoto, T., Ohno, Y., Tsugane, S., Watanabe, S., Shimoyama, M., Tajima, K., Miwa, M., and Shimotohno, K. Multi-step carcinogenesis model for adult T-cell leukemia. *Jpn. J. Cancer Res.*, **80**, 191–195 (1989).
24. Okayama, A., Stuver, S., Iga, M., Okamoto, M., Mueller, N., Matsuoka, M., Yamaguchi, K., Tachibana, N., and Tsubouchi, H. Sequential change of viral markers in seroconverters with community-acquired infection of human T-lymphotropic virus type 1. *J. Infect. Dis.*, **183**, 1031–1037 (2001).
25. Okayama, A., Tachibana, N., Ishihara, S., Nagatomo, Y., Murai, K., Okamoto, M., Shima, T., Sagawa, K., Tsubouchi, H., Stuver, S. O., and Mueller, N. E. Increased expression of interleukin-2 receptor α on peripheral blood mononuclear cells in HTLV-I tax/rex mRNA-positive asymptomatic carriers. *J. Acquir. Immune Defic. Syndr. Hum. Retrovirol.*, **15**, 70–75 (1997).
26. Plancoulaine, S., Buigues, R. P., Murphy, E. L., van Beveren, M., Pouliquen, J. F., Joubert, M., Remy, F., Tuppin, P., Tortevoye, P., de The, G., Moreau, J. P., and Gessain, A. Demographic and familial characteristics of HTLV-1 infection among an isolated, highly endemic population of African origin in French Guiana. *Int. J. Cancer*, **76**, 331–336 (1998).
27. Shioiri, S., Stuver, S., Okayama, A., Murai, K., Shima, T., Tachibana, N., Tsubouchi, H., Essex, M., and Mueller, N. E. Intrafamilial transmission of HTLV-I and its association with anti-Tax antibody in an endemic population in Japan. *Int. J. Cancer*, **75**, 15–18 (1998).
28. Stuver, S., Tachibana, N., Okayama, A., and Mueller, N. Evaluation of morbidity among human T lymphotropic virus type I carriers in Miyazaki, Japan. *J. Infect. Dis.*, **173**, 584–591 (1997).
29. Stuver, S. O., Tachibana, N., Okayama, A., Shioiri, S., Tsunetoshi, Y., Tsuda, K., and Mueller, N. E. Heterosexual transmission of human T cell leukemia/lymphoma virus type 1 among married couples in southwestern Japan: an initial report from the Miyazaki Cohort Study. *J. Infect. Dis.*, **167**, 57–65 (1993).
30. Suzuki, M., Dezzutti, S. C., Okayama, A., Tachibana, N., Tsubouchi, H., Mueller, N. E., and Lal, B. R. Modulation of T-cell responses to a recall antigen in human T-cell leukemia virus type I-infected individuals. *Clin. Diagn. Lab. Immunol.*, **6**, 713–717 (1999).
31. Tachibana, N., Okayama, A., Ishihara, S., Shioiri, S., Murai, K., Tsuda, K., Goya, N., Matsuo, Y., Essex, M., Stuver, S. O., and Mueller, N. E. High HTLV-I proviral DNA

level associated with abnormal lymphocytes in peripheral blood from asymptomatic carriers. *Int. J. Cancer*, **51**, 593–595 (1992).

32. Tajima, K., Kamura, S., Ito, S., Ito, M., Nagatomo, M., Kinoshita, K., and Ikeda, S. Epidemiological features of HTLV-I carriers and incidence of ATL in an ATL-endemic island: a report of the community-based co-operative study in Tsushima, Japan. *Int. J. Cancer*, **40**, 741–746 (1987).
33. Watanabe, T. HTLV-1 associated diseases. *Int. J. Hematol.*, **66**, 257–278 (1997).
34. Wattel, E., Cavrois, M., Gessain, A., and Wain-Hobson, S. Clonal expansion of infected cells: a way of life for HTLV-I. *J. Acquir. Immune Defic. Syndr. Hum. Retrovirol.*, **13** (Suppl. 1), S92-S99 (1996).
35. Wells, S. L., Tachibana, N., Okayama, A., Murai, K., Shioiri, S., Sagawa, K., Katagiri, K., and Mueller, N. The distribution of T-cell subsets among HTLV-I carriers in Japan. *J. Acquir. Immune Defic. Syndr.*, **7**, 509–516 (1994).
36. Yashiki, S., Fujiyoshi, T., Arima, N., Osame, M., Yoshinaga, M., Nagata, Y., Tara, M., Nomura, K., Utsunomiya, A., Hanada, S., Tajima, K., and Sonoda, S. Hla-a*26, hla-b*4002, hla-b*4006, and hla-b*4801 alleles predispose to adult T cell leukemia: the limited recognition of HTLV type 1 tax peptide anchor motifs and epitopes to generate anti-HTLV type 1 tax cd8+ cytotoxic T lymphocytes. *AIDS Res. Hum. Retroviruses*, **17**, 1047–1061 (2001).
37. Yoshida, M. Multiple viral strategies of HTLV-1 for dysregulation of cell growth control. *Annu. Rev. Immunol.*, **19**, 475–496 (2001).

MOLECULAR PATHOLOGY OF HTLV-I-ASSOCIATED MYELOPATHY/TROPICAL SPASTIC PARAPARESIS

Fujio UMEHARA and Mitsuhiro OSAME

*The Third Department of Internal Medicine, Kagoshima University School of Medicine**

Human T lymphotropic virus type I (HTLV-I) is associated with adult T cell leukemia (ATL) and a chronic progressive disease of the central nervous system (CNS) termed HTLV-I-associated myelopathy (HAM)/tropical spastic paraparesis (TSP) (*4, 16*). The pathological analysis indicates that the disease affects the spinal cord, predominantly at the thoracic level. Macroscopic findings revealed atrophy of the spinal cord in the lateral column of the thoracic level. In the spinal cord, degeneration of the lateral corticospinal tract as well as of the spinocerebellar or spinothalamic tract of the lateral column also showed degeneration (Fig. 1A). These lesions are associated with perivascular and parenchymal lymphocytic infiltration with the presence of foamy macrophages, proliferation of astrocytes, and fibrillary gliosis (Fig. 1B) (*7, 13, 14*). There is widespread loss of myelin and axons, particularly in the corticospinal tracts of the spinal cord. Damage is most severe in the lower thoracic to lumbar regions. These findings are consistent with a patient's neurological symptoms of paraparesis, spasticity, hyperreflexia or Babinski's sign. In this chapter, we discuss the immunological and molecular analysis of the CNS lesions of HAM/TSP.

T Cell Subset and Cytokine Expression

Patients whose illness was of short duration (2.5–4.5 years) showed parenchymal lesions with marked inflammatory and degenerative changes in both lateral funiculi (henceforce referred to as active-chronic lesions), where CD4+ cells, CD8+ cells and macrophages were evenly distributed (*21*). Immunohistochemistry showed that inflammatory cytokines, including tumor necrosis factor-α (TNF-α), interleukin-1β (IL-1β) and interferon-γ (IFN-γ) were expressed on perivascular infiltrating macrophages, astrocytes and microglia (*22*). In striking contrast, in patients whose duration of illness was from 8 to 10 years, the spinal cord showed monotonous degeneration of both lateral funiculi with a few inflammatory cells in the subarachnoid and perivascular spaces (henceforce termed inactive-chronic lesions). In inactive-chronic lesions, we noted a predominance of CD8+ cells over CD4+ cells, however, pro-inflammatory cytokine expressions were down-regulated compared with those of active-chronic lesions. These studies suggest that immune responses in the spinal cord lesions of HAM/TSP patients gradually change along with the duration of illness.

* 8-35-1, Sakuragaoka Kagoshima 890-0075, Japan

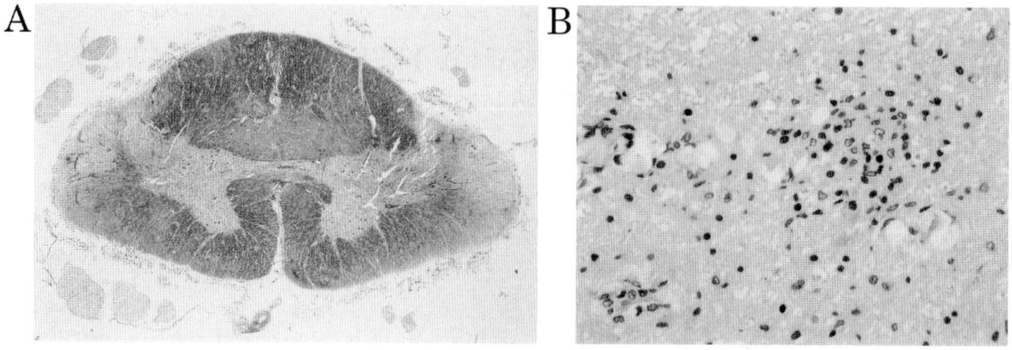

FIG. 1. Pathology of the spinal cord lesions of HAM/TSP
A. Transverse section of the spinal cord at the middle thoracic level. Note the marked myelin loss in both lateral columns. Klüber-Barella staining (kindly provided by Prof. S. Izumo). B. Inflammatory cell infiltration in perivascular space and parenchyma. Hematoxylin-eosin staining.

Role of Macrophages/Microglias

The monocyte/macrophage system has a wide range of immunological effector functions in antigen presentation and the production of cytokines or proteolytic enzymes. Therefore, it is suggested that macrophages/microglias may play an important role in immune reaction of HAM/TSP. We have analyzed the various patterns of macrophage/microglia activation using various monoclonal antibodies recognizing the different stages of activation. Many macrophages/microglias express both MRP8 and MRP14 in active-chronic inflammatory lesions of patients with illness of short duration (*1*). In contrast, MRP-8 and MRP-14 were expressed only in intravascular monocytes/macrophages, but negative in the perivascular space or parenchyma of inactive-chronic lesions. This suggests that many hematogenous macrophages were recruited in the active-chronic lesions, and both macrophages and microglias were chronically activated. In addition, monocyte/macrophage recruitment and activation is also down-regulated along with the duration of illness.

CD8+ Cytotoxic T Lymphocytes (CTLs) Infiltrated the CNS of HAM/TSP

As the pathogenesis of HAM/TSP, CD8+ CTLs against HTLV-I have been considered as the effector cells (*8*). To confirm the role of CD8+ CTLs in the formation of CNS lesions, we analyzed the distribution of TIA-1+ cells in the spinal cord lesions of HAM/TSP. A novel monoclonal antibody (mAb), designated TIA-1 (*2*), recognizes a 15-kD granule associated protein, the expression of which is restricted to CTLs and natural killer (NK) cells. In active-chronic lesions, many TIA-1+ cells were distributed throughout the parenchyma and perivascular cuffs (*23*); dual immunolabeling revealed that about 80% of them coexpressed the CD8 antigen. In contrast, TIA-1+ cells were scarcely observed in inactive-chronic lesions, though CD8+ cells dominated in the parenchyma and perivascular cuffs. The number of TIA-1+ cells correlated with the amount of HTLV-I proviral DNA *in situ* (Fig. 2). The protein TIA-1 has been associated with the induction of apoptosis in target cells (*20*). In active inflammatory lesions we found cells undergoing apoptosis, most of

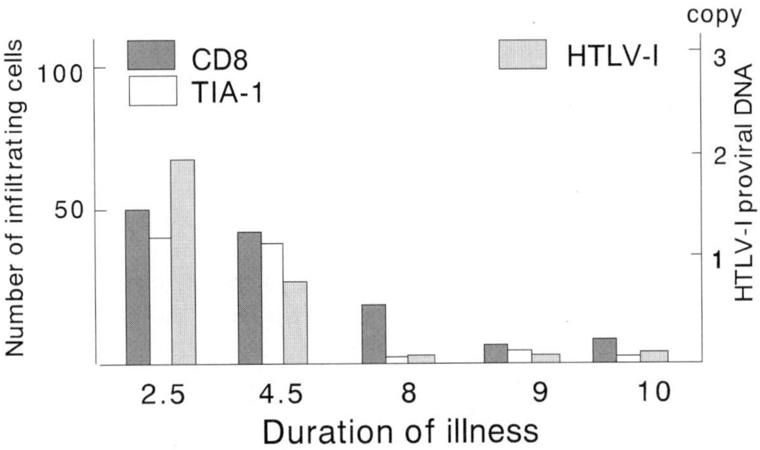

FIG. 2. Morphometric analysis of inflammatory cells in the spinal cord
In active-inflammatory lesions, most CD8+ cells are immunoreactive for TIA-1. The number of CD8+ TIA+ cells is related to the amount of HTLV-I proviral DNA *in situ*. In inactive-lesions which contain a small amount of HTLV-I proviral DNA, the number of CD8+TIA+ cells was markedly decreased.

them identified as helper-inducer CD45RO+ T lymphocytes, which were consistent with *in vivo* cellular tropism of HTLV-I in patients with HAM/TSP (*18*). These findings suggest that HTLV-I specific CD8+ CTL-induced apoptosis of CD4+ T lymphocytes may be one possible mechanism to eliminate HTLV-I infected cells from the CNS.

Fas/FasL Expression

HTLV-I specific CD8+ CTLs are crucial for viral clearance in HTLV-I infection. In general, CD8+ CTLs exert antiviral effector functions via two basic mechanisms: (1) the exocytosis of perforin-containing granules on cognate target cells; and (2) the engagement of Fas on cognate or neighboring target cells by membrane-bound or released FasL. Fas/FasL interaction regulates a major pathway in apoptosis, which may play an important role in mediating both the antiviral effects and the inflammatory process in neurological diseases. The level of soluble Fas is increased in the sera of patients with HAM compared with controls (*6*). The levels of soluble FasL are higher either in the sera (*6*) or the cerebrospinal fluid (CSF) (*19*) of patients in the active stage of HAM. In addition, FasL mRNA expression is up-regulated in peripheral blood T lymphocytes (*9*). In the spinal cord lesions of HAM/TSP, Fas was preferentially expressed on infiltrating T cells in active-chronic lesions (*27*). FasL expression was up-regulated on various cells, mainly microglias/macrophages in these lesions. These findings suggest that macrophages/microglias may play a key role in the elimination of activated effector T cells in the CNS of HAM/TSP. Therefore, increased FasL expression by macrophages/microglias in active-inflammatory lesions of HAM/TSP may represent an immunological response against infiltration of Fas+ T cells in the CNS. In contrast, FasL expression was markedly down-regulated in inactive-chronic lesions of HAM patients who had an illness of long-duration. Taken together, our data suggest that the Fas/FasL system may play a role in the down-regulation of the immune reaction in the CNS of HAM/TSP.

Localization of HTLV-I in the CNS of HAM/TSP

Our series of work suggests that HTLV-I infected CD4+ T lymphocytes enter the CNS driving local expansion of virus specific CD8+ CTLs, which along with cytokine causes pathological changes. Therefore, it was of interest to determine which cells might be targets of CTLs in the CNS. Using semi-quantitative PCR, HTLV-I *pX* and *pol* sequences were found to be increased in the thoracic cord areas where CD4+ cells predominated (*10*). This proviral load decreased with the increased period of the patient's disease which was paralleled by the number of infiltrating CD4+ cells. By a novel *in situ* PCR technique, HTLV-I DNA was localized to inflammatory CD4+ cells (*12*). Using the same CNS samples, an *in situ* hybridization study accurately localized HTLV-I tax mRNA to infiltrating CD4+ T lymphocytes in active lesions in CNS specimens from HAM/TSP patients (*15*). Taken together, these findings suggest that the main harbor of HTLV-I virus may be infiltrating CD4+ T lymphocytes, and transcription of *tax* gene occurred in some of the HTLV-I infected CD4+ T lymphocytes.

However, there are controversial reports which showed that HTLV-I tax mRNA was localized within the neural tissue (some of them astrocytes), but not in perivascular infiltrates (*11*). The reasons for these differences are unknown, but they may be related to variations in samples or detection methods.

T Cell Trafficking into the CNS of HAM/TSP

Leukocyte adhesion molecules to endothelium play an important role in the pathogenesis of inflammatory diseases, including HAM/TSP. The spinal cord lesions of HAM/TSP had greater vascular cell adhesion molecule-1 (VCAM-1) expression on endothelium compared with those of controls (*24*). Infiltrating mononuclear cells, especially perivascular lesions, expressed very late antigen-4 (VLA-4). Monocyte chemoattractant protein-1 (MCP-1) was also up-regulated on perivascular infiltrating cells and vascular endothelium in active-chronic inflammatory lesions of HAM/TSP. These findings suggest that VLA-4/VCAM-1 interaction and MCP-1 are involved in mediating T-lymphocyte/macrophage adhesion and chemotaxis in the CNS inflammatory process (Fig. 3).

After transendothelial migration, T cells/macrophages encounter the extracellular matrix (ECM), must pass through the basement membrane and migrate into the interstitial matrix. Proteolytic disruption of ECM by matrix metalloproteinases (MMPs) is a key process for the damage of the blood brain barrier (BBB). MMPs have been reported to be involved in inflammatory disorders of the CNS. Immunohistochemical studies revealed that collagen IV and decorin immunoreactivity on the basement membrane of CNS parenchymal vessels was partially disrupted where inflammatory mononuclear cells infiltrated active-chronic lesions of HAM/TSP (*25*). In these lesions, MMP-2 (gelatinase A) was immunostained mainly on the surface of foamy macrophages and lymphocytes, whereas MMP-9 (gelatinase B) expression was positive in the intravascular and perivascular mononuclear cells but not on foamy

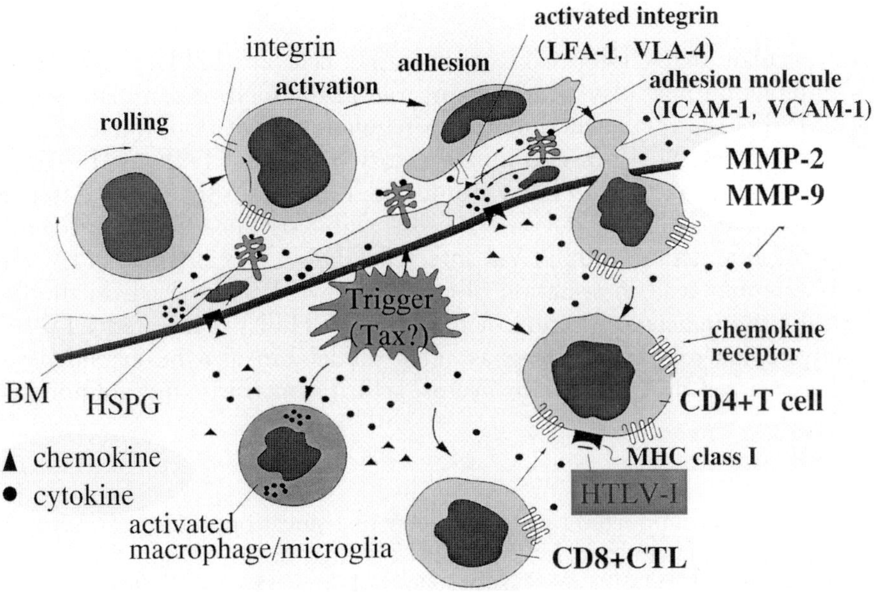

FIG. 3. Scheme of the CNS inflammatory lesions of HAM/TSP
BM, basement membrane; HSPG, heparan sulfate proteoglycan.

macrophages. In contrast, inactive-chronic lesions of the spinal cords of the HAM/TSP contained a much smaller number of MMP-2-positive or MMP-9-positive mononuclear cells than active-chronic lesions. Production levels of MMP-2 and MMP-9 in both sera and CSF were higher in the patients with HAM/TSP than those in non-inflammatory other neurological disease controls (ONDs). Using zymography, proMMP-9 was more frequently detected in the CSF of patients with HAM/TSP than those in ONDs (*5, 25*). Taken together, our data indicate that MMP-2 and MMP-9 may play an important role in the BBB breakdown and tissue remodeling in the CNS of HAM/TSP.

Axonal Degeneration in Spinal Cord Lesions of HAM/TSP

Previous neuropathological studies revealed that both myelin destruction and axonal loss are histological hallmarks of actively inflamed lesions of HAM/TSP. Particularly, axonal loss is responsible for the persistent disability characterized by spastic paraparesis and urinary disturbance. We investigated the localization and extent of β-amyloid precursor protein (APP) immunoreactivity as a sensitive marker for impairment of fast axonal transport in the spinal cords of HAM/TSP (*3, 17*). The results from this study show that APP, used as a marker of early axonal damage in HAM/TSP lesions, is more intensively expressed in areas of active-inflammatory lesions than those of inactive-chronic lesions (*3, 26*). The close localization to the areas containing inflammation (activation of macrophage/microglia) is striking and suggests that axonal damage is closely associated with inflammation in active-chronic lesions.

Conclusion

We speculate that CD8+ CTL versus target cell (i.e., HTLV-I infected CD4+ T cells) interaction may play an important role in the CNS inflammatory process in HAM/TSP. HTLV-I tax will induce up-regulation of various molecules such as adhesion molecules, MMP-9, and inflammatory cytokines (TNF-α, IL-1β, IFN-γ), which result in infiltration of HTLV-I infected T cells into the CNS. After trafficking into the CNS, HTLV-I infected T cells exhibit significant viral antigen expression, and the interaction of HTLV-I infected CD4+ T cells and HTLV-I specific CD8+ CTLs may lead to secretion of cytokines, MMPs and FasL in the CNS. In addition, the spontaneous secretion of IFN-γ from HTLV-I infected CD4+ T cells could activate macrophages/microglias. This model can also be extended to other HTLV-I-associated diseases such as uveitis, arthritis, myositis, and pulmonary alveolitis.

Acknowledgments

This study was supported by the Ministry of Education, Culture, Sports, Science and Technology of Japan, and by the Naito Foundation, the Brain Science Foundation, and the Welfide Medicinal Research Foundation. We thank Professor S. Izumo (Division of Molecular Pathology, Chronic Viral Disease Research Center, Kagoshima University) and many members of the Third Department of Internal Medicine, Kagoshima University, who contributed to the study of HAM/TSP.

REFERENCES

1. Abe, M., Umehara, F., Kubota, R., Moritoyo, T., Izumo, S., and Osame, M. Activation of macrophage/microglias with expression of MRP8 and MRP14 is associated with the lesional activities in the spinal cord lesions in HTLV-I associated myelopathy. *J. Neurol.*, **246**, 358–364 (1999).
2. Anderson, P., Nagler-Anderson, C., O'Brien, C., Levine, H., Watkins, S., Slayter, H. S., Blue, M. L., and Schlossman, S. F. A monoclonal antibody reactive with a 15-kDa cytoplasmic granule-associated protein defines a subpopulation of CD8+ T lymphocytes. *J. Immunol.*, **144**, 574–582 (1990).
3. Gentleman, S. M., Nash, M. J., Sweeting, C. J., Graham, D. I., and Roberts, G. W. β-Amyloid precursor protein (β-APP) as a marker for axonal injury after head injury. *Neurosci. Lett.*, **160**, 139–144 (1993).
4. Gessain, A., Barin, F., Vernant, J. C., Gout, O., Maurs, L., Calender, A., and de-The, G. Antibodies to human T-lymphotropic virus type-I in patients with tropical spastic paraparesis. *Lancet*, **ii**, 407–410 (1985).
5. Giraudon, P., Burt, S., Bernard, A., Thomas, N., and Beeline, M. F. Extracellular matrix-remodeling metalloproteinases and infection of the central nervous system with retrovirus human T-lymphotropic virus type I (HTLV-I). *Prog. Neurobiol.*, **49**, 169–184 (1996).
6. Inoue, A., Koh, C. S., Sakai, T., Yamazaki, M., Yanagisawa, N., Usuku, K., and Osame, M. Detection of the soluble form of the Fas molecule in patients with multiple sclerosis and human T-lymphotropic virus type I-associated myelopathy. *J. Neuroimmunol.*, **75**, 141–146 (1997).

7. Iwasaki, Y. Pathology of chronic myelopathy associated with HTLV-I infection (HAM/TSP). *J. Neurol. Sci.*, **96**, 103–123 (1990).
8. Jacobson, S., Shida, H., McFarlin, D. E., Fauci, A. S., and Koenig, S. Circulating CD8+ cytotoxic T lymphocytes specific for HTLV-I *pX* in patients with HTLV-I associated neurological disease. *Nature*, **348**, 245–248 (1990).
9. Kawahigashi, N., Furukawa, Y., Saito, M., Usuku, K., and Osame, M. Predominant expression of Fas ligand mRNA in CD8+ T lymphocytes in patients with HTLV-I-associated myelopathy. *J. Neuroimmunol.*, **90**, 199–206 (1998).
10. Kubota, R., Umehara, F., Izumo, S., Ijichi, S., Matsumuro, K., Yashiki, S., Fujiyoshi, T., Sonoda, S., and Osame, M. HTLV-I proviral DNA amount correlates with infiltrating CD4+ lymphocytes in the spinal cord from patients with HTLV-I-associated myelopathy. *J. Neuroimmunol.*, **53**, 23–29 (1994).
11. Lehky, T. J., Fox, C. H., Koenig, S., Levin, M. C., Flerlage, N., Izumo, S., Sato, E., Raine, C. S., Osame, M., and Jacobson, S. Detection of human T-lymphotropic virus type I (HTLV-I) tax mRNA in the central nervous system of HTLV-I-associated myelopathy/tropical spastic paraparesis patients by *in situ* hybridization. *Ann. Neurol.*, **37**, 167–175 (1995).
12. Matsuoka, T., Takenouchi, N., Hashimoto, K., Kashio, N., Moritoyo, T., Higuchi, I., Isashiki, Y., Sato, E., Osame, M., and Izumo, S. Perivascular T cells are infected with HTLV-I in the spinal cord lesions with HTLV-I-associated myelopathy/tropical spastic paraparesis: double staining of immunohistochemistry and polymerase chain reaction *in situ* hybridization. *Acta Neuropathol.*, **96**, 334–346 (1998).
13. MoeMoe, Aye., Matsuoka, E., Moritoyo, T., Umehara, F., Suehara, M., Hokezu, Y., Yamanaka, H., Isashiki, Y., Osame, M., and Izumo, S. Histopathological analysis of four autopsy cases of HTLV-I-associated myelopathy/tropical spastic paraparesis: inflammatory changes occur simultaneouly in the entire central nervous system. *Acta Neuropathol.*, **100**, 245–252 (2000).
14. Moore, G. R. W., Traugott, U., Scheinberg, L. C., and Raine, C. S. Tropical spastic paraparesis: a model of virus-induced, cytotoxic T-cell mediated demyelination? *Ann. Neurol.*, **26**, 523–530 (1989).
15. Moritoyo, T., Reinhart, T. A., Moritoyo, H., Sato, E., Izumo, S., Osame, M., and Haase, A. T. Human T-lymphotropic virus type I-associated myelopathy and tax gene expression in CD4+ T lymphocytes. *Ann. Neurol.*, **40**, 84–90 (1996).
16. Osame, M., Matsumoto, M., Usuku, K., Izumo, S., Ijichi, N., Amitani, H., Tara, M., and Igata, A. Chronic progressive myelopathy associated with elevated antibodies to human lymphotropic virus type-I and adult T cell leukemia-like cells. *Ann. Neurol.*, **21**, 117–122 (1987).
17. Povlinshock, J. T. Traumatically induced axonal injury: pathogenesis and pathological implications (Review). *Brain Pathol.*, **2**, 1–12 (1992).
18. Richardson, J. H., Edwards, A. J., Cruickshank, J. K., Rudge, P., and Dalgleish, A. G. *In vivo* cellular tropism of human-T-cell leukemia virus type 1. *J. Virol.*, **64**, 5682–5687 (1990).
19. Saito, M., Nakamura, N., Nagai, M., Shirakawa, K., Sato, H., Kawahigashi, N., Furukawa, Y., Usuku, K., Nakagawa, M., Izumo, S., and Osame, M. Increased levels of soluble Fas ligand in CSF of rapidly progressive HTLV-I-associated myelopathy/tropi-

cal spastic paraparesis. *J. Neuroimmunol.*, **98**, 221–226 (1998).

20. Tian, Q., Streuli, M., Saito, H., Schlossman, S. F., and Anderson, P. A polyadenylate binding protein localized to the granules of cytolytic lymphocytes induces DNA fragmentation in target cells. *Cell*, **67**, 629–639 (1991).
21. Umehara, F., Izumo, S., Nakagawa, M., Ronquillo, A. T., Takahashi, K., Sato, E., and Osame, M. Immunocytochemical analysis of the cellular infiltrate in the spinal cord lesions in HTLV-I-associated myelopathy. *J. Neuropathol. Exp. Neurol.*, **53**, 424–430 (1993).
22. Umehara, F., Izumo, S., Ronquillo, Q. T., Matsumuro, K., and Osame, M. Cytokine expression in the spinal cord lesions in HTLV-I associated myelopathy. *J. Neuropathol. Exp. Neurol.*, **53**, 72–77 (1994).
23. Umehara, F., Nakamura, A., Izumo, S., Kubota, R., Ijichi, S., Kashio, N., Hashimoto, K., Usuku, K., Sato, E., and Osame, M. Apoptosis of T lymphocytes in the spinal cord lesions in HTLV-I-associated myelopathy: a possible mechanism to control viral infection in the central nervous system. *J. Neuropathol. Exp. Neurol.*, **53**, 617–624 (1994).
24. Umehara, F., Izumo, S., Takeya, M., Sato, E., and Osame, M. Expression of adhesion molecules and monocyte chemoattractant protein-1 (MCP-1) in the spinal cord lesions in HTLV-I-associated myelopathy. *Acta Neuropathol.*, **91**, 343–350 (1996).
25. Umehara, F., Okada, Y., Fujimoto, N., Abe, M., Izumo, S., and Osame, M. Expression of matrix metalloproteinases and tissue inhibitors of metalloproteinases in HTLV-I-associated myelopathy. *J. Neuropathol. Exp. Neurol.*, **57**, 839–849 (1998).
26. Umehara, F., Abe, M., Koreeda, Y., Izumo, S., and Osame, M. Axonal damage revealed by accumulation of β-amyloid precursor protein in HTLV-I-associated myelopathy. *J. Neurol. Sci.*, **176**, 95–101 (2000).
27. Umehara, F., Itoh, K., Michizono, K., Abe, M., Izumo, S., and Osame, M. Involvement of Fas/Fas ligand system in the spinal cords of HTLV-I-associated myelopathy. *Acta Neuropathol.*, **8**, 359–364 (2002).

GENETICS AND DYNAMICS OF THE IMMUNE RESPONSE TO HTLV-I

Charles R.M. BANGHAM

*Department of Immunology, Wright-Fleming Institute, Imperial College**

The existence of a remarkably strong cellular immune response to human T cell leukemia virus type I (HTLV-I) in both asymptomatic carriers of the virus and patients with HTLV-I-associated myelopathy/tropical spastic paraparesis (HAM/TSP) implies persistent expression of HTLV-I proteins *in vivo*. Yet it is hard to detect HTLV-I proteins or mRNA in freshly isolated lymphocytes from infected people; and the HTLV-I genome varies relatively little in sequence, implying that the error-prone viral reverse transcriptase contributes little to the maintenance of the high proviral load that is typically observed. The cell-mediated immune response is usually necessary, and often sufficient, to limit the replication of a virus *in vivo* and to prevent disease. But the presence of a dense T lymphocyte infiltrate in the inflammatory lesions in HAM/TSP (and the other inflammatory conditions associated with HTLV-I) suggests that lymphocytes, perhaps including HTLV-I-specific cells, contribute to the tissue damage in these diseases. Because of these seemingly conflicting observations, we have been interested in two fundamental questions: is HTLV-I latent or is it persistently transcribed? And what is the role of the cell-mediated immune response: protective, pathogenic, or both?

I summarize here evidence that the cytotoxic T lymphocyte (CTL) response to HTLV-I is an important determinant of the proviral load and the risk of HAM/TSP. Our observations suggest the following picture of HTLV-I infection: HTLV-I is not latent, but is persistently transcribed *in vivo*; however, the strong CTL response to HTLV-I limits the ability of the virus to complete its replication cycle. I suggest that the efficiency of anti-HTLV-I CTLs is genetically determined and varies between individuals: a person with an efficient specific CTL response successfully controls HTLV-I replication, but in an individual with an inefficient CTL response, the CTLs may contribute to inflammatory tissue damage in diseases such as HAM/TSP. These conclusions suggest an approach to the treatment and prevention of HTLV-I-associated diseases.

The Innate and Acquired Immune Response to Human T Cell Leukemia Virus Type I (HTLV-I)

The specific immune response to viruses (*42*) results from the selective stimulation and growth of clones of B lymphocytes that produce specific antibodies, and T lymphocytes that kill virus-infected host cells. Stimulation of lymphocytes is accom-

* London W2 1PG, UK

plished by potent "antigen presenting cells" known as dendritic cells, and is assisted by soluble cytokines released from antigen-specific helper T lymphocytes. In a non-immune host, only about 1 in 10^5 to10^6 lymphocytes recognizes each antigenic determinant on a viral protein. It therefore takes several (typically 3–7) days for the specific lymphocytes to reach a sufficiently large number to mount an effective defence against the virus. During this early phase of infection, the host relies on the non-antigen-specific innate immune response to restrict virus spread. This innate response to viruses consists chiefly of interferons, which are cytokines that halt viral replication in the cell, and natural killer (NK) cells, which kill host cells that lack surface expression of major histocompatibility complex (MHC) molecules—human leukocyte antigen (HLA) in the human.

Little work has been carried out on the role of interferons on HTLV-I replication (e.g. (*54*)), except in HTLV-I transformed cells; this subject deserves more attention. The importance of NK cells in HTLV-I infection is also poorly understood: Yu *et al.* (*74*) reported that both the frequency and the activity of $CD56^+$ (NK) cells were lower in HAM/TSP (HTLV-I-associated myelopathy/tropical spastic paraparesis) patients than in healthy carriers of HTLV-I, suggesting that a weak NK cell response contributes to the higher proviral load seen in HAM/TSP. But Hanon *et al.* (*19*) found that $CD56^+$ cells caused little death of HTLV-I-Tax-expressing cells in fresh peripheral blood mononuclear cells (PBMCs) within 24 h, whereas $CD8^+$ T lymphocytes (cytotoxic T lymphocytes, or CTLs) efficiently killed the Tax-positive cells. However, no unique surface marker has been found that identifies human NK cells, and many $CD8^+$ T cells have NK-like activity. A full understanding of the role of NK cells in HTLV-I infection will therefore require a thorough phenotypic and functional description of the lymphocyte subpopulations.

The antibody response in HTLV-I infection is directed mainly at the Gag protein in the first 2 months (*40, 45*). Antibodies to the Env protein then appear, and about 50% of individuals also produce detectable antibodies to the Tax protein (*40, 45*). Antibodies usually play little role in defence against persistent, non-cytolytic viruses (*75*), and it is likely that the correlation observed between the specific anti-HTLV-I antibody titre and the proviral load (*45, 46*) passively reflects the viral antigen concentration.

Helper T cells (Th cells), which carry the CD4 protein on their surface, are essential for an efficient antibody and CTL response to most viruses (*42, 56*). $CD4^+$ T cells predominate in active inflammatory lesions in HAM/TSP (*25*), and a genetic haplotype associated with the *DRB1*0101* (*DR1*) allele was found to carry an increased risk of HAM/TSP in southern Japan (*28, 64, 65*). Horiuchi *et al.* (*22*) found that the inflammatory (Th1) type of helper T cells predominated in the circulation in HAM/TSP patients. Recently, we have shown (*18*) that the onset of Tax expression in $CD4^+$ T cells is rapidly followed by the production of interferon-γ in these cells.

These observations suggest that $CD4^+$ Th cells could initiate and maintain inflammatory lesions in HTLV-I-associated diseases. But although a DR1-restricted epitope has been identified in the Env protein (*34, 72*), there has been no systematic study of the frequency or the phenotype of HTLV-I-specific $CD4^+$ T cells in the circulation. We have therefore begun to quantify the frequency of anti-HTLV-I $CD4^+$ T cells, initially concentrating on those specific to Env or Tax, in healthy carriers and

HAM/TSP patients (*16*). We have observed a predominance of the Th1 phenotype (i.e. γ-interferon secretion) among HTLV-I Env- and Tax-specific $CD4^+$ T cells in HTLV-I infection. HAM/TSP patients had a significantly higher frequency of such cells than healthy carriers. We are now extending this study to quantify Th cells specific to all HTLV-I antigens.

Formal tests of the role of $CD4^+$ T cells in HTLV-I-associated inflammatory diseases are, of course, difficult or impossible in humans. However, circumstantial evidence for their role may be gained from studies of phenotype (cytokines or surface markers) or gene expression using DNA microarrays (see below).

The Cytotoxic T Cell Response to Viruses

Cytotoxic T lymphocytes are the main effector agent of the immune response that limits the replication of a virus and eradicates the infection from the host, although the relative importance of humoral and cell-mediated immunity varies among different viral infections (*29*). CTLs accomplish this task either by killing the virus-infected host cell, or by arresting intracellular virus replication with interferons: the relative importance of these two pathways is usually unknown *in vivo*, but there is increasing evidence of the importance of CTL-mediated lysis. In addition to these beneficial effects of $CD8^+$ T cells, the same cells can also secrete inflammatory substances such as interferon-γ (*38*), tumor necrosis factor (TNF)-α and matrix metalloproteinases (*6*) when stimulated by antigens. In all viral infections there is therefore a trade-off between the benefit and the harm caused by virus-specific CTLs. In an acute virus infection, the host may benefit from a vigorous CTL response that quickly eliminates the virus. But in a chronic virus infection, a persistent CTL response can lead to chronic inflammation. It is therefore a matter of critical importance in HTLV-I infection to understand the balance of good and harm caused by anti-HTLV-I CTLs. For these reasons we have studied the CTL response to HTLV-I, concentrating on the following questions:

1. How does HTLV-I persist in the face of a strong CTL response?
2. What determines the risk of inflammatory diseases such as HAM/TSP?
3. Is the net effect of the CTL response to HTLV-I beneficial or harmful?
4. Does the evidence allow us to suggest a logical approach to treatment or prevention of HTLV-I-associated inflammatory disease?

On the basis of experimental evidence gathered over the last 10 years, I suggest preliminary answers to these questions below.

The CTL Response to HTLV-I

CTLs specific to HTLV-I were first identified by Kannagi *et al.* (*31, 32*) in blood samples from patients with adult T cell leukaemia (ATL) in remission. It was found that HTLV-I-infected lymphocytes became susceptible to CTL-mediated lysis before HTLV-I Env protein was detectable on the cell surface. These authors therefore suggested that the CTLs might recognize a viral protein other than Env. This prescient suggestion, which predated the demonstration that class I MHC-restricted CTLs recognize peptides derived from proteins that are degraded in the cytoplasm (*63*), was confirmed in 1990 when Jacobson *et al.* (*26*) showed that the viral transactivator protein Tax was an important antigen for CTLs. There is good agreement that Tax is the dominant target antigen recognized by anti-HTLV-I CTLs (*30, 57*), although activated CTLs specific to other HTLV-I proteins are also detected both among freshly isolated lymphocytes and lymphocytes cultured *in vitro* for up to 2 months (*26, 57*).

An unusual feature of the HTLV-I-specific CTLs is their state of activation: when isolated from fresh blood, the CTLs are able to kill HTLV-I antigen-positive cells immediately, without restimulation *in vitro* with HTLV-I antigens (*26, 57, 58*). Because antigen-specific CTLs are activated by contact with cells that express short-lived complexes of MHC molecules with appropriate antigenic peptides (*62*), the activated state of HTLV-I-specific CTLs strongly implies that these cells have been exposed to newly synthesized HTLV-I proteins within the preceding few days *in vivo*.

The other chief unusual feature of the $CD8^+$ T cell response to HTLV-I is the high frequency of these cells in the PBMCs. $CD8^+$ T cells that recognize a single peptide epitope from HTLV-I Tax often constitute 1% or more of circulating $CD8^+$ T cells in infected people (*7, 10, 14, 17, 28*). High frequencies of virus-specific $CD8^+$ T cells are often found in acute viral infections (*53*), but chronic viral infections are typically accompanied by lower specific $CD8^+$ T cell frequencies. This observation suggested that such cells could cause or contribute to the tissue damage observed in inflammatory diseases such as HAM/TSP. However, since Tax peptide-specific $CD8^+$ cells are also very abundant in many healthy carriers (*10, 28*) and indeed can exceed 2% of PBMCs in such subjects (*28*), it is clear that a high frequency of activated anti-HTLV-I $CD8^+$ T cells is not sufficient to cause HAM/TSP, even if they can contribute to tissue damage in established lesions.

Correlation between Anti-HTLV-I $CD8^+$ T Cell Frequency and HTLV-I Proviral Load

Since CTLs can reduce virus load, it seemed logical to expect an inverse correlation between CTL frequency in PBMCs and the proviral load of HTLV-I, analogous to the negative correlation observed between the frequency of specific $CD8^+$ T cells and plasma virus load of human immunodeficiency virus (HIV)-1 observed by (*52*). However, a simple mathematical analysis suggested that an effective CTL response to a persistent virus could result in a positive correlation between CTL frequency and virus load at equilibrium (*69*). There is now experimental evidence of both positive and negative significant correlations between the specific $CD8^+$ T cell frequency and viral (or proviral) load in both HIV-1 infection (negative, (*52*); positive, (*5*)) and HTLV-I infection (positive, (*36, 69*); negative, Goon, P. *et al.*, unpublished). Because the experimental methods differed between these studies, it is difficult to find a simple explanation for these apparently conflicting observations. Until a convincing explanation is found, it seems clear that the direction of correlation between specific $CD8^+$ T cell frequency and the proviral load is not a useful indicator of the effectiveness of the $CD8^+$ response to a persistent virus at equilibrium.

What Is the Effect of CTLs on HTLV-I?

Although the frequency of HTLV-I-specific $CD8^+$ T cells is often very high in healthy carriers (*10, 28*), the median frequency is about 2–4 times greater in HAM/

TSP patients, although the frequency distributions overlap extensively and the mean (or median) frequency is often not significantly different between these groups of subjects (*10, 28*). We therefore postulated that, if the Tax-specific CTLs exert selection on Tax-expressing lymphocytes *in vivo*, the *tax* gene would be subject to stronger positive selection in HAM/TSP patients than in healthy carriers, because of the higher frequency of CTLs in the HAM/TSP patients. However, the opposite was the case: measurement of the ratio of non-synonymous to synonymous nucleotide changes (the D_n/D_s ratio, (*47*)) showed that *tax* was subject to positive selection in healthy carriers but not in HAM/TSP patients (*48, 49*). Therefore, if the CTLs are indeed responsible for this selection—and no plausible alternative has been suggested—we must conclude that the $CD8^+$ T cells in healthy carriers, despite being slightly lower in frequency, are more effective in exerting selection on HTLV-I than are the $CD8^+$ T cells in HAM/TSP patients. Recent evidence from population genetics (*27, 28*) strongly supports this idea (see below).

If CTLs exert significant selection pressure on the *tax* gene, one would expect to observe immune escape mutations, i.e. variant sequences in known CTL epitopes of Tax that lead to failure of recognition by CTLs. Such variant sequences were observed by Niewiesk *et al.* (*50*): in nine of nine cases, a single naturally-occurring amino acid substitution in a known CTL epitope in Tax led to failure of recognition of that epitope by CTLs from the subject in whom the variant sequence was found, although the wild-type sequence was still recognized. However, these putative CTL escape mutations impaired the transactivation activity of the Tax protein (*50*) (see below).

It was argued above that the activated state of anti-Tax CTLs in fresh PBMCs could only have resulted from recent exposure to Tax protein *in vivo*. However, it has been hard to detect HTLV-I protein in fresh PBMCs (*43*). We therefore developed a sensitive flow cytometric assay of intracellular Tax protein expression (*19*). Using this assay, Hanon *et al.* observed a spontaneous rise in Tax expression in fresh $CD4^+$ T cells between 6 and 24 h after isolation, followed by a fall in Tax expression. Furthermore, the Tax expression increased greatly if $CD8^+$ cells were removed from the PBMCs, or if CTL-mediated lysis was prevented with the perforin inhibitor, concanamycin A. Two main conclusions followed from these experiments: first, all or most of the HTLV-I provirus-containing cells spontaneously expressed HTLV-I Tax within 24 h, and there was no intracellular or serum factor that suppressed this expression. Second, autologous $CD8^+$ CTLs were able to limit Tax expression by killing the infected $CD4^+$ T cells.

But if Tax expression rises rapidly *in vitro*, why is it so low in freshly isolated lymphocytes? We suggest that the reason is very simple (*3*): *in vivo*, $CD8^+$ T cell surveillance of virus-infected cells is much more efficient, i.e. the rate of encounter between $CD4^+$ and $CD8^+$ T cells is much greater, than *in vitro*. Therefore, the life expectancy of a Tax-expressing cell is greater *in vitro* than *in vivo*, and the cell survives long enough to allow Tax expression to rise to a level that is detectable by flow cytometry. This interpretation was supported by a simple experiment ((*20*); Hanon *et al.*, unpublished data). If fresh PBMCs were incubated in a flat-bottomed tissue culture well, Tax expression rose rapidly within 24 h. But if the same number of PBMCs from the same subject were incubated in a round-bottomed tissue culture well, the Tax expression was greatly reduced. We suggest that this reduction in Tax expression was due to the close proximity of cells in the round-bottomed well, which reduced the mean interval between CTL killing events. We excluded the possibility that Tax expression was induced by activation of the T cells by fetal calf serum or heparin.

It remains possible that Tax expression was induced in the PBMCs by the temperature change that resulted from removing the cells from the circulation before putting them in culture. However, the most parsimonious explanation remains that $CD8^+$ T cells are the main factor responsible for limiting Tax expression *in vivo*, as they are *in vitro*.

It has been suggested that the Rex protein of HTLV-I might halt the expression of the HTLV-I provirus *in vivo*, and therefore account for intermittent provirus expression (*73*). However, while Rex expression is necessary for the completion of the replication cycle of HTLV-I, few HTLV-I-expressing cells are likely to survive long enough *in vivo* to express Rex, because most will be killed by CTLs specific to Tax, the first viral protein expressed in the HTLV-I replication cycle. Indeed, I have suggested (*3*) that this provides a simple explanation for the immunodominance of Tax in the anti-HTLV-I CTL response.

Mortreux *et al.* (*44*) reported that, in experimental HTLV-I infection of the squirrel monkey (*Saimiri sciureus*), detectable expression of HTLV-I genes declined at 2–3 weeks after inoculation. The authors concluded that HTLV-I primary infection in squirrel monkeys is a "two step process" consisting of a transient phase of reverse transcription followed by latency. No factor was identified or suggested that might cause such a transient pattern of HTLV-I replication. However, the authors found that the decline in detectable HTLV-I transcription coincided with the development of the immune response to HTLV-I in the inoculated monkeys. The same group had previously reported (*33*) that the squirrel monkey develops a strong CTL response to HTLV-I, chiefly directed against the Tax protein, which therefore resembles the human CTL response to HTLV-I. Again, therefore, the simplest explanation is that the strong CTL response reduces viral gene expression *in vivo* below the limit of detection.

We conclude that it is unnecessary to invoke a factor either inherent in HTLV-I-infected cells or in serum that causes intermittent or transient expression of the HTLV-I provirus *in vivo*.

$CD8^+$ T Cells Are also Infected with HTLV-I in vivo: CTL Fratricide

An unexpected observation was made by Hanon (*20*) in experiments on Tax expression in fresh PBMCs. Between 4% and 28% of the proviral load of HTLV-I was carried not in $CD4^+$ T cells but in $CD8^+$ T cells (*20*). Previous workers had obtained evidence of infection of $CD8^+$ T cells with HTLV-I *in vivo* (*9, 35*), but the results had been widely disregarded, because of the difficulty of excluding contamination of the PCR reaction. Furthermore, the $CD8^+$ T cells that express Tax were then spontaneously and rapidly killed by their "brother" Tax-specific CTLs *in vitro*, and presumably *in vivo*. Such CTL "fratricide" had been demonstrated in artificial systems (*23, 59*), but not among naturally-infected cells.

However, perhaps the most curious observation in these experiments was that, whereas a proportion of HTLV-I-specific CTLs was always infected with HTLV-I, EBV-specific CTLs were never found to be infected. This observation raised the intriguing possibility that antigen-specific recognition enhances, or is required for, transmission of HTLV-I from $CD4^+$ T cell to $CD8^+$ T cell. We are therefore now testing this hypothesis in the laboratory.

Is the Net Effect of HTLV-I-specific CTLs Good or Bad?

Cytotoxic T cells can exert both harmful inflammatory effects and beneficial effects by eliminating virus-infected cells (*29*) (see above). Moreover, the CTL that are most active in killing virus-infected cells might also be expected to exert the strongest harmful effects, i.e. to cause the most inflammation. However, work from the groups of Lanzavecchia and others (*66*) has shown that there is a hierarchy of antigen concentrations required to elicit different responses from a $CD8^+$ T cell. A low antigen concentration (i.e. a low MHC/peptide complex density) on the target cell can elicit a killing response by the CTL, but a higher antigen concentration is required for the induction of cytokine secretion, and a higher still antigen concentration to elicit proliferation of the CTL.

Asquith used these observations to suggest (*1*) that an individual with a strong CTL response to HTLV-I would kill Tax-expressing lymphocytes quickly, and would therefore maintain the equilibrium concentration of Tax protein below the threshold required to elicit inflammatory cytokine secretion by $CD8^+$ T cells. In an individual with a less efficient CTL response, however, there would be a sufficiently high equilibrium concentration of HTLV-I antigens to elicit production of interferon-γ and TNF-α by the $CD8^+$ T cells. On this hypothesis, the net effect of HTLV-I-specific CTLs is beneficial in a strong CTL responder but harmful in a weak responder. It is important to note that a "strong CTL responder" to a virus denotes here a person whose CTLs proliferate and kill rapidly in response to a given quantity of viral antigen.

It was argued above that there is no simple relationship in persistent virus infections at equilibrium between the abundance of virus-specific CTLs and their efficiency. The question arises: what exactly is CTL "efficiency"?

What Is CTL "Efficiency"?

The evidence summarized above has led to the following hypothesis: the efficiency of a person's CTL response to HTLV-I is an important determinant of that person's proviral load and therefore his risk of inflammatory disease. But what is CTL "efficiency" in molecular or functional terms? In 1996 we proposed (*51*) that the efficiency of antiviral CTLs could be defined by two parameters: the rate at which a single CTL kills virus infected cells, and the rate at which it proliferates on contact with a given concentration of viral antigen. We showed (*51*) that if these parameters of CTL efficiency differ between infected individuals then, when equilibrium is reached between the virus replication and the immune response, an efficient ("strong") CTL responder will have a low proviral load whereas an inefficient ("weak") CTL responder will have a high proviral load, although there may be little or no difference in the frequency of virus-specific CTLs between these two individuals. It is important to point out that in an immunologically normal individual one would not expect to find a correlation between the efficiency of the CTL responses to unrelated viruses. That is, CTL efficiency must be separately defined for each virus infection of each individual.

The proviral load in an HTLV-I-infected individual is usually stable over many years (*41*). Furthermore, preliminary evidence ((*61*); Taylor *et al.*, unpublished data) suggests that in HTLV-I, as in other persistent viral infections such as HIV-1, each individual has a characteristic "set point" of proviral load, to which the load returns after an artificial perturbation. We therefore hypothesized that genetically determined differences in the efficiency of the CTL response to HTLV-I contribute to the observed differences in proviral load and disease manifestation among HTLV-I-infected people.

In order to test this hypothesis, we carried out, in collaboration with Professor M. Osame and his colleagues, a population genetic survey in the region of endemic HTLV-I infection in Kagoshima, southern Japan. We drew up a list of candidate genes that were known or suspected either to influence the immune response to viruses, or to be involved in the life cycle of HTLV-I. Because of the evidence on the CTL response to HTLV-I, we placed at the top of the candidate gene list the class I HLA genes, because the HLA proteins are known to control the specificity and activity of CTLs.

The results of this study were clear (*27, 28*). Possession of either the *HLA-A*02* gene or the *HLA-Cw*08* gene was associated with a significantly lower risk of HAM/TSP in HTLV-I-infected people in Kagoshima. These results remained statistically significant even after correction for the large number of polymorphisms tested in the study. But there was a still more powerful indication that this result was not due to chance. Since the likely mechanism of HLA class I-associated protection is efficient CTL-mediated killing of HTLV-I-infected cells, we reasoned that possession of either of the two protective class I HLA alleles should be associated with a significant reduction in the proviral load of HTLV-I, both within the healthy carriers and patients with HAM/TSP. The data on proviral load confirmed this prediction in the case of both *HLA-A*02* gene and the *HLA-Cw*08* gene (*27, 28*). The fact that this reduction in proviral load associated with each allele was observed within the healthy carriers excluded the possibility that the effects of *A*02* and *Cw*08* were artefacts due to an unidentified systematic genetic difference ("genetic stratification") between healthy carriers and HAM/TSP patients.

These results argue strongly in favour of the hypothesis (*3, 51*) (see above) that genetically determined differences in the efficiency or strength of the CTL response to HTLV-I are an important cause of the observed variation in proviral load and disease manifestation among HTLV-I-infected individuals. Because there are differences between endemically infected populations in both host genetic composition and in the dominant HTLV-I variant (see below), the magnitude of effects of individual genes such as *HLA-A*02* is likely to vary between such populations. However, it is very unlikely that the fundamental principles that govern the immune control of HTLV-I will differ: specifically, the principle that an efficient CTL response to HTLV-I is beneficial.

We also observed an increase in susceptibility to HAM/TSP in Kagoshima associated with both *HLA-DR1* (*DRB1*0101*) and *HLA-B*54*. The *HLA-B*54* gene effect is of particular interest, because it suggests that the inflammatory effects of

*B*54*-restricted anti-HTLV-I CD8⁺ T cells predominate over their beneficial effects. We are now investigating this possibility.

The population genetic study also identified effects of polymorphisms in three non-HLA genes. The effect of a polymorphism in the TNF-α promoter (*TNF-863C/A*) is of special interest, because the susceptibility allele (*-863A*) exerted its effect only among HTLV-I-infected subjects with a high proviral load, in excess of 3 proviral copies per 100 PBMCs. A possible interpretation of this observation is that T lymphocytes contribute to inflammatory tissue damage by secreting TNF-α only when the concentration of viral antigens exceeds a threshold value. This behaviour is precisely that predicted by the theory proposed by Asquith (*1*) (see above). Such a threshold effect could account for the observation (*46*) that the relative risk of HAM/TSP in Kagoshima rises steeply when the HTLV-I proviral load exceeds an apparent threshold of about 1 copy per 100 PBMCs (Fig. 1).

With Alun Lloyd we have used logistic and multivariate regression analysis to identify the significant independent determinants of the risk of HAM/TSP and of

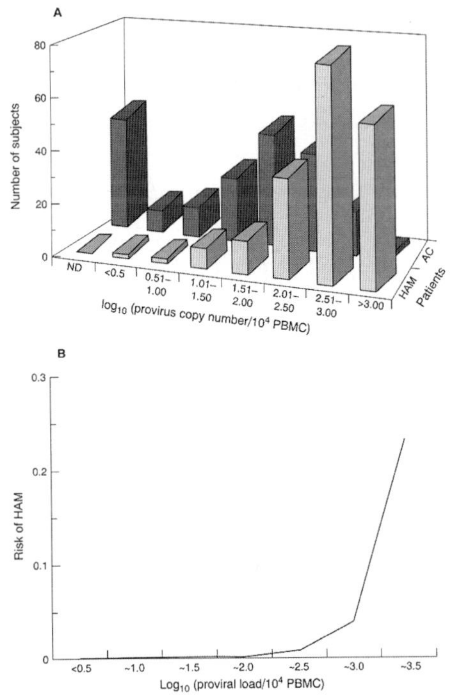

FIG. 1. The risk of HAM/TSP is strongly correlated with HTLV-I proviral load

(A) Distribution of human T lymphotropic virus type I (HTLV-I) proviral load in 202 patients with tropical spastic paraparesis (HAM/TSP) and 200 asymptomatic carriers of HTLV-I (AC) in Kagoshima, southern Japan. The lower limit of detection was one provirus copy/10^4 peripheral blood mononuclear cells (PBMCs). The median proviral load in patients with HAM/TSP was 16 times greater than in asymptomatic carriers. ND, not determined. (B) The risk of HAM/TSP depends on the proviral load of HTLV-I. The vertical axis shows the risk of HAM/TSP at a given $\log_{10}$ (HTLV-I proviral load). The risk remains very low until the proviral load reaches 1% PBMCs; above this apparent threshold, the risk of disease rises rapidly with increasing load. The prevalence of HAM/TSP among the seropositive adult population was assumed to be 0.01. Reproduced (with permission) from Ref. no. *28*.

HTLV-I proviral load in the Kagoshima population. The results (*67*) confirmed that the class I alleles *HLA-A*02* and *-Cw*08* exert the strongest genetic effect on HAM/TSP risk. Interestingly, the results also demonstrated that whereas the protective HLA alleles were the *only* significant determinants of proviral load among the healthy carriers, neither allele was a significant determinant of proviral load in the HAM/TSP patients. This observation suggested that the class I-restricted CTL response limits HTLV-I proviral load and gene expression in strong CTL responders to HTLV-I (i.e. healthy carriers), but not in the weak CTL responders (i.e. a typical HAM/TSP patient). Again, this conclusion is in turn consistent with the analysis made by Asquith (see above (*1*)) that HTLV-I-specific CTLs are beneficial in strong responders but may be harmful in weaker responders.

Molecular and Functional Definition of CTL Efficiency

The above analysis, if correct, suggests that the outcome of HTLV-I infection is largely determined by the efficiency of the CTL response to Tax. Further, the apparent importance of a single type of lymphocyte response to a single antigen suggests that it might be possible to use this system to define "CTL efficiency" in molecular terms and to measure it in functional assays. Experiments are therefore now underway using DNA expression microarrays to identify the gene expression profile of a strong CTL responder to Tax, and assays of CTL-mediated lysis among fresh PBMCs to measure the per-cell rate of lysis effected by CTLs.

HTLV-I Genotype Also Influences the Outcome of Infection

The effects described here do not, of course, exclude the possibility that other factors influence the outcome of HTLV-I infection. Indeed, although no single sequence of HTLV-I is uniquely associated with inflammatory diseases such as HAM/TSP ((*11*); reviewed in (*4*)), Furukawa has shown (*15*) that different strains (subgroups) of HTLV-I are associated with different risks of HAM/TSP in Kagoshima.

How Long Do HTLV-I-infected Cells Live in vivo?

The emerging picture of HTLV-I infection is one of persistent viral gene expression and lysis of infected cells by CTLs. If this picture is accurate, it implies that HTLV-I-infected cells will be rapidly eliminated once they start to express HTLV-I antigens. One estimate of the half-life of a Tax-expressing cell derived from *in vitro* assays was about 0.7 days (*2*): we believe that this should be regarded as an upper-bound estimate of the half-life of such cells *in vivo*, where the mean interval between encounters with a CTL is likely to be shorter than *in vitro* (see above).

Furthermore, evidence from the short-term *in vitro* incubation of fresh unstimulated PBMCs (*19*) (see above) suggests that a high proportion of HTLV-I-provirus-carrying lymphocytes spontaneously start to express HTLV-I antigens within a 24 h period. Since the proviral load of HTLV-I is often of the order of 1–10% of PBMCs, this raises the possibility that there is very rapid turnover of $CD4^+$ T cells *in vivo* in HTLV-I infection. We are now testing this hypothesis using the recently developed technique of *in vivo* cell marking with deuterium-labelled glucose (*39*).

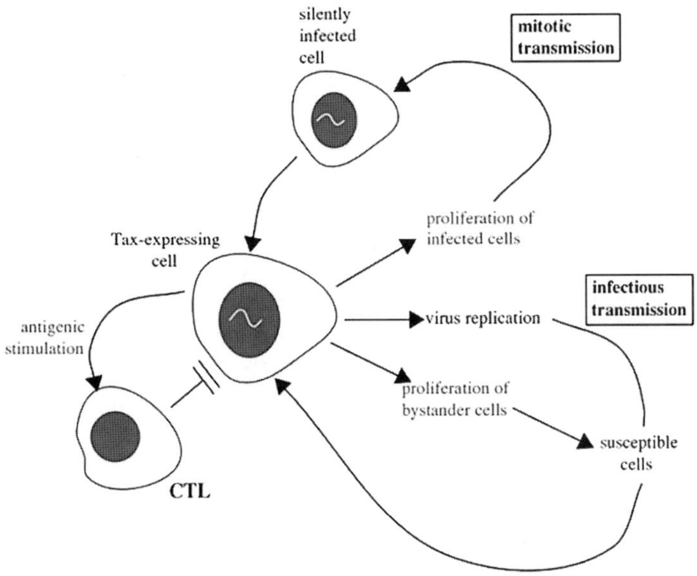

FIG. 2. The Tax-expressing cell occupies a central place in the dynamics of HTLV-I infection
The maintenance of HTLV-I load and the CTL response to HTLV-I. CD4+ T cells that express the HTLV-I Tax protein occupy a central position in the balance between HTLV-I persistence and the immune response. HTLV-I-infected cells drive the proliferation of uninfected bystander lymphocytes. The Tax-expressing cell that is shown in the centre drives both mitotic and infectious spread of the virus, so maintaining the very high proviral load that is observed. Therefore, an individual who makes a strong CTL response to Tax can efficiently limit the proviral load and reduce the risk of associated inflammatory disease such as HAM/TSP. Reproduced (with permission) from Ref. no. *3*.

Mitotic and Infectious Spread of HTLV-I in vivo

For many years it has been recognized that HTLV-I varies much less in sequence than HIV-1, both within and between isolates. Because reverse transcription is accompanied by a high per-cycle mutation rate, whereas genomic DNA replication is about one million times more accurate, the inference has been made (e.g. (*8*)) that HTLV-I persists *in vivo* chiefly by transmission of the provirus from a dividing cell to its two daughters. These two routes of transmission may be called the "infectious" route involving replication via reverse transcriptase, and the "mitotic" route which involves provirus replication by the cell. However, the evidence presented earlier in this chapter favours the view that the HTLV-I provirus is not silent *in vivo*, but on the contrary is frequently expressed. How can we reconcile these interpretations?

Wodarz has made a careful analysis (*68, 70*) of the dynamics of HTLV-I persistence, and has reached three main conclusions. First, mitotic and infectious transmission contribute additively to the basic reproductive ratio of a retrovirus, i.e. the average number of infected cells that are produced by a single HTLV-I-carrying cell. Second, infectious transmission may be required to maintain the infection in an individual, even if mitotic transmission predominates. Third, and most important, when equilibrium is reached between the virus replication and the immune response, most of the newly-produced HTLV-I provirus-carrying cells could arise through mitosis of infected cells, even if infectious spread continues to contribute to (and indeed may be required for) the maintenance of HTLV-I persistence. The interpretation is as follows: in a newly-infected individual, the main mode of HTLV-I transmission is spread of virus particles, i.e. infectious spread. Initially, spread occurs from the inoculated cells to host cells; later, but before the emergence of the immune response, infected host cells undergo mitosis, although infectious spread still contributes significantly to the maintenance of the load. After the cell-mediated response has developed, a cell that starts to express Tax has a much lower (though not zero) chance of surviving long enough to complete the replication cycle of HTLV-I, because it is more likely to be killed by HTLV-I-specific CTLs first. Therefore, although the chance that an individual HTLV-I-infected cell initiates proviral transcription within a specified time remains unchanged throughout the course of infection, the virus is maintained mainly by infectious spread in the early stages of infection, and mainly by mitotic spread later on. Interestingly, the observations of Mortreux *et al.* (*44*) show that HTLV-I expression follows precisely this pattern in primary HTLV-I infection in the squirrel monkey. This hypothesis carries the testable implication that the rate of HTLV-I sequence diversification will be greater in the early stages than the late stages of infection.

Since mitotic spread predominates throughout most of the course of infection of one individual, the exposure of the virus to immune selection by CTLs is limited, the mutation rate of HTLV-I is low, and the resulting sequence diversity is correspondingly small (Fig. 3). This contrasts with the pattern of selection in other retroviral infections such as HIV-1 or feline leukemia virus (FeLV), in which infectious spread is more important in the maintenance of the infection, resulting in a greater mutation rate, greater exposure to immune selection, and therefore greater diversification (*55*)—see Fig. 3.

The Immunodominance of Tax Is a Result of Its Inability to Tolerate Mutation

I suggested above that a simple explanation of the dominance of Tax in the CTL response to HTLV-I is that it is the first viral protein to be expressed: few cells survive long enough to express later HTLV-I proteins and elicit a CTL response to them. But if Tax could evade CTL recognition, then the life expectancy of infected cells would be increased and a more diverse CTL response would be generated. However, both artificial (*60*) and naturally-occurring (*50*) one- or two-amino acid substitutions almost invariably impair or abolish the transactivation activity of the Tax protein. Therefore, CTL escape mutations carry a very high cost to the virus, and can probably be maintained only via mitotic spread. The immunodominance of Tax in the CTL response is thus maintained by its inability to tolerate CTL escape mutations.

Spontaneous Proliferation in HTLV-I-infected Lymphocytes

If PBMCs from an HTLV-I-infected individual are incubated *in vitro* for a few days, the lymphocytes start to proliferate spontaneously, i.e. without addition of exogenous IL-2 or other growth factors. This spontaneous proliferation has often been investigated and described as though it were a single, simple phenomenon. However, it is clear that several independent processes are occurring simultaneously in these mixed PBMC populations:

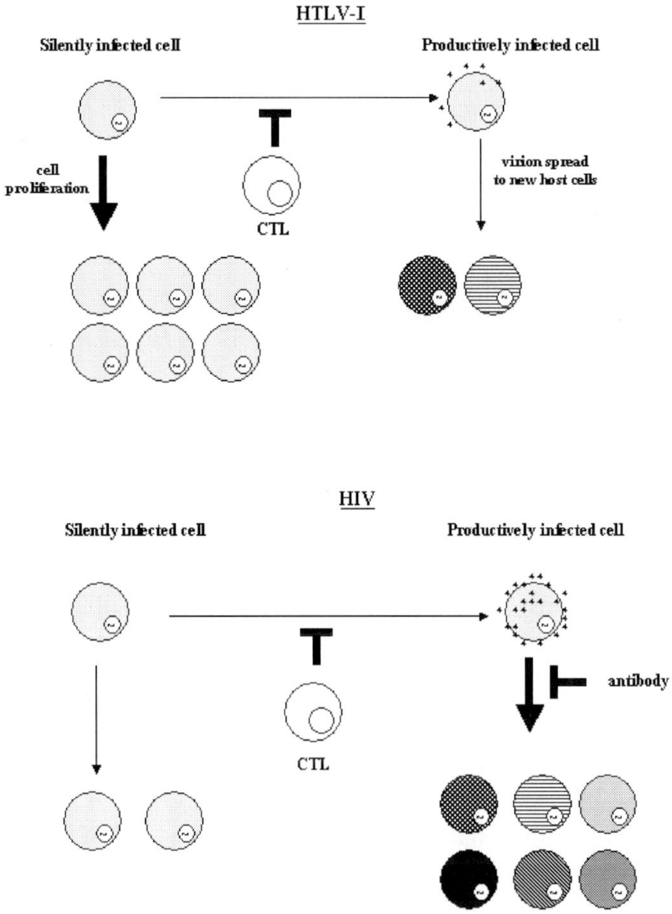

FIG. 3. Immune selection and sequence variation in retrovirus infections
Retroviruses can replicate by two different routes: mitotic division of a provirus-carrying cell replicates the viral genome faithfully, and can occur without viral gene expression. This first "mitotic" route of replication therefore results in a relatively uniform population of virus sequences. The full cycle of virion production requires the action of the highly error-prone viral reverse transcriptase, and the viral gene expression exposes the viral antigens to attack by the host immune system. In this second "infectious" route of retrovirus spread, the frequent reverse transcriptase-generated mutations and immune-mediated selection can result in a population of viruses that is genetically highly diverse. Each shaded pattern represents a cell infected with a different sequence variant of the retrovirus. Reproduced (with permission) from Ref. no. *55*.

1. Initially, the chief subset of lymphocytes that proliferate consists of $CD8^+$ cells (*13*). It is likely that these cells are largely HTLV-I antigen-specific $CD8^+$ cells, stimulated antigenically by the HTLV-I proteins spontaneously expressed by infected T lymphocytes (*19*). $CD4^+$ T cells specific to HTLV-I are also presumably stimulated to proliferate by HTLV-I antigens, but they would not contribute detectably to the phenomenon of spontaneous lymphocyte proliferation because their frequency in fresh PBMCs is too low (*16*).
2. HTLV-I-infected lymphocytes also stimulate proliferation in bystander lymphocytes (*71*), by a process that remains poorly understood but which seems

to require the interaction of CD2 and LFA-3 (*71*), and the costimulatory molecules CD80 and CD86 (*37*) on the surface of the interacting lymphocytes.

3. The spontaneous expression of Tax protein by HTLV-I-infected lymphocytes, both $CD4^+$ and $CD8^+$ (*19, 20*) is also followed by the expression of certain host genes, and the Tax-expressing cells acquire a phenotype typical of activated cells ($CD25^+$, $CD69^+$, $CD71^+$) (*18*). It is therefore probable that Tax-expressing cells proliferate spontaneously, independently of other cells, but this remains to be proved experimentally.

The dynamics of "spontaneous lymphocyte proliferation" are complicated further by the rapid lysis of Tax-expressing cells by specific CTLs (*19*), as described above.

A Putative IL-2 Autocrine Loop Does Not Contribute to the Maintenance of HTLV-I Proviral Load

It has been suggested that an autocrine loop involving IL-2 and the IL-2 receptor (IL-2R) contributes to the maintenance of HTLV-I proviral load, because HTLV-I-infected lymphocytes can express both IL-2 and IL-2R. Lymphocytes that are naturally infected with HTLV-I do indeed express CD25, the high-affinity α chain of the IL-2R, within a few hours of the onset of Tax expression (*18*). However, the same cells did not produce detectable intracellular IL-2 within 24 h of the onset of Tax expression. In a careful *in vitro* study, Inoue *et al.* (*24*) showed that IL-2 expression occurs only at a low level, many hours after Tax protein is expressed. We therefore conclude, in agreement with Hollsberg (*21*) that an IL-2/IL-2R autocrine loop does not normally operate in HTLV-I infection to maintain the proviral load. However, it remains possible that another cytokine such as IL-15 is involved in such an autocrine loop, and that a loop involving IL-2 or IL-15 might contribute to the leukaemogenesis associated with HTLV-I.

Are Antiretroviral Drugs Effective against HTLV-I?

An important implication of the conclusion that HTLV-I persistently replicates *in vivo* is that it should be possible to reduce the proviral load and the risk of inflammatory disease with antiretroviral drugs. Several groups have tested the effects of antiretroviral drugs on HTLV-I-transformed cells *in vivo* or *in vitro*, but there has been relatively little work on their effect on HTLV-I replication in non-transformed cells. Zidovudine (AZT) has some activity against HTLV-I *in vitro* (e.g. (*12*)). Taylor *et al.* (*61*) tested the effect of the nucleoside analogue Lamivudine (3TC) *in vivo*, and showed that the drug typically reduces the proviral load of HTLV-I by a maximum of ten-fold. This result supports the interpretation that infectious spread contributes significantly to the maintenance of HTLV-I proviral load in an infected individual, and strongly encourages further attempts to devise effective antiretroviral treatment. However, as is the case with single-drug treatment of other retroviral infections such as HIV-1, the observed benefit was not sustained. A double-blind, placebo-controlled crossover trial is now in progress (Taylor *et al.*, unpublished), to test the effects of treatment with a combination of Zidovudine and Lamivudine.

The initial results indicated (*61*) that the rate of decline in HTLV-I proviral load during Lamivudine treatment varied considerably between patients. This observation suggests that the relative importance of infectious spread and mitotic spread of HTLV-I varies between individuals. If this suggestion is borne out by future observation, it will be of considerable interest to discover the reasons for such variation, because the variation could account for much of the observed differences in the outcome of HTLV-I infection.

What Are the Implications for Our Understanding of Adult T Cell Leukaemia?

No simple extrapolation can be made, from the arguments presented in this chapter, to increase our understanding of ATL, because the dynamics of the transformed lymphocytes is likely to be highly abnormal. However, the possibility should be considered that immunotherapy could be a useful adjunct to chemotherapy in the treatment of ATL.

Mathematical Analysis of the Dynamics of HTLV-I Replication and the Immune Response

Even the simplest depiction of the equilibrium that is struck between a persistent retrovirus and the host immune response (Fig. 2) makes it evident that intuition and verbal reasoning can lead one astray in trying to explain the experimental data. It is therefore essential to use mathematics in attempts to understand the dynamics of HTLV-I infection, i.e. the interaction of the immune response and HTLV-I and the mechanisms by which the proviral load is maintained. In this work, mathematical consistency must not be regarded as the final criterion of truth, because no experimental data can (*pace* Popper) prove the truth of a theory or a mathematical model. Rather, mathematics serves two main purposes in this context: first, to check that one's reasoning is logical and consistent, and second, to identify the smallest number of plausible biological factors that can explain the experimental observations.

Summary and Conclusions

In conclusion, we can begin to suggest answers to the questions posed earlier in this chapter.

1. How does HTLV-I persist?
    a) HTLV-I is persistently transcriptionally active *in vivo* and is not, as has been believed, latent.
    b) It is unnecessary to postulate an additional factor either in the HTLV-I provirus or the serum that terminates HTLV-I expression *in vivo*, reversibly or irreversibly, because a large and powerful factor that does so has been identified: the abundant, activated HTLV-I-specific CTLs.
    c) The proviral load of HTLV-I is maintained by both infectious spread of HTLV-I from cell to cell, and by mitotic division of HTLV-I provirus-containing cells. Mitotic spread predominates throughout most of the course of infection, despite the continued transcription of the HTLV-I provirus.
2. What determines the risk of HAM/TSP?
    a) The most important predictor of HAM/TSP risk remains the proviral

load (Fig. 1): the reason for this is unknown, because the pathogenetic mechanisms are not known.

b) Host genetic factors, chiefly the HLA genotype, determine the efficiency of the CTL response to HTLV-I: in turn, the CTLs limit the proviral load and therefore the risk of HAM/TSP. CTL "efficiency" is defined here by two parameters: the rate of replication of specific CTL in response to a given antigen load, and the rate at which a specific CTL kills an HTLV-I-infected target cell.

c) The causes of tissue damage in HAM/TSP remain uncertain, but the lesion may be initiated by Th1 type (inflammatory) $CD4^+$ T cells, which are more frequent in the circulation in HAM/TSP patients than in healthy carriers of HTLV-I. $CD4^+$ T cells predominate in early HAM/TSP lesions, but it is not known whether these are Th1 or Th2.

d) Other factors that can influence the risk of HAM/TSP appear to include the following: age, HTLV-I genotype, and host genotype at the *TNF-α* and stromal cell-derived factor 1 (*SDF-1*) loci.

3. Is the CTL response to HTLV-I beneficial or harmful?
   a) The net effect of the CTL response is beneficial in individuals with an efficient CTL response to HTLV-I, as defined in 2b above, i.e. in a typical healthy HTLV-I carrier.
   b) In an individual with an inefficient CTL response to HTLV-I, i.e. a typical HAM/TSP patient, the HTLV-I-specific CTLs may contribute to tissue damage by secreting inflammatory cytokines, such as interferon-γ or TNF-α, or metalloproteinases. These conclusions emphasize the need to understand the causes of CTL efficiency at the molecular and functional levels.
   c) Neither the frequency of virus-specific CTLs nor the correlation between the frequency of such cells and the proviral load is a useful index of the effectiveness of the CTL response against a persistent virus at equilibrium.
4. What are the implications for the treatment of HTLV-I infection?
   a) Treatment with antiretroviral drugs can reduce the proviral load of HTLV-I *in vivo.*
   b) Antiretroviral treatment of a healthy carrier of HTLV-I might be considered if the proviral load is very high or if other risk factors are present.
   c) Antiretroviral treatment of a HAM/TSP patient would be expected to be beneficial, but will not reverse established neurological damage.
   d) An effective anti-HTLV-I vaccine should elicit a CTL response to Tax. A mutated form of the Tax protein could be used, which retains immunogenicity but lacks transactivation activity.
   e) Post-exposure immunization might be appropriate in a healthy carrier with a high proviral load, but would be contraindicated in a HAM/TSP patient because an enhanced CTL response could exacerbate the disease.

REFERENCES

1. Asquith, B. and Bangham, C. R. The role of cytotoxic T lymphocytes in human T-cell lymphotropic virus type 1 infection. *J. Theor. Biol.*, **207**, 65–79 (2000).
2. Asquith, B., Hanon, E., Taylor, G. P., and Bangham, C. R. Is human T-cell lymphotropic virus type I really silent? *Philos. Trans. R. Soc. Lond. B Biol. Sci.*, **355**, 1013–1019 (2000).
3. Bangham, C. R. The immune response to HTLV-I. *Curr. Opin. Immunol.*, **12**, 397–402 (2000).
4. Bangham, C. R. M., Kermode, A. G., Hall, S. E., and Daenke, S. The cytotoxic T-lymphocyte response to HTLV-I: the main determinant of disease? *Semin. Virol.*, **7**, 41–48 (1996).
5. Betts, M. R., Ambrozak, D. R., Douek, D. C., Bonhoeffer, S., Brenchley, J. M., Casazza, J. P., Koup, R. A., and Picker, L. J. Analysis of total human immunodeficiency virus (HIV)-specific CD4+ and CD8+ T-cell responses: relationship to viral load in untreated HIV infection. *J. Virol.*, **75**, 11983–11991 (2001).
6. Biddison, W. E., Kubota, R., Kawanishi, T., Taub, D. D., Cruikshank, W. W., Center, D. M., Connor, E. W., Utz, U., and Jacobson, S. Human T cell leukemia virus type I (HTLV-I)-specific CD8+ CTL clones from patients with HTLV-I-associated neurologic disease secrete proinflammatory cytokines, chemokines, and matrix metalloproteinase. *J. Immunol.*, **159**, 2018–2025 (1997).
7. Bieganowska, K., Hollsberg, P., Buckle, G. J., Lim, D. G., Greten, T. F., Schneck, J., Altman, J. D., Jacobson, S., Ledis, S. L., Hanchard, B., Chin, J., Morgan, O., Roth, P. A., and Hafler, D. A. Direct analysis of viral-specific CD8+ T cells with soluble HLA-A2/Tax11-19 tetramer complexes in patients with human T cell lymphotropic virus-associated myelopathy. *J. Immunol.*, **162**, 1765–1771 (1999).
8. Cavrois, M., Gessain, A., Wain-Hobson, S., and Wattel, E. Proliferation of HTLV-1 infected circulating cells *in vivo* in all asymptomatic carriers and patients with TSP/HAM. *Oncogene*, **12**, 2419–2423 (1996).
9. Cho, I., Sugimoto, M., Mita, S., Tokunaga, M., Imamura, F., and Ando, M. *In vivo* proviral burden and viral RNA expression in T cell subsets of patients with human T lymphotropic virus type-1-associated myelopathy/tropical spastic paraparesis. *Am. J. Trop. Med. Hyg.*, **53**, 412–418 (1995).
10. Daenke, S., Kermode, A. G., Hall, S. E., Taylor, G., Weber, J., Nightingale, S., and Bangham, C. R. High activated and memory cytotoxic T-cell responses to HTLV-1 in healthy carriers and patients with tropical spastic paraparesis. *Virology*, **217**, 139–146 (1996).
11. Daenke, S., Nightingale, S., Cruickshank, J. K., and Bangham, C. R. Sequence variants of human T-cell lymphotropic virus type I from patients with tropical spastic paraparesis and adult T-cell leukemia do not distinguish neurological from leukemic isolates. *J. Virol.*, **64**, 1278–1282 (1990).
12. Derse, D., Hill, S. A., Lloyd, P. A., Chung, H., and Morse, B. A. Examining human T-lymphotropic virus type 1 infection and replication by cell-free infection with recombinant virus vectors. *J. Virol.*, **75**, 8461–8468 (2001).
13. Eiraku, N., Ijichi, S., Yashiki, S., Osame, M., and Sonoda, S. Cell surface phenotype of *in vitro* proliferating lymphocytes in HTLV-I-associated myelopathy (HAM/TSP).

J. Neuroimmunol., **37**, 223–228 (1992).

14. Elovaara, I., Koenig, S., Brewah, A. Y., Woods, R. M., Lehky, T., and Jacobson, S. High human T cell lymphotropic virus type 1 (HTLV-1)-specific precursor cytotoxic T lymphocyte frequencies in patients with HTLV-1-associated neurological disease. *J. Exp. Med.*, **177**, 1567–1573 (1993).
15. Furukawa, Y., Yamashita, M., Usuku, K., Izumo, S., Nakagawa, M., and Osame, M. Phylogenetic subgroups of human T cell lymphotropic virus (HTLV) type I in the tax gene and their association with different risks for HTLV-I-associated myelopathy/tropical spastic paraparesis. *J. Infect. Dis.*, **182**, 1343–1349 (2000).
16. Goon, P. K. C., Hanon, E., Igakura, T., Tanaka, Y., Weber, J. N., Taylor, G. P., and Bangham, C. R. M. High frequencies of Th1 type CD4+ T-cells specific to HTLV-I Env and Tax proteins in patients with HTLV-I-associated myelopathy/tropical spastic paraparesis (HAM/TSP). *Blood*, **99**, 3335–3341 (2002).
17. Greten, T. F., Slansky, J. E., Kubota, R., Soldan, S. S., Jaffee, E. M., Leist, T. P., Pardoll, D. M., Jacobson, S., and Schneck, J. P. Direct visualization of antigen-specific T cells: HTLV-1 Tax11-19-specific CD8+ T cells are activated in peripheral blood and accumulate in cerebrospinal fluid from HAM/TSP patients. *Proc. Natl. Acad. Sci. USA*, **95**, 7568–7573 (1998).
18. Hanon, E., Goon, P., Taylor, G. P., Hasegawa, H., Tanaka, Y., Weber, J. N., and Bangham, C. R. High production of interferon gamma but not interleukin-2 by human T-lymphotropic virus type I-infected peripheral blood mononuclear cells. *Blood*, **98**, 721–726 (2001).
19. Hanon, E., Hall, S., Taylor, G. P., Saito, M., Davis, R., Tanaka, Y., Usuku, K., Osame, M., Weber, J. N., and Bangham, C. R. Abundant tax protein expression in CD4+ T cells infected with human T-cell lymphotropic virus type I (HTLV-I) is prevented by cytotoxic T lymphocytes. *Blood*, **95**, 1386–1392 (2000).
20. Hanon, E., Stinchcombe, J. C., Saito, M., Asquith, B. E., Taylor, G. P., Tanaka, Y., Weber, J. N., Griffiths, G. M., and Bangham, C. R. Fratricide among CD8+ T lymphocytes naturally infected with human T cell lymphotropic virus type I. *Immunity*, **13**, 657–664 (2000).
21. Hollsberg, P. Mechanisms of T-cell activation by human T-cell lymphotropic virus type I. *Microbiol. Mol. Biol. Rev.*, **63**, 308–333 (1999).
22. Horiuchi, I., Kawano, Y., Yamasaki, K., Minohara, M., Furue, M., Taniwaki, T., Miyazaki, T., and Kira, J. Th1 dominance in HAM/TSP and the optico-spinal form of multiple sclerosis versus Th2 dominance in mite antigen-specific IgE myelitis. *J. Neurol. Sci.*, **172**, 17–24 (2000).
23. Huang, J. F., Yang, Y., Sepulveda, H., Shi, W., Hwang, I., Peterson, P. A., Jackson, M. R., Sprent, J., and Cai, Z. TCR-mediated internalization of peptide-MHC complexes acquired by T cells. *Science*, **286**, 952–954 (1999).
24. Inoue, J., Seiki, M., Taniguchi, T., Tsuru, S., and Yoshida, M. Induction of interleukin 2 receptor gene expression by p40x encoded by human T-cell leukemia virus type 1. *EMBO J.*, **5**, 2883–2888 (1986).
25. Iwasaki, Y., Ohara, Y., Kobayashi, I., and Akizuki, S. Infiltration of helper/inducer T lymphocytes heralds central nervous system damage in human T-cell leukemia virus infection. *Am. J. Pathol.*, **140**, 1003–1008 (1992).

26. Jacobson, S., Shida, H., McFarlin, D. E., Fauci, A. S., and Koenig, S. Circulating CD8+ cytotoxic T lymphocytes specific for HTLV-I pX in patients with HTLV-I associated neurological disease. *Nature*, **348**, 245–248 (1990).
27. Jeffery, K. J., Siddiqui, A. A., Bunce, M., Lloyd, A. L., Vine, A. M., Witkover, A. D., Izumo, S., Usuku, K., Welsh, K. I., Osame, M., and Bangham, C. R. The influence of HLA class I alleles and heterozygosity on the outcome of human T cell lymphotropic virus type I infection. *J. Immunol.*, **165**, 7278–7284 (2000).
28. Jeffery, K. J., Usuku, K., Hall, S. E., Matsumoto, W., Taylor, G. P., Procter, J., Bunce, M., Ogg, G. S., Welsh, K. I., Weber, J. N., Lloyd, A. L., Nowak, M. A., Nagai, M., Kodama, D., Izumo, S., Osame, M., and Bangham, C. R. HLA alleles determine human T-lymphotropic virus-I (HTLV-I) proviral load and the risk of HTLV-I-associated myelopathy. *Proc. Natl. Acad. Sci. USA*, **96**, 3848–3853 (1999).
29. Kagi, D., Ledermann, B., Burki, K., Zinkernagel, R. M., and Hengartner, H. Molecular mechanisms of lymphocyte-mediated cytotoxicity and their role in immunological protection and pathogenesis *in vivo*. *Annu. Rev. Immunol.*, **14**, 207–232 (1996).
30. Kannagi, M., Harada, S., Maruyama, I., Inoko, H., Igarashi, H., Kuwashima, G., Sato, S., Morita, M., Kidokoro, M., Sugimoto, M. *et al.* Predominant recognition of human T cell leukemia virus type I (HTLV-I) pX gene products by human CD8+ cytotoxic T cells directed against HTLV-I-infected cells. *Int. Immunol.*, **3**, 761–767 (1991).
31. Kannagi, M., Sugamura, K., Kinoshita, K., Uchino, H., and Hinuma, Y. Specific cytolysis of fresh tumor cells by an autologous killer T cell line derived from an adult T cell leukemia/lymphoma patient. *J. Immunol.*, **133**, 1037–1041 (1984).
32. Kannagi, M., Sugamura, K., Sato, H., Okochi, K., Uchino, H., and Hinuma, Y. Establishment of human cytotoxic T cell lines specific for human adult T cell leukemia virus-bearing cells. *J. Immunol.*, **130**, 2942–2946 (1983).
33. Kazanji, M., Moreau, J. P., Mahieux, R., Bonnemains, B., Bomford, R., Gessain, A., and de The, G. HTLV-I infection in squirrel monkeys (*Saimiri sciureus*) using autologous, homologous, or heterologous HTLV-I-transformed cell lines. *Virology*, **231**, 258–266 (1997).
34. Kitze, B., Usuku, K., Yamano, Y., Yashiki, S., Nakamura, M., Fujiyoshi, T., Izumo, S., Osame, M., and Sonoda, S. Human CD4+ T lymphocytes recognize a highly conserved epitope of human T lymphotropic virus type 1 (HTLV-1) env gp21 restricted by HLA DRB1*0101. *Clin. Exp. Immunol.*, **111**, 278–285 (1998).
35. Koyanagi, Y., Itoyama, Y., Nakamura, N., Takamatsu, K., Kira, J., Iwamasa, T., Goto, I., and Yamamoto, N. *In vivo* infection of human T-cell leukemia virus type I in non-T cells. *Virology*, **196**, 25–33 (1993).
36. Kubota, R., Kawanishi, T., Matsubara, H., Manns, A., and Jacobson, S. HTLV-I specific IFN-gamma+ CD8+ lymphocytes correlate with the proviral load in peripheral blood of infected individuals. *J. Neuroimmunol.*, **102**, 208–215 (2000).
37. Lal, R. B., Rudolph, D. L., Dezzutti, C. S., Linsley, P. S., and Prince, H. E. Costimulatory effects of T cell proliferation during infection with human T lymphotropic virus types I and II are mediated through CD80 and CD86 ligands. *J. Immunol.*, **157**, 1288–1296 (1996).
38. Lin, Y. L. and Askonas, B. A. Biological properties of an influenza A virus-specific

killer T cell clone. Inhibition of virus replication *in vivo* and induction of delayed-type hypersensitivity reactions. *J. Exp. Med.*, **154**, 225–234 (1981).

39. Macallan, D. C., Fullerton, C. A., Neese, R. A., Haddock, K., Park, S. S., and Hellerstein, M. K. Measurement of cell proliferation by labeling of DNA with stable isotope-labeled glucose: studies *in vitro*, in animals, and in humans. *Proc. Natl. Acad. Sci. USA*, **95**, 708–713 (1998).
40. Manns, A., Murphy, E. L., Wilks, R., Haynes, G., Figueroa, J. P., Hanchard, B., Barnett, M., Drummond, J., Waters, D., Cerney, M. *et al.* Detection of early human T-cell lymphotropic virus type I antibody patterns during seroconversion among transfusion recipients. *Blood*, **77**, 896–905 (1991).
41. Matsuzaki, T., Nakagawa, M., Nagai, M., Usuku, K., Higuchi, I., Arimura, K., Kubota, H., Izumo, S., Akiba, S., and Osame, M. HTLV-I proviral load correlates with progression of motor disability in HAM/TSP: analysis of 239 HAM/TSP patients including 64 patients followed up for 10 years. *J. Neurovirol.*, **7**, 228–234 (2001).
42. McMichael, A. J. (1995). Immunology. *In* "Oxford Textbook of Medicine," 3rd Ed., ed. D. J. Weatherall, J. G. G. Ledingham, and D. A. Warrell (1995). Oxford University Press, Oxford.
43. Moritoyo, T., Izumo, S., Moritoyo, H., Tanaka, Y., Kiyomatsu, Y., Nagai, M., Usuku, K., Sorimachi, M., and Osame, M. Detection of human T-lymphotropic virus type I p40tax protein in cerebrospinal fluid cells from patients with human T-lymphotropic virus type I-associated myelopathy/tropical spastic paraparesis. *J. Neurovirol.*, **5**, 241–248 (1999).
44. Mortreux, F., Kazanji, M., Gabet, A. S., de Thoisy, B., and Wattel, E. Two-step nature of human T-cell leukemia virus type 1 replication in experimentally infected squirrel monkeys (*Saimiri sciureus*). *J. Virol.*, **75**, 1083–1089 (2001).
45. Mueller, N. E. and Blattner, W. A. Retroviruses. *In* "Viral Infections of Humans: Epidemiology and Control," 4th Ed., ed. A. S. Evans and R. Kaslow, pp. 785-813 (1997). Plenum Medical Press, New York.
46. Nagai, M., Usuku, K., Matsumoto, W., Kodama, D., Takenouchi, N., Moritoyo, T., Hashiguchi, S., Ichinose, M., Bangham, C. R., Izumo, S., and Osame, M. Analysis of HTLV-I proviral load in 202 HAM/TSP patients and 243 asymptomatic HTLV-I carriers: high proviral load strongly predisposes to HAM/TSP. *J. Neurovirol.*, **4**, 586–593 (1998).
47. Nei, M. and Gojobori, T. Simple methods for estimating the numbers of synonymous and nonsynonymous nucleotide substitutions. *Mol. Biol. Evol.*, **3**, 418–426 (1986).
48. Niewiesk, S. and Bangham, C. R. Evolution in a chronic RNA virus infection: selection on HTLV-I tax protein differs between healthy carriers and patients with tropical spastic paraparesis. *J. Mol. Evol.*, **42**, 452–458 (1996).
49. Niewiesk, S., Daenke, S., Parker, C. E., Taylor, G., Weber, J., Nightingale, S., and Bangham, C. R. The transactivator gene of human T-cell leukemia virus type I is more variable within and between healthy carriers than patients with tropical spastic paraparesis. *J. Virol.*, **68**, 6778–6781 (1994).
50. Niewiesk, S., Daenke, S., Parker, C. E., Taylor, G., Weber, J., Nightingale, S., and Bangham, C. R. Naturally occurring variants of human T-cell leukemia virus type I

Tax protein impair its recognition by cytotoxic T lymphocytes and the transactivation function of Tax. *J. Virol.*, **69**, 2649–2653 (1995).

51. Nowak, M. A. and Bangham, C. R. Population dynamics of immune responses to persistent viruses. *Science*, **272**, 74–79 (1996).
52. Ogg, G. S., Jin, X., Bonhoeffer, S., Dunbar, P. R., Nowak, M. A., Monard, S., Segal, J. P., Cao, Y., Rowland-Jones, S. L., Cerundolo, V., Hurley, A., Markowitz, M., Ho, D. D., Nixon, D. F., and McMichael, A. J. Quantitation of HIV-1-specific cytotoxic T lymphocytes and plasma load of viral RNA. *Science*, **279**, 2103–2106 (1998).
53. Ogg, G. S. and McMichael, A. J. HLA-peptide tetrameric complexes. *Curr. Opin. Immunol.*, **10**, 393–396 (1998).
54. Oka, T., Ohtsuki, Y., Sonobe, H., Furihata, M., and Miyoshi, I. Suppressive effects of interferons on the production and release of human T-lymphotropic virus type-I (HTLV-I). *Arch. Virol.*, **115**, 63–73 (1990).
55. Overbaugh, J. and Bangham, C. R. Selection forces and constraints on retroviral sequence variation. *Science*, **292**, 1106–1109 (2001).
56. Oxenius, A., Zinkernagel, R. M., and Hengartner, H. CD4+ T-cell induction and effector functions: a comparison of immunity against soluble antigens and viral infections. *Adv. Immunol.*, **70**, 313–367 (1998).
57. Parker, C. E., Daenke, S., Nightingale, S., and Bangham, C. R. Activated, HTLV-1-specific cytotoxic T-lymphocytes are found in healthy seropositives as well as in patients with tropical spastic paraparesis. *Virology*, **188**, 628–636 (1992).
58. Parker, C. E., Nightingale, S., Taylor, G. P., Weber, J., and Bangham, C. R. Circulating anti-Tax cytotoxic T lymphocytes from human T-cell leukemia virus type I-infected people, with and without tropical spastic paraparesis, recognize multiple epitopes simultaneously. *J. Virol.*, **68**, 2860–2868 (1994).
59. Raftery, M. J., Behrens, C. K., Muller, A., Krammer, P. H., Walczak, H., and Schonrich, G. Herpes simplex virus type 1 infection of activated cytotoxic T cells: induction of fratricide as a mechanism of viral immune evasion. *J. Exp. Med.*, **190**, 1103–1114 (1999).
60. Smith, M. R. and Greene, W. C. Identification of HTLV-I tax trans-activator mutants exhibiting novel transcriptional phenotypes. *Genes Dev.*, **4**, 1875–1885 (1990).
61. Taylor, G. P., Hall, S. E., Navarrete, S., Michie, C. A., Davis, R., Witkover, A. D., Rossor, M., Nowak, M. A., Rudge, P., Matutes, E., Bangham, C. R., and Weber, J. N. Effect of lamivudine on human T-cell leukemia virus type 1 (HTLV-1) DNA copy number, T-cell phenotype, and anti-tax cytotoxic T-cell frequency in patients with HTLV-1-associated myelopathy. *J. Virol.*, **73**, 10289–10295 (1999).
62. Townsend, A. and Bodmer, H. Antigen recognition by class I-restricted T lymphocytes. *Annu. Rev. Immunol.*, **7**, 601–624 (1989).
63. Townsend, A. R., Rothbard, J., Gotch, F. M., Bahadur, G., Wraith, D., and McMichael, A. J. The epitopes of influenza nucleoprotein recognized by cytotoxic T lymphocytes can be defined with short synthetic peptides. *Cell*, **44**, 959–968 (1986).
64. Usuku, K., Nishizawa, M., Matsuki, K., Tokunaga, K., Takahashi, K., Eiraku, N., Suehara, M., Juji, T., Osame, M., and Tabira, T. Association of a particular amino acid sequence of the HLA-DR beta 1 chain with HTLV-I-associated myelopathy. *Eur. J. Immunol.*, **20**, 1603–1606 (1990).

65. Usuku, K., Sonoda, S., Osame, M., Yashiki, S., Takahashi, K., Matsumoto, M., Sawada, T., Tsuji, K., Tara, M., and Igata, A. HLA haplotype-linked high immune responsiveness against HTLV-I in HTLV-I-associated myelopathy: comparison with adult T-cell leukemia/lymphoma. *Ann. Neurol.*, **23**, S143–S150 (1988).
66. Valitutti, S., Muller, S., Dessing, M., and Lanzavecchia, A. Different responses are elicited in cytotoxic T lymphocytes by different levels of T cell receptor occupancy. *J. Exp. Med.*, **183**, 1917–1921 (1996).
67. Vine, A. M., Witkover, A. D., Lloyd, A. L., Jeffery, K. J. M., Siddiqui, A., Marshall, S. E. F., Bunce, M., Eiraku, N., Izumo, S., Usuku, K., Osame, M., and Bangham, C. R. M. Polygenic control of HTLV-I proviral load and the risk of HTLV-I-associated myelopathy/tropical spastic paraparesis (HAM/TSP). *J. Infect. Dis.*, **186**, 932–939 (2002).
68. Wodarz, D. and Bangham, C. R. Evolutionary dynamics of HTLV-I. *J. Mol. Evol.*, **50**, 448–455 (2000).
69. Wodarz, D., Hall, S. E., Usuku, K., Osame, M., Ogg, G. S., McMichael, A. J., Nowak, M. A., and Bangham, C. R. Cytotoxic T-cell abundance and virus load in human immunodeficiency virus type 1 and human T-cell leukaemia virus type 1. *Proc. R. Soc. Lond. B Biol. Sci.*, **268**, 1215–1221 (2001).
70. Wodarz, D., Nowak, M. A., and Bangham, C. R. The dynamics of HTLV-I and the CTL response. *Immunol. Today*, **20**, 220–227 (1999).
71. Wucherpfennig, K. W., Hollsberg, P., Richardson, J. H., Benjamin, D., and Hafler, D. A. T-cell activation by autologous human T-cell leukemia virus type I-infected T-cell clones. *Proc. Natl. Acad. Sci. USA*, **89**, 2110–2114 (1992).
72. Yamano, Y., Kitze, B., Yashiki, S., Usuku, K., Fujiyoshi, T., Kaminagayoshi, T., Unoki, K., Izumo, S., Osame, M., and Sonoda, S. Preferential recognition of synthetic peptides from HTLV-I gp21 envelope protein by HLA-DRB1 alleles associated with HAM/TSP (HTLV-I-associated myelopathy/tropical spastic paraparesis). *J. Neuroimmunol.*, **76**, 50–60 (1997).
73. Yoshida, M. Multiple viral strategies of HTLV-1 for dysregulation of cell growth control. *Annu. Rev. Immunol.*, **19**, 475–496 (2001).
74. Yu, F., Itoyama, Y., Fujihara, K., and Goto, I. Natural killer (NK) cells in HTLV-I-associated myelopathy/tropical spastic paraparesis-decrease in NK cell subset populations and activity in HTLV-I seropositive individuals. *J. Neuroimmunol.*, **33**, 121–128 (1991).
75. Zinkernagel, R. M. and Hengartner, H. Antiviral immunity. *Immunol. Today*, **18**, 258–260 (1997).

ROLE OF HTLV-I SPECIFIC CTL IN HAM/TSP

Yoshihisa YAMANO, Utano TOMARU, and Steven JACOBSON
*Viral Immunology Section, Neuroimmunology Branch, National Institute of Neurological Disorders and Stroke, National Institutes of Health**

Human T-cell lymphotropic virus type I (HTLV-I)-associated myelopathy/tropical spastic paraparesis (HAM/TSP) is a chronic, progressive neurological disease characterized by marked degeneration of the spinal cord and the presence of infiltrating T cells and macrophages. HAM/TSP patients harbor very high frequencies of HTLV-I infected T cells and HTLV-I specific CD8+ T cells in both peripheral blood and cerebrospinal fluid. The use of novel molecular and immunological methods has improved our understanding of the underlying pathological mechanisms operative in HAM/TSP. Information presented in this review suggests that HTLV-I infected T cells and HTLV-I specific CD8+ T cells accumulate in central nervous system (CNS), in which high cellular immune response continuously driven by this virus may contribute to the inflammatory process within CNS lesions in HAM/TSP patients.

Human T-cell lymphotropic virus type I (HTLV-I) is a human retrovirus which infects approximately 10 million people worldwide (*8*). The majority of infected individuals remain healthy lifelong asymptomatic carriers while approximately 0.25–3% develop a slowly progressive neurological disorder termed HTLV-I-associated myelopathy/tropical spastic paraparesis (HAM/TSP) (*13, 31, 57, 67*). How HTLV-I causes this disease is not completely understood, although virus-host immunological interactions have been suggested to play a role in the pathogenesis of this disorder.

Pathologically, the disease is characterized by perivascular inflammatory cell infiltration with demyelination, which is prominent in the thoracic spinal cord (*1, 28*). Immunologic parameters, including high HTLV-I proviral load (*34, 38, 52, 72*), increased spontaneous lymphoproliferation (*27, 69*), high antibody titers to HTLV-I both in sera and cerebrospinal fluid (CSF) (*56*), and increased cytokine production such as interferon (IFN)-γ and tumor necrosis factor (TNF)-α (*4, 39, 42, 55*) are abnormal in HAM/TSP patients in comparison with HTLV-I infected asymptomatic carriers. In addition, HAM/TSP patients show extraordinarily high levels of circulating CD8+ HTLV-I specific cytotoxic T-lymphocytes (CTL) in peripheral blood and CSF (*10, 29, 36, 39, 53, 71*). Immunohistochemical analysis of the affected spinal cord lesions in the early stage of patients with HAM/TSP reveals the presence of infiltrating CD4+ and CD8+ lymphocytes in which CD8+ cells predominate with duration of illness (*68*). In such lesions, the expression of HLA class I antigens (*47*) and the existence of HTLV-I specific CD8+ CTL have also been shown (*44*). Recent molecular biologic studies have demonstrated HTLV-I genome (*21, 33, 41*)

* Building 10, Room 5B-16, 9000 Rockville Pike, Bethesda, MD 20892, USA

and its expression in the affected lesions (*43, 49*). Collectively, these results support the view that virus-specific CD8+ T-lymphocytes may play a critical role in the immunopathogenesis of HAM/TSP.

In this review, current approaches to understanding the immunopathogenesis of HAM/TSP will be discussed based on newly developed molecular and immunological techniques that have been adapted to studies of HTLV-I proviral load, HTLV-I mRNA load, and HTLV-I Tax specific CD8+ T cells in HAM/TSP. These methods have enabled researchers to study previously inaccessible aspects of this disease and have allowed a more detailed analysis of viral/host immune responses as they relate to disease specificity in HAM/TSP. In particular, the role of HTLV-I specific CD8+ T cell immune responses will be highlighted.

Higher HTLV-I Proviral DNA Load in HAM/TSP than in Asymptomatic Carriers

It has been demonstrated that the HTLV-I proviral load in peripheral blood mononuclear cells (PBMC) of HAM/TSP patients is significantly higher than in HTLV-I asymptomatic carriers (*15, 34, 38, 72*). Recently, using newly established real time quantitative PCR techniques (TaqMan), this observation was strengthened by analyzing HTLV-I proviral DNA in PBMC of more than 400 subjects infected with HTLV-I (*52*). In HAM/TSP patients, HTLV-I proviral load was remarkably (approximately 16-fold) higher than that in asymptomatic carriers. In typical asymptomatic carriers the median HTLV-I proviral load is 0.1–1.0 copies per 100 PBMC, while in HAM/TSP patients it is 5.0–20.0 copies per 100 PBMC but may exceed 60 copies per 100 PBMC. In addition, HTLV-I proviral load of HTLV-I asymptomatic carriers in the families of HAM/TSP patients was higher than those of unrelated asymptomatic carriers suggesting that genetic factors may influence this load (*52, 53*).

Although HTLV-I can infect a wide range of human and non-human cells *in vitro* (*66*), it has been thought to preferentially infect CD4+ T cells *in vivo* (*60*). However, extraordinarily high HTLV-I proviral loads in some HAM/TSP cases led to a hypothesis that, *in vivo*, HTLV-I may also infect non-CD4+ T cells. Recent work using quantitative PCR (TaqMan) indicated that CD8+ T cells were also infected with HTLV-I *in vivo* (*50*) (certain HAM/TSP patients had as high as 60 copies of HTLV-I proviral DNA per 100 isolated CD8+ T cells). Hanon *et al.* (*20*) also showed that in natural HTLV-I infection, a significant proportion of CD8+ T cells were infected with HTLV-I using a sensitive flow cytometric technique. Interestingly, a portion of HTLV-I specific CD8+ CTL were also infected with HTLV-I, and HTLV-I protein expression in naturally infected CD8+ T cells rendered them susceptible to cytolysis mediated by autologous HTLV-I specific CD8+ CTL. HTLV-I infected CD8+ T cells may therefore have a role as a significant virus reservoir *in vivo* and may also serve to drive the high HTLV-I specific immune response observed in patients with HAM/TSP. Although high HTLV-I proviral load seems to be associated with pathogenesis of HAM/TSP, this is clearly not sufficient to cause disease as some patients have low levels of HTLV-I provirus.

Higher HTLV-I mRNA Load in HAM/TSP than in Asymptomatic Carriers

Even though a high proviral DNA load is characteristic of HAM/TSP patients, the expression of HTLV-I in PBMC appears to be very low (*14, 32, 63–65*). These observations have led a number of investigators to consider that HTLV-I may be latent in peripheral blood. However, this suggestion is difficult to reconcile with the observation of highly activated immune responses in HAM/TSP patients such as the existence of high anti-HTLV-I IgM antibodies suggesting continuous immune stimulation *in vivo* (*35, 54*), and the existence of chronically activated HTLV-I Tax specific CD8+ T cells (*51*). Moreover, it has been reported that a significant proportion of HTLV-I infected PBMC are capable of expressing HTLV-I proteins after short-term cultivation *in vitro* (*14, 19*), supporting the hypothesis that HTLV-I is not inherently latent. Recently, using a newly established real time quantitative RT-PCR technique (TaqMan), HTLV-I mRNA load in PBMC of HAM/TSP patients appeared to be significantly higher compared to that in HTLV-I asymptomatic carriers as reported previously (*12*). Although HTLV-I mRNA load is higher in PBMC of HAM/TSP patients, (and if we consider that one provirus DNA is equally expressing mRNA between MT-2 cells and HTLV-I infected PBMCs), it is estimated that 0.004–0.4% of PBMC are expressing HTLV-I mRNA in HAM/TSP patients. Recently, Moritoyo *et al.* (*48*) developed a novel, highly sensitive technique which can visualize and quantify fluorescent positive cells by laser scanning cytometry and have detected HTLV-I Tax protein in PBMC and CSF cells from HAM/TSP patients. In HAM/TSP patients, 0.02–0.54% of PBMC and 0.04–1.16% of CSF cells were positive for the HTLV-I Tax protein. Interestingly, HTLV-I Tax protein expression in CSF cells is higher than that in PBMC and is more frequent in HAM/TSP patients with shorter duration of illness. It is surprising that the percentage of HTLV-I Tax expressing PBMC in HAM/TSP patients measured by laser scanning cytometry is similar to the estimated HTLV-I mRNA load using the quantitative RT-PCR technique. This amount of HTLV-I mRNA and Tax protein expressing PBMC is very low if compared with HTLV-I proviral DNA load, suggesting that the majority of HTLV-I infected cells may be silent in peripheral blood, however, this amount seems to be sufficient to continuously activate the immune system *in vivo.*

HTLV-I Specific CD8+ CTL

Exceptionally high levels of HTLV-I specific CD8+ T cells have been demonstrated directly in PBMC and CSF of HAM/TSP patients (*10, 29, 39, 53, 71*). CTL activity was predominantly restricted to products of the HTLV-I *pX* gene. Although HTLV-I Env, Pol, Rof and Tof (*59*) could also be target proteins of HTLV-I specific CTL, the magnitude of CTL activity to HTLV-I Tax protein was much higher than to other HTLV-I antigens. Specifically, the HTLV-I Tax 11–19 peptide (LLFGYPVYV) was defined as an immunodominant epitope by HLA-A2 restricted CD8+ CTL (*36, 58*). This Tax specific CD8+ T cell has been shown to produce IFN-γ and TNF-α which are proinflammatory and neurotoxic cytokines in HAM/TSP patients (*39*). Recently, HTLV-I Tax peptide-loaded HLA-A2(*0201) dimers and tetramers were developed and used to demonstrate HTLV-I Tax specific HLA-A2-restricted CD8+ cells (*5, 16*). This technique allows not only for the quantification of the frequency of antigen-specific T cells, but also for the direct characterization of these cells. HTLV-I Tax 11–19-specific CD8+ cells from PBMC of HLA-A2 HAM/TSP patients were found to represent an extraordinarily high proportion of the total CD8+ population (as high as 25% of CD8+ cells in some HAM/TSP patients) (*16, 51, 71*) and showed cytotoxicity to HTLV-I Tax expressing target cells (*45*). More than 80% of these cells had a memory and/or effector phenotype and some were activated effector CTL *in vivo*, which contained perforin and readily lysed target cells (*51*). In addition, the recently described cytokine IL-15 may also be an important factor that functions to maintain CD8+ T cells *in vivo* (*3, 37*) since IL-15 mRNA expression was up-regulated in the PBMC from HAM/TSP patients (*2*).

An important question in HTLV-I immunobiology is how HTLV-I specific immune responses are continuously stimulated by HTLV-I *in vivo* in HAM/TSP patients. To date, it has been difficult to detect HTLV-I antigens in lymphocytes *ex vivo*. It has recently been shown that HTLV-I proviral load (DNA) correlated with the HTLV-I mRNA load (as measured by real time RT-PCR techniques (TaqMan)) and the frequency of HTLV-I Tax specific CD8+ T cell (*40, 51, 71*). These results indicate that HAM/TSP patients have more HTLV-I proviral DNA load than asymptomatic carriers, which may be associated with higher mRNA expression resulting in heightened virus-specific immune responses *in vivo*. Furthermore, these results may support the hypothesis that expansion of HTLV-I can occur in the presence of an activated immune system which is trying to control the infection in HAM/TSP patients.

It has been demonstrated that the frequency of HTLV-I specific CD8+ T cells in CSF is much higher than that in PBMC of HAM/TSP patients (*10, 16*). Recently, this accumulation was proved to be an HTLV-I specific phenomenon in HAM/TSP patients by comparing it with the frequency of cytomegalovirus specific CD8+ T cells (*53*). These results suggest that HTLV-I specific CD8+ T cells may preferentially migrate into the CSF from peripheral blood and/or that HTLV-I specific CD8+ T cells may selectively expand in this compartment. These two possibilities are not mutually exclusive. It has been demonstrated that activated T cells can migrate into the central nervous system (CNS) irrespective of antigen specificity, however, only CNS antigen-specific cells are retained (*23*). Recently, the increased expression of T cell activation markers has been demonstrated in HTLV-I Tax specific CD8+ T cells in PBMC of HAM/TSP patients, suggesting that these cells may have a selective advantage to migrate into the CNS of HAM/TSP patients (*51, 53*). Detailed analysis of *in vitro* spontaneous lymphoproliferation (SP) supports the idea that HTLV-I specific CD8+ T cells may selectively expand in CNS. SP was defined as the spontaneous proliferation of PBMC in the absence of exogenous antigens or stimulants *in vitro* (*27, 30*). This SP was reported to be similar to the inflammatory process in the CNS (*25*). The magnitude of the SP is more pronounced in HAM/TSP patients than in asymptomatic HTLV-I carriers. Recent data showed that the percentage of proliferating CD8+ T cells was 2–5 times higher than that of CD4+ T cells, and HTLV-I Tax 11–19-specific CD8+ cells also expanded in SP of PBMC from HAM/TSP patients (*62*). These results may support the hypothesis that chronically activated HTLV-I specific CD8+ T cells accumulate in the CNS and may be stimulated by the continuous expression of the target antigen in the affected lesions.

Model of Immunopathogenesis in HAM/TSP

The results summarized above demonstrate that virus-host immunological interactions play a pivotal role in HAM/TSP and suggest a number of models regarding pathogenesis of this disorder. Three major immunopathogenic models have been considered: 1) cytotoxic model; inflammatory HTLV-I infected T cells in the CNS or CNS resident glial cells may be presenting HTLV-I antigens and become direct targets for lysis by HTLV-I specific CTL (*29*); 2) autoaggressive bystander model; HTLV-I virus-specific cells that have migrated into the CNS that recognize HTLV-I infected cells may result in the release of cytokines and chemokines which may cause damage to nearby cells such as glial and neural cells (*26*); 3) autoimmune model; activated HTLV-I specific T cells cross-react with self antigens in the CNS (*24*) and/or virus specific immune responses lead to epitope spreading and recognition of multiple CNS antigens (*17*). It has not been proven which, if any, of these mechanisms predominate in HAM/TSP or if one or more of these mechanisms act in concert. Based on a large body of evidence and the results presented in this review, it is reasonable to suggest that HTLV-I specific CD8+ T cells contribute to the inflammatory process in CNS lesions of HAM/TSP regardless of which mechanism is involved. HTLV-I specific CD8+ T cells may kill HTLV-I antigen(s) expressing target cells directly by a perforin-dependent mechanism (*19*) as well as by the production of a large amount of

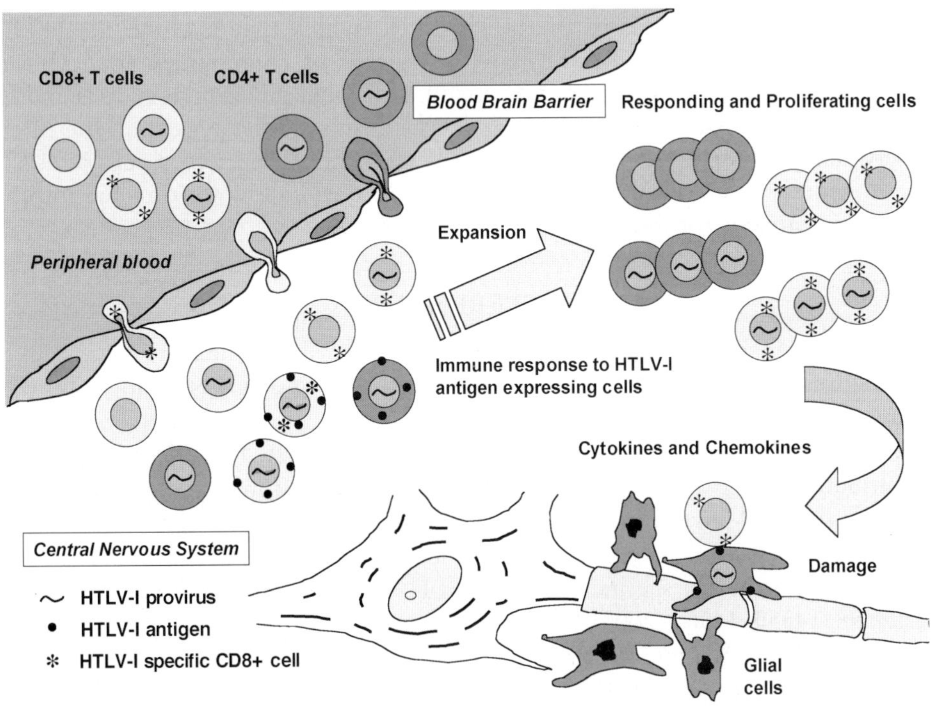

FIG. 1. Model of Immunopathogenesis in HAM/TSP

Activated T cells which may be HTLV-I infected CD4+, CD8+, or antigen-specific T cells migrate across the blood-brain barrier from the peripheral blood to the central nervous system (CNS). HTLV-I specific T cells and soluble immune responses recognize HTLV-I antigen expressing T cells and/or glial cells in the CNS. Recognition by functional HTLV-I specific CD8+ cells (such as CTL) can result in lysis of infected target cells and the release of a cascade of chemokines and cytokines. These molecules may also be produced by inflammatory HTLV-I infected CD4+ and CD8+ T cells expressing HTLV-I antigens. Cytokines such as IL-2 and IL-15 may help bystander T cells to expand and IFN-γ and TNF-α may be associated with damage to resident CNS cells such as glia and neurons.

matrix metalloproteinases (MMP-9), chemoattractants (macrophage inflammatory protein (MIP)-1α and -1β) and pro-inflammatory cytokines (TNF-α and IFN-γ) which can induce human leukocyte antigens (HLA) expression in neuronal cells and damage CNS tissue (*4, 16, 39*). HTLV-I Tax specific CD8+ CTL clones recognize and lysis self-peptide(s) pulsed target cells *in vitro* (*22*) which is consistent with any of the above theories.

Recently, HTLV-I proviral load in CSF of HAM/TSP patients was demonstrated to be more than 2 times higher than that in PBMC using quantitative TaqMan techniques (*53*). This result suggests that HTLV-I infected lymphocytes may preferentially migrate into the CSF from peripheral blood, and/or HTLV-I infected lymphocytes may selectively expand in this compartment. The exaggerated transmigrating activity of HTLV-I infected cells to the CNS tissue has been reported (*6, 18, 46, 61*). Furthermore, expansion of HTLV-I infected T cell clone has also been observed (*7, 9, 11, 70*).

A model for HTLV-I immunopathogenesis is represented in Fig. 1 which encompasses many aspects of HTLV-I T cell tropism and host immune responses to the virus. In the HAM/TSP lesion, HTLV-I infected CD4+ and CD8+ T cells can cross the blood brain barrier. In addition, activated CD8+ antigen-specific cells may also migrate into the CNS. These cells may recognize productively infected cells and respond by either direct lysis of the infected cell (which may also include HTLV-I infected resident glial cells) or through the release of chemokines and cytokines. These molecules can act to recruit and expand additional inflammatory cells and have been shown to be toxic to glial cells (Fig. 1). Intensive studies regarding the interaction between HTLV-I specific CD8+ T cells and HTLV-I infected cells will clarify the pathogenesis of HAM/TSP. This understanding will allow for directed immunotherapeutic strategies for the treatment of this disease.

REFERENCES

1. Akizuki, S., Yoshida, S., and Setoguchi, M. *In* "HTLV-I and the Nervous System," ed. G. C. Roman, J. C. Vernant, and M. Osame, pp. 253–260 (1989). Alan R. Liss, New York.
2. Azimi, N., Mariner, J., Jacobson, S., and Waldmann, T. A. How does interleukin 15 contribute to the pathogenesis of HTLV type 1-associated myelopathy/tropical spastic paraparesis? *AIDS Res. Hum. Retroviruses*, **16**, 1717–1722 (2000).
3. Azimi, N., Nagai, M., Jacobson, S., and Waldmann, T. A. IL-15 plays a major role in the persistence of Tax-specific CD8 cells in HAM/TSP patients. *Proc. Natl. Acad. Sci. USA*, **98**, 14559–14564 (2001).
4. Biddison, W. E., Kubota, R., Kawanishi, T., Taub, D. D., Cruikshank, W. W., Center, D. M., Connor, E. W., Utz, U., and Jacobson, S. Human T cell leukemia virus type I (HTLV-I)-specific CD8+ CTL clones from patients with HTLV-I-associated neurologic disease secrete proinflammatory cytokines, chemokines, and matrix metalloproteinase. *J. Immunol.*, **159**, 2018–2025 (1997).
5. Bieganowska, K., Hollsberg, P., Buckle, G. J., Lim, D. G., Greten, T. F., Schneck, J., Altman, J. D., Jacobson, S., Ledis, S. L., Hanchard, B., Chin, J., Morgan, O., Roth, P. A., and Hafler, D. A. Direct analysis of viral-specific CD8+ T cells with soluble HLA-

A2/Tax11–19 tetramer complexes in patients with human T cell lymphotropic virus-associated myelopathy. *J. Immunol.*, **162**, 1765–1771 (1999).
6. Cavrois, M., Gessain, A., Gout, O., Wain-Hobson, S., and Wattel, E. Common human T cell leukemia virus type 1 (HTLV-1) integration sites in cerebrospinal fluid and blood lymphocytes of patients with HTLV-1-associated myelopathy/tropical spastic paraparesis indicate that HTLV-1 crosses the blood-brain barrier via clonal HTLV-1-infected cells. *J. Infect. Dis.*, **182**, 1044–1050 (2000).
7. Cavrois, M., Leclercq, I., Gout, O., Gessain, A., Wain-Hobson, S., and Wattel, E. Persistent oligoclonal expansion of human T-cell leukemia virus type 1-infected circulating cells in patients with tropical spastic paraparesis/HTLV-1 associated myelopathy. *Oncogene*, **17**, 77–82 (1998).
8. de The, G. and Bomford, R. An HTLV-I vaccine: why, how, for whom? *AIDS Res. Hum. Retroviruses*, **9**, 381–386 (1993).
9. Eiraku, N., Hingorani, R., Ijichi, S., Machigashira, K., Gregersen, P. K., Monteiro, J., Usuku, K., Yashiki, S., Sonoda, S., Osame, M., and Hall, W. W. Clonal expansion within CD4+ and CD8+ T cell subsets in human T lymphotropic virus type I-infected individuals. *J. Immunol.*, **161**, 6674–6680 (1998).
10. Elovaara, I., Koenig, S., Brewah, A. Y., Woods, R. M., Lehky, T., and Jacobson, S. High human T cell lymphotropic virus type 1 (HTLV-1)-specific precursor cytotoxic T lymphocyte frequencies in patients with HTLV-1-associated neurological disease. *J. Exp. Med.*, **177**, 1567–1573 (1993).
11. Furukawa, Y., Fujisawa, J., Osame, M., Toita, M., Sonoda, S., Kubota, R., Ijichi, S., and Yoshida, M. Frequent clonal proliferation of human T-cell leukemia virus type 1 (HTLV-1)-infected T cells in HTLV-1-associated myelopathy (HAM-TSP). *Blood*, **80**, 1012–1016 (1992).
12. Furukawa, Y., Osame, M., Kubota, R., Tara, M., and Yoshida, M. Human T-cell leukemia virus type-1 (HTLV-1) Tax is expressed at the same level in infected cells of HTLV-1-associated myelopathy or tropical spastic paraparesis patients as in asymptomatic carriers but at a lower level in adult T-cell leukemia cells. *Blood*, **1**, 1865–1870 (1995).
13. Gessain, A., Barin, F., Vernant, J. C., Gout, O., Maurs, L., Calender, A., and de The, G. Antibodies to human T-lymphotropic virus type-I in patients with tropical spastic paraparesis. *Lancet*, **ii**, 407–410 (1985).
14. Gessain, A., Saal, F., Giron, M., Lasneret, J., Lagaye, S., Gout, O., De, T. G., Sigaux, F., and Peries, J. Cell surface phenotype and human T lymphotropic virus type 1 antigen expression in 12 T cell lines derived from peripheral blood and cerebrospinal fluid of West Indian, Guyanese and African patients with tropical spastic paraparesis. *J. Gen. Virol.*, **71**, 333–341 (1990).
15. Gessain, A., Saal, F., Gout, O., Daniel, M., Flandrin, G., de The, G., Peries, J., and Sigaux, F. High human T-cell lymphotropic virus type I proviral DNA load with polyclonal integration in peripheral blood mononuclear cells of French West Indian, Guianese, and African patients with tropical spastic paraparesis. *Blood*, **75**, 428–433 (1990).
16. Greten, T. F., Slansky, J. E., Kubota, R., Soldan, S. S., Jaffee, E. M., Leist, T. P., Pardoll, D. M., Jacobson, S., and Schneck, J. P. Direct visualization of antigen-specific T

cells: HTLV-1 Tax11–19-specific CD8(1) T cells are activated in peripheral blood and accumulate in cerebrospinal fluid from HAM/TSP patients. *Proc. Natl. Acad. Sci. USA*, **95**, 7568–7573 (1998).

17. Hafler, D. A. The distinction blurs between an autoimmune versus microbial hypothesis in multiple sclerosis. *J. Clin. Invest.*, **104**, 527–529 (1999).
18. Hanon, E., Goon, P., Taylor, G. P., Hasegawa, H., Tanaka, Y., Weber, J. N., and Bangham, C. R. High production of interferon gamma but not interleukin-2 by human T-lymphotropic virus type I-infected peripheral blood mononuclear cells. *Blood*, **98**, 721–726 (2001).
19. Hanon, E., Hall, S., Taylor, G. P., Saito, M., Davis, R., Tanaka, Y., Usuku, K., Osame, M., Weber, J. N., and Bangham, C. R. Abundant tax protein expression in CD4+ T cells infected with human T-cell lymphotropic virus type I (HTLV-I) is prevented by cytotoxic T lymphocytes. *Blood*, **95**, 1386–1392 (2000).
20. Hanon, E., Stinchcombe, J. C., Saito, M., Asquith, B. E., Taylor, G. P., Tanaka, Y., Weber, J. N., Griffiths, G. M., and Bangham, C. R. Fratricide among CD8(+) T lymphocytes naturally infected with human T cell lymphotropic virus type I. *Immunity*, **13**, 657–664 (2000).
21. Hara, H., Morita, M., Iwaki, T., Hatae, T., Itoyama, Y., Kitamoto, T., Akizuki, S., Goto, I., and Watanabe, T. Detection of human T lymphotropic virus type I (HTLV-I) proviral DNA and analysis of T cell receptor V beta CDR3 sequences in spinal cord lesions of HTLV-I-associated myelopathy/tropical spastic paraparesis. *J. Exp. Med.*, **180**, 831–839 (1994).
22. Hausmann, S., Biddison, W. E., Smith, K. J., Ding, Y. H., Garboczi, D. N., Utz, U., Wiley, D. C., and Wucherpfennig, K. W. Peptide recognition by two HLA-A2/Tax11–19-specific T cell clones in relationship to their MHC/peptide/TCR crystal structures. *J. Immunol.*, **162**, 5389–5397 (1999).
23. Hickey, W. F., Hsu, B. L., and Kimura, H. T-lymphocyte entry into the central nervous system. *J. Neurosci. Res.*, **28**, 254–260 (1991).
24. Hollsberg, P. and Hafler, D. Pathogenesis of diseases induced by human lymphotropic virus type I infection. *N. Engl. J. Med.*, **328**, 1173–1182 (1993).
25. Ijichi, N., Eiraku, N., and Osame, M. *In* "HTLV-I and the Nervous System," ed. G. C. Roman, J. C. Vernant, and M. Osame, pp. 242–259 (1989). Alan R. Liss, New York.
26. Ijichi, S., Izumo, S., Eiraku, N., Machigashira, K., Kubota, R., Nagai, M., Ikegami, N., Kashio, N., Umehara, F., Maruyama, I., and Osame, M. An autoaggressive process against bystander tissues in HTLV-I-infected individuals: a possible pathomechanism of HAM/TSP. *Med. Hypotheses*, **41**, 542–547 (1993).
27. Itoyama, Y., Minato, S., Kira, J., Goto, I., Sato, H., Okochi, K., and Yamamoto, N. Spontaneous proliferation of peripheral blood lymphocytes increased in patients with HTLV-I-associated myelopathy. *Neurology*, **38**, 1302–1307 (1988).
28. Izumo, S., Usuku, K., Osame, M., Machigashira, K., Johnson, M., and Nakagawa, M. *In* "HTLV-I and the Nervous System," ed. G. C. Roman, J. C. Vernant, and M. Osame, pp. 261–267 (1989). Alan R. Liss, New York.
29. Jacobson, S., Shida, H., McFarlin, D. E., Fauci, A. S., and Koenig, S. Circulating CD8+ cytotoxic T lymphocytes specific for HTLV-I pX in patients with HTLV-I

associated neurological disease. *Nature*, **348**, 245–248 (1990).

30. Jacobson, S., Zaninovic, V., Mora, C., Rodgers-Johnson, P., Sheremata, W. A., Gibbs, C. J., Jr., Gajdusek, C., and McFarlin, D. E. Immunological findings in neurological diseases associated with antibodies to HTLV-I: activated lymphocytes in tropical spastic paraparesis. *Ann. Neurol.*, **23**, S196–S200 (1988).
31. Kaplan, J. E., Osame, M., Kubota, H., Igata, A., Nishitani, H., Maeda, Y., Khabbaz, R. F., and Janssen, R. S. The risk of development of HTLV-I-associated myelopathy/tropical spastic paraparesis among persons infected with HTLV-I. *J. Acquir. Immune Defic. Syndr.*, **3**, 1096–1101 (1990).
32. Kinoshita, T., Shimoyama, M., Tobinai, K., Ito, M., Ito, S., Ikeda, S., Tajima, K., Shimotohno, K., and Sugimura, T. Detection of mRNA for the tax1/rex1 gene of human T-cell leukemia virus type I in fresh peripheral blood mononuclear cells of adult T-cell leukemia patients and viral carriers by using the polymerase chain reaction. *Proc. Natl. Acad. Sci. USA*, **86**, 5620–5624 (1989).
33. Kira, J., Itoyama, Y., Koyanagi, Y., Tateishi, J., Kishikawa, M., Akizuki, S., Kobayashi, I., Toki, N., Sueishi, K., and Sato, H. Presence of HTLV-I proviral DNA in central nervous system of patients with HTLV-I-associated myelopathy. *Ann. Neurol.*, **31**, 39–45 (1992).
34. Kira, J., Koyanagi, Y., Yamada, T., Itoyama, Y., Goto, I., Yamamoto, N., Sasaki, H., and Sakaki, Y. Increased HTLV-I proviral DNA in HTLV-I-associated myelopathy: a quantitative polymerase chain reaction study [published erratum appears in *Ann. Neurol.*, 1991, Apr.; **29**(4):363]. *Ann. Neurol.*, **29**, 194–201 (1991).
35. Kira, J., Nakamura, M., Sawada, T., Koyanagi, Y., Ohori, N., Itoyama, Y., Yamamoto, N., Sakaki, Y., and Goto, I. Antibody titers to HTLV-I-p40tax protein and gag-env hybrid protein in HTLV-I-associated myelopathy/tropical spastic paraparesis: correlation with increased HTLV-I proviral DNA load. *J. Neurol. Sci.*, **107**, 98–104 (1992).
36. Koenig, S., Woods, R. M., Brewah, Y. A., Newell, A. J., Jones, G. M., Boone, E., Adelsberger, J. W., Baseler, M. W., Robinson, S. M., and Jacobson, S. Characterization of MHC class I restricted cytotoxic T cell responses to tax in HTLV-1 infected patients with neurologic disease. *J. Immunol.*, **151**, 3874–3883 (1993).
37. Ku, C. C., Murakami, M., Sakamoto, A., Kappler, J., and Marrack, P. Control of homeostasis of CD8+ memory T cells by opposing cytokines. *Science*, **288**, 675–678 (2000).
38. Kubota, R., Fujiyoshi, T., Izumo, S., Yashiki, S., Maruyama, I., Osame, M., and Sonoda, S. Fluctuation of HTLV-I proviral DNA in peripheral blood mononuclear cells of HTLV-I-associated myelopathy. *J. Neuroimmunol.*, **42**, 147–154 (1993).
39. Kubota, R., Kawanishi, T., Matsubara, H., Manns, A., and Jacobson, S. Demonstration of human T lymphotropic virus type I (HTLV-I) tax-specific CD8+ lymphocytes directly in peripheral blood of HTLV-I-associated myelopathy/tropical spastic paraparesis patients by intracellular cytokine detection. *J. Immunol.*, **161**, 482–488 (1998).
40. Kubota, R., Kawanishi, T., Matsubara, H., Manns, A., and Jacobson, S. HTLV-I specific IFN-gamma+ CD8+ lymphocytes correlate with the proviral load in peripheral blood of infected individuals. *J. Neuroimmunol.*, **102**, 208–215 (2000).
41. Kubota, R., Umehara, F., Izumo, S., Ijichi, S., Matsumuro, K., Yashiki, S., Fujiyoshi, T., Sonoda, S., and Osame, M. HTLV-I proviral DNA amount correlates with infil-

vitro cultured peripheral lymphocytes from adult T-cell leukemia patients and healthy virus carriers. *Int. J. Cancer*, **36**, 1–7 (1985).

66. Trejo, S. R. and Ratner, L. The HTLV receptor is a widely expressed protein. *Virology*, **268**, 41–48 (2000).
67. Trujillo, J. M., Concha, M., Munoz, A., Bergonzoli, G., Mora, C., Borrero, I., Gibbs, C. J., and Arango, C. Seroprevalence and cofactors of HTLV-I infection in Tumaco, Colombia. *AIDS Res. Hum. Retroviruses*, **8**, 651–657 (1992).
68. Umehara, F., Izumo, S., Nakagawa, M., Ronquillo, A. T., Takahashi, K., Matsumuro, K., Sato, E., and Osame, M. Immunocytochemical analysis of the cellular infiltrate in the spinal cord lesions in HTLV-I-associated myelopathy. *J. Neuropathol. Exp. Neurol.*, **52**, 424-430 (1993).
69. Usuku, K., Sonoda, S., Osame, M., Yashiki, S., Takahashi, K., Matsumoto, M., Sawada, T., Tsuji, K., Tara, M., and Igata, A. HLA haplotype-linked high immune responsiveness against HTLV-I in HTLV-I-associated myelopathy: comparison with adult T-cell leukemia/lymphoma. *Ann. Neurol.*, **23**, S143–S150 (1988).
70. Wattel, E., Vartanian, J. P., Pannetier, C., and Wain-Hobson, S. Clonal expansion of human T-cell leukemia virus type I-infected cells in asymptomatic and symptomatic carriers without malignancy. *J. Virol.*, **69**, 2863–2868 (1995).
71. Yamano, Y., Nagai, M., Brennan, M., Mora, C. A., Soldan, S. S., Tomaru, U., Takenouchi, N., Izumo, S., Osame, M., and Jacobson, S. Correlation of human T-cell lymphotropic virus type 1 (HTLV-1) mRNA with proviral DNA load, virus-specific CD8(+) T cells, and disease severity in HTLV-1-associated myelopathy (HAM/TSP). *Blood*, **99**, 88–94 (2002).
72. Yoshida, M., Osame, M., Kawai, H., Toita, M., Kuwasaki, N., Nishida, Y., Hiraki, Y., Takahashi, K., Nomura, K., Sonoda, S. *et al.* Increased replication of HTLV-I in HTLV-I-associated myelopathy. *Ann. Neurol.*, **26**, 331–335 (1989).

ANIMAL DISEASE MODEL

PATHOGENESIS OF HAM/TSP IN A RAT MODEL

Takashi YOSHIKI

*Department of Pathology/Pathophysiology, Hokkaido University Graduate School of Medicine**

Among several strains of human T lymphocyte virus type I (HTLV-I) carrier rats, only WKAH rats with $RT1^k$ haplotype of rat major histocompatibility complex developed HAM/TSP (HTLV-I-associated myelophathy/tropical spastic paraparesis)-like myeloneuropathy (HAM rat disease). Symmetrical white matter degeneration of the spinal cord was caused by apoptotic death of the myelin forming cells, oligodendrocytes. Schwann cells of the peripheral nerve were similarly affected. Chronological examination showed that high provirus load and pX expression in the infected microglia/macrophages were evident in the affected spinal cord just before and at the appearance of apoptosis. High tumor necrosis factor (TNF)-α expressions in the affected spinal cord and high amounts of TNF-α in the cerebrospinal fluid were also observed. Moreover, HTLV-I infection increased the sensitivity of oligodendrocytes to TNF-α induced apoptosis. Marked suppression of bcl-2 expression was selectively observed in the isolated oligodendrocytes from the affected spinal cord but not in the microglia/macrophages. Collective evidence suggests that local expansion of HTLV-I provirus with increment of pX expression, followed by high TNF-α expression in the infected microglia/macrohages in the spinal cord leads to apoptosis of nearby oligodendrocytes with bcl-2 suppression, thus a major pathogenetic pathway in HTLV-I induced myeloneuropathy, HAM rat disease of WKAH rats. The HAM rat disease model described here may contribute greatly not only to investigating the etiopathogenesis of HAM/TSP in humans but also to selecting new drugs and effective vaccines for HTLV-I infection.

The first human infectious retrovirus, human T lymphocyte virus type I (HTLV-I) (*13, 21*), is now known to be a causative agent of adult T cell leukemia (ATL) (*2, 22*), HTLV-I-associated myelopathy/tropical spastic paraparesis (HAM/TSP) (*1, 12*), and HTLV-I uveitis (*8*). HTLV-I may also be associated with diverse inflammatory diseases, chronic destructive polyarthritis resembling rheumatoid arthritis (*10*), Sjögren syndrome (*19*), T cell alveolitis (*15*), polymyositis (*9*) and infective dermatitis (*6*). No oncogen is encoded in the virus, and insertional activation of cellular oncogenes cannot account for the malignant transformation. A unique pX region that contains four open reading frames exists in this virus, and one of the products of this gene, p40Tax nuclear phosphoprotein, functions as a transregulator of its own viral and cellular transcription (*23*). To investigate virus-host interactions and pathogenesis

* Sapporo 060-8638, Japan

of ATL and HTLV-I-associated diseases, we first established a rat model of HTLV-I infection (*17, 25*) and a disease model for HAM/TSP in humans (*3*). In our previous reviews (*24, 26*), it was shown that HTLV-I can infect and immortalize rat T cells with helper-inducer phenotype *in vitro* and can be transmitted into different strains of inbred rats. Only the WKAH (Wistar-King-Aptekman/Hokudai) strain rats with HTLV-I carrier develop a chronic progressive myeloneuropathy with hind limb paraparesis resembling HAM/TSP after a long incubation period of 15 months (HAM rat disease). In this review, neuropathological, molecular and chronological findings of the affected spinal cord lesions implicated in the pathogenesis of this rat model are described.

Development of HAM Rat Disease

WKAH rats inoculated with MT-2 cells either at birth or at 3 months of age became seronegative and seropositive HTLV-I carriers, respectively, as determined by the presence of provirus DNAs in peripheral blood mononuclear cells (PBMCs) and serum antibody against HTLV-I (*3, 14*). In the seronegative HTLV-I carrier rats, gait disturbance was evident 15 months after the HTLV-I infection and gradually hind limb paraparesis occurred; some rats had severe paraparesis and urinary incontinence (Fig. 1A). Seropositive carrier rats showed gait disturbance 16 months after the infection, and hind limb paraparesis occurred in some rats at the 17th month and progressed rapidly even in a one month period. Seronegative and seropositive HTLV-I carriers of other strains (F344, ACI, and LEW), WKAH rats inoculated with HTLV-I unrelated HUT-78 cells and untreated WKAH rats at comparable ages showed no clinical symptoms. In a separate experiment, findings in a group of 3-month-old WKAH rats inoculated with 10^8 MT-2 cells were compared with cases of inoculation with 10^7 MT-2 cells to determine if a high dose of MT-2 cell inocula (10^8 cells) accelerated the onset of disease. Rats inoculated with 10^8 MT-2 cells began to develop hind limb paraparesis after age 15 months whereas no rat inoculated with 10^7 MT-2 cells showed clinical symptoms after similar incubation periods. It appears that a high dose of MT-2 cells is more effective for the induction of HAM rat disease and accelerates the onset.

Neuropathology of HAM Rat Disease

1. *Spinal cord lesions (3, 14)*

Chronological histopathological change of the spinal cord in seronegative HTLV-I carriers showed that small demyelinated foci with a cluster of activated microglial cells began to appear in marginal areas of anterolateral funiculi of the upper thoracic cord at age 9 months, 6–7 months before any clinical manifestation. Vacuolation of varying small sizes was noted in demyelinated areas and increased in number and size. The lesion gradually expanded and spread to the marginal areas of mainly anterior and lateral funiculi of the mid-to-upper thoracic cord and lower cervical cord with activated microglia/macrophage infiltration, astrogliosis and remyelination up to 18 months after MT-2 inoculation. Posterior funiculi were also affected but apparently to a lesser extent. The white matter degeneration progressed more slowly after age 18 months. At 19–26 months after MT-2 inoculation, the end-stage of HAM rat

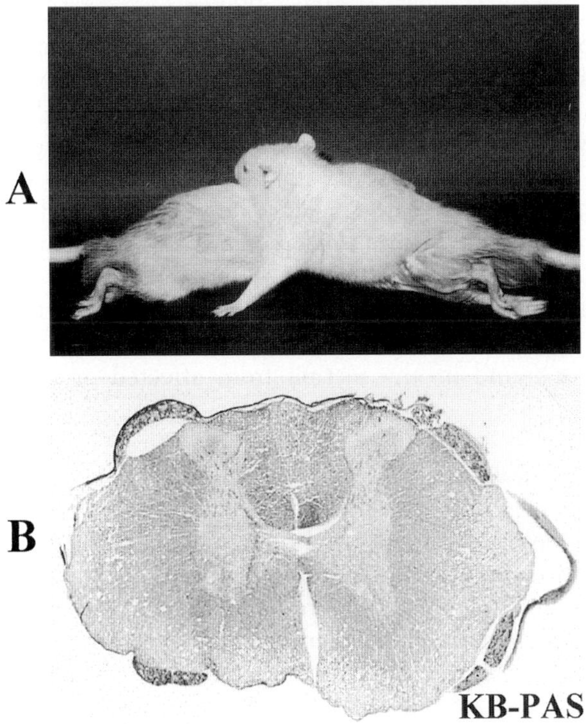

FIG. 1. HTLV-I-associated myelopathy (HAM) rat disease
A. Macroscopic view of two seronegative HAM rats. B. Histopathological findings of HAM rat disease. Symmetrical distribution of the damage in the white matter is shown, as a whole view of the affected thoracic cord (KB-PAS stain).

disease revealed a symmetrical myelin loss in the marginal area at the level of mid-to-lower cervical, thoracic, and upper lumbar cord, and was accentuated in the mid-to-upper thoracic cord. The lesion was characterized by extensive demyelination, vacuolations of various sizes, activated microglia/macrophage infiltration, and astrocytic proliferation (Fig. 1B). The anterior and lateral columns were more severely affected than were the posterior columns. The grade of astroglial proliferation and microglia/macrophage infiltration generally correlated with the severity of myelin loss and vacuolations. Lymphocytic infiltration was absent throughout the disease process.

With high power magnification, apoptotic cells appeared in the affected lesion with a condensed chromatin into the crescent along the nuclear envelope or with a nucleus into a black hole, as early as 7 months after MT-2 inoculation. Some had intact myelinated fibers and others appeared in the center of the demyelinated area or in vacuolated foci of the marginal area. The occurrence was predominant in the anterior and lateral columns of the spinal cord and most frequent in mid-to-upper thoracic spinal cord in rats at age 21 months. Spinal cord lesions in the seropositive HAM rats examined at 18 months after MT-2 inoculation were basically the same as observed in seronegative HAM rats.

Electron microscopically, a considerable number of oligodendrocytes showed morphological characteristics of apoptosis, the nuclear chromatin had collapsed down into peripheral crescents along the nuclear envelope or into a uniformly dense fragmentation with normal-appearing intracellular organelles. Apoptosis of various stages was observed in the affected lesion, and usually occurred in close vicinity to a small demyelinated lesion. Macrophages contained phagocytosed cell fragments or apoptotic bodies, often present in demyelinated lesions or perivascular spaces of the affected lesion. The majority of apoptotic cells was considered to be oligodendrocytes, based on their location, size and organella, some of which produced myelin sheaths. Apoptotic macrophages filled with myelin debris were sometimes present in the demyelinated lesion. Macrophages with undigested myelin were observed not only in demyelinated areas but within the vacuoles delineated with a thin myelin sheath. Axons in close contact with macrophages were generally preserved. Virus particles were never found in any section from affected or nonaffected lesions of the spinal cord.

2. Peripheral nerve lesions (14)

Foamy macrophages and demyelination began to appear at age 15 months, they gradually became widespread and affected the tibial nerve at age 21 months, the end-stage of HAM rat disease. The sciatic nerve had a loss of myelin, scattered vacuolations and foamy macrophages. Axons were relatively well preserved. In the large vacuoles or intramyelinic space, lipid-laden macrophages were often present. Lymphocytic infiltration was not evident. The sciatic nerve was more severely damaged than was the tibial nerve.

Electron microscopically, major neuropathological changes in the peripheral nerves were a reduced number in myelinated fibers, numerous demyelinated axons with macrophage infiltrations and frequent incompletely remyelinated axons. Vacuoles were observed as sharply outlined spaces surrounded by a thin myelin sheath or as periaxonal edema. Occasionally, macrophages containing myelin debris and lipid droplets were present in large vacuoles or near demyelinated axons. In the longitudinal sections, macrophages invaded the periaxonal and intramyelinic spaces, most likely passing through the node of Ranvier. As seen in oligodendrocytes in the affected spinal cord, the nucleus of Schwann cells collapsed into a uniform dense fragmentation with condensed cytoplasm, which suggested apoptotic death of Schwann cells. The axon appeared morphologically intact. Virus particles were not present in any sections from affected or nonaffected lesions of the peripheral nerve.

Chronological Analyses of ED-1 and GFAP-positive Cells (11)

In a quantitative analysis of the time kinetics of ED-1-positive cells, a few ED-1-positive small round cells (activated microglial cells) appeared from 9 months after the infection and gradually increased. After 15 months, both small round and large foamy cells (activated microglias and infiltrating macrophages) rapidly increased (Fig. 2A). In the controls, these cells began to appear the 13th month after the infection and gradually increased to reach half the number in HAM rats. Mean number of ED-1-positive cells in diseased rats after the 18th month was 393±164 (n=11, mean±SD) cells per section, a value significantly higher than that for aged controls (235±38, n=5) (P<0.05, Student's t test). Astrogliosis with a gemistocytic appearance stained with anti-GFAP antibody, which was observed in marginal areas of

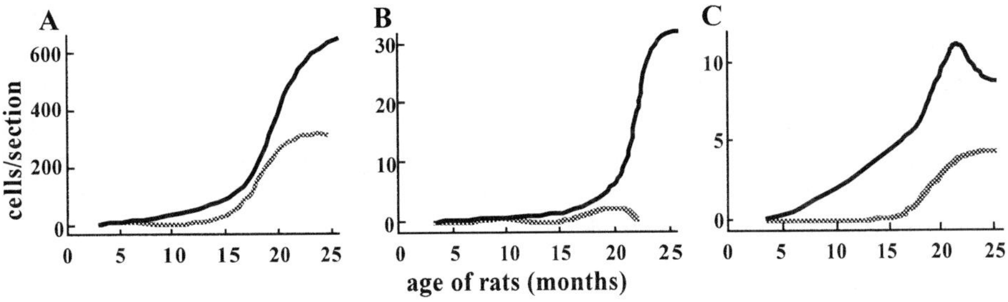

FIG. 2. Chronological change of ED-1-positive cells (A), GFAP-positive astrocytes (B), and TUNEL-positive apoptotic cells (C) in the affected spinal cord of HAM (solid lines) rats and normal (dotted lines) rats

Table I. Cell Type of Apoptotic Cells in the Thoracic Spinal Cord of HAM Rats

Age (months)	No. rats	No. apoptotic cells	Percent of double positive cells[a]		
			MOG	ED-1	GFAP
9–12	5	1.3±0.5	78.8±6.8	—	—
21–26	4	8.5±1.4	61.1±5.2	29.3±8.7	1.1±2.1

[a] TUNEL and immunostaining.

anterior and lateral funiculi, drastically increased 20 months after the infection to a maximum of 31 cells per section in HAM rats, but there were few in the aged controls (Fig. 2B).

Chronological Analysis and Characterization of Cells with DNA Fragmentation

The number of apoptotic cells in thoracic spinal cord of rats with HTLV-I infection was chronologically analyzed using the TUNEL method (*11*). Apoptotic cells first appeared 7 months after HTLV-I infection, and the number increased gradually up to a maximum of 12 cells per section at the 21st month (Fig. 2C). Although apoptotic cells were also present in aged controls, they began to appear in rats at about 18 months and were significantly lower in number than in HAM rats (2.3±1.8 in 5 aged controls and 7.8±2.7 in 10 HAM rats at the late disease phase).

The origin of these apoptotic cells was determined using both TUNEL and immunohistochemical methods. In asymptomatic young carriers at 9 to 12 months after the infection, about 80% of all apoptotic cells were stained with anti-myelin oligodendrocyte glycoprotein (MOG) antibody, but there was no evidence of apoptotic cells with ED-1 or anti-GFAP stains in the serial sections (Table I), indicating that most of the apoptotic cells were oligodendrocytes. While more than 60% of apoptotic cells were also positive with the anti-MOG stain, in HAM rats with overt disease a considerable number of apoptotic cells stained with ED-1. Very few apoptotic astrocytes were evident in HAM rats with overt disease.

In situ Detection of HTLV-I Provirus DNA in the Affected Spinal Cord

To identify infected cells in the affected lesions, *in situ* hybridization of amplified fragments of the provirus DNA by PCR on thin sections was carried out, plus immunohistochemistry on the same sections. The provirus DNA was localized in some microglia/macrophages in the spinal cord lesion (*5*). In addition, the HTLV-I provirus was clearly evident not only in ED-1-negative lymphoid cells but also in ED-1-positive macrophages from lymph nodes. These observations suggest that cells of microglia/macrophage lineage may be one of the dominant viral reservoirs in the spinal cords and lymph nodes in HAM rat disease.

Selective Expression of pX mRNAs in the Affected Spinal Cord and Peripheral Nerve

Next, the presence of proviral DNAs and mRNA transcripts of the *HTLV-I pX* gene were examined in various perfused organ tissues of HAM rats, using PCR and RT-PCR (*18*). HTLV-I proviral DNAs were consistently detected in all tissues examined, including the central nervous system (CNS), although integration in the sciatic nerves could not be examined because of inadequate tissue amounts. In the CNS (n=7), pX mRNAs were selectively expressed in 5 diseased spinal cords and in 3 sciatic nerves but not in other unaffected tissues (cerebra and cerebella) (Fig. 3A). The collective evidence strongly suggests that selective activation of HTLV-I integrated in the target spinal cord and peripheral nerve appears to be causally related to the pathogenesis of HAM rat disease. PX mRNAs were also detected in peripheral blood mononuclear cells (2/6), lymph nodes (4/7), spleens (1/7), hearts (2/7), and livers (1/7) (Table I), although these organs were histologically normal.

Expression of the pX mRNAs in in vitro-separated Glial Cells (4)

The pX mRNAs were detected at least in *in vitro*-purified microglial cells and oligodendrocytes from spinal cords of 9-month-old rats with HTLV-I infection. Astroglias from adult spinal cords were not examined, since they were technically difficult to purify *in vitro*. The ratio of the pX:β-actin in microglial cells was relatively

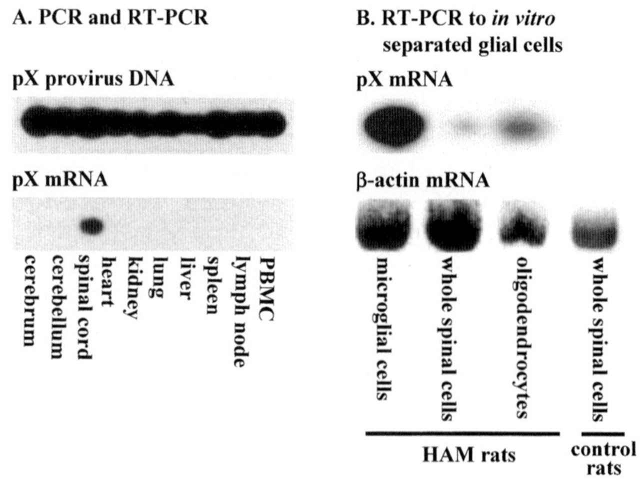

FIG. 3. Local and cell specific expression of pX mRNA in the affected spinal cord of HAM rats

higher than those in oligodendrocytes and whole spinal cord cells before the separation, which was detected by NIH image (Fig. 3B).

Immunohistochemical Detection of Tax Protein in the Affected Spinal Cord

The HTLV-I Tax protein in the spinal cord of HAM rats was immunohistochemically determined using a specific monoclonal antibody, Lt-4, against p40Tax of HTLV-I (*4, 16*). Specimens from rats at 7 (n=4), 9 (n=2), and 12 (n=2) months after the HTLV-I infection were examined. Although the type of glial cells with Lt-4-positive stain could not be identified, the production of p40Tax protein was evident in some glial cells of both white and gray matter from all WKAH rats with HTLV-I infection. Some glial cells in the nucleus and in the cytoplasm were stained. Positive staining was also seen in probable vascular endothelial cells. No staining was observed in either the white or gray matter of the spinal cords of the control rats.

Cytokine Expression in Affected Lesions of HAM Rat Disease

Since the nuclear protein Tax has been reported to transactivate a series of cytokines, mRNA transcripts for several cytokines were determined in RNAs isolated from HAM rat tissues, with emphasis on those from the affected spinal cord and sciatic nerve (*18*). Tumor necrosis factor (TNF)-α mRNAs were detected in affected spinal cords in 7 of 7 HAM rats and in the sciatic nerves in 5 of 7, but far less frequently in age-matched untreated controls (1/3 and 0/3). In the experimental allergic encephalitis (EAE) rats, TNF-α mRNAs were detected in the affected spinal cord (2/3) but not in unaffected sciatic nerves (0/3). Simultaneous investigation of TNF receptor 1 (TNFR1) expression in the spinal cord and peripheral nerve showed that TNFR1 was equally expressed in the spinal cord and sciatic nerve, regardless of health or disease. The levels of TNF-α in colony-stimulating factor (CSF) were elevated significantly in the HAM rats, but not in EAE rats or controls. Serum levels of TNF-α were equally undetectable in rats with HAM or EAE and untreated controls. Interferon (IFN)-γ mRNAs were detected with high frequency in the spinal cord and sciatic nerves of HAM and EAE rats but not in untreated controls. We found that interleukin (IL)-2 mRNAs were present in the affected lesions of all EAE rats but less frequently in HAM rats. This finding may coincide with the absence of T cell infiltration in the affected lesions of HAM rat disease (*3, 14*). mRNAs of IL-1α, IL-1β, and IL–6 were detected equally in the spinal cord and sciatic nerve of HAM and EAE rats and in untreated controls.

Sensitivity of Oligodendrocytes to TNF-α-induced Cell Death

To clarify the susceptibility of spinal oligodendrocytes from HTLV-I-infected rats against TNF-α-induced cell death, *in vitro*-separated oligodendrocytes were cultured under conditions with a series of TNF-α concentrations (*4*). The experiments were done under serum-free conditions and the decision was made 24 h after the TNF-α treatment. Oligodendrocyte death of HTLV-I-infected rats was susceptible to TNF-α treatment and was markedly increased when compared with findings in control rats at the concentration of TNF-α between 5 and 50 ng/ml ($P<0.01$) (Fig.

4). In contrast, the number of non-viable oligodendrocytes of control rats increased only at the concentration of 100 ng/ml and was closer to that for the HTLV-I-infected rats. Oligodendrocyte death of untreated rats by addition of this level of TNF-α concentration was also reported by other investigators (*20*). The TUNEL method was done to detect apoptotic death of oligodendrocytes by TNF-α. The TNF-α treatment significantly increased TUNEL-positive apoptotic oligodendrocytes from rats with HTLV-I infection compared with untreated oligodendrocytes.

Chronological and Semiquantitative Analyses of HTLV-I Provirus DNA and pX mRNA

To determine the possible participation of HTLV-I infection in the pathogenetic process of HAM rat disease, amounts of the provirus DNA and expression levels of the pX mRNA were measured in spinal cords of HTLV-I-infected WKAH rats each month after the infection, using competitive PCR and RT-competitive PCR (Tomaru, U., unpublished data). Although provirus DNA was detected consistently from 4 to 23 months after the infection, the amounts increased to a peak (about 400 molecules in 2 μg of the spinal cord DNA) in the early asymptomatic stage of the disease course, about 7 months after infection (Fig. 5). Relative amounts of pX mRNAs also changed, in a pattern similar to that of provirus DNA from 3 to 18 months; thereafter only minimal levels of pX mRNAs were evident, although provirus DNA was consistently observed at about 200 molecules in 2 μg of the DNA. In this early asymptomatic period, no significant histological findings were observed in spinal cords except for a few apoptotic oligodendrocytes.

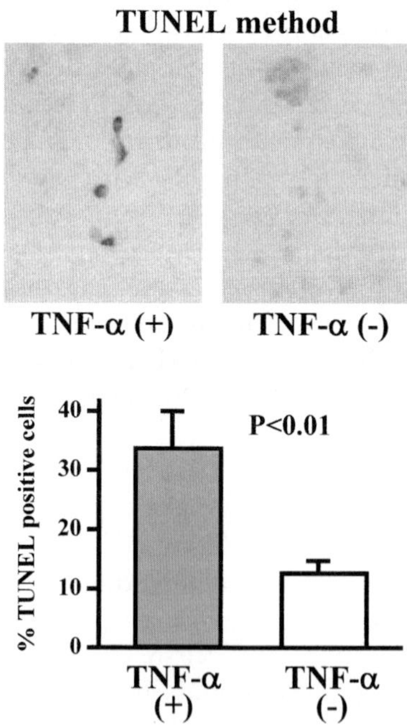

FIG. 4. Induction of apoptosis in the separated oligodendrocytes by TNF-α *in vitro*
Oligodendrocytes from WKAH rats were separated and incubated for 24 h with or without TNF-α (100 ng/ml).

Chronological Analyses of Expression of Apoptosis-related Genes

Since apoptotic death of oligodendrocytes in spinal cords of HAM rat disease is strongly correlated to the development of demyelination (*14*), we examined apoptosis-related host genes, including *TNF-α*, *bcl-2*, *bax* and *p53*, using RT-competitive PCR (Tomaru, U., unpublished data). The expression of TNF-α mRNA increased gradually starting 7 months after the infection, but there was only a minimal level of expression in untreated controls at all ages (Fig. 6). Levels of bcl-2 mRNA expression in spinal cords of HTLV-I-infected WKAH rats were strongly suppressed from 7 to 12 months and the expression gradually recovered after 15 months. In untreated control rats, the expression was consistently high throughout their lifetime. Levels of the bax mRNA expression in HTLV-I-infected WKAH rats changed, with a pattern similar to the findings in case of bcl-2 mRNA. On the other hand, levels of the p53 mRNA expression in HTLV-I-infected WKAH rats were more than twice as high than the findings in control rats at all ages examined.

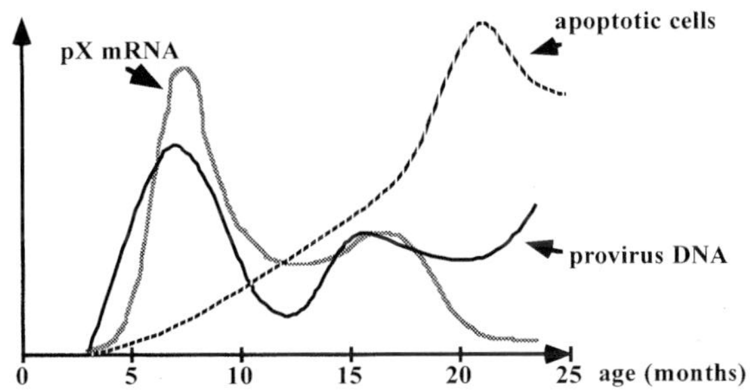

FIG. 5. Chronological relationship among provirus DNA, pX mRNA and apoptosis in the spinal cord of HAM rat disease

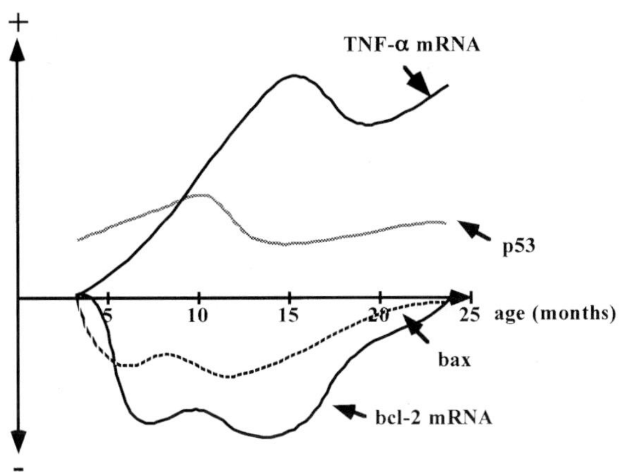

FIG. 6. Chronological relationship among apoptosis-related genes in the spinal cord of HAM rat disease

bcl-2 Expression in Oligodendrocytes and Microglial Cells

Since the bcl-2 is a major anti-apoptotic gene in the CNS (*7*), we examined the expression levels of bcl-2 mRNA in *in vitro*-separated oligodendrocytes and microglial cells of the spinal cords of HTLV-I-infected rats using semi-quantitative RT-PCR assay. In oligodendrocytes separated from the spinal cords of HTLV-I-infected rats 7, 9 and 12 months after the infection, the expression of bcl-2 mRNAs was strongly suppressed compared with findings in 9-month-old control rats (Fig. 7). The expression tended to recover at the late stage of the disease (18 months after the infection). On the other hand, bcl-2 expression was not suppressed in microglial cells separated from the 10-month-old HTLV-I-infected rats but, rather, was enhanced, as compared to findings in the 10-month-old control rats.

Amounts of Provirus DNA and pX, TNF-α and bcl-2 mRNAs in HAM-susceptible WKAH and Other HAM-resistant Strains

Although all rat strains tested were susceptible to HTLV-I infection, HAM rat disease developed only in WKAH strain rats. To investigate the role of HTLV-I infection and apoptosis-related genes in the pathogenesis, we compared the chronological distribution of amounts of provirus DNA and pX, TNF-α and bcl-2 mRNAs in spinal cords between HAM-susceptible WKAH strain and HAM-resistant ACI and LEW strains with HTLV-I infection (Tomaru, U., unpublished data). Using PCR, HTLV-I provirus DNA was detected in spinal cord of all rat strains examined, WKAH (26/26), ACI (12/24) and LEW (10/14). However, the amounts of provirus DNA in the spinal cords of ACI and LEW rats were smaller than that in WKAH rats throughout their lifetime and the provirus expansion observed in WKAH rats did not occur in ACI or LEW rats (Fig. 8). On the other hand, in the cerebrum, (which is unaffected by HTLV-I infection (*3*)), no significant difference in amounts of provi-

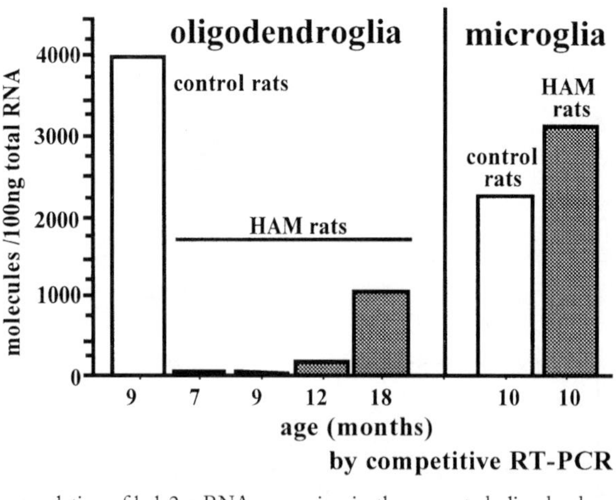

FIG. 7. Down-regulation of bcl-2 mRNA expression in the separated oligodendrocytes *in vitro* but not in the microglial cells

Closed columns indicate the amounts of bcl-2 molecules in the spinal glial cells from HTLV-I carrier WKAH rats and open columns indicate those from the control rats.

rus was observed among these three strains and amounts were less than 100 molecules in 2 μg DNA. In other unaffected tissues such as kidney and spleen, the amounts were also less than 100 molecules, with similar levels among strains. As a result of examination of HTLV-I pX mRNA by RT-PCR, the pX mRNA was selectively expressed in spinal cords of WKAH rats (20/26), but was detected in only 1 of 14 rats in each HAM-resistant strain. The expression levels in the two HAM-resistant rats were under the minimal limit of sensitivity of the RT-competitive PCR we used (under 9 molecules per 100 ng of total RNA) (Fig. 8). Levels of TNF-α mRNA expression in spinal cords of HTLV-I-infected WKAH rats gradually increased from 7 to 23 months after the infection, but those of HTLV-I-infected ACI and LEW rats were consistently low (Fig. 9). Expression of the bcl-2 mRNA in spinal cords of both HTLV-I-infected HAM-resistant strains tended to be higher than findings in untreated controls (Fig. 10). The suppression of bcl-2 mRNAs seen in WKAH rats with HTLV-I infection was not evident in either strain.

Summary and Conclusions

Among lifelong HTLV-I carriers in several rat strains, only WKAH rats with the $RT1^k$ haplotype developed chronic progressive myeloneuropathy, resembling

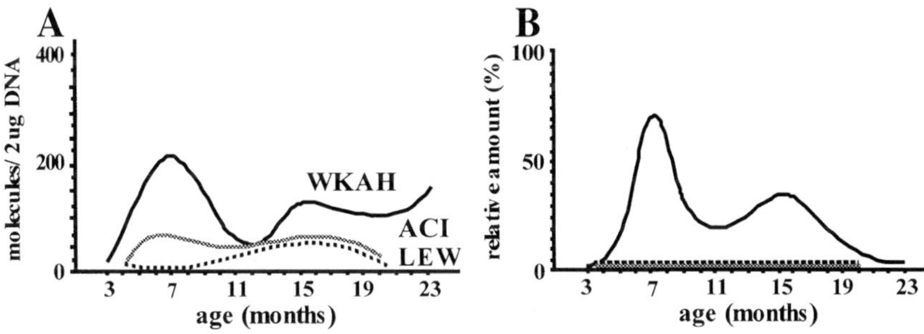

FIG. 8. Chronological analyses of provirus load and amounts of pX mRNAs in HAM-sensitive and -resistant strains of rats

More than 3 rats of each age were used. Amounts were measured by competitive PCR for provirus DNA (A) and competitive RT-PCR for pX mRNAs (B).

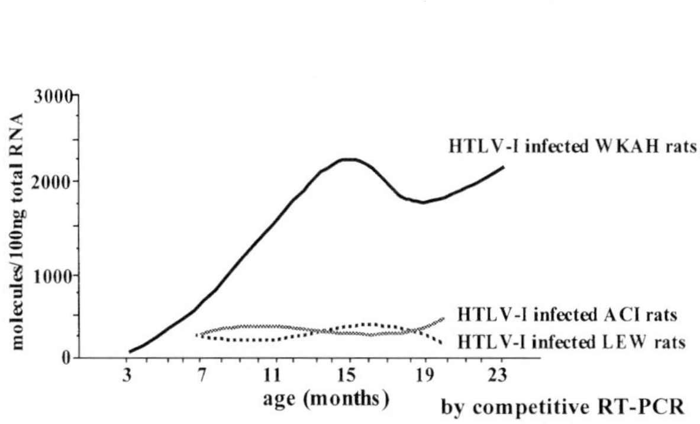

FIG. 9. Chronological analyses of TNF-α mRNA expression in the spinal cords of HAM-sensitive WKAH and HAM-resistant ACI and LEW strains of rats with HTLV-I infection

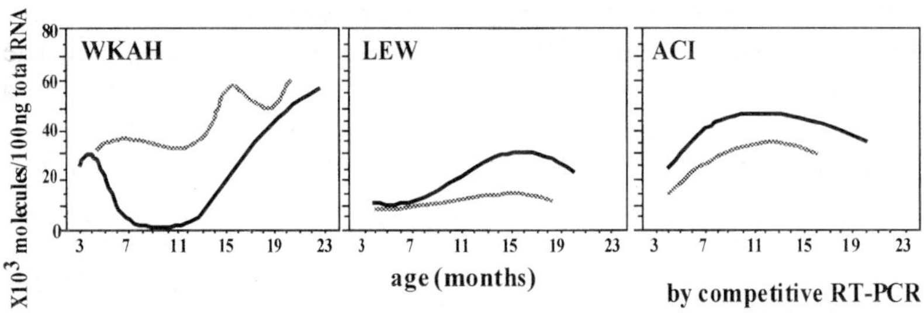

FIG. 10. Chronological analyses of bcl-2 mRNA expression in the spinal cords of HAM-sensitive WKAH and HAM-resistant LEW and ACI strains of rats with HTLV-I infection (solid lines) Dotted lines show age-matched normal rats.

HAM/TSP clinically and histologically in humans, designated as HAM rat disease and having a long incubation period. Neuropathologically, apoptotic death of the myelin forming cells, oligodendrocytes and Schwann cells starting 7 months after the infection appears to be the main cause of HAM rat disease. It was shown that the majority of HTLV-I-infected cells in the spinal cords are resident microglia/macrophages. TNF-α produced by infected microglia/macrophages appears to induce apoptosis of nearby oligodendrocytes in the spinal cord lesions. Moreover, HTLV-I infection increases the sensitivity of oligodendrocytes to TNF-α induced apoptosis. Expansion of HTLV-I provirus and increment of pX expression seen 7 months after the infection are unique events in the spinal cords of HAM-sensitive WKAH rats but not in those of HAM-resistant strain rats. Marked suppression of bcl-2 expression is found in the spinal cords of HAM rats from 7 to 12 months after infection. This suppression is evident in the isolated oligodendrocytes but not in the microglial cells. Suppression of bcl-2 expression is not observed in the spinal cords of HAM-resistant strain rats with HTLV-I infection. Collective evidence suggests that local expansion of HTLV-I provirus with increment of pX, followed by high TNF-α expressions in the infected microglia/macrophages leads to apoptosis of nearby oligodendrocytes with bcl-2 suppression, thus a major pathogenetic pathway in HTLV-I induced myeloneuropathy, HAM rat disease of WKAH rats. This rat model may greatly contribute not only to investigating the etiopathogenesis of HAM/TSP but also to the screening of new drugs and vaccines for HTLV-I infection in humans.

Acknowledgments

The author thanks many colleagues for their collaboration and valuable discussion and Ms. M. Ishizaka for secretarial assistance. This study was supported by grants from the Ministry of Education, Culture, Sports, Science and Technology, and the Ministry of Health, Labour and Welfare of Japan.

REFERENCES

1. Gessain, A., Barin, F., Vernant, J. C., Gout, O., Maurs, L., Calender, A., and de The, G. Antibodies to the human T cell lymphotropic virus type I in patients with tropical spastic paraparesis. *Lancet*, **i**, 407–410 (1985).
2. Hollsberg, P. and Hafler, D. A. Seminars in medicine of Beth Israel Hospital, Boston.

Pathogenesis of diseases induced by human lymphotropic virus type I infection. *N. Engl. J. Med.*, **328**, 1173–1182 (1993).

3. Ishiguro, N., Abe, M., Seto, K., Sakurai, H., Ikeda, H., Wakisaka, A., Togashi, T., Tateno, M., and Yoshiki, T. A rat model of human T lymphocyte virus type I (HTLV-I) infection. 1. Humoral antibody response, provirus integration, and HTLV-I-associated myelopathy/tropical spastic paraparesis-like myelopathy in seronegative HTLV-I carrier rats. *J. Exp. Med.*, **176**, 981–989 (1992).
4. Jiang, X., Ikeda, H., Tomaru, U., Morita, K., Tanaka, Y., and Yoshiki, T. A rat model for human T lymphocyte virus type I-associated myeloneuropathy: down-regulation of bcl-2 expression and increase in sensitivity to TNF-α of the spinal oligodendrocytes. *J. Neuroimmunol.*, **106**, 105–113 (2000).
5. Kasai, T., Ikeda, H., Tomaru, U., Yamashita, I., Ohya, O., Morita, K., Wakisaka, A., Matsuoka, E., Moritoyo, T., Hashimoto, K., Higuchi, I., Izumo, S., Osame, M., and Yoshiki, T. A rat model of human T lymphocyte virus type I (HTLV-I) infection: *in situ* detection of HTLV-I provirus DNA in microglia/macrophages in affected spinal cords of rats with HTLV-I-induced chronic progressive myeloneuropathy. *Acta Neuropathol.*, **97**, 107–112 (1999).
6. LaGrenade, L., Hanchard, B., Fletcher, V., Cranston, B., and Blattner, W. Infective dermatitis of Jamaican children: a marker for HTLV-I infection. *Lancet,* **i**, 1345–1347 (1990).
7. Merry, D. E., Veis, D. J., Hicky, W. F., and Korsmeyer, S. J. bcl-2 protein expression is widespread in the developing nervous system and retained in the adult PNS. *Development*, **120**, 301–311 (1994).
8. Mochizuki, M., Watanabe, T., Yamaguchi, K., Takatsuki, K., Yoshimura, K., Shirao, M., Nakashima, S., Mori, S., Araki, S., and Miyata, N. HTLV-I uveitis: a distinct clinical entity caused by HTLV-I. *Jpn. J. Cancer. Res.*, **83**, 236–239 (1992).
9. Morgan, O. S., Rodgers-Johnson, P., Mora, C., and Char, G. HTLV-I and polymyositis in Jamaica. *Lancet*, **i**, 1184–1187 (1989).
10. Nishioka, K., Maruyama, I., Sato, K., Kitajima, I., Nakajima, Y., and Osame, M. Chronic inflammatory arthropathy associated with HTLV-I. *Lancet*, **i**, 441 (1989).
11. Ohya, O., Ikeda, H., Tomaru, U., Yamashita, I., Kasai, T., Morita, K., Wakisaka, A., and Yoshiki, T. Human T-lymphocyte virus type I (HTLV-I)-induced myeloneuropathy in rats: oligodendrocytes undergo apoptosis in the presence of HTLV-I. *APMIS*, **108**, 459–466 (2000).
12. Osame, M., Usuku, K., Izumo, S., Ijichi, N., Amitani, H., Igata, A., Matsumoto, M., and Tara, M. HTLV-I associated myelopathy, a new clinical entity. *Lancet*, **i**, 1031–1032 (1986).
13. Poiesz, B. J., Ruscetti, F. W., Gazdar, A. F., Bunn, P. A., Minna, J. D., and Gallo, R. C. Detection and isolation of type-C retrovirus particles from fresh and cultured lymphocytes of a patient with cutaneous T-cell lymphoma. *Proc. Natl. Acad. Sci. USA*, **77**, 7415–7419 (1980).
14. Seto, K., Abe, M., Ohya, O., Itakura, O., Ishiguro, N., Ikeda, H., Wakisaka, A., and Yoshiki, T. A rat model of HTLV-I infection: development of chronic progressive myeloneuropathy in seropositive WKAH rats and related apoptosis. *Acta Neuropathol.*, **89**, 483–490 (1995).
15. Sugimoto, M., Nakashima, H., Watanabe, S., Uyama, E., Tanaka, F., Ando, M., Araki,

S., and Kawasaki, S. T lymphocytes alveolitis in HTLV-I-associated myelopathy. *Lancet*, **i**, 1220 (1987).

16. Tanaka, Y., Yoshida, A., Tozawa, H., Shida, H., Nyunoya, H., and Shimotohno, K. Production of a recombinant human T-cell leukemia virus type-I trans-activator (tax1) antigen and its utilization for generation of monoclonal antibodies against various epitopes on the tax1 antigen. *Int. J. Cancer*, **48**, 623–630 (1991).
17. Tateno, M., Kondo, N., Itoh, T., Chubachi, T., Togashi, T., and Yoshiki, T. Rat lymphoid cell lines with human T cell leukemia virus production. I. Biological and serological characterization. *J. Exp. Med.*, **159**, 1105–1116 (1984).
18. Tomaru, U., Ikeda, H., Ohya, O., Abe, M., Kasai, T., Yamashita, I., Morita, K., Wakisaka, A., and Yoshiki, T. Human T lymphocyte virus type I-induced myeloneuropathy in rats: implication of local activation of the pX and tumor necrosis factor-α genes in pathogenesis. *J. Infect. Dis.*, **174**, 318–323 (1996).
19. Vernant, J. C., Buisson, G., Magdeleine, J., De Thore, J., Jouannelle, A., Neisson-Vernant, C., and Monplaisir, N. T lymphocytes alveolitis, tropical spastic paraparesis and Sjögren's syndrome. *Lancet*, **i**, 177 (1988).
20. Wilt, S. G., Milward, E., Zhon, J. M., Nagasato, K., Patton, H., Rusten, R., Grittin, D. E., O'Conner, M., and Dubois-Daleq, M. *In vitro* evidence for a dual role of tumor necrosis factor-α in human immunodeficiency virus type 1 encephalopathy. *Ann. Neurol.*, **37**, 381–394 (1995).
21. Yoshida, M., Miyoshi, I., and Hinuma, Y. Isolation and characterization of retrovirus from cell lines of human adult T-cell leukemia and its implication in disease. *Proc. Natl. Acad. Sci. USA*, **79**, 2031–2035 (1982).
22. Yoshida, M., Seiki, M., Yamaguchi, K., and Takatsuki, K. Monoclonal integration of human T-cell leukemia provirus in all primary tumors of adult T-cell leukemia suggests causative role of human T-cell leukemia virus in the disease. *Proc. Natl. Acad. Sci. USA*, **81**, 2534–2537 (1984).
23. Yoshida, M., Suzuki, T., Fujisawa, J., and Hirai, H. HTLV-I oncoprotein tax and cellular transcription factors. *Curr. Top. Microbiol. Immunol.*, **193**, 79–89 (1995).
24. Yoshiki, T. Chronic progressive myeloneuropathy in WKAH rats induced by HTLV-I infection as an animal model for HAM/TSP in humans. *Intervirology*, **38**, 229–237 (1995).
25. Yoshiki, T., Kondo, N., Chubach, T., Tateno, M., Togashi, T., and Itoh, T. Rat lymphoid cell lines with HTLV-I production. III. Transmission of HTLV-I into rats and analysis of cell surface antigens associated with HTLV-I. *Arch. Virol.*, **97**, 181–196 (1987).
26. Yoshiki, T., Tateno, M., Sakurai, H., Ishiguro, N., Ikeda, H., and Wakisaka, A. A rat model of HTLV-I infection. *In* "Advances in Adult T-Cell Leukemia and HTLV-I Research," ed. K. Takatsuki, Y. Hinuma, and M. Yoshida, Gann Monograph on Cancer Research No. 39, pp. 205–216 (1992). Japan Sci. Soc. Press, Tokyo/CRC Press, USA.

AUTOIMMUNE CHRONIC INFLAMMATORY ARTHROPATHY IN MICE TRANSGENIC FOR THE HTLV-I *tax* GENE

Yoichiro IWAKURA

*Center for Experimental Medicine, Institute of Medical Science, University of Tokyo**

Rheumatoid arthritis (RA) is a serious medical problem, with ~1% of the people in the world affected. The disease is autoimmune in nature and characterized by chronic inflammation of the synovial tissue in multiple joints, which leads to joint destruction. It is remarkable that expression of inflammatory cytokines is augmented in the joints of the patients, although the pathological roles have not been elucidated completely. We recently reported on an inflammatory arthropathy resembling RA that develops in high incidence among transgenic (Tg) mice that carry the *env-pX* region of the human T cell leukemia virus type I (HTLV-I) genome. Autoimmune pathogenesis was suggested in this RA model because: 1) high levels of autoantibodies were detected in the serum, 2) oligoclonal accumulation of T cells was detected in the affected joints, 3) the development of arthritis was suppressed in athymic nude mice, and 4) the disease was transferred by bone-marrow (BM) cell transplantation and suppressed by wild-type BM cell transplantation. We found that cytokine levels including interleukin (IL)-1 were elevated in the joints of these Tg mice. Depletion of IL-1 by gene targeting greatly reduced onset of the disease and T cell proliferative response against synovial components was also reduced, indicating importance of this cytokine in the development of arthritis and autoimmunity. Furthermore, we found that IL-1 receptor antagonist (IL-1Ra)-deficient mice also developed arthritis spontaneously, and autoimmune nature of the disease was suggested. These observations suggest that excess IL-1 signal causes autoimmunity. We show that IL-1 induces expression on T cells of CD40L and OX40 co-signaling molecules, which play important roles in T cell-antigen presenting cell interaction, and activate the immune system non-specifically. In this review, I will discuss pathogenic roles of Tax-induced IL-1 in the development of autoimmunity and arthritis in mouse models.

Rheumatoid arthritis (RA) is one of the most serious medical problems worldwide with approximately 1% of the people of all races in the world affected. It is estimated that approximately 700,000 patients are in Japan. The disease is autoimmune in nature and characterized by chronic inflammation of the synovial tissues in multiple joints that leads to joint destruction, although the etiopathogenesis has not been elucidated completely (*11, 32, 88*).

It is known that people who carry specific major histocompatibility complex (MHC) haplotypes such as HLA-DR4 and DR1 show higher risk for the develop-

* 4-6-1 Shirokanedai, Minato-ku, Tokyo 108-8639, Japan

ment of chronic RA than those who carry other MHC molecules, suggesting that antigen-presenting ability to specific pathogenic antigens correlates to susceptibility to the disease (*22, 40, 117, 140*). On the other hand, infectious agents such as viruses, bacteria, and mycoplasmas are also suggested to be involved in the development of arthritis (*130*). In fact, it is well known that bacteria such as mycobacteria and streptococci can induce arthritis in animals when animals are infected with these bacteria (*15, 70*). It is shown that some bacterial components like HSP65 or proteoglycans are immunologically cross-reactive with surface molecules of the synovial cells (*65*). Super antigens in the bacterial components are also suggested to be involved in the development of autoimmunity (*101*).

Involvement of virus infections is also suggested in many studies. It is known that rubella virus infection can cause arthritis transiently (*115*), and Epstein-Barr (EB) virus (*2*) and parvovirus (*122*) are also suspected to be involved. With regard to this, it was shown that amino acid sequences of the viral components from EB virus have sequence homologies with those of DR chains (*21, 110*). Association of arthritis with retrovirus is also reported (*77*). Nishioka *et al.* first reported a patient with chronic inflammatory arthropathy infected with human T cell leukemia virus type I (HTLV-I) (*96*), and, as I will describe later, we found that HTLV-I could induce arthritis in mice (*60*). Other retroviruses such as human immunodeficiency virus (HIV) (*111*) and caprine arthritis and encephalitis virus (CAEV) (*10, 13, 64*) are also known to induce arthritis.

Various animal disease models have been developed to elucidate pathogenesis of the disease and to evaluate efficacy of possible therapeutics. These RA models include spontaneous models, induced models, and gene manipulated mouse models. The MRL/lpr mouse is known as a model for systemic lupus erythematosus (SLE), and some of these mice also develop arthritis spontaneously (*50*). It is known that these mice carry a mutation in the *fas* gene and T cell apoptosis is impaired in these animals (*134*). Collagen-induced arthritis (CIA) is a typical induced arthritis model in which arthritis is induced in rat, rabbit, mouse, or monkey by injecting type II collagen, one of the major components of the joint (*12, 118, 124*). It is known that injection with either mycobacterium cell wall (complete Freund's adjuvant) (*69, 102*), streptococcal cell wall (*15*), peptide glycans (*70*), or muramyl dipeptide (*71*) also induces acute joint inflammation. It has been shown that molecular mimicry is present between cartilage antigens and a mycobacterium antigen (heat shock protein: HSP65) (*129*) or a streptococcal cell wall antigen (*128*).

Several RA models are produced using transgenic techniques. The transgenic mice carrying a modified human *tumor necrosis factor (TNF)-α* gene under its own promoter, which produces very stable TNF-α mRNA, develop arthritis by 4 weeks of age (*66*). Similar phenotypes were observed in mice lacking TNF AU-rich elements which regulate TNF-α mRNA destabilization (*72*). Since this arthritis develops in the $RAG^{-/-}$ background, the involvement of innate and/or stromal mechanism rather than the autoimmune mechanism is suggested (*72*). As I will describe later, transgenic mice carrying a human *interleukin (IL)-1α* gene (*95*) and IL-1receptor antagonist-deficient mice (*54*) develop arthritis, indicating that IL-1 also plays important roles in development of the symptoms. Furthermore, we also developed a transgenic (Tg) arthritis model which carries the HTLV-I *tax* gene. Although each model has its own merits, the HTLV-I Tg mouse model is important, because epidemiological studies suggest that this virus actually causes RA in humans (*26, 97*).

In spite of enormous efforts, the etiopathogenesis of RA remains mostly unknown. Especially, it is critically important to elucidate mechanisms of the development of autoimmunity, because autoimmunity triggers the development of the disease. In this review, I will discuss pathogenic roles of cytokines, especially of IL-1, in the development of arthritis and autoimmunity caused by HTLV-I.

HTLV-I Tg Mouse Model

HTLV-I is known as the causative agent of adult T cell leukemia (*121, 137*). This virus encodes a transcriptional transactivator, Tax, in the *pX* region, that transactivates transcription from the cognate viral promoter through the 21 bp enhancer (*138*). It was shown that Tax also activates many cellular genes (*137*) including those for cytokines such as IL-1, IL-2, IL-6, nerve growth factor (NGF), transforming growth factor (TGF)-β, and granulocyte macrophage colony-stimulating factor (GM-CSF) (*67, 82, 86, 114, 133*), cytokine receptors such as α-subunit of IL-2 receptor (*42, 56*), and the immediate early transcriptional factors c-fos and c-jun (*34, 35, 59*) through activation of enhancers like the NF-κB-dependent enhancers or serum responsive elements (*138*).

Previously, we found that Tg mice carrying the HTLV-1 *env-pX* region with its own long terminal repeat (LTR) promoter (HTLV-I Tg mice) developed chronic inflammatory polyarthropathy at a high incidence (*60*). The arthritis develops spontaneously in multiple joints as early as 4 weeks of age, and at 3 months of age, 60% (BALB/cAn background) of the mice are affected. The histopathology is very similar to that of RA in humans, showing marked synovial and periarticular inflammation with articular erosion caused by invasion of granulation tissues (*135*). These mice develop autoimmunity with elevated levels of antibodies against immunoglobulin (Ig) G, type II collagen, and HSPs (*58*), and show IgG hyper-gammaglobulinemia in which agalactosylated forms of the Ig carbohydrate chains increase (*27*).

We also produced transgenic mice that expressed the *tax* gene alone under the control of either its own LTR or the CD4 enhancer/promoter and found that both of these mice developed RA-like inflammatory arthropathy similar to HTLV-I Tg mice that carry the LTR-*env-pX*-LTR region, indicating that the *tax* gene is arthritogenic (*49*).

The incidence of arthritis greatly differs among mouse strains (*57*). When HTLV-I Tg mice were backcrossed with different inbred strains, the incidence of arthritis was: 64% and 72% in BALB/cAn ($H\text{-}2^d$), 25% and 46% in C3H/HeN ($H\text{-}2^k$), and 0% and 2% in C57BL/6J ($H\text{-}2^b$) background at 3- and 6-months of age, respectively. Rheumatoid factor (antibodies against IgG) levels in the serum correlated with the susceptibility to the disease, whereas *IL-1β* and *MHC* gene expression were similarly elevated in all these strains, suggesting involvement of immune regulatory genes in this strain difference. However, introduction of the $H\text{-}2^d$ locus into C57BL/6J *pX*-Tg mice did not increase the incidence and substitution of the BALB/cAn *H-2* locus with the $H\text{-}2^b$ did not decrease it. The results indicate that the *H-2* locus is not the major determinant of the disease. Since involvement of multiple genes in RA is also suggested in humans, identification of these susceptibility genes may provide useful information to identify susceptible genes in humans.

Autoimmune pathogenesis was suggested in this model because high levels of autoantibodies were detected in the serum, oligoclonal T cells were accumulated in the joints, and the disease onset was suppressed in athymic *nu/nu* mice (*49, 57, 58, 76*). Furthermore, when bone-marrow (BM) cells from HTLV-I Tg mice were transferred into irradiated non-Tg mice, arthritis developed in these mice (*113*). In contrast, arthritis in HTLV-I Tg mice was completely suppressed by the transplantation of non-Tg BM cells, indicating that the disease is not caused by the abnormal education of T cells in the periphery but by the disorder in the stem cells.

The T cell receptor (TCR) β-chain variable region (Vβ) repertoires in the lymphatic organs are normal in transgenic mice, however, specific Vβ positive T cells are expanded oligoclonally in the affected joints, suggesting that specific antigens, but not superantigens, were involved in the expansion of these T cells (*76*). These expanded T cells had the same TCRs as those of lymph node T cells reactive to type II collagen. Moreover, these mice became susceptible to CIA even on the C3H/HeN background. These observations indicate that endogenous type II collagen is one of the arthritogenic antigens in the joint, suggesting a tolerance break to this antigen in HTLV-I Tg mice.

We found that both TCR Vβ11 and Vβ12 positive populations were greatly reduced in the spleen and lymph modes (LNs) in transgenic mice as in non-transgenic C3H/HeN mice by endogenous mouse mammary tumor virus (MMTV) superantigens, indicating that negative selection in the thymus proceeds normally (*76*). On the other hand, we found that T cells from HTLV-I Tg mice were refractory against anti-Fas antibody treatment (*68*). This finding suggests that this defect of T cells may be involved in the development of autoimmunity in these mice, because Fas-mediated apoptosis of T cells is believed to be important in eliminating autoreactive T cells in the periphery (*90*). In support of this idea, we showed that the incidence increased when these HTLV-I Tg mice were crossed with *lpr/lpr* mice, while it decreased when they were crossed with *fas* Tg mice (*57*). These observations suggest that aberration of Fas-mediated apoptosis of peripheral lymphocytes rather than negative selection in the thymus is involved in the development of autoimmunity in HTLV-I Tg mice. Furthermore, they suggest that administration of Fas ligand or anti-Fas antibody may be effective for the treatment of arthritis.

It seems difficult, however, to explain the development of autoimmunity solely by the defect of Fas-mediated apoptosis, because even *lpr/lpr* mice develop arthritis only on a specific genetic background and never develop arthritis on the C3H/HeN background on which HTLV-I Tg mice develop arthritis. In this context, the finding that Vβ12 positive T cells reactive with type II collagen expanded oligo-clonally in the affected joints (*76*) was interesting, because it suggested that T cells in the affected joints had recovered from the anergic state caused by MMTV superantigens. This tolerance break at the local sites could explain the development of autoimmunity in these HTLV-I Tg mice.

Suppression of the Development of Arthritis in IL-1-deficient Mice

It is remarkable that various proinflammatory cytokine genes including IL-1α, IL-1β, IL-2, IL-6, TNF-α and interferon (IFN)-γ are activated in the transgenic joints (*58*), probably due to the transcriptional transactivating activity of Tax. Among them, IL-1 is especially interesting, because augmentation of *IL-1* gene expression in the joints of RA patients is also reported by many investigators (*31*). Production of IL-1 is also reported in animal models such as antigen-induced arthritis in the rabbit and CIA in the mouse (*81*). Furthermore, correlation of plasma IL-1 levels with disease activity is reported in rheumatoid arthritis (*25*) and a genetic association between juvenile rheumatoid arthritis and an IL-1α polymorphism is reported (*84*).

Arthritogenic activity of IL-1 is demonstrated by the facts that injection of purified rIL-1 into rabbit knee joints induces the accumulation of polymorphonuclear and mononuclear leukocytes in the joint space and the loss of proteoglycan from the articular cartilage (*103*); continuous infusion of human rIL-1α into the rabbit knee joint induces arthritis (*28*), and mouse rIL-1 induces an acute exacerbation of arthritis in rat joints previously injected by group A streptococcal peptidoglycan-polysaccharide (*119*). These observations suggest involvement of IL-1 in the pathogenesis of RA.

Then, we examined the effect of IL-1 deficiency on the development of arthritis in HTLV-I Tg mice and found that development of arthritis was markedly suppressed in IL-1-deficient mice (*112*). The incidence of arthritis was 60% and 80% at 3 months and 6 months of age, respectively, in wild-type mice. In contrast, that in IL-1α/$\beta^{-/-}$ mice is only 10% at 3 months and 30% at 6 months of age. However, the severity score of the affected mice was as high as that in wild-type mice, indicating that IL-1 plays an important role in the initial phase of the disease, but that this pathologic role can be substituted by some other factors in the later stage of the disease.

We also found that the development of CIA was severely suppressed in IL-1-deficient mice: 68% of wild-type mice developed arthritis 10 weeks after type II collagen immunization, whereas none (0%) of the IL-1α/$\beta^{-/-}$ mice developed arthritis (*112*). These results indicate that IL-1 also plays a crucial function in the development of arthritis in this model.

Consistently with our results, it was reported that treatment with anti-IL-1α/β as well as anti-IL-1β antibody ameliorated CIA in mice (*63, 127*). However, in these reports, anti-IL-1α antibody treatment showed no significant protection. It is also reported that local expression of IL-1Ra protein can prevent murine CIA (*4, 126*).

Roles of IL-1 in the Development of Arthritis

It is known that IL-1 is produced by various types of cells, including macrophages, monocytes, and synovial lining cells, and induces inflammation by activating synovial cells, endothelial cells, lymphocytes and macrophages to produce various chemokines, cytokines, and inflammatory mediators (*29*). These inflammatory mediators include IL-1 itself, IL-6, TNF-α, IL-8 and cyclooxygenase (COX)-2, and cause infiltration of inflammatory cells into inflammatory sites, increase the permeability of the blood vessels, and induce fever (*14, 17, 123*). Furthermore, IL-1 promotes synovial cell growth and activates synovial cells and osteoclasts to produce metalloproteases and collagenases that cause erosion of the bone and cartilage of joints (*16*). Thus, it is possible that these activities of IL-1 are involved in the development of arthritis at the effector phase. However, since the severity score of the affected mice in IL-1-deficient HTLV-I Tg mice was similar to wild-type mice, it was suggested that these activities of IL-1 were not necessarily required at the elicitation phase for the development of inflammation in the HTLV-I Tg model (*112*).

Then, we analyzed the effects of IL-1 deficiency on the immune system. In the immune system, IL-1 is known to activate lymphocytes, monocytes, macrophages and NK cell, although the mechanisms have not been elucidated (*18, 19*). When mice were immunized with protein antigens together with IL-1, serum antibody production was enhanced, suggesting that IL-1 has an adjuvant effect (*107, 116*).

Antibody production after immunization with type II collagen in the presence of complete Freund's adjuvant (CFA) in the CIA model was normal in IL-1α/β$^{-/-}$ mice, while spontaneous autoantibody production against type II collagen was suppressed in IL-1α/β$^{-/-}$ HTLV-I Tg mice (*112*). Thus, IL-1 is required for the development of autoantibodies in HTLV-I Tg mice, whereas IL-1 seems not to be required for the production of antibodies against type II collagen in the CIA model. We reported previously that adjuvant abrogated the effects of IL-1 deficiency, because various cytokines such as TNF-α or IL-6, which have similar biological activities as IL-1, were induced by the treatment (*91*). Nonetheless, CIA was severely suppressed in the IL-1-deficient mice. Thus, the function of IL-1 in enhancing antibody production does not seem crucial for the development of arthritis. On the other hand, T cell proliferating response against type II collagen was greatly reduced in both CIA and HTLV-I Tg models, suggesting requirement of IL-1 in T cell activation (*112*). The role of IL-1 in the immune system was further analyzed using IL-1Ra$^{-/-}$ mice.

Development of Autoimmune Arthritis in IL-1Ra Knockout Mouse

IL-1Ra binds IL-1 receptors without exerting agonistic activity, and competes for the binding of both IL-1α and IL-1β (*8, 51*). We have recently generated IL-1Ra-deficient mice and showed that these mice developed chronic inflammatory arthropathy spontaneously (*54*). The incidence of arthritis differs among different genetic backgrounds; the incidence is high in the BALB/c background, but low in the C57BL/6J background, the same tendency that is observed in HTLV-I-induced arthritis. The arthritis starts to develop at 5 weeks of age, and at 12 weeks of age almost all the BALB/c background mice suffered from arthritis. The histopathology of the lesion shows marked synovial and periarticular inflammation with articular erosion caused by invasion of granulation tissues, closely resembling that of RA in humans. Activation of osteoclasts is obvious in the pannus, and infiltration of inflammatory cells mostly consisting of neutrophils are detected in the pannus and the synovial space. Fibrin clots are also detected in the synovial space.

Interestingly, total immunoglobulin levels as well as autoantibody levels against immunoglobulin, type II collagen and dsDNA are elevated in these mice (*54*). When *Rag2* gene in the IL-1Ra$^{-/-}$ mice was deleted, the development of arthritis was completely suppressed (Horai and Iwakura, unpublished observation). Furthermore, when T cells from IL-1Ra knockout mice were transferred to *nu*/*nu* mice, these mice developed arthritis (Horai and Iwakura, unpublished observation). These observations clearly indicate that excess IL-1 signal on T cells causes autoimmunity, resulting in joint specific inflammation and bone destruction.

Recently, it was reported that transgenic mice carrying the human *IL-1α* gene on the C3H/HeN background developed arthritis spontaneously (*95*). The authors suggested, however, that macrophages and polymorphonuclear neutrophils were directly involved in the joint destruction, although they had not examined the possibility that autoimmunity is involved in the pathogenesis.

We then asked why IL-1Ra deficiency causes autoimmunity. We found that IL-1α, IL-1β and IL-1Ra were all constitutively expressed under the physiological conditions in the joints of normal wild-type mice, and the expression of IL-1β, but not IL-1α, increased 2 to 3 times in IL-1Ra knockout mice compared to wild-type mice (*54*). The expression of the other inflammatory cytokines TNF-α and IL-6 was also elevated, indicating regulatory roles of IL-1Ra in the cytokine network. It is thought that this elevation of cytokine expression in IL-1Ra knockout mice is caused by excess IL-1 signal, because IL-1 can induce IL-6, TNF-α, and IL-1 itself (*17, 53*). Thus, in the absence of IL-1Ra, physiological levels of IL-1 induce pathogenic IL-1 responses resulting in the development of autoimmunity. These results have been the first to suggest a link between excessive IL-1 signal and autoimmunity, and that the balance between IL-1 and IL-1Ra is crucial for the homeostasis of the immune system.

T Cell Sensitization with IL-1 through Induction of CD40 Ligand (L) and OX40

Since it is suggested that IL-1 plays a crucial role in activating the immune system, we then analyzed the mechanism. For this, we investigated roles of IL-1 in T cell-dependent (TD) antibody production using gene-targeted mice. Both primary and secondary antibody production against sheep red blood cells (SRBC) were significantly reduced in IL-1α/$\beta^{-/-}$ mice, and were enhanced in IL-1$Ra^{-/-}$ mice (*91, 92*). The intrinsic functions of B cells, T cells, and antigen-presenting cells (APCs) were normal. However, proliferative response and cytokine production of T cells against SRBC through the interaction with APCs were markedly impaired when APCs from IL-1α/$\beta^{-/-}$ mice were used, and enhanced when those from IL-1$Ra^{-/-}$ mice were used. These results indicate that IL-1 produced by APCs promotes the antigen-specific T cell helper function and promotes TD antibody production.

Furthermore, we found that IL-$1^{-/-}$ APCs did not fully activate T cells from DO11.10 mouse, an ovalbumin (OVA)-specific T cell receptor Tg mouse, upon stimulation with OVA, while IL-1$Ra^{-/-}$ APCs were more efficient to activate T cells (*91*). These observations indicate that IL-1 promotes T cell-priming.

In analyzing molecules involved in the interaction between T cell and APC, we found that the expression of co-signaling molecules, CD40L and OX40, on T cells was reduced upon interaction with IL-$1^{-/-}$ APCs and increased upon interaction with IL-1$Ra^{-/-}$ APCs. The expression of these molecules on T cells was induced directly with recombinant IL-1, although the preceding induction of IL-1R on T cells through activation of TCR was required in naive T cells. Furthermore, the reduced antibody production in IL-$1^{-/-}$ mice was recovered by the treatment with agonistic anti-CD40 monoclonal antibody (mAb) both *in vitro* and *in vivo*. These results indicate that IL-1 activates T cells by inducing CD40L and OX40 expression on T cells.

Roles of the CD40L/CD40 and OX40L/OX40 System in the Immune System

The CD40L (CD154) is a type II membrane glycoprotein that is expressed transiently on $CD4^+$ helper T cells after activation. Interactions between CD40L and CD40, its mitogenic receptor on B cells, provide an essential signal for the induction of B cell activation and Ig production. Abnormal CD40L expression causes the Ig class switch defects observed in the X-linked hyper-IgM immunodeficiency syndrome which is characterized by elevated levels of IgM and low levels of IgA, IgG, and IgE, the absence of germinal centers, and the inability to mount TD humoral response (*20, 74, 106*). CD40 is expressed by a large variety of cell types other than B cells, and these include dendritic cells, follicular dendritic cells, monocytes, macrophages, mast cells, fibroblasts, and endothelial cells (*44*). Thus, the CD40-CD40L interaction is also important to produce inflammatory cytokines, chemokines, and IL-12 from APCs, to induce co-signaling molecules, and to activate macrophages, natural killer (NK) cells, and endothelial cells (*30*). Importantly, it is shown that CD40 signal can also induce another co-signaling molecule, OX40L, on APCs (*89*). As a result, CD4 T cell priming is enhanced through this interaction, and T cell-priming to both protein antigens and alloantigens is impaired in CD40L-deficient mice (*45*). This interaction is also important to enhance CD4 and CD8 T cell clonal expansion, and to protect from the death of the activated T cells in the periphery (*83*).

The role of CD40-CD40L signaling in the development of autoimmunity such as collagen-induced arthritis, experimental autoimmune encephalitis (EAE), lupus nephritis, colitis, and oophoritis has been demonstrated in mouse models (*23, 37, 46, 87*). Treatment of mice with anti-CD40L antibodies blocks development of diseases in these models. It is reported that insufficiency of CD40L-CD40 interaction is involved in neonatal transplantation tolerance (*33*), and excess CD40 signal in murine epidermis in CD40L transgenic mice causes chronic skin inflammation and systemic autoimmunity (*85*).

OX40 (CD134) is another important co-signaling molecule that is involved in the activation of T cells and APCs (*79*). OX40 and OX40L is a member of the TNF receptor (R) and TNF families, and is expressed on activated T cells and APCs (*6, 80*), respectively. OX40 is not constitutively expressed, and is induced on naive T cells after stimulation with antigens or infection with HTLV-I (*52*). OX40 is expressed most strongly on CD4 T cell blasts in the T cell zone (*24*). OX40L is also inducible and the CD40 signal is important to induce this molecule on APCs (*89*). OX40L is also expressed on HTLV-I-infected human leukemic T cells (*6*). Several groups have shown that OX40L expressed on APCs can provide costimulation to CD4 T cells (*1, 7, 36, 39, 41*), and promotes T cell survival through induction of Bcl-xL and Bcl-2 (*108*). Activation of CD8 T cells is also reported (*132*), although OX40$^{-/-}$ mice retain primary and memory cytotoxic T cell response after infection with lymphocytic choriomeningitis virus (LCMV) and influenza virus (*73*).

OX40 stimulation of B cells increases immunoglobulin heavy chain mRNA levels and immunoglobulin secretion (*120*), and the OX40-OX40L interaction is necessary for the differentiation of activated B cells into highly Ig-producing cells. It was shown that T cell-priming was impaired in OX40L$^{-/-}$ mice accompanied by a concomitant reduction of both Th1 and Th2 cytokine production (*89*). Antibody production against keyhole limpet hemocyanin (KLH) and cytotoxic T lymphocyte (CTL) induction were suppressed in those mice, although other investigators reported that serum antigen-specific Ig levels were similar to wild-type mice when OX40$^{-/-}$ or OX40L$^{-/-}$ mice were immunized with various TD antigens (*9, 73, 105*).

OX40$^{-/-}$ mice exhibit an impaired contact hypersensitivity response due to defects in T cell-priming and cytokine production (*9*). It is also reported that the

OX40-OX40L system is involved in many diseases such as EAE (*94, 98*), graft-versus-host disease (GVHD) (*75*), colitis (*80*) and asthma (*62*). Thus, OX40-OX40L interaction seems to regulate not only antibody production but also Th1-dependent cellular responses as well as CTL responses.

The Role of IL-1 in the Development of Autoimmunity and Arthritis

Collectively, the role of IL-1 in T cell and APC activation by antigenic stimulation is considered to be as follows. First, when APCs encounter an antigen, these cells produce IL-1 in response to the antigenic stimulation. On the other hand, IL-1R is induced on T cells by the TCR signal upon interaction with APCs. Then, APC-derived IL-1 activates T cells to induce CD40L and OX40, and the CD40L activates APCs and B cells by interacting with constitutively expressed CD40 molecules, leading to the activation of these cells to express OX40L on the surface. This OX40L again activates T cells that are primed to express OX40 on the surface. These T cells include both CD4 T cells and CD8 T cells, and activate both cellular and humoral immunity by producing Th1 as well as Th2 cytokines (Fig. 1).

Thus, over-expression of CD40L and OX40 can explain the development of autoimmunity in animals in which excess IL-1 signal is generated. Actually, our findings that CD40L and OX40 expression on T cells are augmented in HTLV-I Tg mice and that this enhanced expression is normalized when IL-1 is deleted suggest that overproduction of IL-1 causes autoimmunity through induction of CD40L and OX40 molecules on T cells (*112*). Arthritis development in IL-1Ra$^{-/-}$ mice was also

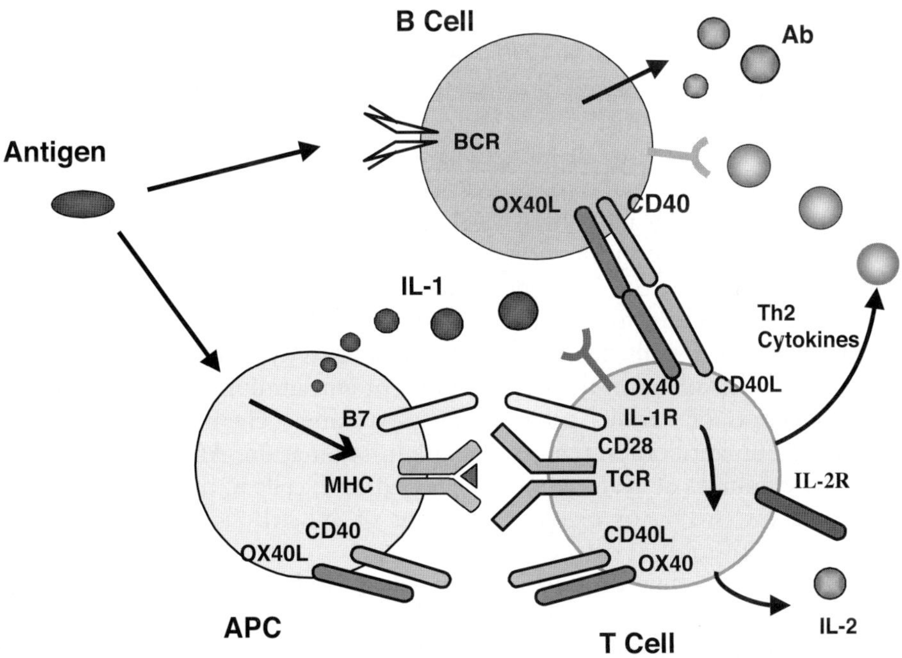

FIG. 1. Activation of the acquired immune system with IL-1

ameliorated when mice were treated with anti-OX40L or anti-CD40L antibodies, suggesting that overexpression of these molecules is responsible for the autoimmunity (Horai *et al.*, unpublished observation). With regard to this, it was reported recently that anti-OX40 antibody treatment was effective to treat CIA model mice (*139*). However, the mechanism seems to be different: development of autoimmunity is suppressed by the treatment in HTLV-I Tg or IL-1Ra$^{-/-}$ mouse models, whereas in the CIA model, T cell-APC interaction upon immunization with collagen is just inhibited.

Involvement of IL-1 in the development of autoimmunity is further supported by the fact that IL-1Ra$^{-/-}$ mice developed CIA even on the C57BL/6J background. Since CIA is never developed in C57BL/6J wild-type mice, these observations suggest that excess IL-1 signal makes the immune system hypersensitive to antigenic challenge (Horai and Saijo, unpublished observation). We have also shown that C3H/HeN mice which are usually resistant to CIA became susceptible to the treatment with type II collagen when they were introduced with HTLV-I *tax* gene (*112*). Consistently with our notion, it is reported that exogenous IL-1 prevents tolerance induction by staphylococcal enterotoxin B (SEB) (*93*), and that signals from an agonistic antibody against OX40 can break an existing state of tolerance in the CD4 T cell compartment (*5*).

It is known that T cells become anergic when these cells encounter antigens without CD28 stimulation (*99*). However, transgenic expression of the B7-1 (CD80) on the islets of Langerhans is not sufficient to abolish the *in vivo* tolerance to islet antigen, and coexpression of the B7-1 and high levels of the class II MHC antigen I-E or TNF-α can induce autoimmune destruction of the β cells of the pancreas (*47, 48*). Our data indicate that IL-1 can activate T cells in the absence of CD28 stimulation, because only the signals from CD3 and IL-1R are enough to activate T cells (*91*). It should be noted that the IL-1 signaling pathway and the CD28 signaling pathway are independent, because CTLA4, a potent inhibitor of the CD28 pathway, could not inhibit IL-1-mediated T cell activation (*91*). Therefore, IL-1 can activate self-reactive T cells which are made anergic in the periphery due to the lack of CD28 costimulation. Since TNF-α can induce IL-1 *in vivo*, the action of TNF-α in abolishing tolerance may actually be mediated by TNF-α-induced IL-1.

Thus, in HTLV-I Tg mice high levels of expression of Tax in synovial cells are thought to directly induce production of inflammatory cytokines, especially of IL-1. These cytokines activate the immune system through induction with CD40L and OX40 on T cells, leading to a tolerance break against synovial antigens. Probably, prolonged survival of T cells due to the reduced sensitivity to Fas-mediated apoptotic signal may also be involved in the development of autoimmunity. This autoimmunity induces production of various inflammatory cytokines, chemokines, inflammatory mediators, and autoantibodies, that exaggerate inflammation. At the same time, these inflammatory cytokines activate synovial cells and/or T cells to produce osteoclast differentiation factor (ODF, or receptor activator of NF-κB ligand: RANKL) (*136*), resulting in the activation of osteoclasts (Fig. 2).

Consistently with our observations, previous studies have also suggested that HTLV-I may be involved in immunological disorders. For example, patients with SLE are found to be carriers of HTLV-I at higher rates than the general population (*104*). HTLV-I has also been suspected of being involved in Sjögren's syndrome

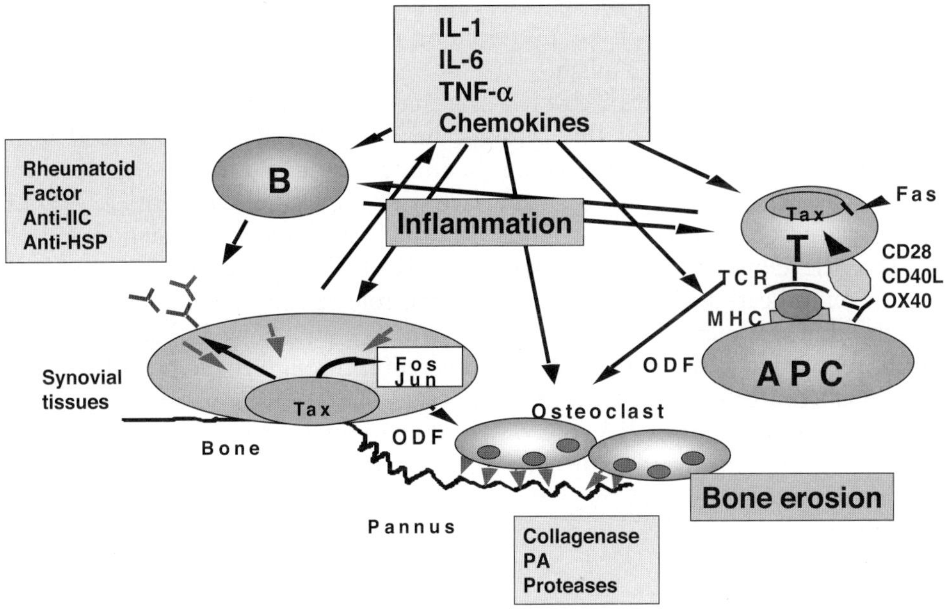

FIG. 2. Pathogenesis of arthritis caused by the HTLV-I *tax* gene

(*131*), and it was shown that transgenic mice carrying the *tax* gene developed an exocrinopathy resembling this syndrome, although there was no data on immunological status reported in those studies (*43*). Elevated immunoglobulin levels (*100*) and increased numbers of autoreactive T cells (*125*) have been reported in patients with HTLV-I-associated myelopathy (HAM), also known as tropical spastic paraparesis (TSP), and augmented immune reactivity against HTLV-I antigens has been implicated in the pathogenesis (*38, 61, 100, 125*). Abnormal immune response against EB virus was reported in patients with adult T cell leukemia (ATL) and healthy HTLV-I carriers (*55*). Other studies have suggested that decreased immune surveillance in HTLV-I carriers may be responsible for their tendency to develop various types of tumors at higher rates than uninfected people (*3*). These observations, however, have not as yet proved a direct pathogenic role of HTLV-I in these immunological disorders. The results presented in the present paper strongly suggest direct involvement of HTLV-I in the induction of these immunological disorders.

It should be noted that the antigenicity of the pathogens is not necessarily cross-reactive with the autoantigens of the host to break the tolerance. In fact, no immunological cross-reactivity was reported between HTLV-I and synovial tissues. The fact that IL-1Ra deficiency causes autoimmunity clearly indicates that excess IL-1 signal is enough to cause autoimmunity. In support of this notion, an enhancing effect of complete Freund's adjuvant (CFA) or inflammation on the development of autoimmunity upon immunization with cross-reactive molecules was reported (*78, 109*). Other than microbial infections, either inflammation caused by mechanical injury or immune reaction, or invasion of a transcriptional transactivator that can activate the *IL-1* genes, or genetic disorder of the IL-1 or its regulatory genes may also cause overproduction or excess signal of this cytokine. Frequent development of arthritis after microbial infections such as mycobacteria or streptococci may be explained in part by this excess-IL-1 mechanism.

Although IL-1 was first discovered as a major mediator of inflammation, it has gradually become evident that this cytokine has numerous functions related to host defense mechanisms, not only regulating the immune system, but also the areas of the neuronal and endocrine systems that interface with the immune system (*17, 18, 123*). Actually, we showed that IL-1 was induced in the brain upon inflammation and associated fever development was not observed in IL-$1^{-/-}$ mice and exaggerated in IL-$1Ra^{-/-}$ mice, indicating that IL-1 plays a crucial role in the development of fever by acting on the neuronal system (*53*). Furthermore, we showed that IL-1 is also involved in the regulation of glucocorticoid secretion through the hypothalamus-pituitary-adrenal cortex axis (*53*). Thus, it is suggested that IL-1 plays critically important roles in the development of RA, not only inducing inflammatory responses but also activating the immune system, the neuronal system and the endocrine system. The control of the production and the activity of this cytokine should be critically important for the treatment of RA.

REFERENCES

1. Akiba, H., Oshima, H., Takeda, K., Atsuta, M., Nakano, H., Nakajima, A., Nohara, C., Yagita, H., and Okumura, K. CD28-independent costimulation of T cells by OX40 ligand and CD70 on activated B cells. *J. Immunol.*, **162**, 7058–7066 (1999).
2. Alspaugh, M. A., Henle, G., Lennette, E. T., and Henle, W. Elevated levels of antibodies to Epstein-Barr virus antigens in sera and synovial fluids of patients with rheumatoid arthritis. *J. Clin. Invest.*, **67**, 1134–1140 (1981).
3. Asou, N., Kumagai, T., Uekihara, S., Ishii, M., Sato, M., Sakai, K., Nishimura, H., Yamaguchi, K., and Takatsuki, K. HTLV-I seroprevalence in patients with malignancy. *Cancer*, **58**, 903–907 (1986).
4. Bakker, A. C., Joosten, L. A., Arntz, O. J., Helsen, M. M., Bendele, A. M., van de Loo, F. A., and van den Berg, W. B. Prevention of murine collagen-induced arthritis in the knee and ipsilateral paw by local expression of human interleukin-1 receptor antagonist protein in the knee. *Arthritis Rheum.*, **40**, 893–900 (1997).
5. Bansal-Pakala, P., Jember, A. G., and Croft, M. Signaling through OX40 (CD134) breaks peripheral T-cell tolerance. *Nat. Med.*, **7**, 907–912 (2001).
6. Baum, P. R., Gayle, R. B., 3rd, Ramsdell, F., Srinivasan, S., Sorensen, R. A., Watson, M. L., Seldin, M. F., Baker, E., Sutherland, G. R., Clifford, K. N. *et al.* Molecular characterization of murine and human OX40/OX40 ligand systems: identification of a human OX40 ligand as the HTLV-1-regulated protein gp34. *EMBO J.*, **13**, 3992–4001 (1994).
7. Calderhead, D. M., Buhlmann, J. E., van den Eertwegh, A. J., Claassen, E., Noelle, R. J., and Fell, H. P. Cloning of mouse Ox40: a T cell activation marker that may mediate T-B cell interactions. *J. Immunol.*, **151**, 5261–5271 (1993).
8. Carter, D. B., Deibel, M. R., Jr., Dunn, C. J., Tomich, C.-S. C., Laborde, A. L., Slightom, J. L., Berger, A. E., Bienkowski, M. J., Sun, F. F., McEwan, R. N., Harris, P. K., Yem, A. W., Waszak, G. A., Chosay, J. G., Sieu, L. C., Hardee, M. M., Zurcher-

Neely, H. A., Reardon, I. M., Heinrikson, R. L., Truesdell, S. E., Shelly, J. A., Eessalu, T. E., Taylor, B. M., and Tracey, D. E. Purification, cloning, expression and biological characterization of an interleukin-1 receptor antagonist protein [see comments]. *Nature,* **344**, 633–638 (1990).

9. Chen, A. I., McAdam, A. J., Buhlmann, J. E., Scott, S., Lupher, M. L., Jr., Greenfield, E. A., Baum, P. R., Fanslow, W. C., Calderhead, D. M., Freeman, G. J., and Sharpe, A. H. Ox40-ligand has a critical costimulatory role in dendritic cell:T cell interactions. *Immunity*, **11**, 689–698 (1999).
10. Clements, J. E., Narayan, O., and Cork, L. C. Biochemical characterization of the virus causing leukoencephalitis and arthritis in goats. *J. Gen. Virol.*, **50**, 423–427 (1980).
11. Cohen, P. L. Systemic autoimmunity. *In* "Fundamental Immunology," 4th Ed., ed. W. E. Paul, pp. 1067–1088 (1999). Lippincott-Raven Publishers, Philadelphia.
12. Courtenay, J. S., Dallman, M. J., Dayan, A. D., Martin, A., and Mosedale, B. Immunisation against heterologous type II collagen induces arthritis in mice. *Nature*, **283**, 666–668 (1980).
13. Crawford, T. B., Adams, D. S., Cheevers, W. P., and Cork, L. C. Chronic arthritis in goats caused by a retrovirus. *Science*, **207**, 997–999 (1980).
14. Davis, P. and MacIntyrek, D. E. Prostaglandins and inflammation. *In* "Inflammation: Basic Principles and Clinical Correlates," ed. J. I. Galin, I. M. Goldstein, and R. Snyderman, pp. 123–137 (1992). Raven Press, Ltd., New York.
15. DeJoy, S. Q., Ferguson, K. M., Sapp, T. M., Zabriskie, J. B., Oronsky, A. L., and Kerwar, S. S. Streptococcal cell wall arthritis. Passive transfer of disease with a T cell line and crossreactivity of streptococcal cell wall antigens with *Mycobacterium tuberculosis*. *J. Exp. Med.*, **170**, 369–382 (1989).
16. DeMarco, D. M. and Zurier, R. B. Cytokines and rheumatology. *In* "Cytokines in Health and Disease," ed. D. G. Remick and J. S. Friedland, pp. 463–485 (1997). Marcel Dekker, Inc., New York.
17. Dinarello, C. A. Biologic basis for interleukin-1 in disease. *Blood*, **87**, 2095–2147 (1996).
18. Dinarello, C. A. Interleukin-1. *Cytokine Growth Factor Rev.*, **8**, 253–265 (1997).
19. Dinarello, C. A. Interleukin-1 and interleukin-1 antagonism. *Blood*, **77**, 1627–1652 (1991).
20. DiSanto, J. P., Bonnefoy, J. Y., Gauchat, J. F., Fischer, A., and de Saint Basile, G. CD40 ligand mutations in x-linked immunodeficiency with hyper-IgM. *Nature*, **361**, 541–543 (1993).
21. Douvas, A. and Sobelman, S. Multiple overlapping homologies between two rheumatoid antigens and immunosuppressive viruses. *Proc. Natl. Acad. Sci. USA*, **88**, 6328–6332 (1991).
22. Duquesnoy, R. J., Marrari, M., Hackbarth, S., and Zeevi, A. Serological and cellular definition of a new HLA-DR associated determinant, MC1, and its association with rheumatoid arthritis. *Hum. Immunol.*, **10**, 165–176 (1984).
23. Durie, F. H., Fava, R. A., Foy, T. M., Aruffo, A., Ledbetter, J. A., and Noelle, R. J. Prevention of collagen-induced arthritis with an antibody to gp39, the ligand for CD40. *Science*, **261**, 1328–1330 (1993).
24. Durkop, H., Latza, U., Himmelreich, P., and Stein, H. Expression of the human OX40

(hOX40) antigen in normal and neoplastic tissues. *Br. J. Haematol.*, **91**, 927–931 (1995).

25. Eastgate, J. A., Symons, J. A., Wood, N. C., Grinlinton, F. M., di Giovine, F. S., and Duff, G. W. Correlation of plasma interleukin 1 levels with disease activity in rheumatoid arthritis. *Lancet*, **ii**, 706–709 (1988).
26. Eguchi, K., Origuchi, T., Takashima, H., Iwata, K., Katamine, S., and Nagataki, S. High seroprevalence of anti-HTLV-I antibody in rheumatoid arthritis. *Arthritis Rheum.*, **39**, 463–466 (1996).
27. Endo, T., Iwakura, Y., and Kobata, A. Structural changes in the N-linked sugar chains of serum immunoglobulin G of HTLV-I transgenic mice. *Biochem. Biophys. Res. Commun.*, **192**, 1004–1010 (1993).
28. Feige, U., Karbowski, A., Rordorf-Adam, C., and Pataki, A. Arthritis induced by continuous infusion of hr-interleukin-1 alpha into the rabbit knee-joint. *Int. J. Tissue React.*, **11**, 225–238 (1989).
29. Feldmann, M., Brennan, F. M., and Maini, R. N. Rheumatoid arthritis. *Cell*, **85**, 307–310 (1996).
30. Ferlin, W. G., von der Weid, T., Cottrez, F., Ferrick, D. A., Coffman, R. L., and Howard, M. C. The induction of a protective response in Leishmania major-infected BALB/c mice with anti-CD40 mAb. *Eur. J. Immunol.*, **28**, 525–531 (1998).
31. Firestein, G. S., Alvaro-Gracia, J. M., Maki, R., and Alvaro-Garcia, J. M. Quantitative analysis of cytokine gene expression in rheumatoid arthritis. *J. Immunol.*, **144**, 3347–3353 (1990).
32. Firestein, G. S. and Zvaifler, N. J. Rheumatoid arthritis. A disease of disordered immunity. *In* "Inflammation: Basic Principles and Clinical Correlates," ed. I. M. G. J. I. Gallin and R. Snyderman, pp. 959–975 (1992). Raven Press, Ltd., New York.
33. Flamand, V., Donckier, V., Demoor, F. X., Le Moine, A., Matthys, P., Vanderhaeghen, M. L., Tagawa, Y., Iwakura, Y., Billiau, A., Abramowicz, D., and Goldman, M. CD40 ligation prevents neonatal induction of transplantation tolerance. *J. Immunol.*, **160**, 4666–4669 (1998).
34. Fujii, M., Niki, T., Mori, T., Matsuda, T., Matsui, M., Nomura, N., and Seiki, M. HTLV-I Tax induces expression of various immediate early serum responsive genes. *Oncogene*, **6**, 1023–1029 (1991).
35. Fujii, M., Sassone-Corsi, P., and Verma, I. M. c-*fos* promoter trans-activation by the tax1 protein of human T-cell leukemia virus type I. *Proc. Natl. Acad. Sci. USA*, **85**, 8526–8530 (1988).
36. Gerloni, M., Xiong, S., Mukerjee, S., Schoenberger, S. P., Croft, M., and Zanetti, M. Functional cooperation between T helper cell determinants. *Proc. Natl. Acad. Sci. USA*, **97**, 13269–13274 (2000).
37. Gerritse, K., Laman, J. D., Noelle, R. J., Aruffo, A., Ledbetter, J. A., Boersma, W. J., and Claassen, E. CD40-CD40 ligand interactions in experimental allergic encephalomyelitis and multiple sclerosis. *Proc. Natl. Acad. Sci. USA*, **93**, 2499–2504 (1996).
38. Gessain, A., Barin, F., Vernant, J. C., Gout, O., Maurs, L., Calender, A., and de The, G. Antibodies to human T-lymphotropic virus type-I in patients with tropical spastic paraparesis. *Lancet*, **ii**, 407–410 (1985).
39. Godfrey, W. R., Fagnoni, F. F., Harara, M. A., Buck, D., and Engleman, E. G. Identi-

fication of a human OX-40 ligand, a costimulator of CD4+ T cells with homology to tumor necrosis factor. *J. Exp. Med.*, **180**, 757–762 (1994).

40. Goronzy, J., Weyand, C. M., and Fathman, C. G. Shared T cell recognition sites on human histocompatibility leukocyte antigen class II molecules of patients with seropositive rheumatoid arthritis. *J. Clin. Invest.*, **77**, 1042–1049 (1986).
41. Gramaglia, I., Weinberg, A. D., Lemon, M., and Croft, M. Ox-40 ligand: a potent costimulatory molecule for sustaining primary CD4 T cell responses. *J. Immunol.*, **161**, 6510–6517 (1998).
42. Green, J. E., Begley, C. G., Wagner, D. K., Waldmann, T. A., and Jay, G. Transactivation of granulocyte-macrophage colony-stimulating factor and the interleukin-2 receptor in transgenic mice carrying the human T-lymphotropic virus type 1 *tax* gene. *Mol. Cell. Biol.*, **9**, 4731–4737 (1989).
43. Green, J. E., Begley, C. G., Wagner, D. K., Waldmann, T. A., and Jay, G. Transactivation of granulocyte-macrophage colony-stimulating factor and the interleukin-2 receptor in transgenic mice carrying the human T-lymphotropic virus type 1 tax gene. *Mol. Cell. Biol.*, **9**, 4731–4737 (1989).
44. Grewal, I. S. and Flavell, R. A. CD40 and CD154 in cell-mediated immunity. *Annu. Rev. Immunol.*, **16**, 111–135 (1998).
45. Grewal, I. S., Xu, J., and Flavell, R. A. Impairment of antigen-specific T-cell priming in mice lacking CD40 ligand. *Nature*, **378**, 617–620 (1995).
46. Griggs, N. D., Agersborg, S. S., Noelle, R. J., Ledbetter, J. A., Linsley, P. S., and Tung, K. S. The relative contribution of the CD28 and gp39 costimulatory pathways in the clonal expansion and pathogenic acquisition of self-reactive T cells. *J. Exp. Med.*, **183**, 801–810 (1996).
47. Guerder, S., Meyerhoff, J., and Flavell, R. The role of the T cell costimulator B7-1 in autoimmunity and the induction and maintenance of tolerance to peripheral antigen. *Immunity*, **1**, 155–166 (1994).
48. Guerder, S., Picarella, D. E., Linsley, P. S., and Flavell, R. A. Costimulator B7-1 confers antigen-presenting-cell function to parenchymal tissue and in conjunction with tumor necrosis factor alpha leads to autoimmunity in transgenic mice. *Proc. Natl. Acad. Sci. USA*, **91**, 5138–5142 (1994).
49. Habu, K., Nakayama-Yamada, J., Asano, M., Saijo, S., Itagaki, K., Horai, R., Yamamoto, H., Sekiguchi, T., Nosaka, T., Hatanaka, M., and Iwakura, Y. The HTLV-I-tax gene is responsible for the development of both inflammatory polyarthropathy resembling rheumatoid arthritis and non-inflammatory ankylotic arthropathy in transgenic mice. *J. Immunol.*, **162**, 2956–2963 (1999).
50. Hang, L., Theofilopoulos, A. N., and Dixon, F. J. A spontaneous rheumatoid arthritis-like disease in MRL/l mice. *J. Exp. Med.*, **155**, 1690–1701 (1982).
51. Hannum, C. H., Wilcox, C. J., Arend, W. P., Joslin, F. G., Dripps, D. J., Heimdal, P. L., Armes, L. G., Sommer, A., Eisenberg, S. P., and Thompson, R. C. Interleukin-1 receptor antagonist activity of a human interleukin-1 inhibitor. *Nature*, **343**, 336–340 (1990).
52. Higashimura, N., Takasawa, N., Tanaka, Y., Nakamura, M., and Sugamura, K. Induction of OX40, a receptor of gp34, on T cells by trans-acting transcriptional activator, Tax, of human T-cell leukemia virus type I. *Jpn. J. Cancer Res.*, **87**, 227–231 (1996).

53. Horai, R., Asano, M., Sudo, K., Kanuka, H., Suzuki, M., Nishihara, M., Takahashi, M., and Iwakura, Y. Production of mice deficient in genes for interleukin (IL)-1α, IL-1β, IL-1α/β, and IL-1 receptor antagonist shows that IL-1β is crucial in turpentine-induced fever development and glucocorticoid secretion. *J. Exp. Med.*, **187**, 1463–1475 (1998).
54. Horai, R., Saijo, S., Tanioka, H., Nakae, S., Sudo, K., Okahara, A., Ikuse, T., Asano, M., and Iwakura, Y. Development of chronic inflammatory arthropathy resembling rheumatoid arthritis in interleukin 1 receptor antagonist-deficient mice [In Process Citation]. *J. Exp. Med.*, **191**, 313–320 (2000).
55. Imai, J. and Hinuma, Y. Epstein-Barr virus-specific antibodies in patients with adult T-cell leukemia (ATL) and healthy ATL virus-carriers. *Int. J. Cancer*, **31**, 197–200 (1983).
56. Inoue, J., Seiki, M., Taniguchi, T., Tsuru, S., and Yoshida, M. Induction of interleukin 2 receptor gene expression by p40x-encoded by human T-cell leukemia virus type I. *EMBO J.*, **5**, 2883–2888 (1986).
57. Iwakura, Y., Itagaki, K., Ishitsuka, C., Yamasaki, Y., Matsuzawa, A., Yonehara, S., Karasawa, S., Ueda, S., and Saijo, S. The development of autoimmune inflammatory arthropathy in mice transgenic for the human T cell leukemia virus type-1 env-pX region is not dependent on H-2 haplotypes and modified by the expression levels of Fas antigen. *J. Immunol.*, **161**, 6592–6598 (1998).
58. Iwakura, Y., Saijo, S., Kioka, Y., Nakayama-Yamada, J., Itagaki, K., Tosu, M., Asano, M., Kanai, Y., and Kakimoto, K. Autoimmunity induction by human T cell leukemia virus type 1 in transgenic mice that develop chronic inflammatory arthropathy resembling rheumatoid arthritis in humans. *J. Immunol.*, **155**, 1588–1598 (1995).
59. Iwakura, Y., Tosu, M., Yoshida, E., Saijo, S., Nakayama-Yamada, J., Itagaki, K., Asano, M., Siomi, H., Hatanaka, M., Takeda, T. *et al.* Augmentation of c-fos and c-jun expression in transgenic mice carrying the human T-cell leukemia virus type-I tax gene. *Virus Genes*, **9**, 161–170 (1995).
60. Iwakura, Y., Tosu, M., Yoshida, E., Takiguchi, M., Sato, K., Kitajima, I., Nishioka, K., Yamamoto, K., Takeda, T., Hatanaka, M., Yamamoto, H., and Sekiguchi, T. Induction of inflammatory arthropathy resembling rheumatoid arthritis in mice transgenic for HTLV-I. *Science*, **253**, 1026–1028 (1991).
61. Jacobson, S., Shida, H., McFarlin, D. E., Fauci, A. S., and Koenig, S. Circulating CD8+ cytotoxic T lymphocytes specific for HTLV-I pX in patients with HTLV-I associated neurological disease. *Nature*, **348**, 245–248 (1990).
62. Jember, A. G., Zuberi, R., Liu, F. T., and Croft, M. Development of allergic inflammation in a murine model of asthma is dependent on the costimulatory receptor OX40. *J. Exp. Med.*, **193**, 387–392 (2001).
63. Joosten, L. A., Helsen, M. M., van de Loo, F. A., and van den Berg, W. B. Anticytokine treatment of established type II collagen-induced arthritis in DBA/1 mice. A comparative study using anti-TNF alpha, anti-IL-1 alpha/beta, and IL-1Ra. *Arthritis Rheum.*, **39**, 797–809 (1996).
64. Jutila, M. A. and Banks, K. L. Increased macrophage division in the synovial fluid of goats infected with caprine arthritis-encephalitis virus. *J. Infect. Dis.*, **157**, 1193–1202 (1988).
65. Kaufman, S. H. Heat shock proteins and the immune response. *Immunol. Today*, **11**,

129–131 (1990).

66. Keffer, J., Probert, L., Cazlaris, H., Georgopoulos, S., Kaslaris, E., Kioussis, D., and Kollias, G. Transgenic mice expressing human tumour necrosis factor: a predictive genetic model of arthritis. *EMBO J.*, **10**, 4025–4031 (1991).
67. Kim, S.-J., Kehrl, J. H., Burton, J., Tendler, C. L., Jeang, K.-T., Danielpour, D., Thevenin, C., Kim, K. Y., Sporn, M. B., and Roberts, A. B. Transactivation of the transforming growth factor 1 (TGF-β1) gene by human T lymphotropic virus type I Tax: a potential mechanism for the increased production of TGF-β1 in adult T cell leukemia. *J. Exp. Med.*, **172**, 121–129 (1990).
68. Kishi, S., Saijo, S., Arai, M., Karasawa, S., Ueda, S., Kannagi, M., Iwakura, Y., Fujii, M., and Yonehara, S. Resistance to Fas-mediated apoptosis of peripheral T cells in human T lymphocyte virus type I (HTLV-I) transgenic mice with autoimmune arthropathy. *J. Exp. Med.*, **186**, 57–64 (1997).
69. Knight, B., Katz, D. R., Isenberg, D. A., Ibrahim, M. A., Le Page, S., Hutchings, P., Schwartz, R. S., and Cooke, A. Induction of adjuvant arthritis in mice. *Clin. Exp. Immunol.*, **90**, 459–465 (1992).
70. Koga, T., Kakimoto, K., Hirofuji, T., Kotani, S., Ohkuni, H., Watanabe, K., Okada, N., Okada, H., Sumiyoshi, A., and Saisho, K. Acute joint inflammation in mice after systemic injection of the cell wall, its peptidoglycan, and chemically defined peptidoglycan subunits from various bacteria. *Infect. Immun.*, **50**, 27–34 (1985).
71. Koga, T., Kakimoto, K., Hirofuji, T., Kotani, S., Sumiyoshi, A., and Saisho, K. Muramyl dipeptide induces acute joint inflammation in the mouse. *Microbiol. Immunol.*, **30**, 717–723 (1986).
72. Kontoyiannis, D., Pasparakis, M., Pizarro, T. T., Cominelli, F., and Kollias, G. Impaired on/off regulation of TNF biosynthesis in mice lacking TNF AU-rich elements: implications for joint and gut-associated immunopathologies. *Immunity*, **10**, 387–398 (1999).
73. Kopf, M., Ruedl, C., Schmitz, N., Gallimore, A., Lefrang, K., Ecabert, B., Odermatt, B., and Bachmann, M. F. OX40-deficient mice are defective in Th cell proliferation but are competent in generating B cell and CTL responses after virus infection. *Immunity*, **11**, 699–708 (1999).
74. Korthauer, U., Graf, D., Mages, H. W., Briere, F., Padayachee, M., Malcolm, S., Ugazio, A. G., Notarangelo, L. D., Levinsky, R. J., and Kroczek, R. A. Defective expression of T-cell CD40 ligand causes X-linked immunodeficiency with hyper-IgM. *Nature*, **361**, 539–541 (1993).
75. Kotani, A., Ishikawa, T., Matsumura, Y., Ichinohe, T., Ohno, H., Hori, T., and Uchiyama, T. Correlation of peripheral blood OX40+ (CD134+) T cells with chronic graft-versus-host disease in patients who underwent allogeneic hematopoietic stem cell transplantation. *Blood*, **98**, 3162–3164 (2001).
76. Kotani, M., Tagawa, Y., and Iwakura, Y. Involvement of autoimmunity against type II collagen in the development of arthritis in mice transgenic for the human T cell leukemia virus type I tax gene. *Eur. J. Immunol.*, **28**, 1–11 (1998).
77. Krieg, A. M. and Steinberg, A. D. Retroviruses and autoimmunity. *J. Autoimmun.*, **3**, 137–166 (1990).
78. Lehmann, P. V., Forsthuber, T., Miller, A., and Sercarz, E. E. Spreading of T-cell

autoimmunity to cryptic determinants of an autoantigen. *Nature*, **358**, 155–157 (1992).

79. Mallett, S., Fossum, S., and Barclay, A. N. Characterization of the MRC OX40 antigen of activated CD4 positive T lymphocytes—a molecule related to nerve growth factor receptor. *EMBO J.*, **9**, 1063–1068 (1990).
80. Malmstrom, V., Shipton, D., Singh, B., Al-Shamkhani, A., Puklavec, M. J., Barclay, A. N., and Powrie, F. CD134L expression on dendritic cells in the mesenteric lymph nodes drives colitis in T cell-restored SCID mice. *J. Immunol.*, **166**, 6972–6981 (2001).
81. Marinova-Mutafchieva, L., Williams, R. O., Mason, L. J., Mauri, C., Feldmann, M., and Maini, R. N. Dynamics of proinflammatory cytokine expression in the joints of mice with collagen-induced arthritis (CIA). *Clin. Exp. Immunol.*, **107**, 507–512 (1997).
82. Maruyama, M., Shibuya, H., Harada, H., Hatakeyama, M., Seiki, M., Fujita, T., Inoue, J., Yoshida, M., and Taniguchi, T. Evidence for aberrant activation of the interleukin-2 autocrine loop by HTLV-1-encoded p40x and T3/Ti complex triggering. *Cell*, **48**, 343–350 (1987).
83. Maxwell, J. R., Campbell, J. D., Kim, C. H., and Vella, A. T. CD40 activation boosts T cell immunity *in vivo* by enhancing T cell clonal expansion and delaying peripheral T cell deletion. *J. Immunol.*, **162**, 2024–2034 (1999).
84. McDowell, T. L., Symons, J. A., Ploski, R., Forre, O., and Duff, G. W. A genetic association between juvenile rheumatoid arthritis and a novel interleukin-1 alpha polymorphism. *Arthritis Rheum.*, **38**, 221–228 (1995).
85. Mehling, A., Loser, K., Varga, G., Metze, D., Luger, T. A., Schwarz, T., Grabbe, S., and Beissert, S. Overexpression of CD40 ligand in murine epidermis results in chronic skin inflammation and systemic autoimmunity. *J. Exp. Med.*, **194**, 615–628 (2001).
86. Miyatake, S., Seiki, M., Yoshida, M., and Arai, K. T-cell activation signals and human T-cell leukemia virus type I-encoded p40x protein activate the mouse granulocyte-macrophage colony-stimulating factor gene through a common DNA element. *Mol. Cell. Biol.*, **8**, 5581–5587 (1988).
87. Mohan, C., Shi, Y., Laman, J. D., and Datta, S. K. Interaction between CD40 and its ligand gp39 in the development of murine lupus nephritis. *J. Immunol.*, **154**, 1470–1480 (1995).
88. Morrow, J., Nelson, J. L., Watts, R., and Isenberg, D. Rheumatoid arthritis. *In* "Autoimmune Rheumatic Diseases," 2nd Ed., pp. 104–146 (1999). Oxford University Press, New York.
89. Murata, K., Ishii, N., Takano, H., Miura, S., Ndhlovu, L. C., Nose, M., Noda, T., and Sugamura, K. Impairment of antigen-presenting cell function in mice lacking expression of OX40 ligand. *J. Exp. Med.*, **191**, 365–374 (2000).
90. Nagata, S. and Golstein, P. The Fas death factor. *Science*, **267**, 1449–1456 (1995).
91. Nakae, S., Asano, M., Horai, R., Sakaguchi, N., and Iwakura, Y. IL-1 enhances T cell-dependent antibody production through induction of CD40 ligand and OX40 on T cells. *J. Immunol.*, **167**, 90–97 (2001).
92. Nakae, S., Horai, R., Asano, M., and Iwakura, I. Interleukin-1β, but not interleukin-1α, is required for T-cell-dependent antibody production. *Immunology*, **104**, 402–409 (2001).
93. Nakata, Y., Matsuda, K., Uzawa, A., Nomura, M., Akashi, M., and Suzuki, G. Administration of recombinant human IL-1 by staphylococcus enterotoxin B prevents toler-

ance induction *in vivo*. *J. Immunol.*, **155**, 4231–4235 (1995).
94. Ndhlovu, L. C., Ishii, N., Murata, K., Sato, T., and Sugamura, K. Critical involvement of OX40 ligand signals in the T cell priming events during experimental autoimmune encephalomyelitis. *J. Immunol.*, **167**, 2991–2999 (2001).
95. Niki, Y., Yamada, H., Seki, S., Kikuchi, T., Takaishi, H., Toyama, Y., Fujikawa, K., and Tada, N. Macrophage- and neutrophil-dominant arthritis in human IL-1 alpha transgenic mice. *J. Clin. Invest.*, **107**, 1127–1135 (2001).
96. Nishioka, K., Maruyama, I., Sato, K., Kitajima, I., Nakajima, Y., and Osame, M. Chronic inflammatory arthropathy associated with HTLV-I. *Lancet*, **i**, 441 (1989).
97. Nishioka, K., Sumida, T., and Hasunuma, T. Human T lymphotropic virus type I in arthropathy and autoimmune disorders. *Arthritis Rheum.*, **39**, 1410–1418 (1996).
98. Nohara, C., Akiba, H., Nakajima, A., Inoue, A., Koh, C. S., Ohshima, H., Yagita, H., Mizuno, Y., and Okumura, K. Amelioration of experimental autoimmune encephalomyelitis with anti-OX40 ligand monoclonal antibody: a critical role for OX40 ligand in migration, but not development, of pathogenic T cells. *J. Immunol.*, **166**, 2108–2115 (2001).
99. Ohashi, P. S., Oehen, S., Buerki, K., Pircher, H., Ohashi, C. T., Odermatt, B., Malissen, B., Zinkernagel, R. M., and Hengartner, H. Ablation of "tolerance" and induction of diabetes by virus infection in viral antigen transgenic mice. *Cell*, **65**, 305–317 (1991).
100. Osame, M., Matsumoto, M., Usuku, K., Izumo, S., Ijichi, N., Amitani, H., Tara, M., and Igata, A. Chronic progressive myelopathy associated with elevated antibodies to human T-lymphotropic virus type I and adult T-cell leukemia-like cells. *Ann. Neurol.*, **21**, 117–122 (1987).
101. Paliard, X., West, S. G., Lafferty, J. A., Clements, J. R., Kappler, J. W., Marrack, P., and Kotzin, B. L. Evidence for the effects of a superantigen in rheumatoid arthritis. *Science*, **253**, 325–329 (1991).
102. Pearson, C. M. Development of arthritis, periarthritis, and periostitis in rats given adjuvants. *Proc. Soc. Exp. Biol. Med.*, **91**, 95–101 (1956).
103. Pettipher, E. R., Higgs, G. A., and Henderson, B. Interleukin 1 induces leukocyte infiltration and cartilage proteoglycan degradation in the synovial joint. *Proc. Natl. Acad. Sci. USA*, **83**, 8749–8753 (1986).
104. Phillips, P. E., Johnston, S. L., Runge, L. A., Moore, J. L., and Poiesz, B. J. High IgM antibody to human T-lymphotropic virus type I in systemic lupus erythematosus. *J. Clin. Immunol.*, **6**, 234–241 (1986).
105. Pippig, S. D., Pena-Rossi, C., Long, J., Godfrey, W. R., Fowell, D. J., Reiner, S. L., Birkeland, M. L., Locksley, R. M., Barclay, A. N., and Killeen, N. Robust B cell immunity but impaired T cell proliferation in the absence of CD134 (OX40) [In Process Citation]. *J. Immunol.*, **163**, 6520–6529 (1999).
106. Ramesh, N., Fuleihan, R., and Geha, R. Molecular pathology of X-linked immunoglobulin deficiency with normal or elevated IgM (HIGMX-1). *Immunol. Rev.*, **138**, 87–104 (1994).
107. Reed, S. G., Pihl, D. L., Conlon, P. J., and Grabstein, K. H. IL-1 as adjuvant. Role of T cells in the augmentation of specific antibody production by recombinant human IL-1 alpha. *J. Immunol.*, **142**, 3129–3133 (1989).

108. Rogers, P. R., Song, J., Gramaglia, I., Killeen, N., and Croft, M. OX40 promotes Bcl-xL and Bcl-2 expression and is essential for long-term survival of CD4 T cells. *Immunity*, **15**, 445–455 (2001).
109. Rose, N. R., Herskowitz, A., Neumann, D. A., and Neu, N. Autoimmune myocarditis: a paradigm of post-infection autoimmune disease. *Immunol. Today*, **9**, 117–120 (1988).
110. Roudier, J., Rhodes, G., Petersen, J., Vaughan, J. H., and Carson, D. A. The Epstein-Barr virus glycoprotein gp110, a molecular link between HLA DR4, HLA DR1, and rheumatoid arthritis. *Scand. J. Immunol.*, **27**, 367–371 (1988).
111. Rynes, R. I., Goldenberg, D. L., DiGiacomo, R., Olson, R., Hussain, M., and Veazey, J. Acquired immunodeficiency syndrome-associated arthritis. *Am. J. Med.*, **84**, 810–816 (1988).
112. Saijo, S., Asano, M., Horai, R., Yamamoto, H., and Iwakura, Y. Suppression of autoimmune arthritis in interleukin-1-deficient mice in which T cell activation is impaired due to low levels of CD40 ligand and OX40 expression on T cells. *Arthritis Rheum.*, **46**, 533–544 (2002).
113. Saijo, S., Kotani, M., Habu, K., Ishitsuka, C., Yamamoto, H., Sekiguchi, T., and Iwakura, Y. Bone marrow-derived cells are responsible for the development of autoimmune arthritis in human T cell leukemia virus type I-transgenic mice and those of normal mice can suppress the disease. *J. Immunol.*, **163**, 5700–5707 (1999).
114. Siekevitz, M., Feinberg, M. B., Holbrook, N., Wong-Staal, F., and Greene, W. C. Activation of interleukin 2 and interleukin 2 receptor (Tac) promoter expression by the trans-activator (*tat*) gene product of human T-cell leukemia virus, type I. *Proc. Natl. Acad. Sci. USA*, **84**, 5389–5393 (1987).
115. Smith, C. A., Petty, R. E., and Tingle, A. J. Rubella virus and arthritis. *Rheum. Dis. Clin. North Am.*, **13**, 265–274 (1987).
116. Staruch, M. J. and Wood, D. D. The adjuvanticity of interleukin 1 *in vivo*. *J. Immunol.*, **130**, 2191–2194 (1983).
117. Stastny, P. Association of the B-cell alloantigen DRw4 with rheumatoid arthritis. *N. Engl. J. Med.*, **298**, 869–871 (1978).
118. Steffen, C., Zeitlhofer, J., Zielinski, C., Endler, A. T., and Menzel, J. Acute autoimmune collagen-induced arthritis in rabbits. *Z. Rheumatol.*, **37**, 275–285 (1978).
119. Stimpson, S. A., Dalldorf, F. G., Otterness, I. G., and Schwab, J. H. Exacerbation of arthritis by IL-1 in rat joints previously injured by peptidoglycan-polysaccharide. *J. Immunol.*, **140**, 2964–2969 (1988).
120. Stuber, E., Neurath, M., Calderhead, D., Fell, H. P., and Strober, W. Cross-linking of OX40 ligand, a member of the TNF/NGF cytokine family, induces proliferation and differentiation in murine splenic B cells. *Immunity*, **2**, 507–521 (1995).
121. Sugamura, K. and Hinuma, Y. Human retroviruses: HTLV-I and HTLV-II. *In* "The Retroviridae," ed. A. Levy, pp. 399–435 (1993). Plenum Press, New York.
122. Takahashi, Y., Murai, C., Shibata, S., Munakata, Y., Ishii, T., Ishii, K., Saitoh, T., Sawai, T., Sugamura, K., and Sasaki, T. Human parvovirus B19 as a causative agent for rheumatoid arthritis. *Proc. Natl. Acad. Sci. USA*, **95**, 8227–8232 (1998).
123. Tocci, M. J. and Schmidt, J. A. Interleukin-1: structure and function. *In* "Cytokines in Health and Disease," 2nd Ed., ed. D. G. Remick and J. S. Friedland, pp. 1–27 (1997).

Marcel Dekker, Inc., New York.

124. Trentham, D. E., Townes, A. S., and Kang, A. H. Autoimmunity to type II collagen an experimental model of arthritis. *J. Exp. Med.*, **146**, 857–868 (1977).

125. Usuku, K., Sonoda, S., Osame, M., Yashiki, S., Takahashi, K., Matsumoto, M., Sawada, T., Tsuji, K., Tara, M., and Igata, A. HLA haplotype-linked high immune responsiveness against HTLV-I in HTLV-I-associated myelopathy: comparison with adult T-cell leukemia/lymphoma. *Ann. Neurol.*, **23**, S143–S150 (1988).

126. van de Loo, F. A. and van den Berg, W. B. Gene therapy for rheumatoid arthritis. Lessons from animal models, including studies on interleukin-4, interleukin-10, and interleukin-1 receptor antagonist as potential disease modulators. *Rheum. Dis. Clin. North Am.*, **28**, 127–149 (2002).

127. van den Berg, W. B., Joosten, L. A., Helsen, M., and van de Loo, F. A. Amelioration of established murine collagen-induced arthritis with anti-IL-1 treatment. *Clin. Exp. Immunol.*, **95**, 237–243 (1994).

128. van den Broek, M. F., Hogervorst, E. J., Van Bruggen, M. C., Van Eden, W., van der Zee, R., and van den Berg, W. B. Protection against streptococcal cell wall-induced arthritis by pretreatment with the 65-kD mycobacterial heat shock protein. *J. Exp. Med.*, **170**, 449–466 (1989).

129. van Eden, W., Holoshitz, J., Nevo, Z., Frenkel, A., Klajman, A., and Cohen, I. R. Arthritis induced by a T-lymphocyte clone that responds to *Mycobacterium tuberculosis* and to cartilage proteoglycans. *Proc. Natl. Acad. Sci. USA*, **82**, 5117–5120 (1985).

130. Venables, P. J. Infection and rheumatoid arthritis. *Curr. Opin. Rheumatol.*, **1**, 15–30 (1989).

131. Vernant, J. C., Buisson, G., Magdeleine, J., De Thore, J., Jouannelle, A., Neisson-Vernant, C., and Monplaisir, N. T-lymphocyte alveolitis, tropical spastic paresis, and Sjogren syndrome. *Lancet*, **i**, 177 (1988).

132. Wang, H. C. and Klein, J. R. Multiple levels of activation of murine CD8(+) intraepithelial lymphocytes defined by OX40 (CD134) expression: effects on cell-mediated cytotoxicity, IFN-gamma, and IL-10 regulation. *J. Immunol.*, **167**, 6717–6723 (2001).

133. Wano, Y., Hattori, T., Matsuoka, M., Takatsuki, K., Chua, A. O., Gubler, U., and Greene, W. C. Interleukin 1 gene expression in adult T cell leukemia. *J. Clin. Invest.*, **80**, 911–916 (1987).

134. Watanabe-Fukunaga, R., Brannan, C. I., Copeland, N. G., Jenkins, N. A., and Nagata, S. Lymphoproliferation disorder in mice explained by defects in Fas antigen that mediates apoptosis. *Nature*, **356**, 314–317 (1992).

135. Yamamoto, H., Sekiguchi, T., Itagaki, K., Saijo, S., and Iwakura, Y. Inflammatory polyarthritis in mice transgenic for human T cell leukemia virus type I. *Arthritis Rheum.*, **36**, 1612–1620 (1993).

136. Yasuda, H., Shima, N., Nakagawa, N., Yamaguchi, K., Kinosaki, M., Mochizuki, S., Tomoyasu, A., Yano, K., Goto, M., Murakami, A., Tsuda, E., Morinaga, T., Higashio, K., Udagawa, N., Takahashi, N., and Suda, T. Osteoclast differentiation factor is a ligand for osteoprotegerin/osteoclastogenesis-inhibitory factor and is identical to TRANCE/RANKL. *Proc. Natl. Acad. Sci. USA*, **95**, 3597–3602 (1998).

137. Yoshida, M. HTLV-1 Tax: regulation of gene expression and disease. *Trends Microbiol.*, **1**, 131–135 (1993).

138. Yoshida, M., Suzuki, T., Fujisawa, J., and Hirai, H. HTLV-I oncoprotein tax and cellular transcription factors. *Curr. Top. Microbiol. Immunol.*, **193**, 131–148 (1995).
139. Yoshioka, T., Nakajima, A., Akiba, H., Ishiwata, T., Asano, G., Yoshino, S., Yagita, H., and Okumura, K. Contribution of OX40/OX40 ligand interaction to the pathogenesis of rheumatoid arthritis. *Eur. J. Immunol.*, **30**, 2815–2823 (2000).
140. Young, A., Jaraquemada, D., Awad, J., Festenstein, H., Corbett, M., Hay, F. C., and Roitt, I. M. Association of HLA-DR4/Dw4 and DR2/Dw2 with radiologic changes in a prospective study of patients with rheumatoid arthritis. Preferential relationship with HLA-Dw rather than HLA-DR specificities. *Arthritis Rheum.*, **27**, 20–25 (1984).

A RAT MODEL OF ATL AND TUMOR-VACCINE

Mari KANNAGI, Shino HANABUCHI, and Takashi OHASHI

*Department of Immunotherapeutics, Tokyo Medical and Dental University Graduate School**

Only a small population of human T cell leukemia virus type I (HTLV-I)-infected individuals develop adult T cell leukemia (ATL) following an incubation period of more than 40 years. This is partly explained by the time taken for clonal evolution of infected cells, and by insufficient host cellular immune responses, as most ATL patients show poor cellular immune responses against HTLV-I. HTLV-I-specific CD8+ cytotoxic T lymphocyte (CTL) in HTLV-I carriers often recognize HTLV-I Tax, a protein capable of transforming cells, and can lyse ATL cells *in vitro*, suggesting that HTLV-I-specific CTL may be an important effector of host immunosurveillance against HTLV-I-induced tumor development. However, it is still unclear whether these CTL actually play major roles against tumor *in vivo* or are merely a consequence of infection.

To understand the mechanisms of HTLV-I leukemogenesis *in vivo*, we designed a series of rat models. In the first model, we investigated *in vivo* conditions resulting in a variety of host immune responses to HTLV-I, and found that the route of mother-to-child transmission seriously affected host immune responses. Secondly, we established rat models in which T cell lymphomas could reproducibly develop following inoculation of syngeneic HTLV-I-immortalized cell lines in euthymic rats by blocking costimulatory signals for T cell activation. Finally, we found that HTLV-I Tax was recognized by CTL in rats as well as in humans, and demonstrated that vaccination using Tax-coding DNA or synthetic peptides corresponding to a CTL-epitope induced strong T cell immune responses, which were able to eradicate ATL-like lymphomas in rats. These findings strongly indicate the contribution of host T cell immune responses *in vivo* and the effectiveness of vaccine against development of HTLV-I-infected lymphoproliferative disease.

Individual Differences in the Host Immunity against Human T Cell Leukemia Virus Type I (HTLV-I)

HTLV-I causes various diseases such as adult T cell leukemia (ATL) (*12, 37, 43*), HTLV-I-associated myelopathy/tropical spastic paraparesis (HAM/TSP) (*5, 34*) and other inflammatory diseases (*5, 12, 34, 37*). A number of studies have shown that different genetic forms of HTLV-I strains could not explain the variety of clinical outcomes of HTLV-I infection (*2, 25*). Instead, the pathogenesis of HTLV-I is more likely to be influenced by the level of immune response to HTLV-I in individual subjects. For example, the efficiency of *in vitro* induction of HTLV-I-specific cytotoxic T

* Tokyo 113-8519, Japan

lymphocytes (CTL) from peripheral blood is low in ATL patients, but high in HAM/TSP patients (*15, 18, 20*). Weakness in the host CTL response against HTLV-I can be one of the risk factors for ATL development, because HTLV-I-infected cells may have a better chance to multiply and autonomously proliferate. However, the exact mechanisms resulting in weak immune responses to HTLV-I in some HTLV-I-infected individuals are still unclear. Immunological tolerance to HTLV-I, HTLV-I-mediated immune suppression, discrete viral expression (*24*), and genetic factors (*47*) may be involved in the mechanisms.

Experimental Alteration of Host Immune Responses against HTLV-I in Rats

Since the major route of natural HTLV-I infection is mother-to-child transmission caused by breastfeeding (*11, 23*), we examined whether this route of HTLV-I-transmission could affect host anti-HTLV-I immune responses (*21*). HTLV-I-producing MT-2 cells were inoculated into immune competent adult rats either orally (p.o), intravenously (i.v.) or intraperitoneally (i.p.). The results are summarized in Table I. All of these animals were persistently infected with HTLV-I. However, none of the p.o. but all i.v. or i.p. inoculated rats produced significant levels of anti-HTLV-I antibodies. Most importantly, T cell responses against HTLV-I measured by a proliferation assay was hardly detectable in p.o. inoculated rats. Although the initial amounts of HTLV-I taken through the intestine of the p.o. inoculated rats must have been smaller than those of i.v. or i.p. inoculated rats, the peripheral proviral load in the p.o inoculated groups was even higher than that in the i.v. or i.p. inoculated rats. A similar HTLV-I-carrier state with low immune responses was observed in rats that were born to and fed by an HTLV-I-infected mother. These observations indicate that oral transmission of HTLV-I could be a reason for immune unresponsiveness against HTLV-I and elevation of viral load in this animal system.

It is of note, however, that there are some differences in HTLV-I infection in rats and humans. In HTLV-I-infected rat T cells, expression of HTLV-I-structural proteins is much lower than human T cells, probably because of dysfunction of Rex in rat cells (*8, 27*). Since most of the antibodies are directed to viral structural proteins, the antibody response to HTLV-I in rats may not be as clear as in human. A similar seronegative HTLV-I-carrier state could be established when rats were i.p. inoculated with HTLV-I-infected rat cells instead of MT-2 cells. However, these rats exhibit proper levels of cellular immune response against HTLV-I, indicating that the scarcity of the viral structural protein could affect humoral but not cellular immunity.

TABLE I. Immune Responses in Rats Persistently Infected with HTLV-I through Various Routes

Route of infection	Serum antibodies to HTLV-I	HTLV-I-specific T cell proliferation	Presence of HTLV-I proviruses
i.p.[a]	+	+	+
i.v.[a]	+	+	+
p.o.[a]	Undetectable	Undetectable	+
Vertical[b]	Undetectable	Undetectable	+

[a] Four-week-old normal rats were inoculated with 2–5×10^7 Mitomycin C-treated MT-2 cells either intraperitoneally (i.p.), intravenously (i.v.), or orally (p.o.).
[b] Vertical infection was established in rats born to and fed by HTLV-I-infected mothers.

Therefore, the weakness in the cellular immunity in p.o. or vertically HTLV-I-infected rats is considered to be a sign of immune tolerance against HTLV-I.

In humans, a number of vertically HTLV-I-infected individuals lack antibody responses to HTLV-I during infancy (*39*). Although cellular immune response to HTLV-I in these individuals has not been investigated intensively, these subjects may have, to a certain extent, immunological tolerance in the early years of their lives.

Animal Models of ATL

To investigate *in vivo* mechanisms of HTLV-I-induced diseases, it is important to develop a suitable animal model. A number of studies have been made to establish animal models. Persistent infection with HTLV-I can be established in rats, rabbits and monkeys, but does not cause any disorders except for HAM/TSP-like disease in the WKAH strain of rats (*13*). Transgenic mice and rats with *tax* gene driven by HTLV-I long terminal repeat (LTR) develop collagen diseases but rarely T cell malignancies (*6, 14, 30, 46*). Tax-transgenic mice and rats with other promoters than HTLV-I LTR develop non-T-cell tumors. These are mammary tumor with H-2k promoter (*45*), large granular lymphocyte (LGL) leukemia with granzyme promoter (*7*), and epithelial thymoma with p56lck promoter (*22*). Frequency of tumor development in tax-transgenic mice increased when c-myc was additionally introduced (*1*). These observations suggest that the presence of tax alone promotes cell proliferation *in vivo* but may not be sufficient to cause T cell malignancy. Some additional events in the infected cells would be required.

Earlier studies indicated uncontrolled proliferation of HTLV-I-transformed T cell lines in newborn rats, nude mice, normal rats and rabbits (*31, 41, 42*). In addition, ATL cells sometimes show cytokine-dependent growth in SCID mice (*4, 26*). These observations suggest, although HTLV-I infection does not cause tumors immediately, HTLV-I-infected cells potentially proliferate *in vivo.* Since host immunity seems to be one of the critical factors for ATL development, we attempted to analyze host immunity in experimental models, in which HTLV-I-infected T cell lymphomas can be reproducibly induced and host anti-tumor immunity can be evaluated.

Establishment of a Nude Rat System of Syngeneic HTLV-I-infected T Cell Lymphomas

To establish an HTLV-I-infected T cell tumor model, we used athymic F344 Jcl *rnu/rnu* (*nu/nu*) and euthymic F344 Jcl *rnu/+* (*nu/+*) rats with a common genetic background (*32*). To shorten the process of clonal evolution of HTLV-I-infected cells, T lymphocytes of *nu/+* rats were transformed with HTLV-I *in vitro.* When the resulting cell line (FPM1) cells were inoculated into *nu/nu* rats, major histocompatibility complex (MHC) of the tumor cells was identical to that of the host. The experimental design and the results are schematically shown in Fig. 1. Although FPM1 cells expressed rat CD4, CD5, CD25, resembling phenotype of human ATL cells and autonomously proliferated *in vitro*, our initial attempts to establish T cell tumors in adult *nu/nu* rats failed. However, subcutaneously (s.c.) inoculated FPM1 cells into newborn *nu/nu* but not *nu/+* rats sporadically formed solid tumors. When the tumor-derived cells were inoculated to the next series of newborn *nu/nu* rats, the fre-

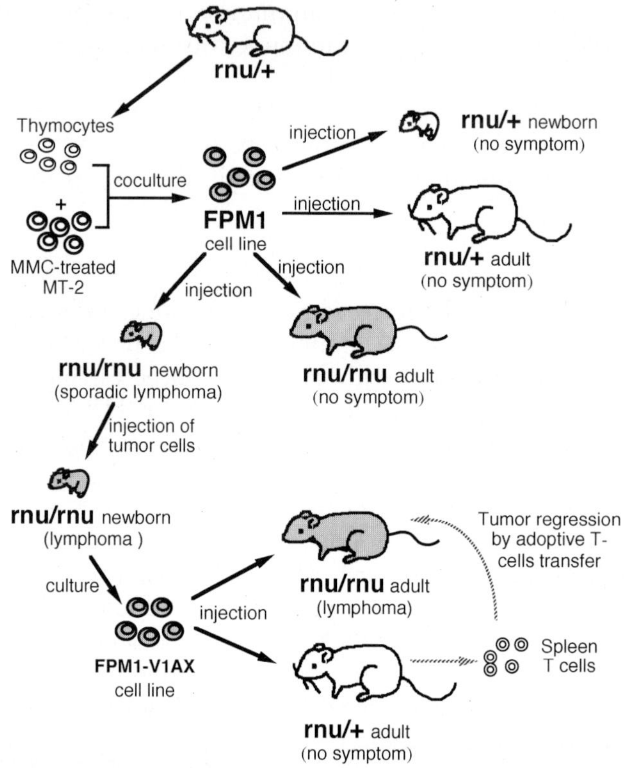

FIG. 1. Scheme of the syngeneic HTLV-I-infected tumor model system in rats

HTLV-I-transformed rat T cell line FPM1 derived from thymocytes of an *rnu*/+ rat did not cause tumors in *rnu*/*rnu* or *rnu*/+ rats except for a newborn *rnu*/*rnu* rat that developed lymphoma. Serial inoculations of the tumor cells in newborn *rnu*/*rnu* rats resulted in establishment of a subline FPM1-V1AX, which could reproducibly cause fatal lymphomas when inoculated in adult *rnu*/*rnu* but not *rnu*/+ rats. The spleen cells from *rnu*/+ rats inoculated with FPM1 or FPM1-V1AX cells could suppress tumor growth in *rnu*/*rnu* rats when adoptively transferred (broken lines).

quency of tumors increased. We isolated a cell line (FPM1-V1AX) from an axillary lymph node of one of these series of rats. Further inoculation of FPM1-V1AX cells into adult *nu*/*nu* rats caused fatal T cell lymphomas with systemic metastases in all the rats tested. Tumor nodules were observed in axillary, inguinal, mesenteric, and mediastinal lymph nodes, lungs, spleen, liver, skin, and retroperitoneal space (Fig. 2). The details of phenotypic progression of the series of subclones of FPM1 during *in vivo* growth are currently under investigation.

In contrast, FPM1-V1AX cells failed to cause such tumors in *nu*/+ rats. Furthermore, T cells isolated from spleens of these rats were able to suppress FPM1-V1AX tumor growth in *nu*/*nu* rats when adoptively transferred (*32*). These results strongly suggested the critical role of T cells in the rejection of HTLV-I-infected tumor cells.

The importance of T cell immunity on HTLV-I leukemogenesis *in vivo* was also indicated in another rat model system, in which otherwise resistant euthymic rats develop HTLV-I-infected T cell lymphomas by treatment with anti-CD80 and -CD86 monoclonal antibodies (*9*). As these antibodies block costimulatory signals

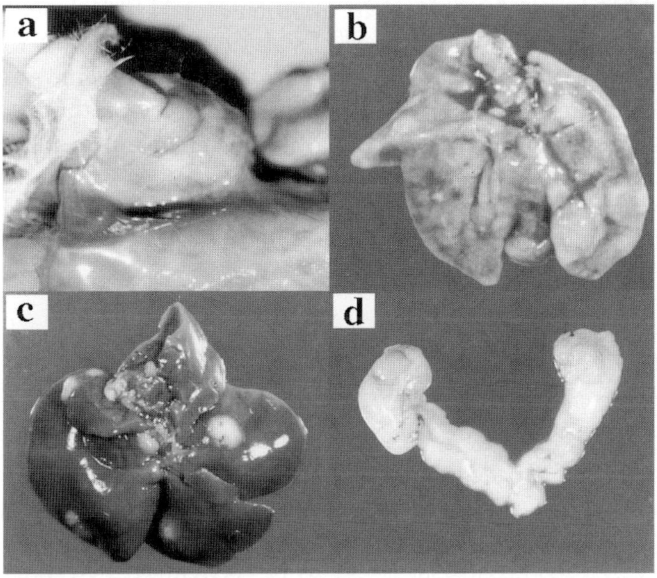

FIG. 2. Systemic lymphomas in *nu/nu* rats inoculated with FPM1-V1AX cells
Pictures indicate macroscopic appearance of tumor nodules in axillary lymph node (a), lung (b), liver (c), and ovary (d).

for T cell activation, HTLV-I-specific T cell immune responses were inhibited. Failure of HTLV-I-specific CTL and T cell proliferative responses in these rats were observed only during antibody treatment and inversely correlated with tumor development, again strongly indicating the contribution of HTLV-I-specific T cells in anti-tumor immunity.

Characterization of Anti-tumor Immunity in a Rat Model of ATL

Since euthymic rats were resistant to HTLV-I-infected lymphomas in the rat system as described above, anti-tumor immunity of *nu*/+ rats inoculated with FPM1-V1AX was characterized (*10*). We first assessed which subsets of immune effectors are required for direct tumor regression, and found both $CD4^+$ and $CD8^+$ T cells exhibited anti-tumor effects on the growth of HTLV-I-infected tumors as well as unfractionated T cells. Both of these fractions exhibited a high level of specific cytotoxicity against FPM1-V1AX. Analysis of the target antigens using recombinant vaccinia viruses revealed that HTLV-I Tax-expressing target cells were most susceptible to the CTL. In addition, HTLV-I envelope but not gag-expressing target cells were also susceptible to a lesser degree. Long-term culture of these HTLV-I-specific CTL with periodic *in vitro* stimulation with FPM1-V1AX resulted in a $CD8^+$ CTL line exclusively recognizing HTLV-I Tax. Further characterization revealed that such HTLV-I Tax-specific CTL line dominantly recognized Tax180-188 epitope that was restricted by rat MHC class I molecule, $RT1.A^l$ (*10*). These results indicate that HTLV-I Tax is one of the major target antigens recognized by CTL *in vivo* in *nu*/+ rats inoculated with syngeneic HTLV-I-infected cells. It is intriguing that both in human and rat HTLV-I-carriers, a major target antigen recognized by HTLV-I-specific CTL is HTLV-I Tax (*15, 16*), a critical protein for HTLV-I-oncogenesis (*48*). It is also clear that sufficient amounts of HTLV-I Tax to maintain CTL responses are expressed *in vivo*. In this rat model, the role of the Tax-specific CTL was further examined for their ability to protect against HTLV-I-infected tumor development *in vivo*. When the long-term-cultured $CD8^+$ CTL line shown above to recognize Tax180-188 was used for passive immunization, significant suppression of tumor growth was observed in *nu/nu* rats that were inoculated with FPM1-V1AX cells. Thus, the HTLV-I Tax-specific CTL contributed to regression of HTLV-I-infected tumor growth *in vivo*.

Tax-directed Anti-tumor Vaccine in a Rat Model of ATL

1. DNA vaccine

The fact that Tax-specific CTL could protect animals from HTLV-I-infected tumors encouraged our proceeding with vaccine studies to see whether induction of Tax-specific immune responses by vaccines may elicit anti-tumor therapeutic or prophylactic effects. To induce anti-tumor immune responses, appropriate tumor antigens should be given to the host in a way to be recognized by cellular immunity. In particular, to evoke $CD8^+$ CTL responses, presentation of processed peptides by MHC class I as well as class II antigens by antigen-presenting cells is essential. Because of the preference of the intrinsic pathway of antigen presentation by MHC class I, we chose DNA vaccine (*33*) and peptide-based vaccines corresponding to CTL epitopes (*10*) (Fig. 3).

DNA vaccine is known to induce not only antibody formation but also both MHC-I and -II restricted T cell responses (*36, 40, 44*). Initially, we investigated the therapeutic effect of Tax-coding DNA vaccine against ATL-like lymphoproliferative disease. As a Tax-coding DNA, we used pβMT-2Tax plasmids containing the entire length of wild-type tax cDNA driven by β-actin promoter (*29*). For vaccination, gold particles coated with the plasmids were shot by a gene gun into the skin of *nu/+* rats twice at a 1-week interval (*33*). Two weeks after the priming, we found that spleen T

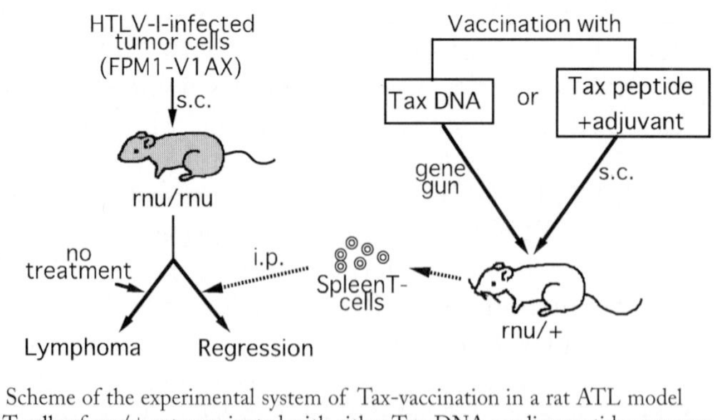

FIG. 3. Scheme of the experimental system of Tax-vaccination in a rat ATL model
Spleen T cells of *rnu/+* rats vaccinated with either Tax-DNA or oligopeptides were capable of protecting *rnu/rnu* rats from FPM1-V1AX lymphoma development when adoptively transferred (broken lines). For DNA vaccine, Tax-encoding plasmids were inoculated into the skin using a gene gun. For peptide vaccine, synthetic oligopeptides corresponding to a dominant CTL epitope in Tax were s.c. injected together with ISS-ODN as an adjuvant.

cells isolated from the vaccinated rats contained CTL capable of specifically killing autologous Tax-expressing cells. However, the sera from these rats did not contain any detectable levels of antibodies specific for HTLV-I Tax during the period tested when analyzed on western blotting assay.

When the spleen T cells isolated from the rats vaccinated with Tax-coding DNA were i.p. transferred into *nu/nu* rats that had been s.c. inoculated with FPM1-V1AX cells, tumors grew very slowly and were completely diminished in 2 weeks. Thus, Tax-DNA-vaccination induced efficient anti-tumor effects *in vivo.* However, the use of wild-type Tax may also induce unwanted effects *in vivo* because of its potential oncogenicity. We therefore used mutant Tax #410 (*29*), which lacks transforming activities but was capable of expressing Tax protein, for DNA-vaccination in the same animal system. In this experiment, we found that DNA vaccine with Tax #410 induced Tax-specific CTL activities and showed inhibitory effects *in vivo* as well as wild type Tax-vaccination. This indicates that the use of mutant Tax DNA vaccine may elicit anti-tumor effects with lower risks induced by Tax functions.

2. *Peptide vaccine*

Peptide-based vaccine corresponding to CTL epitopes that directly bind to MHC molecules is an alternative approach (*28*). Since we identified the major Tax-specific $CD8^+$ CTL epitope Tax180-188 in *nu/+* rats as described above, Tax180-188 peptides were used for peptide-vaccination (*10*). Immunostimulatory DNA sequences encoding-oligodeoxynucleotides (*38*) was employed as an adjuvant for the peptide-based vaccine. When *nu/+* rats were immunized with Tax180-188 peptide mixed with adjuvants, spleen T cells from these rats exhibited significant levels of anti-tumor effects *in vivo* in *nu/nu* rats that were inoculated with FPM1-V1AX cells. However, vaccination with Tax180-188 alone or adjuvants alone failed to induce such anti-tumor immune responses. Thus, vaccination with Tax180-188 peptide together with adjuvants effectively elicited T cell immunity that protected against HTLV-I-infected tumor *in vivo.*

Prospect of Immuno-prophylaxis and -therpy for ATL

The relationship between host immunity and HTLV-I leukemogenesis was investigated in the series of rat models presented here. These studies demonstrated: i) insufficient host anti-HTLV-I immunity associated with elevated viral load in oral infection, ii) contribution of Tax-specific CTL to anti-tumor surveillance, and iii) therapeutic potential of vaccine inducing these CTL in HTLV-I-induced tumor. These correspond with the findings in humans such as the insufficiency of Tax-specific CTL in ATL patients and the susceptibility of ATL cells to these CTL *in vitro* (*17, 20*), and encourage clinical application of vaccination against ATL. Several human CTL epitopes in Tax have already been identified and could be the candidates for peptide vaccine (*3, 19, 35*).

However, there are still many unknown mechanisms in ATL development in humans. Although immune unresponsiveness at early ages in mother-to-child HTLV-I infection could be reproduced in rat models, most human HTLV-I-carriers seem to acquire HTLV-I-specific immunity sometime later in their lives. It is not clear whether anti-HTLV-I cellular immunity in ATL patients has been at low levels until disease development or obtained once but suppressed again by other mechanisms. A longitudinal epidemiological study to dissect the relevance between cellular immunity and ATL development would be required to clarify this point. Development of appropriate adjuvant is also necessary in order to brake immune suppression in ATL patients.

The status of HTLV-I-expression *in vivo* is also obscure. It is well documented that the expression of viral proteins is repressed *in vivo* and induced *in vitro* in several hours (*17, 18, 20*). In the present rat study, HTLV-I-infected FPM1-V1AX cells did not grow in immunocompetent hosts and predominantly expressed Tax proteins, although expression of the other HTLV-I structural proteins was repressed (*27*). These cells might more likely have been at a stage much earlier than ATL development. Therefore, it is controversial to what extent Tax-specific CTL would have a direct therapeutic effect against full-blown ATL. Nevertheless, we still expect some prophylactic effects by CTL in a certain period in the development of ATL, because Tax-specific CTL are actually detected in HTLV-I carriers, which means that Tax is actually expressed in HTLV-I-infected cells at a certain stage of HTLV-I infection.

Conclusions

In conclusion, Tax-specific CTL contributed to anti-tumor surveillance *in vivo* and DNA- and peptide-based vaccines inducing these CTL protected hosts from HTLV-I-induced lymphoproliferative disease in rat models. For clinical application of vaccination against ATL, further studies are required, especially on an adjuvant to evoke cellular immunity, viral expression in ATL cells, and the strategy for gating the risk group for ATL among HTLV-I carriers.

REFERENCES

1. Benvenisty, N., Ornitz, D. M., Bennett, G. L., Sahagan, B. G., Kuo, A., Cardiff, R. D., and Leder, P. Brain tumours and lymphomas in transgenic mice that carry HTLV-I LTR/c-myc and Ig/tax genes. *Oncogene*, **7**, 2399–2405 (1992).
2. Daenke, S., Nightingale, S., Cruickshank, J. K., and Bangham, C. R. Sequence variants of human T-cell lymphotropic virus type I from patients with tropical spastic paraparesis and adult T-cell leukemia do not distinguish neurological from leukemic isolates. *J. Virol.*, **64**, 1278–1282 (1990).
3. Elovaara, I., Koenig, S., Brewah, A. Y., Woods, R. M., Lehky, T., and Jacobson, S. High human T cell lymphotropic virus type 1 (HTLV-1)-specific precursor cytotoxic T lymphocyte frequencies in patients with HTLV-1-associated neurological disease. *J. Exp. Med.*, **177**, 1567–1573 (1993).
4. Feuer, G., Zack, J. A., Harrington, W. J., Jr., Valderama, R., Rosenblatt, J. D., Wachsman, W., Baird, S. M., and Chen, I. S. Establishment of human T-cell leukemia virus type I T-cell lymphomas in severe combined immunodeficient mice. *Blood*, **82**, 722–731 (1993).
5. Gessain, A., Barin, F., Vernant, J. C., Gout, O., Maurs, L., Calender, A., and de The, G. Antibodies to human T-lymphotropic virus type-I in patients with tropical spastic paraparesis. *Lancet*, **ii**, 407–410 (1985).

6. Green, J. E., Hinrichs, S. H., Vogel, J., and Jay, G. Exocrinopathy resembling Sjögren's syndrome in HTLV-1 tax transgenic mice. *Nature*, **341**, 72–74 (1989).
7. Grossman, W. J., Kimata, J. T., Wong, F. H., Zutter, M., Ley, T. J., and Ratner, L. Development of leukemia in mice transgenic for the tax gene of human T-cell leukemia virus type I. *Proc. Natl. Acad. Sci. USA*, **92**, 1057–1061 (1995).
8. Hakata, Y., Yamada, M., and Shida, H. Rat CRM1 is responsible for the poor activity of human T-cell leukemia virus type 1 rex protein in rat cells. *J. Virol.*, **75**, 11515–11525 (2001).
9. Hanabuchi, S., Ohashi, T., Koya, Y., Kato, H., Takemura, F., Hirokawa, K., Yoshiki, T., Yagita, H., Okumura, K., and Kannagi, M. Development of human T-cell leukemia virus type 1-transformed tumors in rats following suppression of T-cell immunity by CD80 and CD86 blockade. *J. Virol.*, **74**, 428-435 (2000).
10. Hanabuchi, S., Ohashi, T., Koya, Y., Kato, H., Hasegawa, A., Takemura, F., Masuda, T., and Kannagi, M. Regression of human T-cell leukemia virus type I (HTLV-I)-lymphomas in a rat model: peptide-induced T-cell immunity. *J. Natl. Cancer Inst.*, **93**, 1775–1783 (2001).
11. Hino, S., Yamaguchi, K., Katamine, S., Sugiyama, H., Amagasaki, T., Kinoshita, K., Yoshida, Y., Doi, H., Tsuji, Y., and Miyamoto, T. Mother-to-child transmission of human T-cell leukemia virus type-I. *Jpn. J. Cancer Res.*, **76**, 474–480 (1985).
12. Hinuma, Y., Nagata, K., Hanaoka, M., Nakai, M., Matsumoto, T., Kinoshita, K. I., Shirakawa, S., and Miyoshi, I. Adult T-cell leukemia: antigen in an ATL cell line and detection of antibodies to the antigen in human sera. *Proc. Natl. Acad. Sci. USA*, **78**, 6476–6480 (1981).
13. Ishiguro, N., Abe, M., Seto, K., Sakurai, H., Ikeda, H., Wakisaka, A., Togashi, T., Tateno, M., and Yoshiki, T. A rat model of human T lymphocyte virus type I (HTLV-I) infection. 1. Humoral antibody response, provirus integration, and HTLV-I-associated myelopathy/tropical spastic paraparesis-like myelopathy in seronegative HTLV-I carrier rats. *J. Exp. Med.*, **176**, 981–989 (1992).
14. Iwakura, Y., Tosu, M., Yoshida, E., Takiguchi, M., Sato, K., Kitajima, I., Nishioka, K., Yamamoto, K., Takeda, T., Hatanaka, M., Yamamoto, H., and Sekiguchi, T. Induction of inflammatory arthropathy resembling rheumatoid arthritis in mice transgenic for HTLV-I. *Science*, **253**, 1026–1028 (1991).
15. Jacobson, S., Shida, H., McFarlin, D. E., Fauci, A. S., and Koenig, S. Circulating CD8+ cytotoxic T lymphocytes specific for HTLV-I pX in patients with HTLV-I associated neurological disease. *Nature*, **348**, 245–248 (1990).
16. Kannagi, M., Harada, S., Maruyama, I., Inoko, H., Igarashi, H., Kuwashima, G., Sato, S., Morita, M., Kidokoro, M., Sugimoto, M., Funahashi, S., Osame, M., and Shida, H. Predominant recognition of human T cell leukemia virus type I (HTLV-I) pX gene products by human CD8+ cytotoxic T cells directed against HTLV-I-infected cells. *Int. Immunol.*, **3**, 761–767 (1991).
17. Kannagi, M., Matsushita, S., and Harada, S. Expression of the target antigen for cytotoxic T lymphocytes on adult T-cell-leukemia cells. *Int. J. Cancer*, **54**, 582–588 (1993).
18. Kannagi, M., Matsushita, S., Shida, H., and Harada, S. Cytotoxic T cell response and expression of the target antigen in HTLV-I infection. *Leukemia*, **8**, S54–S59 (1994).
19. Kannagi, M., Shida, H., Igarashi, H., Kuruma, K., Murai, H., Aono, Y., Maruyama, I.,

Osame, M., Hattori, T., Inoko, H., and Harada, S. Target epitope in the Tax protein of human T-cell leukemia virus type I recognized by class I major histocompatibility complex-restricted cytotoxic T cells. *J. Virol.*, **66**, 2928–2933 (1992).

20. Kannagi, M., Sugamura, K., Kinoshita, K., Uchino, H., and Hinuma, Y. Specific cytolysis of fresh tumor cells by an autologous killer T cell line derived from an adult T cell leukemia/lymphoma patient. *J. Immunol.*, **133**, 1037–1041 (1984).
21. Kato, H., Koya, Y., Ohashi, T., Hanabuchi, S., Takemura, F., Fujii, M., Tsujimoto, H., Hasegawa, A., and Kannagi, M. Oral administration of human T-cell leukemia virus type 1 induces immune unresponsiveness with persistent infection in adult rats. *J. Virol.*, **72**, 7289–7293 (1998).
22. Kikuchi, K. Development of thymic tumor in HTLV-I transgenic rats under control of a lymphoid tissue specific promoter. *Hokkaido Igaku Zasshi*, **72**, 545–564 (1997) (in Japanese).
23. Kinoshita, K., Amagasaki, T., Hino, S., Doi, H., Yamanouchi, K., Ban, N., Momita, S., Ikeda, S., Kamihira, S., Ichimaru, M., Katamine, S., Miyamoto, T., Tsujii, Y., Ishimaru, T., Yamabe, T., Ito, M., Kamura, S., and Tsuda, T. Milk-borne transmission of HTLV-I from carrier mothers to their children. *Jpn. J. Cancer Res.*, **78**, 674–680 (1987).
24. Kinoshita, T., Shimoyama, M., Tobinai, K., Ito, M., Ito, S., Ikeda, S., Tajima, K., Shimotohno, K., and Sugimura, T. Detection of mRNA for the tax1/rex1 gene of human T-cell leukemia virus type I in fresh peripheral blood mononuclear cells of adult T-cell leukemia patients and viral carriers by using the polymerase chain reaction. *Proc. Natl. Acad. Sci. USA*, **86**, 5620–5624 (1989).
25. Kinoshita, T., Tsujimoto, A., and Shimotohno, K. Sequence variations in LTR and env regions of HTLV-I do not discriminate between the virus from patients with HTLV-I-associated myelopathy and adult T-cell leukemia. *Int. J. Cancer*, **47**, 491–495 (1991).
26. Kondo, A., Imada, K., Hattori, T., Yamabe, H., Tanaka, T., Miyasaka, M., Okuma, M., and Uchiyama, T. A model of *in vivo* cell proliferation of adult T-cell leukemia. *Blood*, **82**, 2501–2509 (1993).
27. Koya, Y., Ohashi, T., Kato, H., Hanabuchi, S., Tsukahara, T., Takemura, F., Etoh, K., Matsuoka, M., Fujii, M., and Kannagi, M. Establishment of a seronegative human T-cell leukemia virus type 1 (HTLV-1) carrier state in rats inoculated with a syngeneic HTLV-1-immortalized T-cell line preferentially expressing Tax. *J. Virol.*, **73**, 6436–6443 (1999).
28. Marchand, M., Weynants, P., Rankin, E., Arienti, F., Belli, F., Parmiani, G. *et al.* Tumor regression responses in melanoma patients treated with a peptide encoded by gene MAGE-3. *Int. J. Cancer*, **63**, 883–885 (1995).
29. Matsumoto, K., Shibata, H., Fujisawa, J. I., Inoue, H., Hakura, A., Tsukahara, T., and Fujii, M. Human T-cell leukemia virus type 1 Tax protein transforms rat fibroblasts via two distinct pathways. *J. Virol.*, **71**, 4445–4451 (1997).
30. Nerenberg, M., Hinrichs, S. H., Reynolds, R. K., Khoury, G., and Jay, G. The tat gene of human T-lymphotropic virus type 1 induces mesenchymal tumors in transgenic mice. *Science*, **237**, 1324–1329 (1987).
31. Ogawa, K., Matsuda, S., and Seto, A. Induction of leukemic infiltration by allogeneic transfer of HTLV-I-transformed T cells in rabbits. *Leuk. Res.*, **13**, 399–406 (1989).
32. Ohashi, T., Hanabuchi, S., Kato, H., Koya, Y., Takemura, F., Hirokawa, K., Yoshiki, T.,

Tanaka, Y., Fujii, M., and Kannagi, M. Induction of adult T-cell leukemia-like lymphoproliferative disease and its inhibition by adoptive immunotherapy in T-cell-deficient nude rats inoculated with syngeneic human T-cell leukemia virus type 1-immortalized cells. *J. Virol.*, **73**, 6031–6040 (1999).

33. Ohashi, T., Hanabuchi, S., Kato, H., Tateno, H., Takemura, F., Tsukahara, T., Koya, Y., Hasegawa, A., Masuda, T., and Kannagi, M. Prevention of adult T-cell leukemia-like lymphoproliferative disease in rats by adoptively transferred T cells from a donor immunized with human T-cell leukemia virus type 1 Tax-coding DNA vaccine. *J. Virol.*, **74**, 9610–9616 (2000).
34. Osame, M., Usuku, K., Izumo, S., Ijichi, N., Amitani, H., Igata, A., Matsumoto, M., and Tara, M. HTLV-I associated myelopathy, a new clinical entity. *Lancet*, **i**, 1031–1032 (1986).
35. Parker, C. E., Nightingale, S., Taylor, G. P., Weber, J., and Bangham, C. R. Circulating anti-Tax cytotoxic T lymphocytes from human T-cell leukemia virus type I-infected people, with and without tropical spastic paraparesis, recognize multiple epitopes simultaneously. *J. Virol.*, **68**, 2860–2868 (1994).
36. Pertmer, T. M., Eisenbraun, M. D., McCabe, D., Prayaga, S. K., Fuller, D. H., and Haynes, J. R. Gene gun-based nucleic acid immunization: elicitation of humoral and cytotoxic T lymphocyte responses following epidermal delivery of nanogram quantities of DNA. *Vaccine*, **13**, 1427–1430 (1995).
37. Poiesz, B. J., Ruscetti, F. W., Gazdar, A. F., Bunn, P. A., Minna, J. D., and Gallo, R. C. Detection and isolation of type C retrovirus particles from fresh and cultured lymphocytes of a patient with cutaneous T-cell lymphoma. *Proc. Natl. Acad. Sci. USA*, **77**, 7415–7419 (1980).
38. Roman, M., Martin-Orozco, E., Goodman, J. S., Nguyen, M. D., Sato, Y., Ronaghy, A. *et al.* Immunostimulatory DNA sequences function as T helper-1-promoting adjuvants. *Nat. Med.*, **3**, 849–854 (1997).
39. Saito, S., Ando, Y., Furuki, K., Kakimoto, K., Tanigawa, T., Moriyama, I., Ichijo, M., Nakamura, M., Ohtani, K., and Sugamura, K. Detection of HTLV-I genome in seronegative infants born to HTLV-I seropositive mothers by polymerase chain reaction. *Jpn. J. Cancer Res.*, **80**, 808–812 (1989).
40. Schirmbeck, R., Bohm, W., Ando, K., Chisari, F. V., and Reimann, J. Nucleic acid vaccination primes hepatitis B virus surface antigen-specific cytotoxic T lymphocytes in nonresponder mice. *J Virol.*, **69**, 5929–5934 (1995).
41. Simpson, R. M., Zhao, T. M., Hubbard, B. S., Sawasdikosol, S., and Kindt, T. J. Experimental acute adult T cell leukemia-lymphoma is associated with thymic atrophy in human T cell leukemia virus type I infection. *Lab. Invest.*, **74**, 696–710 (1996).
42. Tateno, M., Kondo, N., Itoh, T., Chubachi, T., Togashi, T., and Yoshiki, T. Rat lymphoid cell lines with human T cell leukemia virus production. I. Biological and serological characterization. *J. Exp. Med.*, **159**, 1105–1116 (1984).
43. Uchiyama, T., Yodoi, J., Sagawa, K., Takatsuki, K., and Uchino, H. Adult T-cell leukemia: clinical and hematologic features of 16 cases. *Blood*, **50**, 481–492 (1977).
44. Ulmer, J. B., Donnelly, J. J., Parker, S. E., Rhodes, G. H., Felgner, P. L., Dwarki, V. J., Gromkowski, S. H., Deck, R. R., DeWitt, C. M., Friedman, A. *et al.* Heterologous protection against influenza by injection of DNA encoding a viral protein. *Science*, **256**,

1745–1749 (1993).

45. Yamada, S., Ikeda, H., Yamazaki, H., Shikishima, H., Kikuchi, K., Wakisaka, A., Kasai, N., Shimotohno, K., and Yoshiki, T. Cytokine-producing mammary carcinomas in transgenic rats carrying the pX gene of human T-lymphotropic virus type I. *Cancer Res.*, **55**, 2524–2527 (1995).
46. Yamazaki, H., Ikeda, H., Ishizu, A., Nakamaru, Y., Sugaya, T., Kikuchi, K., Yamada, S., Wakisaka, A., Kasai, N., Koike, T., Hatanaka, M., and Yoshiki, T. A wide spectrum of collagen vascular and autoimmune diseases in transgenic rats carrying the env-pX gene of human T lymphocyte virus type I. *Int. Immunol.*, **9**, 339–346 (1997).
47. Yashiki, S., Fujiyoshi, T., Arima, N., Osame, M., Yoshinaga, M., Nagata, Y., Tara, M., Nomura, K., Utunomiya, A., Hanada, S., Tajima, K., and Sonoda, S. HLA-A26, HLA-B4002, HLA-B4006, and HLA-B4801 alleles predispose to adult T cell leukemia: the limited recognition of HTLV type 1 Tax peptide anchor motifs and epitopes to generate anti-HTLV type 1 Tax CD8+ cytotoxic T lymphocytes. *AIDS Res. Hum. Retroviruses*, **17**, 1047–1061 (2001).
48. Yoshida, M., Suzuki, T., Fujisawa, J., and Hirai, H. HTLV-1 oncoprotein tax and cellular transcription factors. *Curr. Top. Microbiol. Immunol.*, **193**, 79–89 (1995).

TUMORIGENESIS WITH TRANSGENIC EXPRESSION OF HUMAN T/CELL LEUKEMIA VIRUS Tax

Lee RATNER

*Washington University School of Medicine**

Understanding of the pathogenesis of HTLV-I infection and analysis of the effects of treatments and vaccines is limited by the lack of suitable animal model systems permissive for virus replication and pathogenic effects. Tax transgenic mice provide the most useful animal model to date for studies of the step-wise events involved in tumorigenesis. The breadth of mouse genetic models, extensive panels of mouse monoclonal antibodies, and advances in mouse tumor biology and transplantation provide powerful tools for understanding the role of Tax in lymphoproliferative disease.

Human T cell leukemia virus type I (HTLV-I) is responsible for the development of adult T cell leukemia/lymphoma (ATLL) (*6, 36*). It is an oncoretrovirus that has neither transduced a homologue of a cellular oncogene nor integrated adjacent to a cellular oncogene, but it expresses a viral oncogene, *Tax*, distinct from cellular genes (*42*). Tax has multiple functions, and thus resembles oncogenes found in transforming DNA viruses, like polyoma large T antigen and adenoviral E1A protein (*51*). It has promiscuous transcriptional *trans*-activating effects, functioning through cellular transcription factors, including cAMP-response element binding proteins (CREB), nuclear factor kappa B (NFκB), serum response factor (SRF), zinc-finger proteins Sp1 and Egr1, and *trans*-repressing effects through basic helix-loop-helix proteins (*17*). Tax also has post-transcriptional effects through protein interactions with cell cycle inhibitor p16, cyclin D3, and mitotic arrest protein, MAD1, and by activation of a protein kinase(s) that phosphorylates and inactivates the p53 tumor suppressor protein (*18, 23, 26, 29, 41, 44*).

Several experimental findings provided circumstantial evidence that Tax is an oncoprotein. Tax is capable of cooperating with Ras in transforming primary rat embryo fibroblasts (*35*). Tax expression from herpes or retrovirus vectors is capable of immortalization of primary human CD4+ lymphocytes (*9*). Tax is required for the transforming activity of HTLV-II (*38*). Furthermore, frequent deletions occur in the HTLV-I genome in ATLL patients, but the *tax* gene is retained (*46, 47*). However, the most convincing evidence for the role of Tax as an oncoprotein has been provided from studies of murine transgenic models.

Tax Transgenic Models

Several different murine Tax transgenic models have been established. Expression of Tax from the viral promoter resulted in mesenchymal tumors, bone abnormal-

* Box 8069, 660 S Euclid Ave., St. Louis, MO 63110, USA

ities, exocrinopathy, growth retardation, thymic atrophy, inflammatory arthropathy, and premature mortality (*13, 15, 27, 39*). Mesenchymal tumors were also found with expression of Tax under the control of the Thy1.2 promoter (*28*), whereas thymic atrophy was seen with Tax expression from the mouse mammary tumor virus promoter and the SV40 promoter fused to the immunoglobulin enhancer (*7*). Arthritis developed in mice with Tax expressed from the CD4 promoter (*12*).

Tax expression from the granzyme B promoter resulted in highly restricted expression in natural killer cells and activated CD4+ and CD8+ lymphocytes (*10*). Transgenic animals from two distinct founders developed large granular cell lymphoma-leukemia with 100% penetrance among their progeny (Fig. 1). Within the first 3–9 mos of life the animals developed "peripheral" tumors on their tails, ears, limbs, or nose, at sites of injury resulting from tail DNA analysis, ear clipping, or fighting. Large granular lymphoid cells were also seen in the blood at this time, but

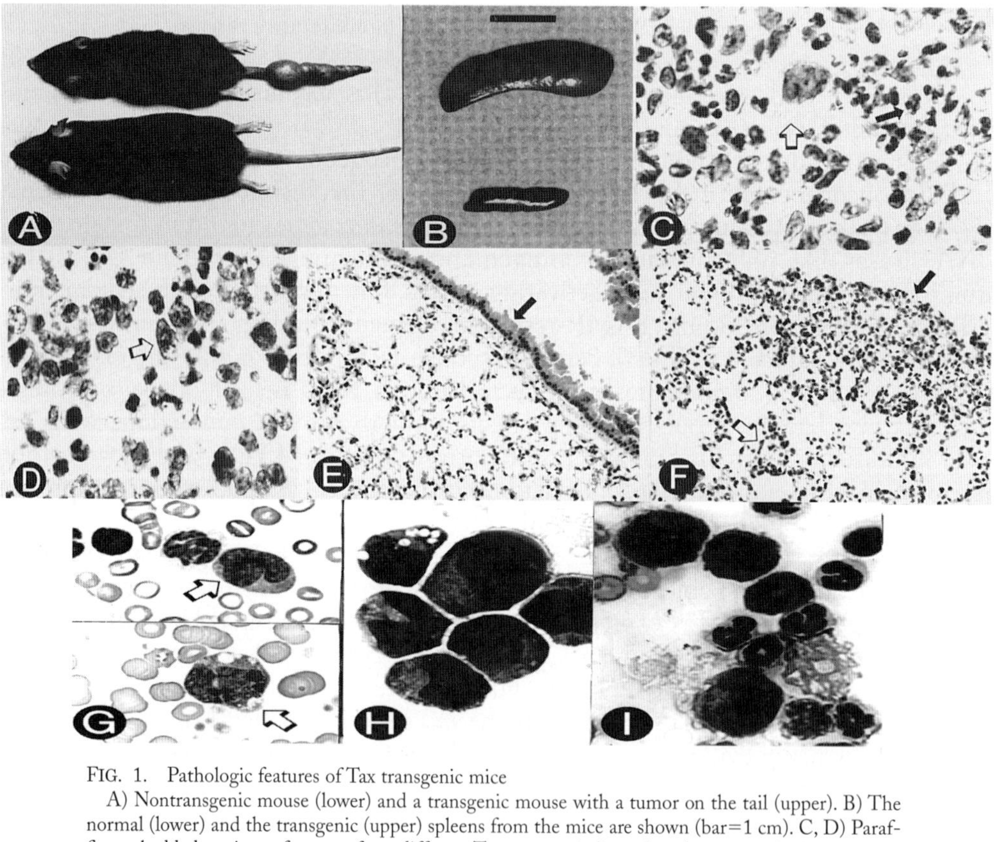

FIG. 1. Pathologic features of Tax transgenic mice
A) Nontransgenic mouse (lower) and a transgenic mouse with a tumor on the tail (upper). B) The normal (lower) and the transgenic (upper) spleens from the mice are shown (bar=1 cm). C, D) Paraffin-embedded sections of tumors from different Tax transgenic lines show large granular lymphocytes (LGLs, open arrow) and infiltrating polymorphonuclear leukocytes (solid arrow) (×720). E, F) Sections of lung from a normal and transgenic mouse, respectively. The transgenic mouse lung exhibits malignant LGLs infiltrating into the bronchus epithelium (solid arrows) and alveolar septa (open arrow) (×360). G) Peripheral blood smears from transgenic animals show circulating LGLs (open arrow) with convoluted nuclei and granular/vacuolar cytoplasm (×2,100). H) LGL cells from an LGL cell line (×2,100). I) Bone marrow smear from transgenic animal shows predominance of LGL cells and myeloid cells. Modified and reprinted with permission from (*10*).

subsequently expanded to represent up to 70% of nucleated cells in the blood and 30–40% of the cells in the bone marrow. Large granular lymphoid cells were also found to diffusely infiltrate the liver, spleen, mesenteric nodes, and lung. In addition, neutrophils were found infiltrating tumors, and the animals exhibited neutrophilia.

The large granular lymphoid cells expressed high levels of FcγR (CD16/CD32), activation marker CD44 (pgp-1), and variable levels of Thy1.2. B cell antigens (CD45R/B220 and surface immunoglobulin) were negative, as were CD3, CD4, CD8, and TCR, on most tumors, although occasional CD8 tumors were identified. These findings suggested that the tumor cells were derived from T-NK cell precursors. Further evidence in support of this cell type derivation was the antibody-dependent cellular cytotoxic activity of these tumor cells towards P815 cells.

Cell lines derived from the Tax transgenic tumors had a similar morphology and cell surface phenotype as the primary tumor cells. Inoculation of SCID mice with these cell lines resulted in a similar malignant phenotype (*11*). Furthermore, it was found that the cell lines and the primary tumors expressed several cellular proteins that are induced by Tax through the NFκB pathway. These included granulocyte macrophage colony-stimulating factor (GM-CSF), interferon-γ, and IL-15 (Fig. 2). We hypothesize that GM-CSF induction is responsible for neutrophilia and neutro-

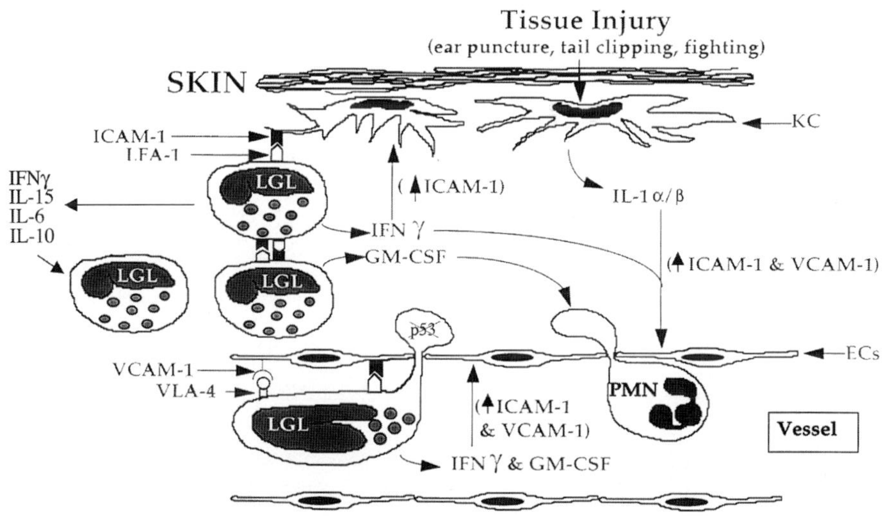

FIG. 2. Hypothetic model of the role of cytokines in tax transgenic mice

This model suggests that there is a need for an initial injury to peripheral skin tissue (e.g. ear puncture, tail clipping, fighting). This injury starts an inflammatory response by skin epithelial cells (e.g. keratinocytes) that release inflammatory cytokines such as IL-1. These cytokines act on endothelial cells (ECs) to cause extravasation of circulating LGLs and neutrophils (polymorphonuclear leukocyte (PMNs)) by upregulating adhesion molecules such as intercellular adhesion molecule (ICAM)-1 and vascular endothelial adhesion molecule (VCAM)-1 on ECs. Circulating LGLs also constitutively express adhesion molecules on their surface (leukocyte function-associated antigen (LFA)-1 and very late antigen (VLA)-4) which promote their binding to activated ECs and extravasation. These LGLs are then activated and release cytokines such as GM-CSF, which cause neutrophilia and PMN extravasation into tumors. In addition, LGLs produce large quantities of IFN-γ that could initiate and exacerbate this inflammatory process by upregulating expression of adhesion molecules such as ICAM-1 and VCAM-1. These cells also release IL-6, -10, and -15, and together with IFN-γ could play an autocrine or paracrine role in tumor cell proliferation or survival. Modified and reprinted with permission from (*11*).

phil infiltration into tumors. Interferon-γ could be a marker of NK-T cell tumors, an anti-apoptotic factor that promotes tumor cell survival, or a component of the immune response to the tumors (*2, 3, 43*). IL-15 may represent an autocrine or paracrine proliferation factor for these tumors (*8, 21, 22, 25*). Moreover, these tumors and cell lines expressed VLA-4, LFA-1, and ICAM-1 which were presumed to mediate adhesion and extravasation of tumor cells at sites of tissue injury.

Tax Activation of NFκB

The tumor cells and cell lines were found to exhibit high levels of NFκB activity as measured by electrophoretic mobility shift analysis and an oligonucleotide-based ELISA (Fig. 3) (*33*). In contrast, uninvolved tissues (e.g. kidney, heart) from transgenic animals and lymphoid tissues from non-transgenic animals showed much lower levels of NFκB activity. Inhibitors of NFκB expression, including sodium salicylate and cyclopentenone prostaglandins were found to block NFκB expression in the tumor cells and cell lines. Moreover, these inhibitors blocked the ability of the tumor cells to proliferate spontaneously and indefinitely when cultured *ex vivo.* In addition, treatment of tumor cell lines or tumor cells with sodium salicylate and cyclopentenone prostaglandins resulted in apoptosis. These findings suggest that the NFκB pathway is important for proliferation of Tax induced tumors, and for their resistance to inducers of apoptosis. Studies are underway to examine the effects on Tax induced

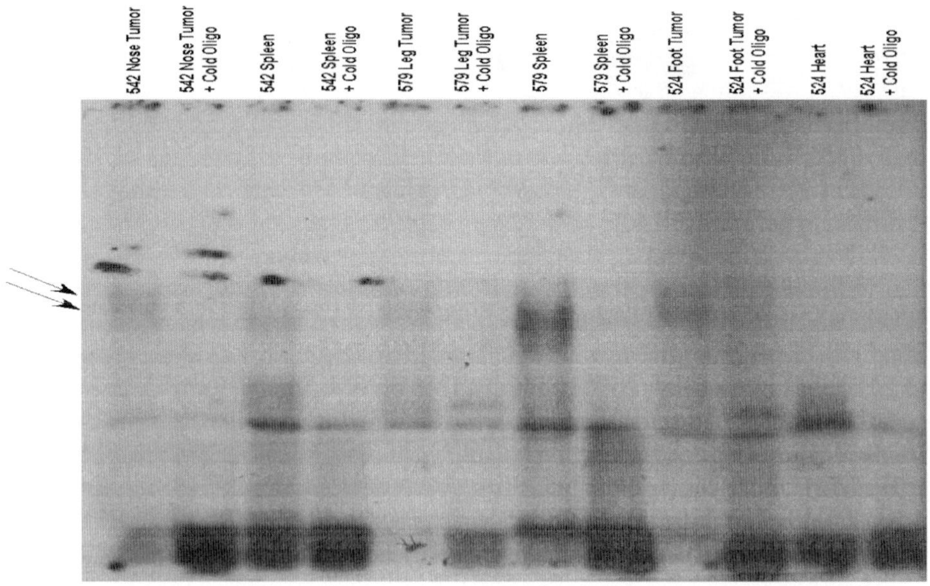

FIG. 3. NFκB activation in Tax transgenic mouse tumors

Nuclear extracts were prepared from Tax transgenic tissues and representative results from 3 animals (#524, #542, and #579) are shown. Electrophoretic mobility gel-shift assays were performed in the absence or presence of 100-fold excess of unlabeled oligonucleotide (indicated as "+ Cold Oligo"). The positions of the retarded NFκB DNA bands are indicated by the arrows on the left, indicating the positions of DNA bound to the p65-p50 heterodimer and p50-p50 homodimer of NFκB. Reprinted with permission from (*33*).

tumors *in vivo* as a result of inhibiting NFκB activation. For this purpose, we will examine both genetic and pharmacological mechanisms for disruption of the NFκB pathway. In addition, we are trying to pinpoint the NFκB target genes that mediate tumorigenesis. Of particular interest is the expression of NFκB-responsive genes that alter T cell proliferation and/or apoptosis, e.g. interferon-γ, IL-15, Bcl-XL and TRAIL (*1, 19, 33, 37*). HTLV-I resembles numerous other viruses in its exploitation of the NFκB pathway (*14*).

Tax and p53

In order to further assess determinants of proliferation and apoptosis of Tax induced tumors, we assessed p53 tumor suppressor protein structure and function (*32*). This tumor suppressor protein was of particular interest since it is mutated in more than 50% of all human malignancies (*4, 20*), it is mutated in about 30% of ATL cases and 50% of bovine leukemia virus induced tumors (*5, 45*), and it is functionally inactivated by Tax through phosphorylation of two critical serine residues (*30, 31*). Mutations in the DNA binding domain of p53 were determined by PCR and by single strand conformation polymorphism analysis of DNA from Tax induced tumors (*32*). Mutations were identified most frequently at 4 specific residues in 3 of 7 tumors arising on the tail and 10 of 11 tumors that had involved visceral sites. These findings suggested that loss of p53 activity was not required for tumor formation, but was associated with tumor dissemination. We further addressed this issue by breeding Tax transgenic mice with p53 knock-out mice. The tumor-free survival of the different mouse strains is shown in Fig. 4A. The p53–/– mice with or without a *Tax* transgene showed a similar incidence of tumors that were T cell lymphomas, and distinct in histological, clinical, and cell surface phenotype from tumors seen in other Tax trans-

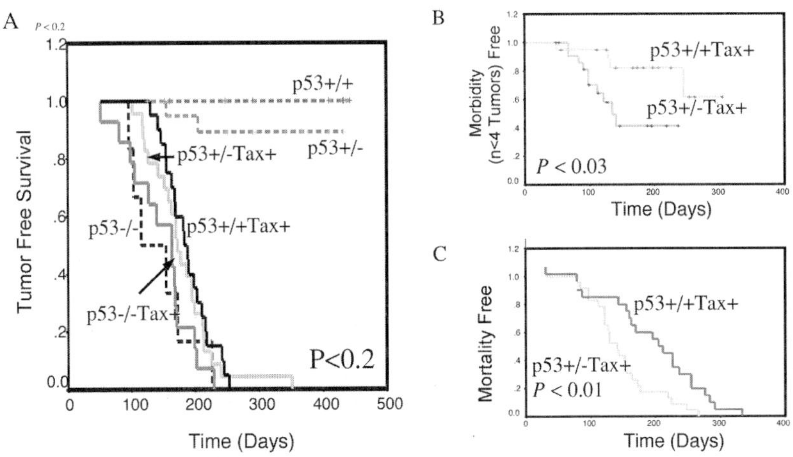

FIG. 4. Tumor initiation and progression in progeny from breeding Tax transgenic mice (solid lines) with p53–/– mice

A) The percent tumor-free survival is plotted as a function of age. Animals were monitored weekly and except for p53–/–Tax+ (*n*=14) and p53–/– (*n*=5) mice, there were at least 20 mice in each group. B) The percent animals with fewer than 4 tumors and C) the percent animals alive from the time of tumor initiation are shown. Reprinted with permission from (*32*).

genic mice (*16*). The non-transgenic mice that were p53+/– or p53+/+ developed very few tumors. The Tax transgenic mice that were p53+/– or p53+/+ developed tumors at a similar incidence (Fig. 4A) with similar gross, microscopic, and cell surface characteristics. However, the p53+/–Tax+ mice differed from the p53+/+Tax+ mice in that there was a more rapid development of multiple tumors, visceral involvement, and death (Fig. 4, B and C). These findings suggest that tumors arising in p53+/– mice are biologically different from those arising in p53+/+ mice. Thus, there may be a difference in tumor cell proliferation rate, sensitivity to apoptosis, invasiveness, angiogenesis, and/or gene expression profiles, depending on the background level of p53 expression. Each of these possibilities is currently under study. Similar results have been found in other transgenic models with loss of p53 or other tumor suppressors (e.g. phosphatase and tensin homologue (PTEN)) (*24, 50*).

Tax induced tumors were further evaluated for p53-dependent activities (*32*). Activation of p53-dependent promoters, p21, bax, and mdm2 after γ-irradiation were intact. Transcriptional activation of other p53-dependent promoters by γ-irradiation or other inducers of p53 are currently underway. In contrast, Tax induced tumors were resistant to induction of apoptosis by γ-irradiation. This could represent selective regulation of p53 actions, as has been described with DNA-dependent protein kinase knock-out mice (*49*). However, recent studies suggest that Tax regulation of the ataxia telangiectasis mutated protein (ATM) may be more important in controlling p53 function (*48*). Alternatively, Tax may disrupt the apoptosis pathway downstream of p53 due to effects on caspases, members of the bcl2 family, mitochondrial integrity, or apoptosis inhibitors. Studies of apoptosis induction by other intrinsic or extrinsic signals is also currently being examined.

We also analyzed cell cycle regulation in Tax induced tumors (*34*). We found that cyclin-dependent kinase (Cdk) 4 and 6 activity was elevated in Tax induced tumors, although Cdk inhibitors p15 and p16 were expressed normally and were not mutated. These *in vivo* findings reaffirm studies demonstrating effects of Tax on Cdk's in cultured cells (*40, 41*).

Acknowledgments

I thank my colleagues for their experimental and intellectual contributions to this work, in particular J. T. Kimata, W. Grossman, T. Portis, J. Harding, S. Mitra-Kaushik, and N. Vander Heyden. The work was supported by Public Health Service grants.

REFERENCES

1. Azimi, N., Brown, K., Bamford, R. N., Tagaya, Y., Siebenlist, U., and Waldmann, T. A. Human T cell lymphotropic virus type I Tax protein trans-activates interleukin 15 gene transcription through an NF-κB site. *Proc. Natl. Acad. Sci. USA*, **95**, 2452–2457 (1999).
2. Beatty, G. L. and Paterson, Y. IFN-gamma can promote tumor evasion of the immune system *in vivo* by down-regulating cellular levels of an endogenous tumor antigen. *J. Immunol.*, **165**, 5502–5508 (2000).
3. Binder, C. K. and Griffin, D. E. Interferon-gamma-mediated site-specific clearance of alphavirus from CNS neurons. *Science*, **293**, 303–306 (2001).

4. Bullock, A. N. and Fersht, A. R. Rescuing the function of mutant p53. *Nat. Rev. Cancer*, **1**, 68–76 (2001).
5. Cesarman, E., Chadburn, A., Inghirami, G., Gaidano, G., and Knowles, D. M. Structural and functional analysis of oncogenes and tumor suppressor genes in adult T-cell leukemia/lymphoma shows frequent p53 mutation. *Blood*, **80**, 3205–3216 (1992).
6. Franchini, G. Molecular mechanisms of human T-cell leukemia/lymphotropic virus type I infection. *Blood*, **86**, 3619–3639 (1995).
7. Furuta, Y., Aizawa, S., Suda, Y., Ikawa, Y., Kishimoto, H., Asano, Y., Tada, T., Hikikoshi, A., Yoshida, M., and Seiki, M. Thymic atrophy characteristic in transgenic mice that harbor pX gene of human T-cell leukemia virus type I. *J. Virol.*, **63**, 3185–3189 (1989).
8. Fehniger, T. A., Suzuki, K., Ponnappan, A., VanDeusen, J. B., Cooper, M. A., Florea, S. M., Freud, A. G., Robinson, M. L., Durbin, J., and Caligiuri, M. A. Fatal leukemia in interleukin 15 transgenic mice follows early expansions in natural killer and memory phenotype CD8+ T cells. *J. Exp. Med.*, **193**, 219–231 (2001).
9. Grassmann, R., Berchtold, S., and Radant, I. Role of human T-cell leukemia virus type 1 X region proteins in immortalization of primary human lymphocytes in culture. *J. Virol.*, **66**, 4570–4575 (1992).
10. Grossman, W. J., Kimata, J. T., Wong, F.-H., Zutter, M., Ley, T. J., and Ratner, L. Development of leukemia in mice transgenic for the tax gene of human T-cell leukemia virus type I. *Proc. Natl. Acad. Sci. USA*, **92**, 1057–1061 (1995).
11. Grossman, W. J. and Ratner, L. Cytokine expression and tumorigenicity of large granular lymphocytic leukemia cells from mice transgenic for the tax gene of human T-cell leukemia virus type I. *Blood*, **90**, 783–794 (1997).
12. Habu, K., Nakayama-Yamada, J., Asano, M., Saijo, S., Itagaki, K., Horai, R., Yamamoto, H., Sekiguchi, T., Nosaka, T., Hatanaka, M., and Iwakura, Y. The human T-cell leukemia virus type I-tax gene is responsible for the development of both inflammatory polyarthropathy resembling rheumatoid arthritis and noninflammatory ankylotic arthropathy in transgenic mice. *J. Immunol.*, **162**, 2956–2963 (1999).
13. Hinrichs, S. H., Nerenberg, M., Reynolds, R. K., Khoury, G., and Jay, G. A transgenic mouse model for human neurofibromatosis. *Science*, **237**, 1340–1343 (1987).
14. Hiscott, J., Kwon, H., and Genin, P. Hostile takeovers: viral appropriation of the NF-κB pathway. *J. Clin. Invest.*, **107**, 143–151 (2001).
15. Iwakura, Y., Saijo, S., Kioka, Y., Nakayama-Yamada, J., Itagaki, K., Tosu, M., Asano, M., Kanai, Y., and Kakimoto, K. Induction of inflammatory arthropathy resembling rheumatoid arthritis in mice transgenic for HTLV-1. *Science*, **253**, 1026–1028 (1991).
16. Jacks, T., Remington, L., Williams, B. O., Schmitt, E. M., Hatachmi, S., Bronson, R. T., and Weinberg, R. A. Tumor spectrum analysis in p53-mutant mice. *Curr. Biol.*, **4**, 1–7 (1994).
17. Jeang, K. T., Widen, S. G., Semmes, O. J., and Wilson, S. H. HTLV-I trans-activator protein, tax, is a trans-repressor of the human beta-polymerase gene. *Science*, **247**, 1082–1084 (1990).
18. Jin, D. Y., Spencer, F., and Jeang, K. T. Human T cell leukemia virus type 1 oncoprotein Tax targets the human mitotic checkpoint protein MAD1. *Cell*, **93**, 81–91 (1998).
19. Khoshnan, A., Tindell, C., Laux, I., Bae, D., Bennett, B., and Nel, A. E. The NF-κB

cascade is important in Bcl-xL expression and for the anti-apoptotic effects of the CD28 receptor in primary human CD4+ lymphocytes. *J. Immunol.*, **165**, 1743–1754 (2000).

20. Levine, A. J. p53, the cellular gatekeeper for growth and division. *Cell*, **88**, 323–331 (1997).
21. Li, X. C., Demirci, D., Ferrari-Lacraz, S., Groves, C., Coyle, A., Malek, T. R., and Strom, T. B. IL-15 and IL-2: a matter of life and death for T cells *in vivo*. *Nat. Med.*, **7**, 114–118 (2001).
22. Lodolce, P. P., Boone, D. L., Chai, S., Swain, R. E., Dassopoulos, T., Trettin, S., and Ma, A. IL-15 receptor maintains lymphoid homeostasis by supporting lymphocyte homing and proliferation. *Immunity*, **9**, 669–676 (1998).
23. Low, K. G., Dorner, L. F., Fernando, D. B., Grossman, J., Jeang, K. T., and Comb, M. J. Human T-cell leukemia virus type I tax releases cell cycle arrest induced by p16 (INK4a). *J. Virol.*, **71**, 1956–1962 (1997).
24. Lu, X., Magrane, G., Yin, C., Louis, D. N., Gray, J., and vanDyke, T. Selective inactivation of p53 facilitates mouse epithelial tumor progression without chromosomal instability. *Mol. Cell. Biol.*, **21**, 6017–6030 (2001).
25. Marks-Konczalik, J., Dubois, S., Losi, J. M., Sabzevari, H., Yamada, N., Feigenbaum, L., Waldmann, T. A., and Tagaya, Y. IL-2-induced activation-induced cell death is inhibited in IL-15 transgenic mice. *Proc. Natl. Acad. Sci. USA*, **97**, 11445–11450 (2000).
26. Mesnard, J. M. and Devaux, C. Multiple control levels of cell proliferation by human T-cell leukemia virus type 1 Tax protein. *Virology*, **257**, 277–284 (1999).
27. Nerenberg, M., Hinrichs, S. H., Reynolds, R. K., Khoury, G., and Jay, G. That tat gene of human T-lymphotropic virus type 1 induces mesenchymal tumors in transgenic mice. *Science*, **237**, 1324–1329 (1987).
28. Nerenberg, M. I., Minor, T., Price, J., Ernst, D. N., Shinohara, T., and Schwarz, H. Transgenic thymocytes are refractory to transformation by the human T-cell leukemia virus type I tax gene. *J. Virol.*, **65**, 3349–3353 (1991).
29. Neuveut, C., Low, K. G., Maldarelli, F., Schmitt, I., Majone, F., Grassmann, R., and Jeang, K. T. Human T-cell leukemia virus type 1 Tax and cell cycle progression: role of cyclin D-cdk and p110Rb. *Mol. Cell. Biol.*, **18**, 3620–3632 (1998).
30. Pise-Masison, C. A., Mahieux, R., Jiang, H., Ashcroft, M., Radonovich, M., DuGuillerm, C., and Brady, J. N. Inactivation of p53 by human T-cell lymphotropic virus type 1 requires activation of the NF-κB pathway and is dependent on p53 phosphorylation. *Mol. Cell. Biol.*, **10**, 3377–3386 (2000).
31. Pise-Masison, C. A., Radonovich, M., Sakaguchi, K., Appella, E., and Brady, J. Phosphorylation of p53: a novel pathway for p53 inactivation in human T-cell lymphotropic virus type 1-transformed cells. *J. Virol.*, **72**, 6348–6355 (1998).
32. Portis, T., Grossman, W. J., Harding, J. C., Hess, J. L., and Ratner, L. Analysis of p53 inactivation in a human T-cell leukemia virus type 1 tax transgenic mouse model. *J. Virol.*, **75**, 2185-2193 (2001).
33. Portis, T., Harding, J. C., and Ratner, L. The contribution of NF-κB activity to spontaneous proliferation and resistance to apoptosis in human T-cell leukemia virus type 1 Tax-induced tumors. *Blood*, **98**, 1200–1208 (2001).
34. Portis, T., Harding, J. C., and Ratner, L. Enhanced cyclin-dependent kinase activity by

human T-cell leukemia virus type I (HTLV-I) tax in a mouse model of tumorigenesis. *Leukemia* (2002). submitted.

35. Pozzatti, R., Vogel, J., and Jay, G. The human T-lymphotropic virus type I tax gene can cooperative with the ras oncogene to induce neoplastic transformation of cells. *Mol. Cell. Biol.*, **10**, 413–417 (1990).
36. Ratner, L. and Poiesz, B. J. Leukemias associated with human T-cell lymphotropic virus type I in a non-endemic region. *Medicine*, **67**, 401–422 (1988).
37. Rivera-Walsh, I., Waterfield, M., Xiao, G., Fong, A., and Sun, S.-C. NF-κB signaling pathway governs TRAIL gene expression and human T-cell leukemia virus-I tax-induced T-cell death. *J. Biol. Chem.*, **276**, 40385–40388 (2001).
38. Ross, T. M., Pettiford, S. M., and Green, P. L. The tax gene of human T-cell leukemia virus type 2 is essential for transformation of human T lymphocytes. *J. Virol.*, **70**, 5194–5202 (1996).
39. Ruddle, N. H., Li, C. B., Horne, W. C., Santiago, P., Troiano, N., Jay, G., Horowitz, M., and Baron, R. Mice transgenic for HTLV-I LTR-tax exhibit tax expression in bone, skeletal alterations, and high bone turnover. *Virology*, **197**, 196–204 (1993).
40. Santiago, F., Clark, E., Chong, S., Molina, C., Mozafari, F., Mahieux, R., Fujii, M., Azimi, N., and Kashanchi, K. Transcription up-regulation of the cyclin D2 gene and acquisition of new cyclin-dependent kinase partners in human T-cell leukemia virus type I-infected cells. *J. Virol.*, **73**, 9917–9927 (1999).
41. Schmitt, I., Rosin, O., Rohwer, P., Gossen, M., and Grassmann, R. Stimulation of cyclin-dependent kinase activity and G1- to S-phase transition in human lymphocytes by the human T-cell leukemia/lymphotropic virus type I Tax protein. *J. Virol.*, **72**, 633–640 (1998).
42. Seiki, M., Eddy, R., Shows, T. B., and Yoshida, M. Nonspecific integration of the HTLV provirus genome into adult T-cell leukaemia cells. *Nature*, **309**, 640–642 (1984).
43. Shankaran, V., Ikdea, H., Bruce, A. T., White, J. M., Swanson, P. E., Old, L. J., and Schreiber, R. D. IFNgamma and lymphocytes prevent primary tumour development and shape tumour immunogenicity. *Nature*, **410**, 1107–1111 (2001).
44. Suzuki, T., Narita, T., Uchida-Toita, M., and Yoshida, M. Down-regulation of the INK4 family of cyclin-dependent kinase inhibitors by tax proteins of HTLV-I through two distinct mechanisms. *Virology*, **259**, 384–391 (1999).
45. Tajima, S., Zhuang, W. Z., Kato, M. V., Okada, K., Ikawa, Y., and Aida, Y. Function and conformation of wild-type p53 protein are influenced by mutations in bovine leukemia virus-induced B-cell lymphosarcoma. *Virology*, **243**, 735–746 (1998).
46. Tamiya, S., Matsuoka, M., Etoh, K., Watanabe, T., Kamihira, S., Yamaguchi, K., and Takatsuki, K. Two types of defective human T-lymphotropic virus type I provirus in adult T-cell leukemia. *Blood*, **88**, 3065–3073 (1996).
47. Tsukasaki, K., Tsushima, H., Yamamura, M., Hata, T., Murata, K., Maeda, T., Atogami, S., Sohda, H., Momita, S., Ideda, S., Katamine, S., Yamada, Y., Kamihira, S., and Tomonaga, M. Integration patterns of HTLV-I provirus in relation to the clinical course of ATL: frequent clonal change at crisis from indolent disease. *Blood*, **89**, 948–956 (1997).
48. Van, P. L., Yim, K.-W., Jin, D.-Y., Dapolito, D., Kurimasa, A., and Jeang, K.-T. Genetic evidence for a role for ATM in functional interaction between human T-cell

leukemia virus type 1 tax and p53. *J. Virol.*, **75**, 396–407 (2001).

49. Wang, S., Guo, M., Ouyang, H., Li, X., Cordon-Cardo, C., Kurimasa, A., Chen, D. J., Fuks, Z., Ling, C. C., and Li, G. C. The catalytic subunit of DNA-dependent protein kinase selectively regulates p53-dependent apoptosis but not cell cycle arrest. *Proc. Natl. Acad. Sci. USA*, **97**, 1584–1588 (2000).
50. Wong, K. S., Lik, Y.-J., Howard, J., and Ben-David, Y. Loss of p53 in F-MuLV induced-erythroleukemias accelerates the acquisition of mutational events that confers immortality and growth factor independence. *Oncogene*, **18**, 5525–5534 (1999).
51. Yoshida, M. Multiple viral strategies of HTLV-1 for dysregulation of cell growth control. *Annu. Rev. Immunol.*, **19**, 475–496 (2001).

PREVENTION AND THERAPEUTIC APPROACH

PRIMARY PREVENTION OF ADULT T CELL LEUKEMIA (ATL) IN NAGASAKI, JAPAN BY REFRAINING FROM BREAST-FEEDING

Shigeo HINO

*Department of Virology, Faculty of Medicine, Tottori University**

Breast-feeding plays an important role in mother-to-child transmission of human T lymphotropic virus type 1 (HTLV-1), the causative agent of adult T cell leukemia (ATL). The influence of refraining from breast-feeding by carrier mothers on HTLV-1 transmission has been evaluated in a prefecture-wide intervention study designated the ATL Prevention Program, Nagasaki (APP). Since 1987, over 100,000 pregnant women in the third trimester were screened for anti-HTLV-1 antibody in Nagasaki, Japan. The seropositive mothers were asked to refrain from breast-feeding and to bring in their children for check-ups of their antibody status. Children with a follow-up period of at least 2 years were analyzed. Without any intervention, the prevalence of antibody-positive women decreased significantly, from 8% for those born in 1945 to below 2% for those born in 1975. Among the children born to carrier mothers, 18.4% of those breast-fed for a long-term were infected with HTLV-1, in contrast to 3.2% of those bottle-fed ($P<0.001$). The incidence in children breast-fed for 1–5 months was 11.4%, which is significantly higher than that of bottle-fed children ($P=0.032$). Considering the trend toward decrease in maternal prevalence, the prevalence in pregnant women born in 1987 is estimated to be 1.0%. The results confirm that breast-feeding plays a major role in the mother-to-child transmission of HTLV-1. They also suggest that the APP has prevented approximately 800 new infections in the study period and is expected to prevent the development of 40 cases of ATL in future.

Adult T cell leukemia (ATL) develops specifically in carriers of human T lymphotropic virus type 1 (HTLV-1) (*10, 20*). ATL has been a significant health problem in the endemic areas for HTLV-1 including the southwestern part of Japan. For example, the annual number of reported ATL cases in Nagasaki Prefecture, approximately 70 cases, corresponds to 10% of ATL deaths in Japan, while the prefecture (1.5 million) only houses 1% of the entire population of Japan (*6, 20*). The ATL mortality corresponds to nearly 1% of total mortality in Nagasaki, and to mortality from all other types of leukemia and lymphoma combined (*6*). Although recent advances in leukemia therapy can induce complete cures in various types of leukemia, only limited cases of ATL have been successfully treated (*2, 16, 21, 24*).

The incidence of ATL among HTLV-1 carriers is approximately 1/1,000 per year among adult carriers (*21*). These carriers include those who acquired HTLV-1 in adult life by transfusion and sexual transmission. In all probability, ATL develops only

*Yonago 680-8503, Japan

in carriers who had acquired HTLV-1 early in life, since most mothers of ATL cases were found to be carriers. There have been no confirmed cases of ATL patients who acquired HTLV-1 in adult life except for a few cases under intensive myelosuppressive therapy directed for preoccupying leukemia or bone marrow transplantation (*5*). Therefore, the incidence of ATL among carriers infected early in life is expected to be ca. 5%.

In the literature as of 1985, it was evident that the mother-to-child route plays a major role in the HTLV-1 transmission in endemic areas. The prevalence in children born to carrier mothers was significantly higher than that in the general child population in the same area, and the prevalence in the mothers of carriers was close to one (*9*). Evidence of intrauterine infection was negative from a survey of cord blood samples obtained from babies born to carrier mothers using IgM antibody and HTLV-1 antigen after short-term culture as markers (*9, 15, 19*), although a few controversial reports have been published (*18*). Furthermore, common marmosets orally inoculated with HTLV-1 bearing cultured cells or breast-milk of carrier mothers were infected with HTLV-1 (*13, 26*).

To confirm the milk-borne hypothesis, an intervention by shutting off the breast-feeding by these mothers was inevitable. However, stopping the breast-feeding by these mothers might induce some medical, familial or socially adverse effects. In order to test the feasibility of the intervention, a pilot intervention study was started in 1985: the antibody-positive women were asked to nurse their expected babies by bottle-feeding instead of breast-feeding and to let us follow the antibody status of their babies. The pilot study suggested that the intervention program in Nagasaki Prefecture was acceptable for carrier women and the community.

ATL Prevention Program, Nagasaki (APP)

The prefecture-wide intervention program called the ATL Prevention Program,

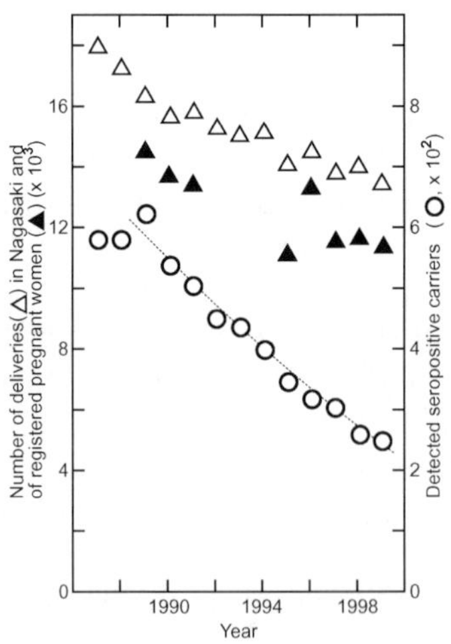

FIG. 1. Number of deliveries in Nagasaki Prefecture (excluding Kami-Goto and Tsushima areas) (open triangles), and numbers of pregnant women registered (closed triangles) and of seropositive women screened (circles) each year

Nagasaki (APP) was commenced in the middle of 1987. The program was promoted under the auspices of the Nagasaki University School of Medicine, the Nagasaki Prefectural Government, and Nagasaki Prefecture Branches of the Japan Maternity Protection Association and of the Japanese Association of Pediatricians. The committee of medical ethics in Nagasaki University Medical School approved this program. The Tsushima and Kami-Goto areas in Nagasaki Prefecture, which accounted for 5% of the population were excluded from the survey, since a separate study group covered these areas.

Pregnant women in the third trimester were screened for anti-HTLV-1 antibody at either commercial laboratories or at the Department of Molecular Microbiology and Immunology (formerly the Department of Bacteriology) of Nagasaki University, using a passive particle agglutination test (Serodia ATL [until 1989] or Serodia HTLV [thereafter], FujiRebio, Tokyo, Japan). Informed consent was obtained from each subject. The positive and undetermined samples were collected at the Department of Bacteriology and used for the confirmation test. In this test, an indirect immunofluorescence test with MT-2 cells as the target was used in addition to the particle agglutination test. For undetermined samples, a western blotting (FujiRebio) was supplemented (*7, 9*). The antibody-positive pregnant women were advised to refrain from breast-feeding to minimize the transmission of HTLV-1 to their children. Approximately 90% of these women agreed, and bromocriptine was administered to suppress breast-milk secretion and thus minimize their distress. A fraction of the mothers chose breast-feeding for a short period, after the media reported the procedure was to decrease the mother-to-child transmission of HTLV-1 (*22, 23*).

Antibody-positive mothers were instructed to bring their children into the Department of Pediatrics, Nagasaki University 6, 12, 18, 24, and 36 months after birth for the examination of antibody status. Children born before the registration of maternal antibody status were eliminated. In total, 859 prospective children born to antibody-positive mothers were followed for at least 24 months. Children were grouped into three categories by nutrition status: 1) those nursed completely with complete bottle-feeding, 2) full-term breast-feeding ($\geq$10 months) and 3) short-

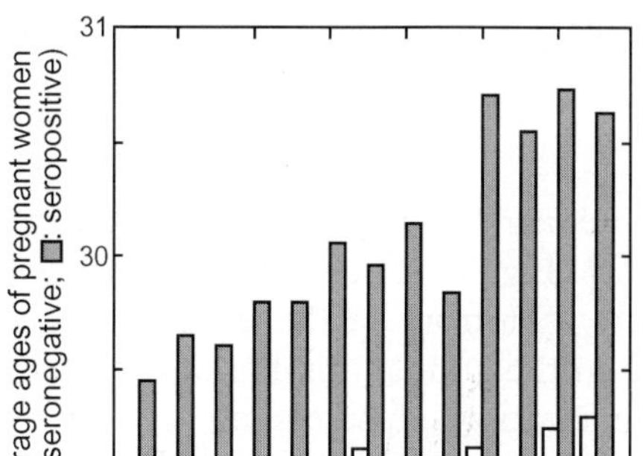

FIG. 2. Average age of seronegative (open bars) and seropositive (shaded bars) pregnant women each year

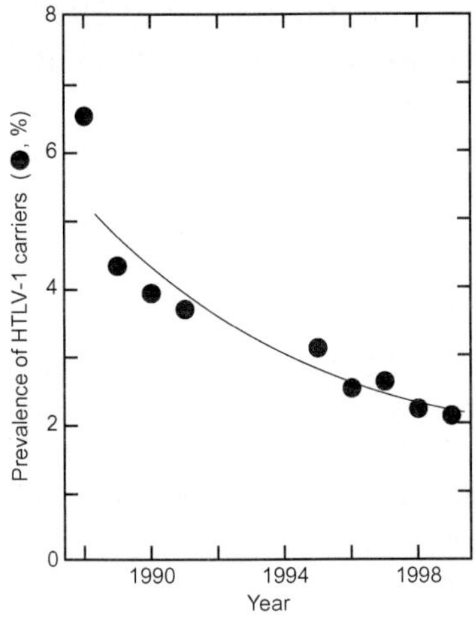

FIG. 3. Yearly declining curve in the prevalence of HTLV-1 carriers among pregnant women in Nagasaki

TABLE I. Prevalence of Pregnant Women Stratified by the Birth Year

Birth year[a]	Positives	Tested	Carrier rate[b] (%)	Carrier rate[b] Log
1943–1947	8	112	7.1	0.85
1948–1952	85	1,380	6.2	0.79
1953–1957	368	7,321	5.0	0.70
1958–1962	837	20,431	4.1	0.61
1963–1967	686	22,281	3.1	0.49
1968–1972	333	16,155	2.1	0.31
1973–1977	119	6,630	1.8	0.25
1978–1982	13	929	1.4	0.15

[a] Birth years of pregnant women were stratified in every five years for women registered in 1979–1981 and 1997–1999.
[b] Regression line: $y=-0.0212\ (x-1900)+1.842$ (x: birth year, y: log[carrier rate]), $r^2=0.886$.

term (1 to 5 months) breast-feeding. After eliminating children which could not be assigned to any of these groups, a total of 820 children were analyzed.

Pregnant Women Enrolled

More than 100,000 pregnant women were enrolled in the APP during the 12-year period of 1987 to 1999 (Fig. 1). The number of deliveries in Nagasaki gradually decreased by approximately 25% from 18,000 to 13,500 according to the Prefecture census. The steady change in the annual number of detected seropositive women suggested the continuous penetration of screening in the field. The discontinuously higher positive rates due to partial registration failure of the screened seronegative women in 1988 and 1992–1994 than those in the adjacent years were eliminated from Fig. 1.

The average age of seropositive women increased from 29.5 years to 30.5 years, while the average age of seronegative women stayed almost constant at 29.0–29.2

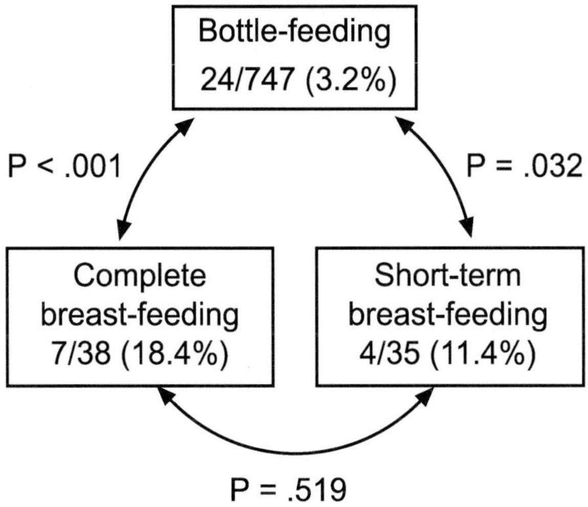

FIG. 4. Difference in the rates of mother-to-child transmission of HTLV-1 by feeding methods: bottle-, breast- and short-term breast-feeding

years throughout the observation period (Fig. 2). The average age of seropositive women was significantly higher than that of seronegative women by unpaired *t* test ($P<0.001$ each year, data not shown). Thus, the difference in average age of seropositive and seronegative women increased.

In addition to the 25% drop in annual deliveries, the observed carrier rate each year diminished even more profoundly from approximately 5% to 2% (Fig. 3). Because of these two factors, the registered number of seropositive women per year dropped by two-thirds, from approximately 600 in the late 1980s to 200 in the late 1990s (Fig. 1). This decrease had nothing to do with APP, however, since these women had been born well before the commencement of the program.

When the pregnant women were stratified by birth year, the prevalence decreased linearly after logarithmic conversion (Table I). The results were also consistent with a previous observation that the prevalence of HTLV-1 carriers is dependent on the birth year cohort (*19*). Given 29 years for the average age of the pregnant women, i.e., the generation period, the slope of the regression line suggested that the seroprevalence would decline to 24% in a generation. The results were consistent with the previous observation that HTLV-1 was transmitted to breast-fed babies born to carrier mothers at a rate of ca. 20% (*8, 9*). A study in Jamaica also reported a transmission rate of 23% in breast-fed children (*25*). Extrapolation of the linear regression line suggests that the prevalence has dropped to 1.0% for females born in the year 1987.

Period of Seroconversion in Children

There has been no difference in the transmission rate between boys and girls in any of the groups, including the bottle-fed group by Fisher's exact test (15/374 [4.0%] and 9/371 [2.4%], respectively [$P=0.299$]). The average age of mothers was not significantly different for infected and uninfected children. Among 35 children who

TABLE II. Basis for Calculation of Expected Bulk Carrier Rate in the Year of 2010

Age range	Median birth year	Calculated carrier rate (Log %)[a]	Calculated carrier rate (%)	Frequency (%)[b]	Effective carrier rate (%)
44–48	1964	0.485	3.06	0.2	0.01
39–43	1969	0.379	2.39	2.2	0.05
34–38	1974	0.273	1.88	13.8	0.26
29–33	1979	−0.038	0.92	38.4	0.35
24–28	1984	−0.227	0.59	35.5	0.21
19–23	1989	−0.416	0.38	9.5	0.04
−18	1994	−0.605	0.25	0.4	0.00
Total					0.92

[a] Calculated by the following equations. Women born by 1987, $y=-0.0212\ (x-1900)+1.842$; women born after 1987, $y=-0.0378\ (x-1987)$.
[b] Calculated from the histogram of age distribution which was unchanged by the test year.

were finally diagnosed as seroconverted, there was no data on the serum status for five of them until the age of 24 months. Only one of the remaining 30 children (3%) seroconverted between 24 and 36 months of age and all others had become seropositive by the age of 24 months. Thus, the appropriate period to diagnose the serological status of children was found to be at the age of 24 months. These results were consistent with a previous report that none of over 50 children who had been seronegative at the age of three seroconverted thereafter (*14*).

Effect of Complete Bottle-feeding vs. Full-term Breast-feeding

Of 747 children nursed with complete bottle-feeding, only 24 seroconverted (3.2%) (Fig. 4). In contrast, of 38 children breast-fed for a full-term, seven seroconverted (18.4%). The incidence of HTLV-1 transmission in the bottle-fed children was significantly lower than that in the breast-fed children for a full-term ($P<0.001$, relative risk [RR]: 0.174 with 95% confidence interval [95%CI]: 0.080–0.379). The results suggest a reduction of 80% in HTLV-1 transmission by refraining from breast-feeding, and confirmed the importance of breast-milk transmission in the endemicity of HTLV-1.

However, to maintain the endemicity of HTLV-1 for a long period, the transmission rate per generation should be one. How HTLV-1 was maintained in a community before World War II remains unsolved. Introduction of the Western life style after the war might have altered the ancient higher transmission rate. Rapid fading of HTLV-1 among Japanese emigrants in Hawaii could be explained by this hypothesis (*27*). Another factor which might have contributed to the high prevalence of women born in the 1960's include the historical practice that transfusions of 200–400 ml were frequently given pregnant women to treat iron deficiency anemia.

The transmission route of HTLV-1 for children nursed with complete bottle-feeding has not been fully elucidated. However, our previous reports suggested that intrapartum transmission would be more likely than the intrauterine transmission. In addition to an earlier study on cord blood samples of babies born to carrier mothers with IgM antibody and HTLV-1 antigen after culture as markers (*9*), the cord blood samples were tested by PCR for the presence of HTLV-1 harboring cells. No babies of PCR-positive in their cord blood became carriers, and all the carrier children tested were PCR negative in their cord blood (*11, 12*). The elective cesarean section has been suggested to reduce the incidence of intrapartum transmission in cases of human immunodeficiency virus (HIV) (*1*) and hepatitis C virus (HCV) (*3, 17*). However, it seems hard to justify this maneuver because the expected incidence of ATL in bottle-fed children born to carrier mothers is less than 0.2%.

Effect of Short-term Breast-feeding vs. Complete Bottle-feeding

While our study was in progress, Takahashi *et al.* reported that breast-feeding for no more than 6 months significantly decreased the incidence of HTLV-1 transmission from HTLV-1 carrier mothers, from 14.4% (20/139 cases) to 4.4% (4/90 cases) (P=0.018) (*22*). Furthermore, they also stated that children breast-fed for a short-term have no higher risk than those bottle-fed, although the number of seropositive

TABLE III. Prevalence of HTLV-1 Carriers among Pregnant Women, Observed and Expected

Year	Carrier rate (%)	
	Observed	Expected
1985		4.7
1987	10.5[a]	
1988	6.5[a]	
1989	4.3	
1990	3.9	3.7
1991	3.7	
1992	3.9[a]	
1993	4.5[a]	
1994	4.4[a]	
1995	3.1	2.9
1996	2.5	
1997	2.6	
1998	2.2	
1999	2.1	
2000		2.2
2005		1.5
2010		0.92
2015		0.56
2020		0.37

[a] Prevalences observed in these years were deviated higher than actual because of failures in the registration of seronegative carriers.

TABLE IV. Estimation of Bulk Transmission Rate after the Commencement of APP

Group	Fraction (%)	Transmission rate	Net prevalence
Bottle-feeding	66	0.03	2.0
Breast-feeding			
full-term	12	0.20	2.4
short-term	7	0.10	0.7
Unregistered	15	0.20	3.0
Total			8.1

children in their study was too small for statistical analysis. However, their report let obstetricians and pregnant women in our study field have the option of "short-term" breast-feeding, which allowed us to analyze its effect. Among 35 children breast-fed for a short-term, four (11.4%) seroconverted (Fig. 4). The incidence of HTLV-1 transmission in the bottle-fed group was also significantly lower than that of this group (*P*=0.032, RR: 0.281, 95%CI: 0.103–0.767). The current results show that the risk of HTLV-1 transmission in children breast-fed for a short-term was significantly higher than that in bottle-fed children. Nevertheless, we do not intend to recommend complete bottle-feeding in developing countries where the nutritional advantage of breast-feeding and the risk of other infantile infections are more common (*4*). Although the incidence of HTLV-1 transmission of the breast-fed group for a short-term was approximately two-thirds that of the full-term, the difference was not significantly lower in this study (*P*=0.519). However, when the elder siblings of the study group were included, the difference came close to being statistically significant (27/150 or 18.0% vs. 7/82 or 8.5%, *P*=0.055, RR: 0.474, 95%CI: 0.216–1.042).

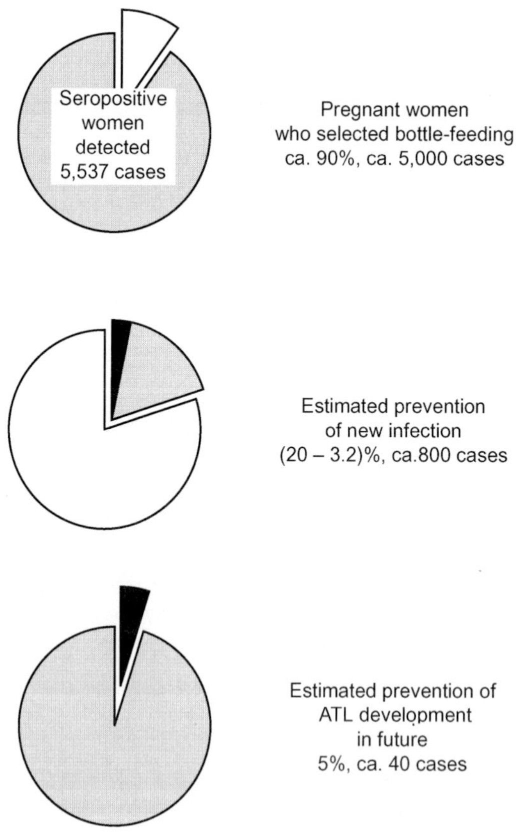

FIG. 5. Estimated numbers of preventions of new HTLV-1 transmissions and of ATL developments in future by APP during the years 1987 through 1999

Prevalence of Pregnant Women in Future

To estimate the seroprevalence of pregnant women in future, the estimated prevalence in a given year was computed back from the age-dependent histogram of pregnant women and the birth-year dependent prevalence in each age group using the following equation:

$$EP = \Sigma BP \times Fr$$

where *EP*: estimated prevalence in a given year (%), *BP*: birth-year specific prevalence (%), *Fr*: fraction of pregnant women in each birth-year cohort.

Table II shows an example estimating the prevalence in the year 2010 using the above equation. The prevalence thus expected is consistent with that observed until 1999 (Table III). In the near future, girls born after the commencement of APP will become pregnant women. Since not all of the pregnant women enrolled and chose bottle-feeding, the adjustment was made. In 1996–1999, 15% of all pregnant women in the area were unregistered. Of the remaining, 78% of mothers chose the bottle-feeding, 8% short-term and 14% full-term breast-feeding. For calculation purposes, all of the unregistered children were assumed to have received breast-feeding. Using the specific transmission rates for each feeding group, a bulk transmission under the APP was computed as 8% (Table IV). With this bulk transmission rate for mothers born after 1987, the prevalence in future was computed: in the year 2020 the expected prevalence was calculated as 0.37% (Table II), and that of women born in 2010 as 0.14% (Table III).

In total, 5,537 pregnant women were found to be seropositive for HTLV-1 during this period. Using 20% as the transmission rate for full-term breast-feeding and 5% for disease incidence, we estimate that approximately 800 new infections and 40 future ATL cases were prevented by APP (Fig. 5). In the decade of the 2010s, HTLV-1 carriers within the pregnant women population will be well below 1% in this field, i.e. less than 100 per year in this area. If the annual number of pregnant carriers becomes less than 100, the expected annual incidence of ATL for children born in the area will be below one, even if breast-feeding is restarted. This would mean virtual extinction of ATL in the Nagasaki area in the near future.

Acknowledgments

The author is deeply indebted to the families of carrier mothers for their participation, to obstetricians and pediatricians in the study field for their enthusiastic efforts to take care of carrier mothers and their children, and to the Nagasaki University School of Medicine, Nagasaki Prefectural Government, Nagasaki Prefecture Branches of the Japan Maternity Protection Association and of the Japanese Association of Pediatricians for assistance. He also thanks Emeritus Professors of Nagasaki University: Tsutome Miyamoto, Yoshiro Tsuji, and Toru Yamabe for their efforts in establishing APP. The study was partly supported by Grants-in-Aid of special cancer research by the Ministry of Education, Culture, Sports, Science and Technology of Japan.

REFERENCES

1. Anonymous. The mode of delivery and the risk of vertical transmission of human

immunodeficiency virus type 1—a meta-analysis of 15 prospective cohort studies. The International Perinatal HIV Group. *N. Engl. J. Med.*, **340**, 977–987 (1999).

2. Bazarbachi, A. and Hermine, O. Treatment with a combination of zidovudine and alpha-interferon in naive and pretreated adult T-cell leukemia/lymphoma patients. *J. Acquir. Immune Defic. Syndr. Hum. Retrovirol.*, **13** (Suppl. 1), S186–S190 (1996).
3. Gibb, D. M., Goodall, R. L., Dunn, D. T., Healy, M., Neave, P., Cafferkey, M., and Butler, K. Mother-to-child transmission of hepatitis C virus: evidence for preventable peripartum transmission. *Lancet*, **356**, 904–907 (2000).
4. Hanchard, B. Adult T-cell leukemia/lymphoma in Jamaica: 1986–1995. *J. Acquir. Immune Defic. Syndr. Hum. Retrovirol.*, **13** (Suppl. 1), S20–S25 (1996).
5. Hino, S. ATL development after adult infection of HTLV-I? *Jpn. J. Cancer Res.*, **80**, 1016 (1989).
6. Hino, S., Katamine, S., Kawase, K., Miyamoto, T., Doi, H., Tsuji, Y., and Yamabe, T. Intervention of maternal transmission of HTLV-1 in Nagasaki, Japan. *Leukemia*, **8** (Suppl. 1), S68–S70 (1994).
7. Hino, S., Katamine, S., Miyata, H., Tsuji, Y., Yamabe, T., and Miyamoto, T. Primary prevention of HTLV-1 in Japan. *Leukemia*, **11**, 357–359 (1997).
8. Hino, S., Sugiyama, H., Doi, H., Ishimaru, T., Yamabe, T., Tsuji, Y., and Miyamoto, T. Breaking the cycle of HTLV-I transmission via carrier mothers' milk. *Lancet*, **ii**, 158–159 (1987).
9. Hino, S., Yamaguchi, K., Katamine, S., Sugiyama, H., Amagasaki, T., Kinoshita, K., Yoshida, Y., Doi, H., Tsuji, Y., and Miyamoto, T. Mother-to-child transmission of human T-cell leukemia virus type-I. *Jpn. J. Cancer Res.*, **76**, 474–480 (1985).
10. Hinuma, Y., Nagata, K., Hanaoka, M., Nakai, M., Matsumoto, T., Kinoshita, K. I., Shirakawa, S., and Miyoshi, I. Adult T-cell leukemia: antigen in an ATL cell line and detection of antibodies to the antigen in human sera. *Proc. Natl. Acad. Sci. USA*, **78**, 6476–6480 (1981).
11. Katamine, S., Moriuchi, R., Yamamoto, T., Terada, K., Eguchi, K., Tsuji, Y., Yamabe, T., Miyamoto, T., and Hino, S. HTLV-1 proviral DNA in umbilical cord blood of babies born to carrier mothers. *Lancet*, **343**, 1326–1327 (1994).
12. Kawase, K., Katamine, S., Moriuchi, R., Miyamoto, T., Kubota, K., Igarashi, H., Doi, H., Tsuji, Y., Yamabe, T., and Hino, S. Maternal transmission of HTLV-1 other than through breast milk: discrepancy between the polymerase chain reaction positivity of cord blood samples for HTLV-1 and the subsequent seropositivity of individuals. *Jpn. J. Cancer Res.*, **83**, 968–977 (1992).
13. Kinoshita, K., Yamanouchi, K., Ikeda, S., Momita, S., Amagasaki, T., Soda, H., Ichimaru, M., Moriuchi, R., Katamine, S., Miyamoto, T., and Hino, S. Oral infection of a common marmoset with human T-cell leukemia virus type-I (HTLV-I) by inoculating fresh human milk of HTLV-I carrier mothers. *Jpn. J. Cancer Res.*, **76**, 1147–1153 (1985).
14. Kusuhara, K., Sonoda, S., Takahashi, K., Tokugawa, K., Fukushige, J., and Ueda, K. Mother-to-child transmission of human T-cell leukemia virus type I (HTLV-I): a fifteen-year follow-up study in Okinawa, Japan. *Int. J. Cancer*, **40**, 755–757 (1987).
15. Nakano, S., Ando, Y., Saito, K., Moriyama, I., Ichijo, M., Toyama, T., Sugamura, K., Imai, J., and Hinuma, Y. Primary infection of Japanese infants with adult T-cell leu-

kaemia-associated retrovirus (ATLV): evidence for viral transmission from mothers to children. *J. Infect.*, **12**, 205–212 (1986).
16. Obama, K., Tara, M., Sao, H., Taji, H., Morishima, Y., Mougi, H., Maruyama, Y., and Osame, M. Allogenic bone marrow transplantation as a treatment for adult T-cell leukemia. *Int. J. Hematol.*, **69**, 203–205 (1999).
17. Okamoto, M., Nagata, I., Murakami, J., Kaji, S., Iitsuka, T., Hoshika, T., Matsuda, R., Tazawa, Y., Shiraki, K., and Hino, S. Prospective reevaluation of risk factors in mother-to-child transmission of hepatitis C virus: high virus load, vaginal delivery, and negative anti-NS4 antibody. *J. Infect. Dis.*, **182**, 1511–1514 (2000).
18. Satow, Y., Hashido, M., Ishikawa, K., Honda, H., Mizuno, M., Kawana, T., and Hayami, M. Detection of HTLV-I antigen in peripheral and cord blood lymphocytes from carrier mothers. *Lancet*, **338**, 915–916 (1991).
19. Sugiyama, H., Doi, H., Yamaguchi, K., Tsuji, Y., Miyamoto, T., and Hino, S. Significance of postnatal mother-to-child transmission of human T-lymphotropic virus type-I on the development of adult T-cell leukemia/lymphoma. *J. Med. Virol.*, **20**, 253–260 (1986).
20. Tajima, K. The 4th nation-wide study of adult T-cell leukemia/lymphoma (ATL) in Japan: estimates of risk of ATL and its geographical and clinical features. The T- and B-cell Malignancy Study Group. *Int. J. Cancer*, **45**, 237–243 (1990).
21. Tajima, K., Amakawa, R., Uehira, K., Matsumoto, N., Shimizu, T., Miyazaki, Y., Fujimoto, M., Kishimoto, Y., and Fukuhara, S. Adult T-cell leukemia successfully treated with allogeneic bone marrow transplantation. *Int. J. Hematol.*, **71**, 290–293 (2000).
22. Takahashi, K., Takezaki, T., Oki, T., Kawakami, K., Yashiki, S., Fujiyoshi, T., Usuku, K., Mueller, N., Osame, M., Miyata, K. *et al.* The Mother-to-Child Transmission Study Group. Inhibitory effect of maternal antibody on mother-to-child transmission of human T-lymphotropic virus type I. *Int. J. Cancer*, **49**, 673–677 (1991).
23. Takezaki, T., Tajima, K., Ito, M., Ito, S., Kinoshita, K., Tachibana, K., and Matsushita, Y. The Tsushima ATL Study Group. Short-term breast-feeding may reduce the risk of vertical transmission of HTLV-I. *Leukemia*, **11**, 360–362 (1997).
24. Waldmann, T. A. T-cell receptors for cytokines: targets for immunotherapy of leukemia/lymphoma. *Ann. Oncol.*, **11**, 1101–1106 (2000).
25. Wiktor, S. Z., Pate, E. J., Murphy, E. L., Palker, T. J., Champegnie, E., Ramlal, A., Cranston, B., Hanchard, B., and Blattner, W. A. Mother-to-child transmission of human T-cell lymphotropic virus type I (HTLV-I) in Jamaica: association with antibodies to envelope glycoprotein (gp46) epitopes. *J. Acquir. Immune Defic. Syndr.*, **6**, 1162–1167 (1993).
26. Yamanouchi, K., Kinoshita, K., Moriuchi, R., Katamine, S., Amagasaki, T., Ikeda, S., Ichimaru, M., Miyamoto, T., and Hino, S. Oral transmission of human T-cell leukemia virus type-I into a common marmoset (*Callithrix jacchus*) as an experimental model for milk-borne transmission. *Jpn. J. Cancer Res.*, **76**, 481–487 (1985).
27. Yanagihara, E. T., Nakamura, J., Kimura, L., and Oishi, N. Retrovirus-associated adult T-cell leukemia-lymphoma: an epidemiologic study of five cases among Hawaii-born offspring of migrant Japanese. *Hematol. Oncol.*, **7**, 181–188 (1989).

ALLOGENEIC HEMATOPOIETIC STEM CELL TRANSPLANTATION FOR ATL

Takayuki ISHIKAWA, Tatsuo ICHINOHE, Kazunori IMADA, and Takashi UCHIYAMA

*Department of Hematology/Oncology, Graduate School of Medicine, Kyoto University**

Aggressive type adult T cell leukemia/lymphoma (ATL) has a very poor prognosis. Chemotherapy can induce long-term remission in only a small percentage of patients due to the intrinsic chemoresistant nature of the disease and to its associated immunodeficiency. Although allogeneic stem cell transplantation (allo-SCT) has been the treatment of choice for otherwise incurable hematological malignancies, allo-SCT was not performed in many patients with ATL until the recent achievement of reduction of transplantation-related toxicity. Recent reports on allo-SCT for relatively young patients with ATL from human leukocyte antigen (HLA)-matched sibling donors are promising, especially if their diseases were well controlled. Unfortunately, many patients with ATL can not receive conventional allo-SCT because of their advanced age, the absence of HLA-matched donors, or failure to induce remission. As the curative potential of allo-SCT is attractive, we have began a clinical trial to apply allo-SCT with reduced intensity conditioning (RI-SCT) to elderly patients with (partially) histocompatibility-matched donors. In RI-SCT, the cure of malignancy is entirely dependent on graft-versus malignancy (GVM) effects, and little information is available regarding the susceptibility of ATL to these effects. Our experience to date has been limited, and therefore we can not define the applicability or feasibility of RI-SCT for ATL. However, we realized that RI-SCT would not be sufficient to cure aggressive ATL. To cure chemotherapy-resistant ATL, successive immune cell therapies after RI-SCT seem necessary.

Adult T cell leukemia/lymphoma (ATL) exhibits diverse clinical features (*21*). It is generally classified into four clinical subtypes: acute, chronic, smoldering, and lymphomatous (*14*). Clinically, acute and lymphomatous types are highly aggressive. Patients often exhibit fever, skin rash, dyspnea, abdominal fullness, thirst and drowsiness. Tumor cells have a strong tendency to spread into various organs including the skin, lung and liver (*9*). Hypercalcemia may develop abruptly, which sometimes results in acute renal failure or coma. In addition, immunity against several pathogens is always reduced in patients with ATL. Serious infections such as *Pneumocystis carinii* pneumonia, invasive aspergillosis, and cytomegalovirus (CMV) infection are frequently encountered, especially after chemotherapy. Treatment of the aggressive form of ATL remains unsatisfactory. Only a small proportion of ATL patients can be

* 54 Kawaracho, Shogoin, Sakyo-ku, Kyoto 606-8507, Japan

induced into sustained remission. Yamada *et al.* have recently reported that combination chemotherapy, designed exclusively for ATL, achieved a complete remission rate of 35% and the median survival time was 13 months (*23*). Discouraging data has also been reported for high dose chemoradiotherapy followed by autologous stem cell transplantation, which has shown to be effective for patients with recurrent lymphoma, due to relapse of ATL and fatal opportunistic infection (*20*).

Until recently, allogeneic stem cell transplantation (allo-SCT) was not regarded as appropriate treatment for ATL because of the poor chemosensitivity, immunological fragility, and advanced age of onset of this disease; the mean age of onset of ATL has been reported to be 55 years (*18*).

Recent advances in graft-versus host disease (GVHD) prophylaxis, management of opportunistic infection, and supportive care have successfully extended the indication of allo-SCT. In the hope of potential immune-mediated eradication of tumor cells, allo-SCT is now performed for ATL patients.

Early Trials on Allo-SCT for ATL

In 1987, Sobue first reported use of allogeneic bone marrow transplantation (allo-BMT) for an ATL patient (*16*). This patient suffered from CMV pneumonia, which occurred soon after transplantation, and was the cause of death 205 days after allo-BMT. In 1994, Ljungman reported T cell-depleted allo-BMT for another ATL patient (*11*). After receiving allo-BMT from a human T cell leukemia virus type 1 (HTLV-1) negative sibling, the patient remained well without disease for 11 months, after which she developed fatal encephalitis caused by an undefined pathogen. HTLV-1 was detectable in autopsy material from the brain; it was also isolated from post-transplant peripheral blood lymphocytes, which showed complete donor chimerism. Autopsies in both patients revealed no residual tumor. Allo-SCT has been shown able to induce durable remission in these highly chemotherapy-resistant patients with ATL. However, these reports highlighted some concerns about allo-SCT for ATL. That is, the immune status in ATL patients receiving allo-SCT might be markedly suppressed for an extended period. In addition, donor-derived lymphocytes might be infected with HTLV-1, which reside anywhere within the patient, and infected lymphocytes might cause HTLV-1-associated diseases.

Recent Experience with Allo-SCT

Recent advances in GVHD prophylaxis and supportive care have decreased the early mortality after allo-SCT (*17*). In addition, increasing evidence has convinced us that allo-SCT is not merely a supportive measure for patients receiving myeloablative conditioning, but the immune-mediated graft-versus malignancy (GVM) effect is essential to prevent relapse (*2*). In other words, conditioning with high-dose chemoradiotherapy before allo-SCT need not eradicate all disease, which would be hardly expected for ATL patients, but a donor-derived immune system introduced into the patient cures the disease. In 1996, Borg first reported successful allo-SCT for an ATL patient (*3*). According to this report, a 41-year-old woman with acute type ATL in durable remission received allo-BMT from an HLA-matched, HTLV-1-negative sibling. She was alive and well 23 months after transplantation, and repetitive PCR-based analyses detected no HTLV-1 proviral DNA in her post-transplant peripheral blood or bone marrow cells. This encouraging report led to increased use of allo-SCT in Japan.

In our institute, two patients in remission have been treated with a conventional myeloablative conditioning regimen (Table I). Early clinical courses after allo-SCT were uneventful with rapid engraftment and controllable acute GVHD. No signs of veno-occlusive disease nor CMV infection were found, and they were discharged on day 64 and 59 after allo-SCT, respectively. In patient 1, the careful termination of FK506 was followed by abrupt development of chronic GVHD with respiratory distress, and she died of respiratory failure on day 298 without the recurrence of ATL. The clinical picture after first discharge in patient 2 was complicated (Fig. 1). Right-sided oculomotor and trochlear nerve palsy developed on day 100. Spinal tap revealed moderate pleocytosis with predominant lymphocytes. Cultivation, PCR-based assay, and flow cytometry did not reveal any pathogens or tumor cells. Bone marrow examination also showed complete donor chimerism, and no residual tumor cells were found. Under a diagnosis of meningitis of unknown etiology, cyclosporin was rapidly tapered off, and her eyeball movement improved 2 weeks later. She then developed extensive chronic GVHD, which required resumption of cyclosporin administration. In accordance with the improved chronic GVHD, multiple bone tumors composed of ATL cells, developed (day 210). Radiotherapy coupled with the re-discontinuation of cyclosporin was markedly effective in tumor regression. It is worth mentioning that radiotherapy was restricted to the bones susceptible to fracture. She was diagnosed as being in complete remission on day 270. In spite of extensive chronic GVHD, which caused severe anorexia, she showed enlarged cervical lymph nodes and smear prepara-

TABLE I. Patient Characteristics and Outcome of ATL Patients Who Received Conventional Allo-SCT in Our Institute

Patient	1	2
Age/Sex	46/M	51/F
Subtype	lymphomatous	acute
Initial presentation	gastric involvement	hypercalcemia
SIL-2R at Dx (U/ml)	7,250	74,500
Dx to SCT (m)	13	5
Status at SCT	PR	CR
Donor relation	MSD	MUD
HTLV-1 status	negative	negative
Stem cell source	peripheral blood	bone marrow
Preparative regimen	CY+TBI (12 Gy)	CY+TBI (12 Gy)
GVHD prophylaxis	CSA+MTX	FK506+MTX
Acute GVHD	grade 1, skin	grade 1, skin
Chronic GVHD	extensive (skin, liver, lung)	extensive (skin, liver, gut)
Complications	ADV-cystitis, ARDS	meningitis
Recurrence	no	yes (day 201)
Current status	dead (ARDS)	alive
Survival from SCT (d)	298	422
HTLV-1 after SCT	provirus +	provirus +

SIL-2R, soluble interleukin-2 receptor; Dx, diagnosis; PR, partial remission; CR, complete remission; MSD, matched sibling donor; MUD, matched unrelated donor; CY, cyclophosphamide; TBI, total body irradiation; GVHD, graft-versus host disease; CSA, cyclosporin A; MTX, methotrexate; ADV, adenovirus; ARDS, adult respiratory distress syndrome.

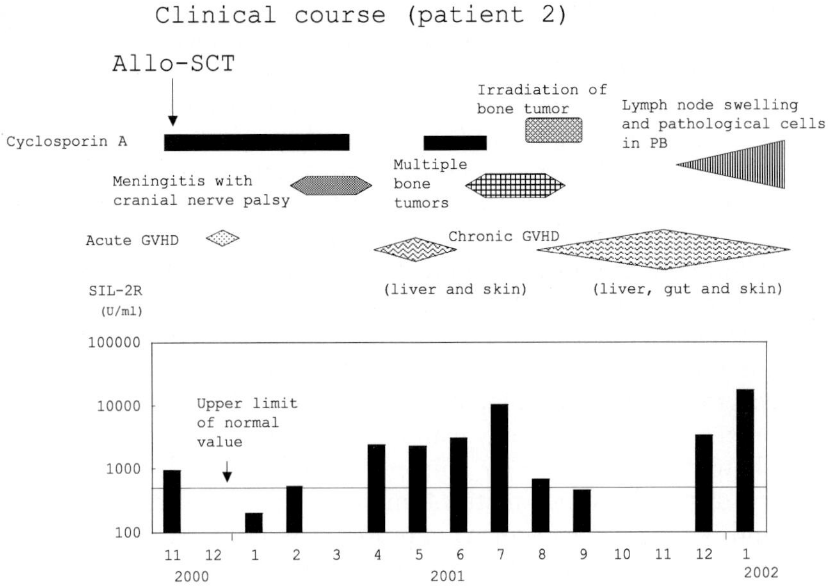

FIG. 1. Clinical course of patient 2 in our institute PB=peripheral blood.

TABLE II. Allogeneic Transplantation for ATL Patients—Characteristics of Reported Cases

Patient					Donor		Transplantation procedure		
No.	Age/ Sex	Subtype	Status at SCT	Dx to SCT (m)		HTLV-1	Stem cell	Conditioning	GVHD prophylaxis
1	33/M	chronic	PR	7.3	MSD	–	BM	CA/CY/TBI	CSA/MTX
2	37/M	acute	PR	9.8	MSD	+	PBPC	ETP/CY/TBI	FK/MTX
3	38/F	acute	CR	3.5	MSD	+	BM	CA/CY/TBI	CSA/MTX
4	41/F	acute	CR	6.0	MSD	–	BM	DNR/CY/TBI	CSA/MTX
5	41/F	acute	CR	5.2	MSD	–	BM	CY/TBI	CSA/MTX
6	42/M	acute	CR	8.7	MSD	–	BM	MEL/TBI	CSA/MTX
7	44/M	acute	PR	6.8	MSD	–	BM	ETP/CA/TBI	CSA/MTX
8	45/M	acute	CR	6.8	MUD	–	BM	MEL/TBI	FK
9	47/F	acute	PR	3.5	MSD	–	BM	ETP/CY/TBI	CSA/MTX
10	47/M	acute	NR	3.0	MSD	–	BM	MCNU/CY/TBI	CSA/MTX
11	48/M	acute	NR	20.1	MSD	–	BM	ETP/CA/TBI	CSA/MTX
12	51/M	lymphomatous	PR	8.9	MSD	+	BM+PBPC	CY/TBI	CSA/MTX

BM, bone marrow; PBPC, peripheral blood progenitor cells; CA, cytosine arabinoside; ETP, etoposide; DNR, daunorubicin; MEL, melphalan; MCNU, ranimustine. For other abbreviations see Table I.

tions of her peripheral blood showed the emergence of ATL cells on day 330. She is being treated with donor lymphocyte infusion now.

Over the last 3 years, three reports on 11 Japanese patients with ATL have been published (*13, 19, 22*), and several conference abstracts suggested that, to date, more than 30 patients have received allo-SCT. Clinical characteristics and outcome of the recently reported cases are summarized in Tables II and III. Allo-SCT from a human leukocyte antigen (HLA)-matched related donor seems to be of benefit to patients

TABLE III. Allogeneic Transplantation for ATL Patients—Outcomes of Reported Cases

Patient No.	GVHD Acute	GVHD Chronic	Relapse (m)	Survival after SCT (m)	Status	Provirus	Reference
1	II	limited	no	32.4+	alive	negative	12, 14
2	0	extensive	no	31.5+	alive	n.d.	14
3	0	n.e.	yes (3.9)	10.1	dead (relapse)	n.d.	12, 14
4	0	extensive	no	23+	alive	negative	11
5	II	no	no	34.4+	alive	n.d.	14
6	II	no	no	19.5+	alive	positive	14
7	0	no	no	28.7+	alive	n.d.	14
8	III	extensive	no	15.5	dead (RRT)	positive	14
9	0	n.e.	yes (3.7)	11.8+	alive	n.d.	14
10	II	no	no	24+	alive	positive	13
11	IV	n.e.	no	4.1	dead (GVHD)	n.d.	14
12	I	extensive	no	9.3	dead (IP)	n.d.	14

n.e., not evaluable; RRT, regimen-related toxicity; IP, interstitial pneumonia.

with well-controlled and indolent disease. Early concerns regarding the possibility of increases in serious opportunistic infections were shown to be unfounded. Because of the risk of occurrence of HTLV-1-associated diseases, further studies in large numbers of patients with longer follow-up are necessary. At least in some patients, allo-SCT may not eradicate HTLV-1 even after tumor cells have disappeared.

Novel Approach to Expand the Application of Allo-SCT

As conventional allo-SCT was designed to eradicate all tumor cells with myeloablative conditioning composed of maximally tolerated doses of high-dose chemotherapy and radiotherapy, the desirable effects are often offset by its overwhelming toxicity in patients over the age of 50 years (*6*). Less than one third of ATL patients are young enough to safely receive conventional allo-SCT (*8*). Moreover, over half of the patients lack HLA-matched sibling donors, and many can achieve only transient disease control, for the duration is too short to allow preparation of allo-SCT from an unrelated donor. Thus, only a small percentage of patients are candidates for conventional allo-SCT. As the curative potential of allo-SCT is attractive for otherwise incurable diseases such as ATL, several attempts to expand the application of allo-SCT are currently underway. One of these approaches utilizes a reduced-intensity conditioning regimen.

Allo-SCT with reduced-intensity conditioning, or RI-SCT, has provided new possibilities for patients who were not eligible for conventional allo-SCT because of age or medical infirmity (*7, 15*). This less toxic, non-myeloablative conditioning regimen was designed not to eradicate all tumor cells but to provide sufficient immunosuppression for the engraftment of allogeneic lymphocytes and hematopoietic stem cells. Tumor eradication is mediated by donor-derived T lymphocytes. Early trials of RI-SCT for patients with hematological malignancies were reported to be successful in achieving durable remission and in reducing transplant-related mortality (*10, 15*).

As the success of RI-SCT relies entirely on immune-mediated graft-versus malignancy (GVM) effects, the indication of RI-SCT is limited to malignancies susceptible to GVM effects. Data concerning the susceptibility of ATL to GVM effects are limited. However, judging from the reports of favorable results of allo-SCT as well as our observations listed in Fig. 1, GVM effects seem to operate on ATL.

Preliminary Results with RI-SCT for ATL

Since 1996, when trials of RI-SCT were first reported at a hematology meeting, RI-SCT has become one of the most exciting developments in the treatment of malignancies. The steady increase in the number of diseases being treated by this method has resulted in an increasing number of preparative regimens. For successful application of RI-SCT to malignant diseases, establishment of stable donor chimerism and suppression of tumor growth are both required. Fludarabine, a new purine analogue, has been regarded as a key drug for the engraftment of donor lymphohematopoietic cells. In RI-SCT from an HLA-identical sibling, complete donor chimerism could be induced with a fludarabine-based preparative regimen coupled with post-transplant calcineurin inhibitor (*12*). With regard to the latter issue, an optimal regimen has yet to be defined (*5*). It is generally believed that several weeks to months are required for the development of an efficient GVM effect. In slow-growing malignancies such as indolent lymphoma or for patients with sustained remission, we could wait for several months. However, the clinical aggressiveness and transient disease control of most patients with ATL does not allow us to wait. The addition of anticancer drugs to preparative regimens is necessary for short-term disease control in patients with ATL. We are currently engaged in a pilot trial to determine whether RI-SCT is beneficial in providing sufficient disease control for hematological malignancies including ATL. Patients who could tolerate conventional allo-SCT were excluded. As RI-SCT was reported to reduce the severity of acute GVHD, we decided to apply RI-SCT to patients for whom partially mismatched related donors

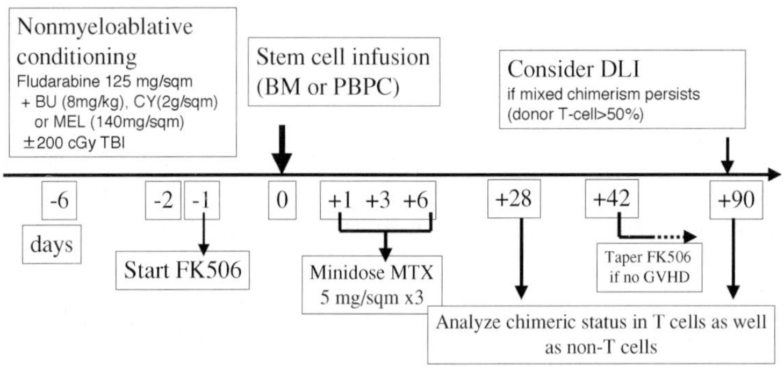

FIG. 2. Schema for RI-SCT against ATL patients

From 6 days before allogeneic stem cell infusion, fludarabine coupled with busulphan (BU), cyclophosphamide (CY) or melphalan (MEL) are administered. If the donor is other than an HLA-identical sibling, 200 cGy total body irradiation (TBI) is added. Donor-derived stem cells are collected from bone marrow (BM) or peripheral blood (PBPC) and infused. A short course of immunosuppressive drugs comprised of FK506 and methotrexate (MTX) is given to prevent graft-versus host disease (GVHD) and graft failure. After transplantation, the status of lymphohematopoietic chimerism is analyzed repeatedly. In cases in which over one third of the T lymphocytes at 90 days after transplantation are of recipient origin, we consider donor lymphocyte infusion to prevent graft rejection or relapse. FK506 would be tapered off rapidly if no sign of GVHD were found at 42 days after transplantation.

were available. The extension of the eligibility of donors has also widened the application of RI-SCT. An outline of our RI-SCT method is illustrated in Fig. 2. As graft rejection has been reported to occur frequently when hematopoietic stem cells from a donor other than an HLA-matched sibling is used (*4*), we intensified pre-transplant immunosuppression by addition of low-dose total body irradiation in such instances. As for anti-cancer drugs before RI-SCT, we used oral busulphan or intravenous cyclophosphamide for patients with durable remission, and intravenous melphalan for others. For post-transplant immunosuppression, a combination of FK506 and small doses of methotrexate was introduced. To date, 26 patients have received RI-SCT in our institute, and 3 of them had ATL. The overall

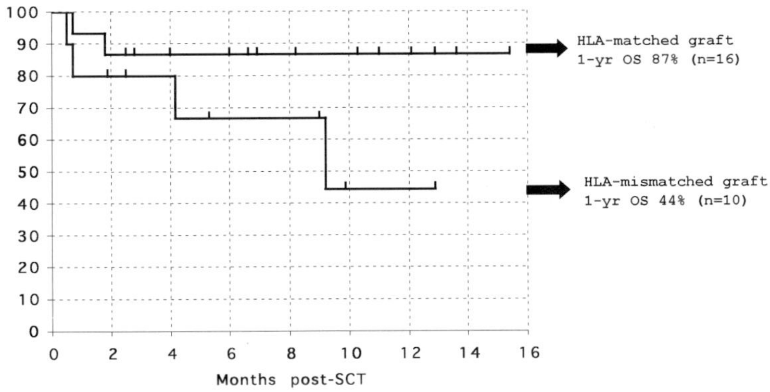

FIG. 3. Overall survival of patients according to HLA compatibility with donor Median follow-up is 190 days.

TABLE IV. Patient Characteristics and Outcome of ATL Patients Who Received RI-SCT in Our Institute

Patient	3	4	5
Age/Sex	53/M	58/M	59/M
Subtype	acute	acute	acute
Initial presentation	leukocytosis	hypercalcemia	hypercalcemia
SIL-2R at diagnosis	26,600	65,000	64,000
Dx to SCT (m)	3	4	3
Status at SCT	PR	PD	PD
Donor relatioon	PMRD	PMRD	MSD
HLA disparity	3 (HVG), 0 (GVH)	2 (HVG), 1 (GVH)	0
HTLV-1 status	negative	positive	positive
Stem cell source	peripheral blood	peripheral blood	bone marrow
Preparative regimen	FLU+MEL+2 Gy TBI	FLU+MEL+2 Gy TBI	FLU+MEL
GVHD prophylaxis	FK506+MTX	FK506+MTX	FK506
Acute GVHD	grade 2, gut	not evaluable	no
Chronic GVHD	extensive (skin, liver)	not evaluable	not evaluable
Complications	engraftment fever	sepsis	not significant
Recurrence	no	no	yes (day 27)
Current status	alive	dead (sepsis)	dead (recurrence)
Survival from SCT (d)	130	20	51
HTLV-l after SCT	provirus +	not evaluable	not evaluable

PD, progressive disease; PMRD, partially mismatched related donor; HVG, host-versus graft direction; GVH, graft-versus host direction; FLU, fludarabine. For other abbreviations see Tables I–III.

results are shown in Fig. 3. The 1-year estimated overall survival was 87% for patients who received HLA-matched grafts. This result was comparable to those of previous reports. None of the 3 patients with ATL were in sustained remission at the time of RI-SCT and all were treated with the melphalan-based regimen. Initial trials with refractory disease (patients 4 and 5) were discouraging (Table IV). In patient 4, prolonged neutropenia before RI-SCT resulted in uncontrollable infection, and early relapse in patient 5 indicated that it is necessary to find a way to hasten the development of GVM effects in refractory ATL patients. For chemotherapy-sensitive patients, RI-SCT might afford the possibility of long-term disease control as shown in patient 3.

Summary and Future Perspectives

Allogeneic transplantation has saved the lives of many patients with hematological malignancies, which would have otherwise been fatal. It is not surprising that ATL was regarded as unsuitable for allo-SCT. It has to be mentioned that, until recently, many patients with refractory non-Hodgkin's lymphoma, Hodgkin's disease, and chronic lymphocytic leukemia did not receive allo-SCT because of excessive transplant-related mortality (*1*). Over the last decade, substantial improvements have been made in lowering transplant-related toxicity, and the range of applicable diseases for allo-SCT has been extended. At present, allo-SCT from an HLA-matched sibling donor is being used for relatively young patients with ATL, and the outcome seems to be promising for those with well-controlled disease.

The development of RI-SCT has markedly raised the upper age limit at which allo-SCT can be tolerated. To extend the application of allo-SCT for elderly patients with ATL, we have tried to cure them with RI-SCT. Although our experience to date has been limited, relapse of malignancy seems to be a major obstacle. It is recognized that it takes several months for the development of an efficient GVM effect. In addition, greater GVM effects are generally accompanied by more GVHD, and we have no way to dissociate the GVM effect from GVHD. There are two different approaches to make more ATL patients free of disease through allo-SCT. One is to develop novel agents that effectively suppress tumor growth for a long period, such as anti-CD20 antibody for B cell lymphoma or tyrosine-kinase inhibitors for chronic myelogenous leukemia. The other is to hasten the occurrence of GVM effects. *In vitro* induction of cytotoxic T cells against tumor antigens or alloantigens exclusively expressed on lymphohematopoietic cells would enable us to better control tumor growth without severe GVHD.

REFERENCES

1. Akpek, G., Ambinder, R. F., Piantadosi, S., Abrams, R. A., Brodsky, R. A., Vogelsang, G. B., Zahurak, M. L., Fuller, D., Miller, C. B., Noga, S. J., Fuchs, E., Flinn, I. W., O'Donnell, P., Seifter, E. J., Mann, R. B., and Jones, R. J. Long-term results of blood and marrow transplantation for Hodgkin's lymphoma. *J. Clin. Oncol.*, **19**, 4314–4321 (2001).
2. Appelbaum, F. R. Haematopoietic cell transplantation as immunotherapy. *Nature*, **411**, 385–389 (2001).

3. Borg, A., Yin, J. A., Johnson, P. R., Tosswill, J., Saunders, M., and Morris, D. Successful treatment of HTLV-1-associated acute adult T-cell leukaemia lymphoma by allogeneic bone marrow transplantation. *Br. J. Haematol.*, **94**, 713–715 (1996).
4. Bornhauser, M., Thiede, C., Platzbecker, U., Jenke, A., Helwig, A., Plettig, R., Freiberg-Richter, J., Rollig, C., Geissler, G., Lutterbeck, K., Oelschlagel, U., and Ehninger, G. Dose-reduced conditioning and allogeneic hematopoietic stem cell transplantation from unrelated donors in 42 patients. *Clin. Cancer Res.*, **7**, 2254–2262 (2001).
5. Champlin, R., Khouri, I., Shimoni, A., Gajewski, J., Kornblau, S., Molldrem, J., Ueno, N., Giralt, S., and Anderlini, P. Harnessing graft-versus-malignancy: non-myeloablative preparative regimens for allogeneic haematopoietic transplantation, an evolving strategy for adoptive immunotherapy. *Br. J. Haematol.*, **111**, 18–29 (2000).
6. Du, W., Dansey, R., Abella, E. M., Baynes, R., Peters, W. P., Klein, J., Akhtar, A., Cherednikova, L., and Karanes, C. Successful allogeneic bone marrow transplantation in selected patients over 50 years of age—a single institution's experience. *Bone Marrow Transplant.*, **21**, 1043–1047 (1998).
7. Giralt, S., Estey, E., Albitar, M., van Besien, K., Rondon, G., Anderlini, P., O'Brien, S., Khouri, I., Gajewski, J., Mehra, R., Claxton, D., Andersson, B., Beran, M., Przepiorka, D., Koller, C., Kornblau, S., Korbling, M., Keating, M., Kantarjian, H., and Champlin, R. Engraftment of allogeneic hematopoietic progenitor cells with purine analog-containing chemotherapy: harnessing graft-versus-leukemia without myeloablative therapy. *Blood*, **89**, 4531–4536 (1997).
8. Hirose, K. and Tajima, K. The 8th nation-wide study of adult T-cell leukemia/lymphoma (ATL) in Japan: the T- and B-cell Malignancy Study Group. *Gann no Rinsyou*, **44**, 381–397 (1998) (in Japanese).
9. Kawano, F., Yamaguchi, K., Nishimura, H., Tsuda, H., and Takatsuki, K. Variation in the clinical courses of adult T-cell leukemia. *Cancer*, **55**, 851–856 (1985).
10. Khouri, I. F., Keating, M., Korbling, M., Przepiorka, D., Anderlini, P., O'Brien, S., Giralt, S., Ippoliti, C., von Wolff, B., Gajewski, J., Donato, M., Claxton, D., Ueno, N., Andersson, B., Gee, A., and Champlin, R. Transplant-lite: induction of graft-versus-malignancy using fludarabine-based nonablative chemotherapy and allogeneic blood progenitor-cell transplantation as treatment for lymphoid malignancies. *J. Clin. Oncol.*, **16**, 2817–2824 (1998).
11. Ljungman, P., Lawler, M., Asjo, B., Bogdanovic, G., Karlsson, K., Malm, C., McCann, S. R., Ringden, O., and Gahrton, G. Infection of donor lymphocytes with human T lymphotrophic virus type I (HTLV-I) following allogeneic bone marrow transplantation for HTLV-I positive adult T-cell leukaemia. *Br. J. Haematol.*, **88**, 403–405 (1994).
12. McSweeney, P. A., Niederwieser, D., Shizuru, J. A., Sandmaier, B. M., Molina, A. J., Maloney, D. G., Chauncey, T. R., Gooley, T. A., Hegenbart, U., Nash, R.A., Radich, J., Wagner, J. L., Minor, S., Appelbaum, F. R., Bensinger, W. I., Bryant, E., Flowers, M. E., Georges, G. E., Grumet, F. C., Kiem, H. P., Torok-Storb, B., Yu, C., Blume, K. G., and Storb, R. F. Hematopoietic cell transplantation in older patients with hematologic malignancies: replacing high-dose cytotoxic therapy with graft-versus-tumor effects. *Blood*, **97**, 3390–4000 (2001).

13. Obama, K., Tara, M., Sao, H., Taji, H., Morishima, Y., Mougi, H., Maruyama, Y., and Osame, M. Allogeneic bone marrow transplantation as a treatment for adult T-cell leukemia. *Int. J. Hematol.*, **69**, 203–205 (1999).
14. Shimoyama, M. Diagnostic criteria and classification of clinical subtypes of adult T-cell leukaemia-lymphoma. A report from the Lymphoma Study Group (1984–1987). *Br. J. Haematol.*, **79**, 428–437 (1991).
15. Slavin, S., Nagler, A., Naparstek, E., Kapelushnik, Y., Aker, M., Cividalli, G., Varadi, G., Kirschbaum, M., Ackerstein, A., Samuel, S., Amar, A., Brautbar, C., Ben-Tal, O., Eldor, A., and Or, R. Nonmyeloablative stem cell transplantation and cell therapy as an alternative to conventional bone marrow transplantation with lethal cytoreduction for the treatment of malignant and nonmalignant hematologic diseases. *Blood*, **91**, 756–763 (1998).
16. Sobue, R., Yamauchi, T., Miyamura, K., Sao, H., Tahara, T., Yoshikawa, H., Yoshikawa, S., Morishima, Y., and Kodera, Y. Treatment of adult T cell leukemia with megadose cyclophosphamide and total body irradiation followed by allogeneic bone marrow transplantation. *Bone Marrow Transplant.*, **4**, 441–444 (1987).
17. Stockerl-Goldstein, K. E. and Blume, K. G. A decade of progress in allogeneic hematopoietic cell transplantation: 1990–2000. *Adv. Cancer Res.*, **81**, 1–59 (2001).
18. Tajima, K. The 4th nation-wide study of adult T-cell leukemia/lymphoma (ATL) in Japan: estimates of risk of ATL and its geographical and clinical features. The T- and B-cell Malignancy Study Group. *Int. J. Cancer*, **45**, 237–243 (1990).
19. Tajima, K., Amakawa, R., Uehira, K., Matsumoto, N., Shimizu, T., Miyazaki, Y., Fujimoto, M., Kishimoto, Y., and Fukuhara, S. Adult T-cell leukemia successfully treated with allogeneic bone marrow transplantation. *Int. J. Hematol.*, **71**, 290–293 (2000).
20. Tsukasaki, K., Maeda, T., Arimura, K., Taguchi, J., Fukushima, T., Miyazaki, Y., Moriuchi, Y., Kuriyama, K., Yamada, Y., and Tomonaga, M. Poor outcome of autologous stem cell transplantation for adult T cell leukemia/lymphoma: a case report and review of the literature. *Bone Marrow Transplant.*, **23**, 87–89 (1999).
21. Uchiyama, T., Yodoi, J., Sagawa, K., Takatsuki, K., and Uchino, H. Adult T-cell leukemia: clinical and hematologic features of 16 cases. *Blood*, **50**, 481–492 (1977).
22. Utsunomiya, A., Miyazaki, Y., Takatsuka, Y., Hanada, S., Uozumi, K., Yashiki, S., Tara, M., Kawano, F., Saburi, Y., Kikuchi, H., Hara, M., Sao, H., Morishima, Y., Kodera, Y., Sonoda, S., and Tomonaga, M. Improved outcome of adult T cell leukemia/lymphoma with allogeneic hematopoietic stem cell transplantation. *Bone Marrow Transplant.*, **27**, 15–20 (2001).
23. Yamada, Y., Tomonaga, M., Fukuda, H., Hanada, S., Utsunomiya, A., Tara, M., Sano, M., Ikeda, S., Takatsuki, K., Kozuru, M., Araki, K., Kawano, F., Niimi, M., Tobinai, K., Hotta, T., and Shimoyama, M. A new G-CSF-supported combination chemotherapy, LSG15, for adult T-cell leukaemia-lymphoma: Japan Clinical Oncology Group Study 9303. *Br. J. Haematol.*, **113**, 375–382 (2001).

CHEMOTHERAPY OF ATL

Kensei TOBINAI

*Hematology Division, National Cancer Center Hospital**

Since 1978 the Lymphoma Study Group of the Japan Clinical Oncology Group (JCOG-LSG) has conducted six consecutive clinical trials against aggressive adult T-cell leukemia-lymphoma (ATL). ATL patients were included in the three earlier trials for advanced non-Hodgkin's lymphoma (NHL) [JCOG7801 (1978–80), JCOG8101 (1981–83) and JCOG 8701 (1987–90)]. The 4-year survival rate of ATL patients in the JCOG-8101 and JCOG8701 was only 8% and 12%, respectively, and prognostic factor analyses of both studies revealed that a diagnosis of ATL was significantly associated with shortened survival. Thus, the following three trials have been conducted exclusively for ATL: JCOG9109 (1991–93), JCOG9303 (1994–96) and JCOG9801 (1998–). Based on the promising results of deoxycoformycin (DCF) in ATL patients, a phase II study of DCF-containing combination chemotherapy was conducted (JCOG9109). However, the clinical outcome of ATL patients was not improved: complete response (CR) rate, 28% (17/60); median survival time (MST), 7 months; 2-year survival rate, 17%. In 1994, a phase II study of a granulocyte colony-stimulating factor (G-CSF)-supported multi-agent chemotherapy (JCOG9303) was initiated. In this protocol, LSG15, ranimustine and carboplatin were included to overcome the multidrug resistance of ATL cells. The overall response rate was 81% (75/93), and the CR rate was 36%. The MST was 13 months and 2-year survival rate was 31%. The therapeutic results of JCOG9303 were superior to those obtained by the previous JCOG-LSG trials. To confirm whether the LSG15, a G-CSF-supported multi-agent treatment strategy, is a new standard in ATL chemotherapy, a randomized phase III study (JCOG9801) comparing it with biweekly CHOP is going on.

Despite the consecutive clinical trials by the JCOG-LSG to develop more effective chemotherapy regimens, most ATL patients are still incurable with current chemotherapies. Further efforts to incorporate new innovative treatment modalities are needed such as new anticancer agents, monoclonal antibody therapies and allogeneic hematopoietic stem cell transplantation (HSCT) including non-myeloablative HSCT.

Adult T-cell leukemia-lymphoma (ATL) was first recognized in Japan by Takatsuki and his colleagues in 1977 (*15, 25*). The disease was characterized as a leukemia or lymphoma derived from peripheral T-cells, generalized lymphadenopathy, hepatosplenomegaly, hypercalcemia and skin involvement. After a novel retrovirus,

* 5-1-1 Tsukiji, Chuo-ku, Tokyo 104-0045, Japan

human T-cell leukemia virus type I (HTLV-I), was revealed to be closely associated with ATL, it was found that ATL shows a marked diversity in its clinical manifestations. ATL cases have been subdivided into four clinical subtypes: acute, lymphoma, chronic and smoldering types. Based on a nationwide survey of 854 patients with ATL who were newly diagnosed between 1983 and 1987 in Japan, the Lymphoma Study Group proposed the diagnostic criteria for these four clinical subtypes (Table I) (*11*).

Clinical Manifestations

The leukemic subtypes include the acute type with a rapidly progressive clinical course and most of the characteristic features of ATL: generalized lymphadenopathy, hepatomegaly, splenomegaly, skin involvement, hypercalcemia, and organ infiltration (lung, gastrointestinal tract, etc.). The symptoms and signs include abdominal pain, diarrhea, ascites, pleural effusion, cough, sputum, and chest X-ray abnormalities. The smoldering type shows an indolent clinical course and only a small percentage of leukemic cells, but may also include skin involvement. The chronic type, with a high percentage of leukemic cells, is occasionally associated with skin involvement, lymphadenopathy, and hepatosplenomegaly, and also shows an indolent clinical course. The lymphoma type (non-leukemic type), is comprised of patients who present with the manifestations of non-Hodgkin's lymphoma (NHL) without circulating malignant cells in the peripheral blood. When ATL patients are staged according to the

TABLE I. Diagnostic Criteria for Clinical Subtypes of ATL[a]

	Smoldering	Chronic	Lymphoma	Acute
Anti-HTLV-I antibody	+	+	+	+
Lymphocyte ($\times 10^3/\mu l$)	<4	≥4[d]	<4	b
Abnormal T lymphocytes	≥5%[f]	+[e]	≤1%	+[e]
"Flower cells" with T-cell marker	c	c	No	+
LDH	≤1.5 N	≤2 N	b	b
Corrected Ca^{2+} (mEq/liter)	<5.5	<5.5	b	b
Histologically proven lymphadenopathy	No	b	+	b
Tumor lesion				
Skin and/or lung	b	b	b	b
Lymph node	No	b	Yes	b
Liver	No	b	b	b
Spleen	No	b	b	b
Central nervous system	No	b	b	
Bone	No	No	b	b
Ascites	No	No	b	b
Pleural effusion	No	No	b	b
Gastrointestinal tract	No	No	b	b

[a] From Shimoyama, M. and members of the Lymphoma Study Group (1984–1987): Diagnostic criteria and classification of clinical subtypes of adult T-cell leukemia-lymphoma. *Br. J. Haematol.*, **79**, 428 (1991) with permission.
[b] No essential qualification except terms required for other subtype(s).
[c] Typical "flower cells" may be seen occasionally.
[d] Accompanied by T-lymphocytosis ($3.5\times 10^3/\mu l$ or more).
[e] If abnormal T-lymphocytes are less than 5% in peripheral blood, histologically proven tumor lesion is required.
[f] Histologically proven skin and/or pulmonary lesion(s) is required if there are fewer than 5% abnormal T-lymphocytes in peripheral blood.
Abbreviations: ATL, adult T-cell leukemia-lymphoma; HTLV-I, human T-cell leukemia virus type I; LDH, lactate dehydrogenase; N, normal upper limit.

Ann Arbor staging classification, which has been widely utilized for staging of patients with Hodgkin's disease or NHL, most ATL patients are categorized as stage IV, because leukemic cells are recognized even in clinically indolent forms such as the smoldering type and chronic type. Therefore, in ATL, the clinical subtype is more important than the Ann Arbor stage for predicting prognosis and deciding treatment in each patient.

ATL, particularly the aggressive forms (acute and lymphoma types), has been found to infiltrate the stomach in 29% and the intestine in 25% of patients at autopsy (*26*). The involvement may be focal as an isolated gastric lesion or diffuse involvement of the entire gastrointestinal tract. Extensive infiltration in the intestine can lead to moderate to severe diarrhea and malabsorption. ATL patients suffer from a variety of abdominal symptoms, such as nausea, vomiting, abdominal fullness and diarrhea, which may be attributable to the infiltration by neoplastic cells, but because of the associated immunodeficiency, various opportunistic infections such as *Strongyloidiasis* may be complicated.

Hepatic involvement of ATL cells may be found in up to one fourth of the patients with acute and lymphomatous ATL and is not infrequently manifested by jaundice and hepatic transaminase elevations. One hundred eleven patients with acute-type or lymphoma-type ATL were analyzed for hepatic involvement, compared with 106 patients with NHL other than ATL (*29*). There were more frequent palpable hepatomegaly, elevations of total bilirubin, hepatic transaminase, lactic dehydrogenase (LDH), and alkaline phosphatase values than other NHL patients ($P < 0.0001$). Autopsy liver samples disclosed that the portal area was most frequently infiltrated with ATL cells. A tendency for ATL patients with impaired hepatic function to have shorter survival was noted (*29*).

Pulmonary complications, common in ATL, are due to leukemic infiltration in one half of the patients and to infections with a variety of bacterial and opportunistic organisms in the other half (*31*). Twenty-six percent of 854 Japanese patients with ATL had active infections at diagnosis (*11*). The incidence was highest in the chronic and smoldering types (36%) and lower in the acute (27%) and lymphoma subtypes (11%). The encountered infections were bacterial (pneumonias, sepsis, and tuberculosis) in 43%, fungal in 31%, protozoal in 18%, and viral in 8% of ATL patients. The immunodeficiency at presentation in ATL can be exacerbated by the neutropenia produced by cytotoxic chemotherapy, leading to an extremely high risk of infection throughout therapy. Infections are responsible for the patient's death in about half of the cases.

Central nervous system (CNS) involvement occurs in approximately 10% of patients with ATL. Teshima *et al.* identified 15 instances of CNS involvement in 10 of 99 patients with ATL (*16*). Leptomeningeal involvement was present in nine of the 10 patients, intracerebral infiltration was noted in three, and the spinal cord was involved in two. The initial symptoms included muscle weakness (47%), altered mental status (47%), paresthesia (40%), headache (33%), and urinary incontinence (27%). Signs included nuchal rigidity (33%) and cranial nerve palsies (13%). Hyponatremia secondary to the syndrome of inappropriate secretion of antidiuretic hormone was observed in four patients. CNS involvement frequently occurs in association with a systemic progression of disease; only rarely is it involved at presentation or as the only site of relapse.

Laboratory Findings

Laboratory findings also depend on the clinical subtype of ATL (Table I) (*11*). Leukocytosis is found in patients with the acute or chronic type at presentation, exhibiting characteristic atypical lymphoid cells with marked polymorphic nuclei, so-called "flower cells." Although not all patients present with a leukemic feature, peripheral blood involvement develops in most patients at some time during their courses. Most patients with the acute or lymphoma type have elevated serum LDH levels.

The most striking laboratory finding in ATL patients is hypercalcemia, which was found in 32% of Japanese patients with ATL (*11*). Lytic bone lesions have been described in some patients; however, examinations of bone obtained at autopsy or from bone marrow biopsies usually reveal activated osteoclasts with increased bone resorption; infiltrating neoplastic T-cells are rarely found. ATL patients have low phosphates, hypercalciuria, high levels of nephrogenous cAMP, and low levels of 1,25-dihydroxyvitamin D. This pattern suggests the presence of humoral hypercalcemia of malignancy. The following factors have been suggested to have roles in the hypercalcemia in ATL patients; parathyroid hormone-related peptide (PTHrP), osteoclast-activating factor, tumor necrosis factors (TNF-α and TNF-β), interleukin-1α (IL-1α), IL-1β and overexpression of the receptor activator of nuclear factor κB (*RANK*) ligand gene.

Elevated serum levels of soluble IL-2 receptor (sIL-2R) in ATL patients, especially in those with acute- or lymphoma-type, have been noted by several investigators. The serum level of sIL-2R is thought to be one of the useful markers for evaluating the clinical aggressiveness of the disease and for monitoring the response to therapy in ATL patients.

Histopathology and Diagnosis of ATL

The circulating cells in the peripheral blood have markedly pleomorphic nuclei with an irregular nuclear contour, the so-called "flower cells," which is characteristic of ATL. The swollen lymph nodes in ATL patients show diffuse NHL of various histopathological subtypes, including pleomorphic, large cell, mixed cell, or medium-sized cell types. The pleomorphic pattern, i.e., a mixture of various-sized lymphoma cells from small cells to giant cells, and nuclear polymorphism are recognized. No correlation between histopathological subtype and clinical behavior has been recognized.

Skin is frequently involved by ATL cells. Generalized nodular or papulonodular eruptions are common in ATL patients, however, tumorous lesions are also recognized in some patients. Erythematous plaque formation and sometimes nodular tumors are other cutaneous manifestations. Histologically, diffuse or patchy infiltration of atypical lymphoid cells, usually small or medium in size with polymorphic nuclear contours in the upper dermis, sometimes with an intraepidermal infiltration is noted. Large nuclear cells with highly irregular or cerebriform features are intermingled in some cases.

One of the difficult issues in the diagnosis of ATL is the relationship with other peripheral T-cell malignancies not associated with HTLV-I. The clinical diagnosis of ATL is suspected by the unique combination of its clinical and pathologic features.

One of the T-cell malignancies that is likely to be confused with ATL is mycosis fungoides-Sezary syndrome (MF/SS). Since cutaneous involvement in ATL is frequent, the distinction of smoldering-type ATL from MF/SS is often difficult based on the clinical manifestations alone. In the differential diagnosis of ATL and other peripheral T-cell malignancies, HTLV-I virology such as the detection of serum antibody to HTLV-I and the molecular detection of the monoclonal integration of HTLV-I proviral DNA is most important. Various kinds of serological assays have been used including the immunofluorescence assay, particle agglutination assay, enzyme-linked immunosorbent assay (ELISA) and western blot assay. Generally, a particle agglutination assay or ELISA is useful as a screening test, and western blot assay is used for the confirmation of the presence of serum antibody to HTLV-I. The demonstration of the monoclonal integration of HTLV-I proviral DNA by Southern blot analysis can lead to a definite diagnosis of ATL.

Clinical Course and Treatment

ATL most often pursues the prototypic acute course originally described by Takatsuki and his colleagues (*15*, *25*), however, approximately one fourth of the patients show a more indolent course (chronic and smoldering types), with disease predominantly limited to the peripheral blood and/or skin. These patients may experience multiple infections but can remain remarkably free of disease progression for many years (*11*). These indolent diseases frequently progress to full-blown acute or lymphomatous ATL, an event that is sometimes called the "crisis." There have been reports that various kinds of infectious episodes may predispose to the transformation from an indolent to an aggressive disease course.

Most patients with ATL are not curable with current treatment modalities, even at the limited stage of disease. In addition, no treatment has been shown to prevent disease progression to a more aggressive disease. Thus, patients with chronic- or smoldering-type ATL should be carefully watched for the development of infectious complications and for signs of disease progression to acute or lymphomatous ATL.

Most previously untreated patients with aggressive forms (acute or lymphoma type) of ATL die within weeks or months of the diagnosis without treatment. The treatment of patients with acute or lymphomatous ATL has not been very successful.

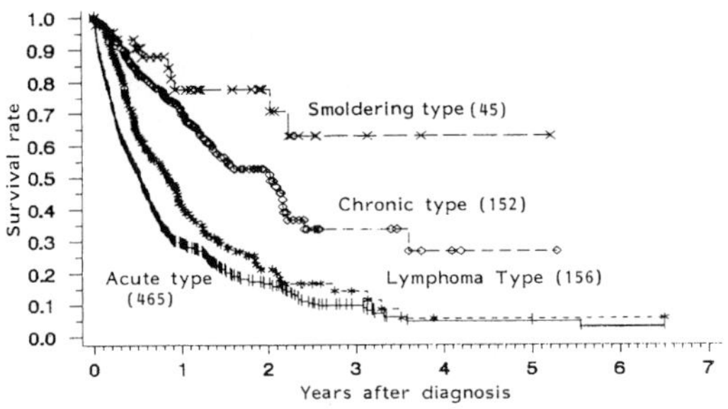

FIG. 1. Survival curves of 818 patients with ATL in Japan, according to four clinical subtypes determined by the diagnostic criteria

Numbers in parentheses indicate number of patients. From Shimoyama, M. and members of the Lymphoma Study Group [1984–1987]. *Br. J. Haematol.*, **79**, 428–437 (1991) with permission.

TABLE II. Significant Prognostic Factors of Each Clinical Subtype of ATL Based on a Multivariate Analysis[a]

Poor prognostic factor	All cases	Subtype			
		Acute	Lymphoma	Chronic	Smoldering
Involved lesion ≥4	0.0016				
Hypercalcemia	<0.0001				
Age ≥40	<0.0001	0.0002			
Performance status 2–4[b]	<0.001	<0.0001	<0.0001		
High LDH	<0.0001	<0.0001		<0.0001	
High BUN		0.0011		0.0065	0.0047
Low albumin				<0.0001	

[a] From Lymphoma Study Group: Major prognostic factors of patients with adult T-cell lenkemia-lymphoma: a cooperative study. *Leuk. Res.*, **15**, 81 (1991). Copyright 1991 Elsevier Science, with permission.
[b] Performance status grade was defined by WHO criteria.
Abbreviations: ATL, adult T-cell leukemia-lymphoma; LDH, lactate dehydrogenase; BUN, blood urea nitrogen; WHO, World Health Organization.

Figure 1 shows the survival curves of 818 ATL patients in Japan according to the four clinical subtypes (*11*). Eighty-five percent of the patients received chemotherapy with one of a variety of different regimens. Most of the patients with the smoldering type lived well without chemotherapy for a long time. Approximately two-thirds of the chronic-type patients died within about 2.5 years from diagnosis. Patients with the lymphoma type had poor prognoses, with a median survival time of 10.2 months. The most aggressive type of ATL was acute-type, with a median survival time of 6.2 months. The projected 4-year survival rates of patients with the lymphoma and acute types were only 5%. The clinical subtype clearly determines the prognosis of each patient, suggesting that it can be used as a prognostic indicator in ATL patients (*11*).

The major prognostic factors for each clinical subtype of ATL are summarized in Table II (*6, 11*). In all cases of ATL, high age (≥40 years), poor performance status (≥2), high LDH, and hypercalcemia and four or more involved lesions were unfavorable prognostic factors. For patients with chronic-type ATL, the major prognostic factors were the serum values of LDH, albumin and blood urea nitrogen (BUN). Patients with chronic-type ATL and normal values for the three factors showed a prognosis as good as those with smoldering-type ATL, and accounted for 30% of the patients with the chronic type. Thus, patients with the favorable chronic type with normal LDH, albumin and BUN values need not be treated immediately and can be placed on follow-up without treatment, whereas patients with the unfavorable chronic type having an abnormal value in at least one of the three factors would be targets for cytotoxic chemotherapy.

Clinical Trials for ATL by the Lymphoma Study Group of the Japan Clinical Oncology Group (JCOG-LSG)

Six chemotherapy trials have been consecutively conducted by the JCOG-LSG since 1978, focusing on ATL (*5, 12, 13, 21, 22, 30*). The first trial, called LSG1 protocol (1978–1980), utilized "VEPA" therapy, which consisted of vincristine (VCR), cyclophosphamide (CPA), prednisolone (PSL), and doxorubicin (DOX). In this study, patients with NHL at an advanced stage, including ATL, were enrolled. The complete response (CR) rate was lowest (18%) in ATL, intermediate (36%) in peripheral T-lymphoma other than ATL, and highest (64%) in B-lymphoma (*5*). Between 1981 and 1983, the JCOG-LSG conducted a randomized phase III trial, LSG1-VEPA versus LSG2-VEPA-M (VEPA+methotrexate) against advanced NHL including ATL (*12, 13*). Patients' sera were examined for anti-HTLV-I antibody to distinguish ATL from peripheral non-ATL T-lymphoma (PNTL) (*14*). The CR rate of LSG2-VEPA-M for ATL (37%) was higher than that of LSG1-VEPA (17%) (P=0.09). In the LSG1/LSG2 trial, however, the CR rate was significantly lower for ATL than for B-lymphoma and PNTL (P<0.001). The median survival time (MST) of the 54 patients with ATL treated with LSG1/LSG2 were six months, and the estimated four-year survival rate was only 8% (*12, 13*). These results indicate that the first-generation combination chemotherapy was not very effective against ATL.

In 1987, against advanced aggressive NHL including ATL, the JCOG-LSG initiated a combination phase II study (JCOG8701) of a new second-generation combination chemotherapy called LSG4, consisting of three different regimens: (1) VEPA-B, i.e., VCR, CPA, PSL, DOX and bleomycin (BLM); (2) M-FEPA, i.e., methotrexate (MTX), vindesine (VDS), CPA, PSL and DOX; (3) VEPP-B, i.e., VCR, etoposide (ETP), procarbazine (PCZ), PSL and BLM (*21*). The overall CR rate (72%) of the LSG4 protocol in aggressive NHL patients was significantly higher than that of the LSG1/LSG2 trial (57%) (P<0.05). The CR rate for ATL was improved from 28% (LSG1/LSG2) to 43% (LSG4). However, the CR rate was significantly lower in ATL than in B-lymphoma and PNTL (P<0.01). The patients with ATL still showed a poor prognosis with an MST of eight months and a four-year survival rate of 12%. However, the continued CR rate was increased to 12% (5/43) as compared to 4% (2/54) in the LSG1/LSG2 trial (P=0.13). A multivariate analysis of the 267 eligible patients with advanced aggressive NHL who were treated with the LSG4 protocol demonstrated that the clinical diagnosis of ATL was the most significant unfavorable prognostic factor (relative risk 3,185; P=0.0001) for the aggressive NHL patients (*21*).

The disappointing results against ATL patients with conventional chemotherapies have led to the search for new active agents.

2′-Deoxycoformycin (DCF), which is an irreversible inhibitor of adenosine deaminase, has been shown to be effective in a number of lymphoid malignancies. Based on the promising results of some single-institute studies of DCF against ATL, multicenter phase I and phase II studies of DCF as a single agent were conducted against ATL in Japan (*10, 20*). The phase II study of DCF revealed a response rate of 32% (10/31) in relapsed or refractory ATL, using the weekly intravenous administration of 5 mg/m^2. Two patients achieved CR and eight achieved PR (*10*).

These encouraging results prompted us to conduct a DCF-containing combination phase II trial (JCOG9109; LSG11 protocol) as initial chemotherapy for ATL (*22*). In this trial, ATL patients with acute, lymphoma and unfavorable chronic types were eligible. Between 1991 and 1993, 62 previously untreated patients with ATL (34 patients of acute, 21 of lymphoma and seven of chronic types) were enrolled, but two of them were ineligible because they were judged to be of the favorable chronic type. VCR 1 mg/m^2 intravenously (i.v.) on days 1 and 8, DOX 40 mg/m^2 i.v. on day 1, ETP 100 mg/m^2 i.v. on days 1–3, PSL 40 mg/m^2 *per os* (p.o.) on days 1 and 2 and

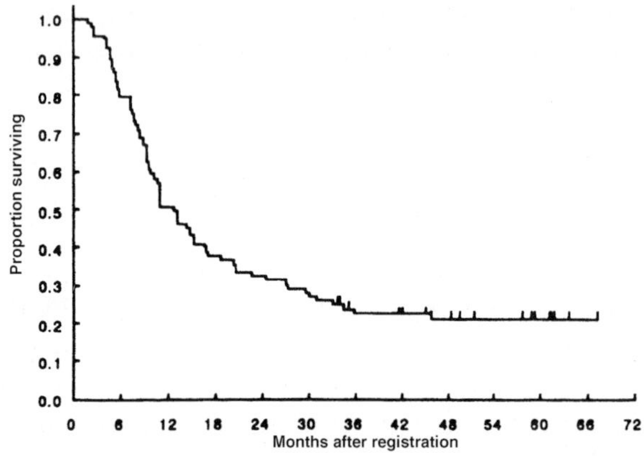

FIG. 2. Kaplan-Meier estimate of the overall survival (OS) for the 93 eligible patients with aggressive ATL

OS was defined as the time from registration until death from any cause or until the last follow-up evaluation for patients who were still alive (20 patients). From Yamada, Y. *et al. Br. J. Haematol.*, **113**, 375–382 (2001) with permission.

DCF 5 mg/m^2 i.v. on days 8, 15 and 22 were administered every 28 days for 10 cycles unless disease progression occurred. Among the 61 patients evaluable for toxicity, four patients (7%) died of fatal infections [two sepsis and two cytomegalovirus (CMV) pneumonia]. Other fatal non-hematological toxicities did not occur. In the 60 eligible patients, there were 17 CR [28%; 95% confidence interval (CI) 19–41%] and 14 PR (overall response rate 52%; 95% CI 39–64%). After a median observation time of 27 months, the MST was 7.4 months, and the estimated 2-year survival was 17%, data which were identical to those of the 43 ATL patients who were treated with the previous LSG4 protocol (JCOG8701) (*21, 22*). The following conclusions were obtained from the JCOG9109 study. First, ATL patients who were treated with a DCF-containing five-drug regimen (the LSG11 protocol) showed survival data comparable to those treated with a nine-drug second-generation combination chemotherapy (the LSG4 protocol). Second, the prognosis of the ATL patients was still poor even though they were treated with a DCF-containing combination chemotherapy.

In 1994, based on the promising results of a single-institute feasibility study, the JCOG-LSG initiated a new multiagent combination phase II trial (JCOG9303; LSG15 protocol); an eight-drug regimen consisting of VCR, CPA, DOX, PSL, nimustine (MCNU), VDS, ETP, and carboplatin (CBDCA), for untreated patients with ATL (*30*). In this trial, the elevation of relative dose intensity was attempted with the prophylactic use of granulocyte colony-stimulating factor (G-CSF). In addition, non-cross-resistant agents such as MCNU and CBDCA were incorporated to overcome multidrug resistance. Ninety-six previously untreated patients with aggressive ATL were enrolled: acute type (58), lymphoma type (28) and unfavorable chronic type (10). Eighty-one percent of the 93 eligible patients responded [75/93, 95% CI 71–88%], with 33 patients obtaining CR (35%) and 42 obtaining PR (45%). Patients with lymphoma type showed a better CR rate (67%, 18/27) than acute-type patients (20%, 11/56) and unfavorable chronic-type patients (40%, 4/10). Overall survival of 93 eligible patients at two years was estimated to be 31% (Fig. 2). The MST after

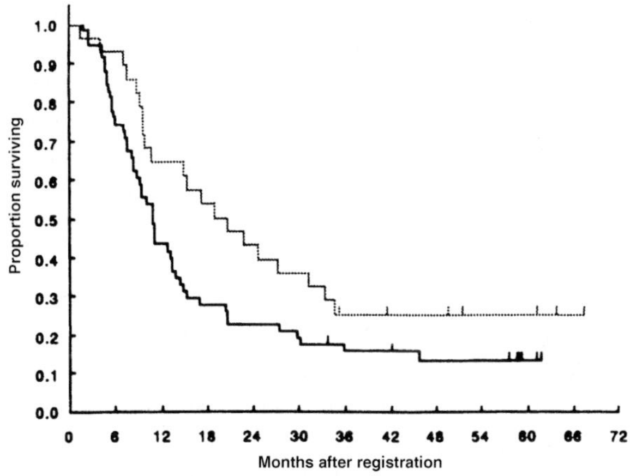

FIG. 3. Survival by clinical subtypes in the JCOG9303 study
Survival curves of acute-type ATL and lymphoma-type ATL are shown. A trend towards better survival for lymphoma-type patients compared with acute-type patients was observed (hazard ratio, 1.65). From Yamada, Y. *et al. Br. J. Haematol.*, **113**, 375–382 (2001) with permission.

registration was 13 months and the median follow-up duration of the 20 surviving patients was 4.2 years (range 2.8–5.6). Survival curves by clinical subtypes are shown in Fig. 3. A trend towards better survival for lymphoma-type patients (MST, 20 months) compared with acute-type patients (MST, 11 months) was recognized (hazard ratio, 1.65). These results may suggest that lymphoma type is the most appropriate target of the LSG15 protocol. Grade 4 hematological toxicities of neutropenia and thrombocytopenia were observed in 65% and 53% of the patients respectively, but grade 4 non-hematological toxicity was observed in only one patient. We concluded that LSG15 protocol is feasible with mild non-hematological toxicity and improved the clinical outcome of ATL patients (*30*). To confirm whether the LSG15 is a new standard for the treatment of aggressive ATL patients, JCOG-LSG is conducting a randomized phase III study comparing the LSG15 with biweekly CHOP (CPA, DOX, VCR, PSL), eight cycles of CHOP every two weeks. In both regimens, elevation of relative dose-intensity is attempted with the prophylactic use of G-CSF.

New Agents for ATL under Investigation

In addition to DCF, several types of new agents against ATL have been investigated.

Sobuzoxane (MST-16) is a bis(2,6-dioxopiperazine) analogue that inhibits topoisomerase II. Sobuzoxane is bioavailable following oral administration and has myelosuppression as its dose-limiting toxicity. In a phase I-II study of sobuzoxane, which was administered orally at 1,200–2,800 mg/day for seven days, the response rate was reported to be 44% of 23 patients with ATL, including two CR and eight PR (*7*). Unfortunately, the duration of remission was generally short (on the order of a few months).

Irinotecan hydrochloride (CPT-11) is a semisynthetic analogue of camptothecin with inhibitory activity against topoisomerase I. Preclinical studies of CPT-11 have suggested a lack of cross-resistance between topoisomerase I inhibitors and other anticancer agents. It has been demonstrated that CPT-11 has definitive activity against various kinds of solid tumors, including colorectal cancer and lung cancer. Multicenter, early and late phase II studies of CPT-11 were conducted against relapsed or refractory NHL in Japan (*8, 24*). The study revealed nine CR and 17 PR [response rate 38% (26/69)], using a weekly intravenous administration of 40 mg/m^2/day for three consecutive days. Among them, five (38%) of the 13 patients with ATL who were assessable for therapeutic efficacy showed a response to CPT-11 (1 CR and 4 PR) (*24*). The major toxicities of CPT-11 were leukopenia, diarrhea and nausea/vomiting. To develop a new effective chemotherapy regimen against NHL and ATL, we conducted two kinds of phase I/II study of CPT-11 in combination with CBDCA or ETP for relapsed or refractory NHL (*9, 17*). However, in both studies, dose-escalation was halted because of hematological toxicity in combination with CBDCA and hepatotoxocity in combination with ETP. We concluded that CBDCA and ETP are not recommendable agents in combination with CPT-11 for relapsed or refractory lymphoid malignancies.

Based on the preliminary documentation of the efficacy of interferon-α against ATL, two kinds of phase II trials of high-dose interferon-α (intravenous administration and subcutaneous administration) were conducted; however, the results were not impressive. In 1995, Gill and his colleagues in the United States reported that 11 of the 19 patients with acute or lymphomatous ATL achieved major responses (five CR and six PR) by the combination therapy of interferon-α and zidovudine (*1*). The therapeutic efficacy of this combination was also observed by a French group; the major objective responses were obtained in all five ATL patients (four with acute type and one with smoldering type) treated with this combination (*2*). Although the results of the combination with interferon-α and zidovudine are encouraging, the overall survival of previously untreated ATL patients was rather short (4.8 months) (*1, 18*) compared to those in the JCOG-LSG (7–13 months) (*12, 13, 21, 22, 30*). Furthermore, the CR-rate by the use of interferon-α and zidovudine in previously untreated patients (25%, 3/12) was not superior to the CR-rates in those treated with the JCOG8701, JCOG9109 and JCOG9303 (28–42%) (*1, 18, 21, 22, 30*). Thus, it is unclear whether this combination has a role in the frontline treatment of aggressive ATL. Alternatively, the combination may be one of the promising salvage treatments for ATL patients who do not show responses to cytotoxic chemotherapy, because four of the seven patients with no response to prior cytotoxic chemotherapy showed major responses to the combination (*1*). To evaluate the optimal role of this combination in the treatment of ATL, further large-scale studies are needed.

Cladribine (2-chlorodeoxyadenosine, 2-CdA) is a chlorinated purine analogue that is resistant to degradation by adenosine deaminase. Cladribine has been found to be effective against various indolent B-cell malignancies such as hairy cell leukemia, B-chronic lymphocytic leukemia and indolent B-lymphoma. It is known that deoxycytidine kinase is rich in T-cells, and an *in vitro* study showed the sensitivity of T-lymphoblastoid cell lines to cladribine. In addition, cladribine has been reported to be effective against cutaneous T-cell lymphoma; therefore, it could have some benefit in the treatment of ATL. In the Japanese phase I study of cladribine against relapsed or refratory lymphoid malignancies, one refractory ATL patient after receiving sobuzoxane achieved PR (*19*). Based on this encouraging result, a multiinstitutional phase II study of cladribine against ATL was conducted in Japan (*23*). Between July 1996 and December 1998, 16 patients with relapsed or refractory ATL were enrolled in the phase II study of cladribine against ATL, and 15 of them were eligible. Cladribine was administered at 0.09 mg/kg/day by 7-day continuous intravenous infusion every 28 days up to 3 courses. However, the interim analysis revealed that only one of the 15 eligible patients showed PR [overall response rate (ORR) 7%; 90% CI 0–28%]. As the upper limit of the 90% CI of the ORR did not reach 30%, the independent monitoring committee advised us to discontinue the patient enrollment. We concluded that cladribine showed objective response in a fraction of ATL patients, however, its role in the treatment of ATL is limited with this dose and schedule.

Since most ATL cells express the α-chain of IL-2R (CD25), Waldmann and his colleagues have treated ATL patients with monoclonal antibodies to CD25 (*27*). Anti-Tac (anti-CD25) has been shown to prevent the growth of certain cell lines *in vitro*, even in the absence of complement, by blocking IL-2 from gaining access to its receptor. Six (32%) of 19 patients who were treated with anti-Tac showed PR (four patients), or CR (two patients), lasting from nine weeks to more than three years. One of the significant impediments to this approach is that a quantity of soluble IL-2R is shed by the tumor cells into the circulation. The soluble IL-2R can bind to anti-Tac and inhibit binding to the tumor cell.

Other strategies using the IL-2R as a target for the treatment of ATL are conjugation with an immunotoxin (*Pseudomonas* exotoxin) or radioisotope (90-yttrium). Anti-Tac coupled with *Pseudomonas* exotoxin, which inhibits protein synthesis, has been administered to patients with ATL (*27*). Although relatively low doses were used, tumor regressions were observed in some patients. The action of immunotoxins depends on the expression of the target antigen on all malignant cells and on the cell's ability to internalize the antigen-antibody-complex containing the toxin. To circumvent facts that not all malignant cells express the target antigen, and not all cells internalize bound substances, radiolabeled monoclonal antibodies (radioimmunoconjugates) were developed. Radioimmunoconjugates have the advantage of killing adjacent antigen-negative neoplastic cells or cells that fail to internalize the antigen-antibody complex. Waldmann and his colleagues developed a stable conjugate of anti-Tac with 90-yttrium (*28*). They have treated 18 ATL patients with this radioimmunoconjugate. Among the 16 ATL patients who received 5- to 15-mCi doses, nine (56%) showed major objective responses (two CR and seven PR). The response duration was longer than the previous results with unconjugated anti-Tac antibody. Grade 3 or more toxicities were largely limited to hematological toxicities. They concluded that radioimmunotherapy with ^{90}Y-labeled anti-Tac may provide a useful approach for the treatment of ATL (*28*).

One of the potentially promising strategies for developing a new treatment against ATL is to circumvent drug resistance (*3, 4*). Kuwazuru *et al.* analyzed the expression of P-glycoprotein (P-gp) in samples from 25 ATL patients, by immunoblotting with a monoclonal antibody against P-gp (*3*). Neoplastic cells from all six patients at the relapsed stage were P-gp positive. More importantly, neoplastic cells from eight of 20 ATL patients expressed P-gp at the initial presentation. These results suggest that the expression of multidrug-resistant (*mdr1*)/P-gp may be correlated with the refractory nature of ATL cells to cytotoxic chemotherapy. Subsequently, Lau *et al.* reported the results of their investigation of the presence of an active multidrug-resistance phenotype in freshly isolated peripheral blood mononuclear cells from HTLV-I-infected subjects, including asymptomatic HTLV-I carriers, patients with tropical spastic paraparesis/HTLV-I associated myelopathy (TSP/HAM), and those with ATL (*4*). Significant P-gp-mediated efflux activity and enhanced *mdr1* mRNA expression were observed in CD3+ T-cell populations from nine of 10 subjects. Furthermore, it was found that *mdr1* gene promoter is transcriptionally activated by the HTLV-I *tax* protein. These observations suggest the possibility of new chemotherapeutic strategies against ATL through the use of P-gp inhibitors.

Despite the various clinical trials to find a more effective chemotherapy regimen, including those conducted by the JCOG-LSG, most ATL patients are still incurable with currently available chemotherapies. Further efforts to incorporate new innovative treatment modalities are needed such as new anticancer agents, molecular-targetting therapies monoclonal antibody therapies and allogeneic hematopoietic stem cell transplantation (HSCT) including non-myeloablative HSCT.

Acknowledgments

The clinical trials by the JCOG-LSG were supported by Grants-in-Aid for Cancer Research (2S-1, 5S-1, 8S-1, 11S-1, 1-1 and 4-5) and for a New 10-Year Strategy for Cancer Control from the Ministry of Health, Labour and Welfare, Japan. The author is grateful to all investigators who participated in the multicenter trials by the JCOG-LSG.

REFERENCES

1. Gill, P. S., Harrington, W., Kaplan, M. H., Ribeiro, R. C., Bennett, J. M., Liebman, H. A., Bernstein-Singer, M., Espina, B. M., Cabral, L., and Allen, S. Treatment of adult T-cell leukemia-lymphoma with a combination of interferon alfa and zidovudine. *N. Engl. J. Med.*, **332**, 1744–1748 (1995).
2. Hermine, O., Blouscary, D., Gessain, A., Turlure, P., Leblond, V., Franck, N., Buzyn-Veil, A., Rio, B., Macintyre, E., and Dreyfus, F. Treatment of adult T-cell leukemia-lymphoma with zidovudine and interferon alfa. *N. Engl. J. Med.*, **332**, 1749–1751 (1995).
3. Kuwazuru, Y., Hanada, S., Furukawa, T., Yoshimura, A., Sumizawa, T., Utsunomiya, A., Ishibashi, K., Saito, T., Uozumi, K., and Maruyama, M. Expression of p-glycoprotein in adult T-cell leukemia cells. *Blood*, **76**, 2065–2071 (1990).
4. Lau, A., Nightingale, S., Taylor, G. P., Gant, T. W., and Cann, A. J. Enhanced MDR1 gene expression in human T-cell leukemia virus-I-infected patients offers new prospects for therapy. *Blood*, **91**, 2467–2474 (1998).
5. Lymphoma Study Group. Final results of cooperative study of VEPA [vincristine, cyclophosphamide (Endoxan), prednisolone and adriamycin] therapy in advanced adult non-Hodgkin's lymphoma: relation between T- or B-cell phenotype and response. *Jpn. J. Clin. Oncol.*, **12**, 227–238 (1982).
6. Lymphoma Study Group. Major prognostic factors of patients with adult T-cell leukemia-lymphoma: a cooperative study. *Leuk. Res.*, **15**, 81–90 (1991).
7. Ohno, R., Masaoka, T., Shirakawa, S., Sakamoto, S., Hirano, M., Hanada, S., Yasunaga, K., Yokomaku, S., Mitomo, Y., and Nagai, K. Treatment of adult T-cell leukemia/lymphoma with MST-16, a new oral antitumor drug and a derivative of bis(2,6-diox-

opiperazine). *Cancer*, **71**, 2217–2221 (1993).
8. Ohno, R., Okada, K., Masaoka, T., Kuramoto, A., Arima, T., Yoshida, Y., Ariyoshi, H., Ichimaru, M., Sakai, Y., and Oguro, M. An early phase II study of CPT-11: a new derivative of camptothecin, for the treatment of leukemia and lymphoma. *J. Clin. Oncol.*, **8**, 1907–1912 (1990).
9. Ohtsu, T., Sasaki, Y., Igarashi, T., Murayama, T., Kobayashi, Y., and Tobinai, K. Unexpected hepatotoxicities in patients with non-Hodgkin's lymphoma treated with irinotecan (CPT-11) and etoposide. *Jpn. J. Clin. Oncol.*, **28**, 502–506 (1998).
10. Shimoyama, M. Chemotherapy of ATL. *In* "Adult T-cell Leukaemia," ed. K. Takatsuki, pp. 221–237 (1994). Oxford University Press, Oxford.
11. Shimoyama, M. and members of the Lymphoma Study Group (1984–1987). Diagnostic criteria and classification of clinical subtypes of adult T-cell leukemia-lymphoma. *Br. J. Haematol.*, **79**, 428–437 (1991).
12. Shimoyama, M., Ota, K., Kikuchi, M., Yunoki, K., Konda, S., Takatsuki, K., Ichimaru, M., Ogawa, M., Kimura, I., Tominaga, S., Tsugane, S., Taguchi, H., Minato, K., Takenaka, T., Tobinai, K., Kurita, S., Oyama, A., Hisano, S., Kozuru, M., Matsumoto, M., Nomura, K., Takiguchi, T., Sugai, S., Yamaguchi, K., Hattori, T., Kinoshita, K., Tajima, K., Suemasu, K. for the Lymphoma Study Group. Chemotherapeutic results and prognostic factors of patients with advanced non-Hodgkin's lymphoma treated with VEPA or VEPA-M. *J. Clin. Oncol.*, **6**, 128–141 (1988).
13. Shimoyama, M., Ota, K., Kikuchi, M., Yunoki, K., Konda, S., Takatsuki, K., Ichimaru, M., Tominaga, S., Tsugane, S., Minato, K., Tobinai, K., Oyama, A., Hisano, S., Matsu-moto, M., Takiguchi, T., Yamaguchi, K., Kinoshita, K., Tajima, K., Suemasu, K. for the Lymphoma Study Group (1981–1983). Major prognostic factors of adult patients with advanced T-cell lymphoma/leukemia. *J. Clin. Oncol.*, **6**, 1088–1097 (1988).
14. Shimoyama, M., Oyama, A., Tajima, K., Tobinai, K., Minato, K., Takenaka, T., Konda, C., Takeyama, K., Kohno, A., and Narabayashi, M. Differences in clinicopathological characteristics and major prognostic factors between B-lymphoma and peripheral T-lymphoma excluding adult T-cell leukemia/lymphoma. *Leuk. Lymphoma*, **10**, 335–342 (1993).
15. Takatsuki, K., Uchiyama, T., Sagawa, K., and Yodoi, J. Adult T cell leukemia in Japan. *In* "Topics in Hematology," ed. S. Seno, F. Takaku, and S. Irino, pp. 73–77 (1977). Excerpta Medica, Amsterdam.
16. Teshima, T., Akashi, K., Shibuya, T., Taniguchi, S., Okamura, T., Harada, M., Sumida, I., Hanada, M., and Niho, Y. Central nervous system involvement in adult T-cell leukemia/lymphoma. *Cancer*, **65**, 327–332 (1990).
17. Tobinai, K., Hotta, T., Saito, H., Ohnishi, K., Ohno, R., Ogura, M., Ariyoshi, Y., Takeyama, K., Kobayashi, T., Ohashi, Y., and Shirakawa, S. Combination phase I/II study of irinotecan hydrochloride and carboplatin in relapsed or refractory non-Hodgkin's lymphoma. *Jpn. J. Clin. Oncol.*, **26**, 455–460 (1996).
18. Tobinai, K., Kobayashi, Y., Shimoyama, M. (letter), Gill, P. S., Harrington, W., and Levine, A. M. (the authors' reply). Interferon alfa and zidovudine in adult T-cell leukemia-lymphoma (Correspondence). *N. Engl. J. Med.*, **333**, 1285–1286 (1995).
19. Tobinai, K., Ogura, M., Hotta, T., Kobayashi, Y., Narabayashi, M., Suzuki, R., Kinoshita, T., Kozuru, M., Uike, N., and Ohashi, Y. Phase I study of cladribine (2-chlorodeoxyadenosine) in lymphoid malignancies. *Jpn. J. Clin. Oncol.*, **27**, 146–153

(1997).
20. Tobinai, K., Shimoyama, M., Inoue, S., Takayasu, S., Mikuni, C., Kozuru, M., Oda, S., and Nakajima, H. Phase I study of YK-176 (2′-deoxycoformycin) in patients with adult T-cell leukemia-lymphoma. *Jpn. J. Clin. Oncol.*, **22**, 164–171 (1992).
21. Tobinai, K., Shimoyama, M., Minato, K., Shirakawa, S., Hotta, T., Fukuhara, S., Suzuki, H., Ichimaru, M., Sai, T., Konda, S., Mikuni, C., Toki, H., Kozuru, M., Deura, K., Takatsuki, K. for the Lymphoma Study Group, JCOG (1987–91). Japan Clinical Oncology Group phase II trial of second-generation "LSG4 protocol" in aggressive T- and B-lymphoma: a new predictive model for T- and B-lymphoma (Abstract). *Proc. Am. Soc. Clin. Oncol.*, **13**, 378 (1994).
22. Tobinai, K., Shimoyama, M., Tajima, K., Kozuru, M., Tomonaga, M., Araki, K., Kasai, M., Takatsuki, K., Tara, M., Hotta, T., Mikuni, C., members of the 30 institutions for the JCOG-Lymphoma Study Group. Deoxycoformycin-containing combination chemotherapy for adult T-cell leukemia-lymphoma: Japan Clinical Oncology Group (JCOG) Study #9109 (Abstract). *Proc. Am. Soc. Clin. Oncol.*, **14**, 390 (1995).
23. Tobinai, K., Uike, N., Saburi, Y., Ohashi, Y. and members for the Cladribine/ATL Study Group. Phase II study of cladribine in adult T-cell leukemia-lymphoma. *Proc. Am. Soc. Clin. Oncol.*, **19**, Abstr. #149 (2000).
24. Tsuda, H., Takatsuki, K., Ohno, R., Masaoka, T., Okada, K., Shirakawa, S., Ohashi, Y., and Ota, K. Treatment of adult T-cell leukemia-lymphoma with irinotecan hydrochloride. *Br. J. Cancer*, **70**, 771–774 (1994).
25. Uchiyama, T., Yodoi, J., Sagawa, K., Takatsuki, K., and Uchino, H. Adult T-cell leukemia: clinical and hematologic features of 16 cases. *Blood*, **50**, 481–492 (1977).
26. Utsunomiya, A., Hanada, S., Terada, A., Kodama, M., Uematsu, T., Tsukasa, S., Hashimoto, S., and Tokunaga, M. Adult T-cell leukemia with leukemia cell infiltration into the gastrointestinal tract. *Cancer*, **61**, 824–828 (1988).
27. Waldmann, T. A. Multichain interleukin-2 receptor: a target for immunotherapy in lymphoma. *J. Natl. Cancer Inst.*, **81**, 914–923 (1989).
28. Waldmann, T. A., White, J. D., Carrasquillo, J. A., Reynolds, J. C., Paik, C. H., Gansow, O. A., Brechbiel, M. W., Jaffe, E. S., Fleisher, T. A., and Goldman, C. K. Radioimmunotherapy of interleukin-2Ra-expressing adult T-cell leukemia with yttrium-90-labeled anti-Tac. *Blood*, **86**, 4063–4075 (1995).
29. Yamada, Y., Kamihira, S., Murata, K., Yamamura, M., Maeda, T., Tsukasaki, K., Jubash, T., Atogami, S., Sohda, H., Taguchi, T., and Tomonaga, M. Frequent hepatic involvement in adult T-cell leukemia: comparison with non-Hodgkin's lymphoma. *Leuk. Lymphoma*, **26**, 327–335 (1997).
30. Yamada, Y., Tomonaga, M., Fukuda, H., Hanada, S., Utsunomiya, A., Tara, M., Sano, M., Ikeda, S., Takatsuki, K., Kozuru, M., Araki, K., Kawano, F., Niimi, M., Tobinai, K., Hotta, T., Shimoyama, M. and members of the JCOG-Lymphoma Study Group (LSG) (1994–1996). A new G-CSF-supported combination chemotherapy, LSG15, for adult T-cell leukemia-lymphoma (ATL): Japan Clinical Oncology Group (JCOG) Study 9303. *Br. J. Haematol.*, **113**, 375–382 (2001).
31. Yoshioka, R., Yamaguchi, K., Yoshinaga, T., and Takatsuki, K. Pulmonary complications in patients with adult T-cell leukemia. *Cancer*, **55**, 2491–2494 (1985).

EPIDEMIOLOGY

ETHNOEPIDEMIOLOGY OF ATL/HTLV IN MONGOLOIDS WITH SPECIAL REFERENCE TO MONGOLOID DISPERSAL

Kazuo TAJIMA and Toshiro TAKEZAKI

*Division of Epidemiology and Prevention, Aichi Cancer Center Research Institute**

The origins of human T-cell leukemia virus type I and type II (HTLV-I/II) in the world are not well understood, but are probably linked to a history of human dispersal with geographical clustering due to the natural transmission routes of HTLV. From the previous comprehensive research on genetic characteristics of Mongoloids in America, the Southern Japanese are similar to those of Andean highlanders with HTLV-I, but exclusively different from those of lowlanders with HTLV-II. To clarify the common origin of HTLV-I in Asia and Andes, we analyzed HTLV-I provirus DNA from Andean mummies about 1,500 years old. In 104 mummy bone marrow specimens, one out of two human *β-globin* gene positives produced bands of 159 bp and 157 bp of *HTLV-I-pX* and *-LTR* (long terminal repeat) genes. Nucleotide sequences of ancient HTLV-I-LTR clones isolated from mummy bone marrow were similar to those in contemporary Andeans and Japanese. To explain the unique distribution of HTLV-I/II among Mongoloids, one hypothetical model of migration history of HTLV-I and -II carriers in Pan-Pacific areas can be explored. Because of the similarity of genetic background between Southern Japanese and Andeans in South America, a paradoxical hypothesis which has not found general acceptance amongst anthropologists and archeologists seems plausible: HTLV-I-carrying people arrived in the Andes through a transpacific route after migration of HTLV-II-carrying Asians to the South American continent.

To confirm the geographic and ethnic segregation of human T-cell leukemia virus type I (HTLV-I) (*11, 23*) and type II (HTLV-II) (*14*) carriers in native Mongoloids on Asian and American continents, we have conducted a series of sero-epidemiological studies of HTLV during the last 10 years (*3, 5–7, 19, 35*). HTLV-I carriers are mainly distributed among Japanese in Asia and Andeans in South America. The geographic distribution of the foci of HTLV-I is different from those of HTLV-II (*26, 27*). HTLV-I is prevalent worldwide in Southwestern Japanese, native Andeans and North Iranians, as well as Central Africans, and African descendants in the Caribbean and South America (*4, 7, 8*). On the other hand, HTLV-II is known to be prevalent mainly in some native populations in North, Central and South America. To explain the unique distribution of HTLV-I/II among Mongoloids, one hypothetical model of migration history of HTLV-I and -II carriers can be explored.

* 1-1 Kanokoden, Chikusa-ku, Nagoya 464-8681, Japan

General Epidemiologic Features of HTLV-I/II

According to evidence obtained from a biennial nationwide survey of cases of adult T-cell leukemia/lymphoma (ATL) (*34*) carried out in Japan since 1986, the annual incidence is estimated to be 700 within approximately 1.2 million HTLV-I carriers in Japan (*30, 32, 33*). The affected individuals mainly inhabit the southern and northern ATL endemic areas. Among 3,499 ATL patients newly diagnosed during the 10 years from 1988 to 1997 (*33*), approximately 70% were born in the southwestern districts of Kyushu and Shikoku, and 10% in the northern districts of Hokkaido and Tohoku. The main reasons for this unique geographical clustering of ATL in Japan might be linked to a history of early Japanese migration with geographical clustering of HTLV-I carriers in the Japanese archipelago due to the natural transmission routes of mother-to-child through breast milk and man-to-woman through semen (*15, 20, 29*). In the ATL endemic areas in Japan a systematic prevention trial against HTLV-I transmission through breast milk from HTLV-I carrying mothers to newborn babies, e.g., strong recommendation to pregnant women with HTLV-I to completely stop breast feeding and/or to raise their babies by short-term breast feeding (>6 months) (*10, 31*).

In East Asian countries other than Japan, only a hunting-gathering people, the Aeta in the Philippines, have a conspicuously high rate of HTLV-I carriers (*13*). A recent study showed more than 10% of the Mashadi Jews in Northern Iran to carry the virus (*18*). In other Central Asian countries, sporadic distributions of HTLV-I

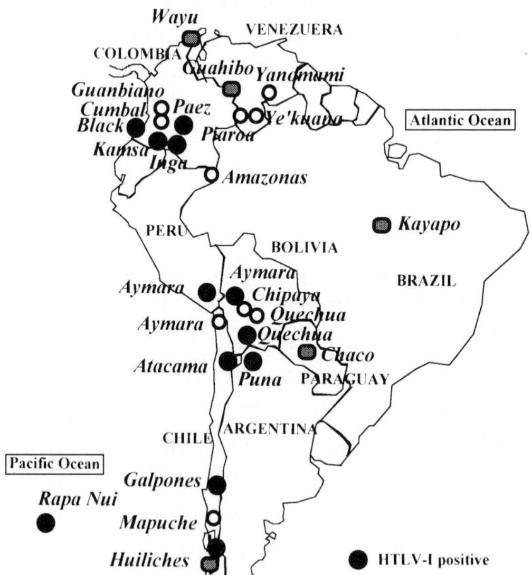

FIG. 1. Geographical distribution of positive anti-HTLV-I/II among South American native groups

Location and name of ethnic group are corresponding to the ethnicity and locality in Table I cited from Fujiyoshi *et al.* (*7*).

carriers and patients with ATL were disclosed recently among Chinese on the east coast of China, Singapore and Hong Kong, Koreans in Korea, Indians in Singapore and India, and Jewish in Israel and Eskimos in Alaska (*26*). We found no evidence of any HTLV-II clustering areas nor ethnic groups in Asian countries.

Since 1989 we have collected a total of 2,289 samples from 26 different ethnic groups in South America (*7*). HTLV-I positive cases were found in indigenous groups in the Andes highlands, while HTLV-II positive cases were broadly detected in lowland ethnic groups (Fig. 1). Incidence rates of HTLV-I and HTLV-II carriers varied with each ethnic group ranging from 0.8% to 27.3% for HTLV-I seropositivity and 1.4% to 57.9% for HTLV-II seropositivity (Table I). These results revealed that HTLV-II was broadly distributed among native populations on an Atlantic coast-Orinoco-Amazon-Grand Chaco-Patagonia axis, but that HTLV-I was clustered only among limited native groups in Andes areas. Thus, the characteristic geographical distribution of HTLV-I/II showed mutually exclusive foci in the South American continent, that is, HTLV-I foci in the Andes highlands and HTLV-II foci in broad lowlands from the north to southernmost South America.

Contrasting HTLV-I Related Diseases

The annual incidence rate of ATL in endemic areas of Japan is approximately 1/1,000 in male adult HTLV-I carriers and 1/2,000 in females over 40 years old of age (*30, 32*), not very different from findings in Jamaica. HTLV-I associated myelopathy/tropical spastic paraparesis (HAM/TSP) is the second most common disease caused by HTLV-I infection (*8, 22*). The main immunological features of HAM/TSP are the presence of high titers of anti-HTLV-I antibodies in serum and cerebrospinal fluid, a high proviral load in the peripheral blood cells, and the presence of cytotoxic T-cells which recognize epitopes of the *tax* gene product (*22*). A small number of circulating abnormal flower-like lymphocytes is observed in many patients, which may cause confusion with the chronic type of ATL. The pathogenesis of HAM/TSP is uncertain, but some specific type of human leukocyte antigen (HLA) variability, associated with viral load and the immune response, is considered as a potential factor determining susceptibility to disease development (*25*).

A previous study showed that most ATL patients have specific HLA haplotypes found only in a minority Japanese subpopulation (*32*), while HAM/TSP patients share common HLA haplotypes appearing in the majority of Japanese (*25*). In Japan the number of cases of ATL exceeds that of HAM/TSP, but HAM/TSP is more prevalent in Chile (Table II) (*1, 2, 21, 33*). In both countries, the ratio of males to females is more than 1.2 for ATL patients and less than 0.5 for HAM/TSP cases. The previous case reports of ATL and HAM/TSP in Chile indicated that the risk of HAM/TSP there is relatively higher than that in Japan, even though many more cases of ATL were reported in Japan. Such a geographical discrepancy between ATL and HAM/TSP might be explained by a different genetic background between Japanese and Chilean carriers of HTLV-I. All Japanese carriers of HTLV-I at risk of HTLV-I related diseases are Mongoloids, whereas recent Chilean carriers are of mixed Mongoloid and Caucasoid stock. Thus present-day Chilean carriers of HTLV-I share a genetic background with Caucasoid people which is very different from that of the original South American Mongoloids.

TABLE I. Summary Results of Serological Tests for HTLV-I/II in South American Native Groups

Country	Ethnicity	Locality (Area)	Date (Mo/Yr) of sampling	Number of subjects	Age range (years)	Number of seropositives HTLV-I (%)	HTLV-II (%)	M/F[f] ratio
Colombia	Paez[a]	Paez (Andes)	08/1989	32	18–53	2 (6.3%)	0	5/27
	Guambiano[a]	Silvia (Andes)	08/1989	53	18–53	0	0	20/33
	Inga[d]	Aponte (Andes)	08/1990	48	15–65	1 (2.1%)	0	15/33
	Kamsa[b]	Sibundoy (Andes)	08/1991	59	11–71	4 (6.8%)	0	7/52
	Cumbal[d]	Cumbal (Andes)	08/1991	55	16–60	0	0	14/41
	Guahibo[d]	Puerto carreno (Orinoco)	08/1992	92	10–65	0	29 (31.5%)	44/48
	Wayu[d]	Guajira (Atlantic Coast)	08/1991 & 08/1992	123	8–80	0	5 (4.1%)	46/77
	Amazonas[a]	Leticia & Narino (Amazon)	11/1989	85	13–79	0	0	33/52
	Amazonas[e]	Leticia & La Pedrera (Amazon)	08/1990	80	14–62	0	0	39/41
	Black[e]	Guapi & Timbiqui (Pacific Coast)	08/1990	270	14–80	17 (6.3%)	0	90/180
Venezuela[e]	Yanomami	Majawana *et al.* (Gyana)	10/1993	129	10–60	0	0	74/55
	Ye'kuana	Chajorana (Orinoco)	10/1993	40	10–60	0	0	12/28
	Piaroa	Manapiare (Orinoco)	10/1993	30	14–60	0	0	13/17
Brazil[e]	Kayapo	Xingu (Amazonas)	09/1990	19	8–74	0	11 (57.9%)	NS
Peru[e]	Aymara	Carumas (Andes)	08/1997	28	19–74	1 (3.6%)	0	15/13
	Aymara	San Cristobal *et al.* (Andes)	08/1997	34	18–56	0	0	16/18
Bolivia[e]	Aymara	Huachacalla (Andes)	08/1996 & 08/1997	32	20–78	6 (18.8%)	0	18/14
	Aymara	Escara (Andes)	08/1996	7	48–77	1 (14.0%)	0	6/1
	Aymara	Achacachi (Andes)	08/1995	49	13–75	0	0	17/32
	Aymara	Sabaya *et al.* (Andes)	08/1997	12	35–90	0	0	4/8
	Aymara	Llica (Andes)	08/1997	51	6–78	2 (4.0%)	0	23/28
	Quechua	San Juan (Andes)	08/1997	26	12–76	3 (11.5%)	0	9/17
	Quechua	Copacabana (Andes)	08/1997	20	12–77	0	0	8/12
	Quechua	Alota (Andes)	08/1997	29	17–74	0	0	8/21
	Quechua	Quetena Chico (Andes)	08/1998	10	30–80	0	0	4/6
	Quechua	San Pablo de Lipez (Andes)	08/1998	11	32–65	3 (27.3%)	0	4/87
	Quechua	Candelaria (Andes valley)	08/1995	38	14–80	0	0	14/24
	Chipaya	Chipaya (Andes)	08/1999	9	20–60	0	0	6/3
Argentina[e]	Puna	Susques (Andes)	06/1997	40	18–56	1 (2.5%)	0	19/21
	Puna	Cochinoca(Andes)	10/1997	48	16–88	1 (2.1%)	0	13/35
Paraguay[e]	Chaco	Colonias Mennonitas (Chaco)	03/1998	146	14–75	0	24 (16.4%)	71/75
Chile	Aymara[e]	Putre *et al.* (Andes)	12/1992	57	18–82	0	0	24/33
	Atacama[e]	Toconao (Andes)	11/1992	29	19–78	4 (13.8%)	0	10/19
	Atacama[e]	San Pedro (Andes)	11/1992	44	19–89	5 (11.4%)	0	13/31
	Atacama[e]	Caspana *et al.* (Andes)	11/1992	87	17–89	0	0	30/57
	Mapuche[c]	Los Galpones *et al.* (Andes plain)	01/1991	133	12–81	1 (0.8%)	0	54/79
	Pehuenche[c]	Icalma (Andes)	01/1991	10	18–63	0	0	3/7
	Huiliches[c]	Chiloe Island (Pacific coast)	01/1991	70	13–83	1 (1.4%)	1 (1.4%)	29/41
	Alacalfes[e]	Wellington Island (Pacific coast)	11/1993	23	14–73	0	8 (34.8%)	9/14
	Yaganes[e]	Navarino Island (Patagonia)	12/1993	22	13–70	0	2 (9.1%)	9/13
	Rapa Nui[e]	Easter Island (Pacific ocean)	09/1996	132	7–86	1 (0.8%)	0	44/88

[a] Data reported by Zamora *et al.* (*35*).
[b] Data reported by Fujiyama *et al.* (*5*).
[c] Data reported by Cartier *et al.* (*3*).
[d] Data reported by Zaninovic *et al.* (*36*).
[e] Data reported by Fujiyoshi *et al.* (*7*).
[f] Male/Female ratios (NS: M/F was not specified).

Ancient HTLV-I Provirus DNA in Andean Mummies

The Museo Archeologia San Miguel (MASM) around Azapa valley and the Museo Padre le Paige (MPP) in Atacama desert (Fig. 2) preserve mummies excavated from ruins of cemeteries in North Chile. Each mummy was examined for age by carbon-14 (C^{14}) dating and body surfaces were thoroughly inspected for exogenous contamination, especially with human blood. We succeeded in extracting DNA from bone marrow materials in a couple of mummies out of the 104 selected (*16*). We isolated the DNA by employing a Geneclean Kit for Ancient DNA and general PCR amplification processing. The DNA samples were subjected to PCR amplification of human *β-globin*, *HTLV-I-pX* and *HTLV-I-LTR* (long terminal repeat) genes using two sets of primers according to a standard protocol. In parallel with mummy DNA

TABLE II. Comparison of Clinico-epidemiological Features of Adult T-cell Leukemia/Lymphoma (ATL) and HTLV-I Associated Myelopathy (HAM/TSP) between Japan and Chile

Comparative item	ATL		HAM/TSP	
	Japan[a]	Chile[b]	Japan[a]	Chile[b]
Period of registration	1988–97	1989–98	1987–94	1987–90
Number of cases	3,499	26	1,103	45
Average annual cases reported	350	2.6	137.9	11.1
Average age at onset (Years)	58.5	50.0	45.2	47.2
(Range)	(19–89)	(24–78)	(5–80)	(20–67)
Sex ratio (Male/Female)	1.22	1.00	0.46	0.45

[a] New cases were collected by nationwide registry from hospitals. Cited from the T- and B-cell Malignancy Study Group, 2001; Osame *et al.*, 1994.
[b] Reported cases rather than nationwide study from references. Cited from Cabrera *et al.*, 1998; Cartier *et al.*, 1992.

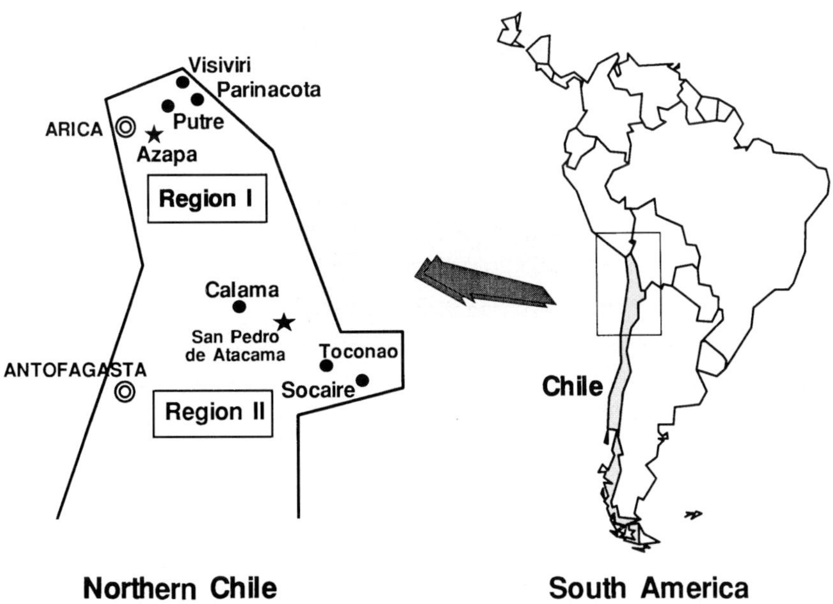

FIG. 2. Map of Northern Chile (1st and 2nd regions) showing locations in Table III
∗Site of mummy museum.

TABLE III. Antibody Profiles of Contemporary HTLV-I Carriers in Toconao and San Pedro de Atacama (SPA), Northern Chile

Number	Age Sex	Locality	PA[a] test	ELISA[b] (A)	IF[c] (MT-1)	HTLV-I/II WB[d]						
						p19	p24	p28	p53	rgp21	rgp46-I	rgp46-II
CHI-383	54F	Toconao	×128	0.124	<×10	+	+	+	+	+	+	−
CHI-384	49F	Toconao	>×512	2.747	×40	+	+	+	+	+	+	−
CHI-399	59F	Toconao	>×512	>3.000	×80	++	+	++	++	++	+	−
CHI-408	67M	Toconao	×64	>3.000	×80	++	+	+	+	+	+	−
CHI-413	89M	SPA	>×512	>3.000	×80	++	+	+	+	+	+	−
CHI-430	68F	SPA	>×512	>3.000	×80	++	++	++	++	++	++	−
CHI-431	43F	SPA	×512	2.248	×20	+	+	+	+	+	+	−
CHI-444	36F	SPA	>×512	>3.000	×80	++	++	+	+	++	+	−
CHI-430	68F	SPA	×256	0.955	<×10	+	+	+	+	+	+	−

[a] PA: Particle agglutination (Fujirebio, Tokyo).
[b] ELISA: Eitest ATL (Cut off point=0.123, Eizai, Tokyo).
[c] IF: Immunofluorescence test with MT-1 cells.
[d] HTLV-I/II WB: HTLV-I/II blot 2.4 (Diagnostic Biotechnology, Singapore).

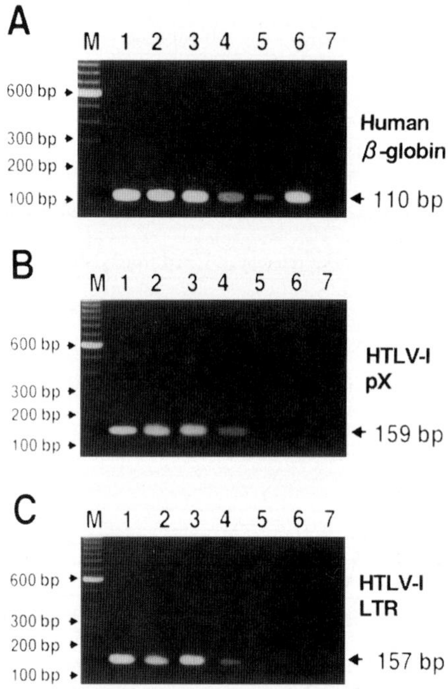

FIG. 3. PCR amplification of β-globin DNA (A), HTLV-I-pX DNA (B) and HTLV-I-LTR DNA (C) cited from Li *et al.* (*16*)

Lane M, molecular markers; lane 1, Japanese HTLV-I seropositive PBL (KAG-130); lane 2, Chilean HTLV-I seropositive PBL (CHI-430); lane 3, Chilean HTLV-I seropositive PBL (CHI-383); lane 4, SP-2 mummy bone marrow; lane 5, SP-11 mummy bone marrow; lane 6, Japanese HTLV-I seronegative PBL (KAG-991); lane 7, control without DNA sample.

amplification, 3 samples from contemporary Japanese and Chilean HTLV-I DNA were similarly amplified for human *β-globin*, *HTLV-I-pX* and *HTLV-I-LTR* genes.

We performed two independent PCR reactions for all DNA samples to detect sequence changes during PCR amplification. DNA samples extracted from all 104

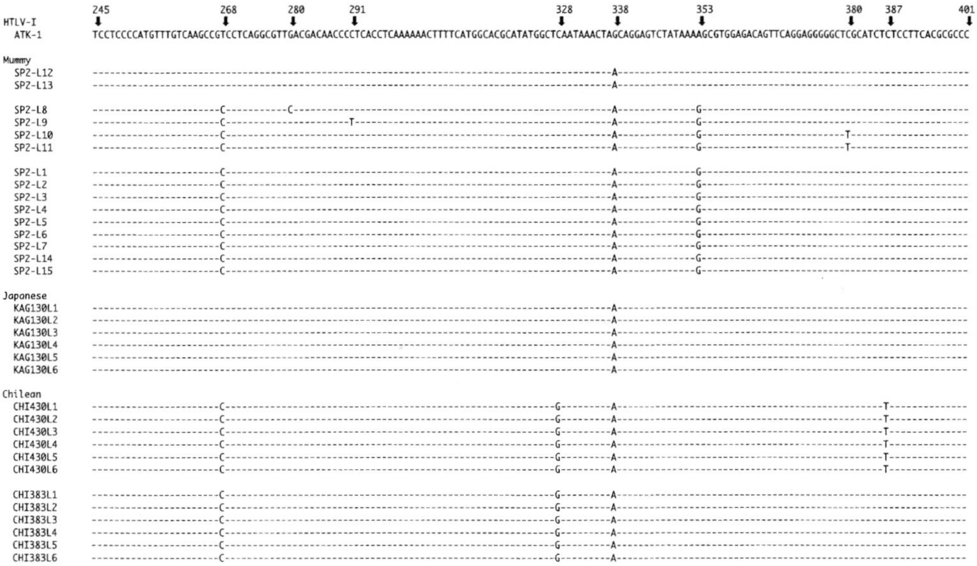

FIG. 4. Nucleotide sequences of PCR amplified HTLV-I-LTR DNA cited from Li *et al.* (*16*)
Nucleotide positions 245–401 (157 bp) were aligned by sequencing 15 SP-2 clones (SP2-L1–L15), 6 Japanese clones (KAG130-L1–L6) and 12 Chilean clones (CHI430-L1–L6, CHI383-L1–L6). The nucleotide sequences of the standard HTLV-I (ATK-1) (*24*) clones are shown in the top and bottom lines. SP-2 clones were categorized by the nucleotide substitution referring to the ATK-1 LTR sequence (see text); one substitution (category I), random substitutions (category II) and three substitutions at the 268, 338 and 353 base positions (category III). Chilean HTLV-I-LTR clones showed two unique substitutions at the 328 and 387 base positions as compared with the ATK-1 and Japanese HTLV-I-LTR sequences.

mummy bone marrow specimens and 3 contemporary DNA samples of peripheral blood lymphocytes (PBL) from two Chileans (Table III) and one Japanese HTLV-I seropositive individuals were amplified by PCR with 2 sets of primers for *β-globin* and *HTLV-I* genes. Two of the mummy DNAs from San Pedro de Atacama (SP-2 and SP-11) yielded bands of 110 base pair (bp) length corresponding with the PCR product of the human *β-globin* gene (Fig. 3A). One out of two (SP-2) produced bands of 159 bp and 157 bp length corresponding with the *HTLV-I-pX* and the *-LTR* genes (Fig. 3, B and C). The DNA samples from the remaining 102 cases of mummy bone marrow failed to provide any amplified sequences of β-globin and HTLV-I.

All the 9 clones except one exhibited exactly the same sequences in the 110 bp of β-globin DNA as that of the contemporary DNA clones. These results confirmed the DNA extracts from mummy bone marrow to be of human origin. In the next, we picked up a total 40 clones of the HTLV-I-pX DNA from the SP-2 mummy and two-thirds of the 22 SP-2 HTLV-I-pX DNA clones from mummy bone marrow showed the same sequence as the standard HTLV-I clone (ATK-1) (*24*) in the 159 bp of the gene, while the other 7 clones displayed two point substitutions. Finally, we picked up a total 33 clones of the HTLV-I-LTR DNA from the SP-2 mummy and contemporary PBL of Japanese and Chilean donors. Two of the 15 clones of the SP-2 HTLV-I-LTR DNA from mummy bone marrow (SP2-L12L–13) showed the same sequence as the contemporary Japanese HTLV-I-LTR clones (KAG-130), but

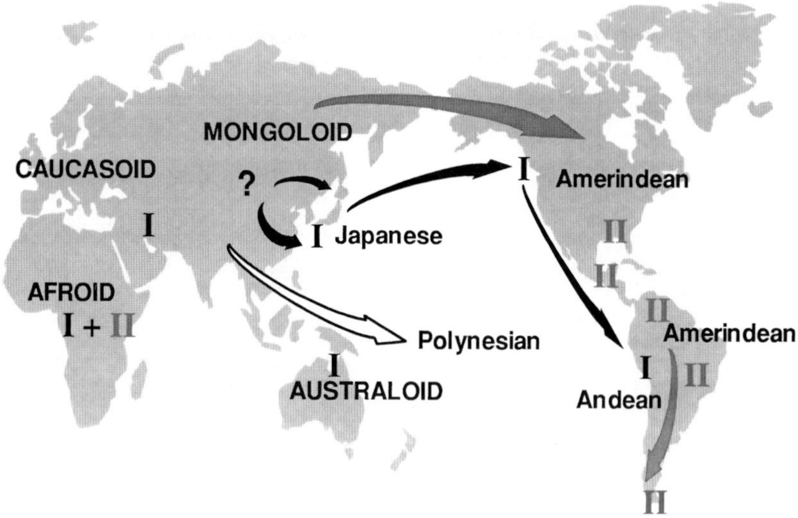

FIG. 5. Hypothetical migration routes of Asian Mongoloids from Asia to the American continent

the other 9 clones of SP-2 showed 2 point substitutions (Fig. 4). Twelve clones of the contemporary Chilean HTLV-I-LTR DNA (CHI-430 and CHI-383) showed 2 point substitutions from the Japanese HTLV-I-LTR DNA but showed one additional point substitution from the ATK-1 clone.

Recent studies of the phylogeny of HTLV-I focusing on genetic variation in the LTR have allowed classification into three major lineages: Melanesian, Central African and Cosmopolitan. The Cosmopolitan lineage can be further divided into four subgroups: Transcontinental, Japanese, West African and North African (*19*). The HTLV-I of contemporary Andeans from Northern Chile belongs to the Transcontinental subtype, this also being found in some Japanese subpopulations including the Ainu. To throw light on the unsolved issue of the origin of HTLV-I in Andes and Asian Mongoloids, we investigated ancient DNA of mummies from San Pedro de Atacama which were well preserved. The ancestral HTLV-I DNA sequence was likely preserved more than 1,000 years by virtue of the low frequency of HTLV-I replication *in vivo*, that is down-regulated by the specific function of HTLV-I *rex* gene. Taking these results together, the HTLV-I provirus of the SP-2 mummy might be the aboriginal HTLV-I prevailing among Mongoloid populations in Asia and Andes at least 1,500 years ago (*28*).

Asian Mongoloid Dispersal to South America

HTLV-I carriers in the Andes highlands were found to be remarkable with respect to ethnic clusters in Inga, Kamsa, Aymara, Quechua, Atakama, and Puna, distributed from Southern Colombia to Southeast Peru, Central-southwestern Bolivia, Northern Chile and Northern Argentina. Phylogenetic studies of HTLV-I isolates of Andean ethnic groups have shown that the native Amerindians harbor HTLV-I isolates of the transcontinental subgroup (subtype A), and there is some relation between the Japanese HTLV-I foci and Andes HTLV-I carriers. It is thus suggested that all South American HTLV-I carriers may have originated in the Andes high lands, genetically derived from a Mongoloid lineage carrying the Transcontinental subtype A of HTLV-I provirus DNA.

HTLV-II foci were found in the Brazilian Kayapo tribe, Paraguayan Chacos, Chilean Alacalfes and Yaganes, in line with a previous report of natural endemic foci in indigenous South American groups on the Atlantic coast and in the Orinoco lowlands of Colombia (*7*). Thus there is a clear geographic independence of HTLV-I and HTLV-II, the former mainly infecting Andes ethnic groups, and the latter non-Andean lowland ethnic groups. Furthermore, the incidence rate for HTLV-II carriers is generally higher than that for HTLV-I carriers (Table I), again suggesting independent and isolated populations.

HTLV-I and HTLV-II foci exhibit geographically and ethnically independent distributions in the Andes highlands and the Orinoco lowlands, and a common HLA haplotype is associated with HTLV-I carriers in both Andes Indians and Japanese HTLV-I endemic populations, in contrast to the completely independent HLA haplotypes associated in HTLV-II carriers specific to Orinoco lowland and North American Indian groups (*25*). Our data on geographical distribution of HTLV-I/II carriers in the world suggest that HTLV-I and HTLV-II evolved among Mongoloids and were transmitted independently to people on the American continent by historical dispersal of ancestral Mongoloids who migrated from Asia.

To explain this unique distribution of HTLV-I, one hypothetical model of the migration of HTLV-I and -II carriers can be generated as follows (Fig. 5). HTLV-II carriers after the Asian migration arrived or the American continent and spread HTLV-II through Pan-American areas other than the Andes. After migration of HTLV-II-carrying Asians, HTLV-I carriers newly arrived in Andes areas through transpacific routes. As Meggers first pointed out more than 30 years ago (*17*), this hypothesis seems plausible because of similarity of genetic background between Southern Japanese and Andeans in South America. However, it has not been accepted by all anthropologists and archeologists. To clarify the relationship between Asian and Andean HTLV-I carriers, further comprehensive information on the HTLV-I distribution in Pan-Pacific areas may give a crucial hint to the relationship between Japanese and Andean ancestors and the history of their dispersal and settlements in each area.

Acknowledgments

We are most grateful to Dr. Shunro Sonoda, Professor, Department of Virology, Faculty of Medicine, Kagoshima University, for his lengthy collaboration with our ethnoepidemiologic study on ATL and HTLV-I among Mongoloids in South America. This work was supported mainly by Grants-in-Aid from Japan's Ministry of Education, Culture, Sports, Science and Technology International Scientific Research Program (1989–1998), the Japan Society for the Promotion of Science (1999–2001) and the Princess Takamatsu Cancer Research Fund.

REFERENCES

1. Cabrera, M. E., Labra, S., Ford, A. M., and Matutes, E. Adult T-cell leukaemia/lymphoma in Chile. *Estud. Atacamenos*, **15**, 59–62 (1998).

2. Cartier, L., Araya, F., Castillo, J. L., Ruiz, F., Gormaz, A., and Tajima, K. Progressive spastic paraparesis associated with human T-cell leukemia virus type I (HTLV-I). *Intern. Med.*, **31**, 1257–1261 (1992).
3. Cartier, L., Araya, F., Castillo, J. L., Zaninovicm, V., Hayami, M., Miura, T., Imai, J., Sonoda, S., Shiraki, H., Miyamoto, K., and Tajima, K. Southernmost carriers of HTLV-I/II in the world. *Jpn. J. Cancer Res.*, **84**, 1–3 (1993).
4. Duenas-Barjas, E., Bernal, J. E., Vaught, D. R., Briceno, I., Duran, C., Yanagihara, R., and Gajdusek, C. Coexistence of human T-lymphotropic virus types I and II among the Wayuu Indians from the Guajira region of Colombia. *AIDS Res. Hum. Retroviruses*, **8**, 1851–1855 (1992).
5. Fujiyama, C., Fujiyoshi, T., Miura, T., Yashiki, S., Matsumoto, D., Zaninovic, V., Blanco, A., Harrington, W., Jr., Byrnes, J. J., Hayami, M., Tajima, K., and Sonoda, S. A new endemic focus of human T lymphotropic virus type II carriers among Orinoco natives in Colombia. *J. Infect. Dis.*, **168**, 1075–1077 (1993).
6. Fujiyoshi, T., Yashiki, S., Fujiyama, C., Kuwayama, M., Miyashita, H., Ohnishi, H., Blank, M., Zaninovic, V., Cartier, L., Blank, A., Byrnes, J. J., Harrington, W., Jr., Miura, T., Hayami, M., Tajima, K., and Sonoda, S. Ethnic segregation of HTLV-I and HTLV-II carriers among South American native Indians. *Int. J. Cancer*, **63**, 510–515 (1995).
7. Fujiyoshi, T., Li, H.-C., Lou, H.-L., Yashiki, S., Karino, S., Zanimovic, V., Guillen, S., Gamacho, M., Andrade, R., Hurtado, L. V., Hurtado, L., Damiani, E., Cartier, L., Dipierri, J. E., Hayami, M., Sonoda, S., and Tajima, K. Characteristic distribution of HTLV type I and HTLV type II carriers among native groups in South America. *AIDS Res. Hum. Retroviruses*, **15**, 1235–1239 (1999).
8. Gessain, A., Barin, F., Vernant, J. C., Gout, O., Maurs, L., Calender, A., and de-The, G. Antibody to human T-lymphotropic virus type-I in patients with tropical spastic paraparesis. *Lancet*, **ii**, 407–409 (1985).
9. Gessain, A., Mauclere, P., Proment, A., Biglione, M., Hesran, J. Y. L., Tekaia, F., Millan, J., and de-The, G. Isolation and molecular characterization of a human T-cell lymphotropic virus type II (HTLV-II), subtype B, from a healthy pygmy living in a remote area of Cameroon: an ancient origin for HTLV-II in Africa. *Proc. Natl. Acad. Sci. USA*, **92**, 4041–4045 (1995).
10. Hino, S., Katamine, S., Miyata, H., Tsuji, Y., Yamabe, T., and Miyamoto, T. Primary prevention of HTLV-1 in Japan. *Leukemia*, **11**, S57–S59 (1997).
11. Hinuma, Y., Nagata, K., Hanaoka, M., Nakai, M., Matsumoto, T., Kinoshita, K., Shirakawa, S., and Miyoshi, I. Adult T-cell leukemia: antigen in an ATL cell line and detection of antibodies to the antigen in human sera. *Proc. Natl. Acad. Sci. USA*, **78**, 6476–6480 (1981).
12. Ido, E., Yamashita, M., and Hayami, M. Ethnic variation of HTLV in the world. *Gann Monogr. Cancer Res.*, **44**, 107–120 (1996).
13. Ishida, T., Yamamoto, K., Omoto, K., Iwanaga, M., Osato, T., and Hinuma, Y. Prevalence of a human retrovirus in native Japanese: evidence for a possible ancient origin. *J. Infect.*, **11**, 153–157 (1985).
14. Kalyanaraman, V. S., Sarngadharan, M. G., Robert-Guroff, M., Miyoshi, I., Blayney, D., Golde, D., and Gallo, R. C. A new subtype of human T-cell leukemia virus

(HTLV-II) associated with a T-cell variant of hairy cell leukemia *Science*, **218**, 571–573 (1982).

15. Kinoshita, K., Hino, S., Amagasaki, T., Ikeda, S., Yamada, Y., Suzuyama, J., Momita, S., Toriya, K., Kamihira, S., and Ichimaru, M. Demonstration of adult T-cell leukemia virus antigen in milk from three sero-positive mothers. *Jpn. J. Cancer Res.*, **75**, 103–105 (1984).
16. Li, H.-C., Fujiyoshi, T., Lou, H., Yashiki, S., Sonoda, S., Cartier, L., Nunez, L., Munoz, I., Horai, S., and Tajima, K. The presence of ancient human T-cell lymphotropic virus type I provirus DNA in an Andean mummy. *Nat. Med.*, **5**, 1428–1432 (1999).
17. Meggers, B. J. Archeological evidence for transpacific voyages from Asia since 6000 BP. *Estud. Atacamenos*, **15**, 107–124 (1998).
18. Meytes, D., Schochat, B., Lee, H., Nadel, G., Sidi, Y., Cerney, M., Swanson, P., Shaklai, M., Kilim, Y., Elgat, M., Ghin, E., Danon, Y., and Rosenblatt, J. D. Serological and molecular survey for HTLV-I infection in a high-risk Middle Eastern group. *Lancet*, **336**, 1533–1535 (1990).
19. Miura, T., Yamashita, M., Vladimir, Z., Cartier, L., Takehisa, J., Igarash, T., Ido, E., Fojiyoshi, T., Sonoda, S., Tajima, K., and Hayami, M. Molecular phylogeny of human T-cell leukemia virus type I and II of Amerindians in Colombia and Chile. *J. Mol. Evol.*, **44**, S76–S82 (1997).
20. Nakano, S., Ando, Y., Ichijo, M., Moriyama, I., Saito, S., Sugamura, K., and Hinuma, Y. Search for possible routes of vertical and horizontal transmission of adult T-cell leukemia virus. *Jpn. J. Cancer Res.*, **75**, 1044–1045 (1984).
21. Osame, M., Nakagawa, M., Kubota, H., and Takamori, M. The second nationwide survey of HTLV-I-associated myelopathy (HAM) in Japan. *Annual Report in 1994 on Neuroimmunological Research Committee of the Ministry of Health and Welfare of Japan*, 11–13 (1994).
22. Osame, M., Usuku, K., Izumo, S., Ijichi, N., Amitani, H., Igata, A., Matsumoto, M., and Tara, M. HTLV-I-associated myelopathy, a new clinical entity. *Lancet*, **i**, 1031–1032 (1986).
23. Poiesz, B. J., Ruscetti. F. W., Gazdar, A. F., Bunn, P. A., Minna, J. D., and Gallo, R. C. Detection and isolation of type-C retrovirus particles from fresh and .cultured lymphocytes of patients with cutaneous T-cell lymphoma. *Proc. Natl. Acad. Sci. USA*, **77**, 7415–7419 (1980).
24. Seiki, M., Hattori, S., Hirayama, Y., and Yoshida, M. Human adult T-cell leukemia virus: complete nucleotide sequence of the provirus genome integrated in leukemia cell DNA. *Proc. Natl. Acad. Sci. USA,* **80**, 3618–3622 (1983).
25. Sonoda, S. and Fujiyoshi, T. HTLV-I infection and HLA. *Gann Monogr. Cancer Res.*, **44**, 207–217 (1996).
26. Tajima, K. HTLV-I/II related diseases with special reference to its distribution among Mongoloids. *Gann Monogr. Cancer Res.*, **44**, 207–217 (1996).
27. Tajima, K. and Hinuma, Y. Epidemiology of HTLV-I/II in Japan and the world. *Gann Monogr. Cancer Res.*, **39**, 129–149 (1992).
28. Tajima, K. and Sonoda, S. Origins of HTLV-1 in South America. *Nat. Med.*, **6**, 232–233 (2000).
29. Tajima, K., Tominaga, S., Suchi, T., Kawagoe, T., Komoda, A., Hinuma, Y., Oda, T.,

and Fujita, K. Epidemiological analysis of the distribution of antibody to adult T-cell leukemia-virus-associated antigen: possible horizontal transmission of adult T-cell leukemia virus. *Jpn. J. Cancer Res.*, **73**, 893–901 (1982).

30. Tajima, K. The T- and B-cell Malignancy Study Group and Co-authors. The 4th nation-wide study of adult T-cell leukemia/lymphoma (ATL) in Japan: estimates of risk of ATL and its geographical and clinical features. *Int. J. Cancer*, **45**, 237–243 (1990).
31. Takezaki, T., Tajima, K., Ito, M., Ito, S., Kinoshita, I., Tachibana, K., Yamashita, Y., The Tsushima ATL Study Group. Short-term breast-feeding may reduce the risk of vertical transmission of HTLV-I. *Leukemia*, **11** (Suppl.), 60–61 (1997).
32. The T- and B-cell Malignancy Study Group. The third nation-wide study on adult T-cell leukemia/lymphoma (ATL) in Japan: characteristic patterns of HLA antigen and HTLV-I infection in ATL patients and their relatives. *Int. J. Cancer*, **41**, 505–512 (1988).
33. The T- and B-cell Malignancy Study Group. A report from the 8th nationwide study of adult T-cell leukemia/lymphoma (ATL in Japan). *Jpn. J. Cancer Clin.*, **44**, 381–397 (2001).
34. Uchiyama, T., Yodoi, J., Sagawa, K., Takatsuki, K., and Uchino, H. Adult T-cell leukemia: clinical and hematologic features of 16 cases. *Blood*, **50**, 481–492 (1977).
35. Zamora, T., Zaninovic, V., Kajiwara, M., Komoda, H., Hayami, M., and Tajima, K. Antibody to HTLV-I in indigenous inhabitants of the Andes and Amazon regions in Colombia. *Jpn. J. Cancer Res.*, **81**, 715–719 (1990).
36. Zaninovic, V., Sanzon, F., Lopez, F., Velandia, G., Blank, A., Blank, M., Fujiyama, C., Yashiki, S., Matsumoto, D., Katahira, Y., Miyashita, H., Fujiyoshi, T., Chan, L., Sawada, T., Miura, T., Hayami, M., Tajima, K., and Sonoda, S. Geographic independence of HTLV-I and HTLV-II foci in the Andes Highland, the Atlantic Coast, and the Orinoco of Colombia. *AIDS Res. Hum. Retroviruses*, **10**, 97–101 (1994).

GENETIC RISK OF DISEASE DEVELOPMENT AMONG HTLV-I CARRIERS

Shunro SONODA

*Department of Island Medicine, Faculty of Medicine, Kagoshima University**

Genetic risk for adult T-cell leukemia (ATL) has been implicated by ethnic and familial segregation of ATL patients from human T-lymphotropic virus type I (HTLV)-I-associated myelopathy/tropical spastic paraparesis (HAM/TSP). In order to clarify the genetic risk for ATL, we characterized human leukocyte antigen (HLA) class I alleles of ATL patients and analyzed the anchor motifs of HTLV-I peptides binding to HLA class I molecules using 291 lines of anti-HTLV-I $CD8^+$ cytotoxic T-lymphocytes (CTL) generated *in vitro* with a total of 165 synthetic peptides for HTLV-I Tax and Env proteins. Allele frequencies of HLA-A*26, B*4002, B*4006 and B*4801 were significantly higher in ATL patients than HAM/TSP patients and asymptomatic HTLV-I carriers in the southern Japan. $CD8^+$ CTL analysis revealed the HTLV-I Tax peptide sequence to completely lack anchor motifs of peptides binding to HLA-A*26, B*4002 and B*4006 molecules but to possess one anchor motif for HLA-B*4801, while the HTLV-I Env peptide sequence had many anchor motifs for HLA-A*26, B*4002, B*4006 and B*4801 molecules. Most ATL patients featured heterozygous HLA class I alleles composed of HLA-A*26, B*4002, B*4006 and B*4801, with a lower number of HTLV-I Tax peptide anchor motifs and epitopes generating anti-HTLV-I Tax $CD8^+$ CTL than individuals possessing other HLA alleles. Relationship between Tax epitope and ATL incidence was verified by the significantly lower number of HTLV-I Tax epitopes in ATL patients than in asymptomatic HTLV-I carriers ($P>0.006$) or late onset ATL patients ($P>0.002$). These results indicate that HLA-A*26, B*4002, B*4006 and B*4801 alleles predispose to ATL due to the limited recognition of HTLV-I Tax peptide anchor motifs and epitopes capable of generating anti-HTLV-I Tax $CD8^+$ CTL. These results suggest the presence of genetic risk for ATL among Japanese subpopulations. Based on the knowledge of HLA-binding peptide anchor motifs, HTLV-I Tax peptide vaccine for ATL prevention is discussed.

The human T-lymphotropic virus type I (HTLV-I) found worldwide causes adult T-cell leukemia (ATL) and non-leukemic diseases such as HTLV-I-associated myelopathy/tropical spastic paraparesis (HAM/TSP) among HTLV-I-infected people in the southern Japan, the Caribbean basin, South America, Iran, Melanesia and Equatorial Africa (*27*). The fact that an identical species of HTLV-I causes two clinically independent diseases, ATL and HAM/TSP (*35*), suggests the presence of segregate host factors determining disease risk. This hypothesis is supported by evidence

* 8-35-1 Sakuragaoka, Kagoshima 890-8520, Japan

that ATL patients possess characteristic human leukocyte antigen (HLA) haplotypes whose genetic lineages are distinct from those of HAM/TSP patients (*24, 25, 31*), the latter showing increased replication of peripheral blood T-cells infected with HTLV-I (*29, 34*) and a high immune responsiveness of $CD8^+$ cytotoxic T-lymphocytes (CTL) against HTLV-I Tax antigens (*10, 13, 20*). It is known that HLA alleles are individually different in human hosts, and that they determine levels of $CD8^+$ CTL responses to HTLV-I and other retrovirus infections (*3, 6, 10, 13, 17, 20, 30, 32*). It is thus suggested that the genetic prediction for ATL and HAM/TSP is linked to the $CD8^+$ CTL response to HTLV-I viral antigens as well as HTLV-I-infected $CD4^+$ T-cell antigens in the context of HLA class I molecules.

In order to clarify genetic risk factors being susceptible to ATL, we characterized HLA class I alleles of ATL patients, comparing their HLA class I alleles with those of HAM/TSP and asymptomatic HTLV-I carriers as well as healthy controls, and then we analyzed HTLV-I epitopes of $CD8^+$ CTL and the anchor motifs of HTLV-I peptides binding to HLA class I molecules. This article reports evidences of a unique immunogenetic background predisposing to ATL, featuring HLA class I molecules whose binding peptide anchor motifs are incompatible to recognize HTLV-I Tax peptides and incapable to generate anti-HTLV-I Tax $CD8^+$ CTL.

Genetic Background of HTLV-I Carriers in Japan

Genetic clusters of HTLV-I carriers in Japan have been explained by the anthropological postulate of aboriginal Japanese who might have carried HTLV-I and migrated from the Asian Continent in prehistoric time (*8, 9*). To clarify the genetic

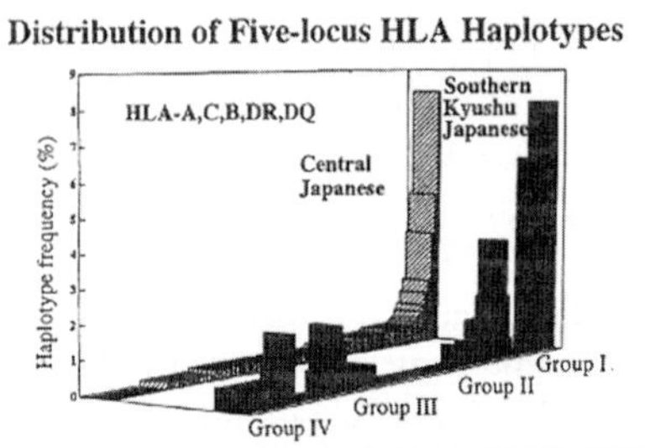

	HLA haplotypes	Southern Kyushu Japanese (N=162) HF(%)	Central Japanese (N=1786) HF(%)
Group I	A24Cw- B52DR15DQ1	8.0	8.3
	A24Cw1B54DR4 DQ4	6.2	1.6
	A33Cw- B44DR13DQ1	4.9	4.9
	A24Cw7 B7 DR1 DQ1	3.1	3.6
Group II	A24Cw1B59DR4 DQ4	3.7	0.6
	A2 Cw1B54DR4 DQ4	1.9	0.5
	A26Cw3B62DR15DQ1	1.2	0.5
	A26Cw3B61DR9 DQ3	1.2	0.5
Group III	A2 Cw3B61DR9 DQ3	1.9	0.3
	A26Cw3B62DR14DQ7	1.9	0.2
	A2 Cw3B61DR14DQ1	1.2	0.3
Group IV	A26Cw3B61DR14DQ1	1.9	<0.2
	A24Cw- B51DR14DQ1	1.2	<0.2

FIG. 1. Distribution of HLA haplotypes frequencies in southern Kyushu Japanese and the central Japanese populations

Genetic subpopulations of Japanese were found in southern Kyushu, Japan (modified from Ref. *26*).

background of Japanese HTLV-I carriers, we investigated HLA haplotypes of the southern Japanese subpopulations in Kyushu, where 10–20% of the population is infected with HTLV-I. Their HLA haplotypes were determined by familial analysis of HLA antigens and compared with the known HLA haplotypes of the central Japanese population (*26*). The distribution of HLA haplotypes was remarkably different between the southern Kyushu Japanese and the central Japanese populations (Fig. 1). The southern Kyushu Japanese were composed of four subgroups: one group corresponded to the major Japanese population and the other three were minor genetic groups among the central Japanese population. The southern Kyushu Japanese were disparate from the central Japanese in genetic makeup whose HLA haplotypes appeared to determine genetic susceptibility to HTLV-I infection and carrier state.

Segregation of ATL and HAM/TSP

A nationwide survey of ATL and HAM/TSP in Japan revealed that there is much greater incidence of the former in Japan (Fig. 2) (*19, 28*). In contrast to Japanese, the Chilean mestizo population showed higher incidence of HAM/TSP than ATL (*1, 2*), while the Caribbean black population appeared to have almost the same number with the two diseases (personal communication). HAM/TSP patients were prevalent in the Mudunga tribe, a minor ethnic group of the Zairean native population (*15*). These ethnic variations in the incidence of the two diseases suggest that there may be ethnically-determined genetic factors. An ethnicity-based genetic factor(s) would be consistent with the epidemiological findings of Japanese ATL and HAM/TSP patients whose occurrence was segregated in both family and population (*18, 23*). The segregation between the diseases was further investigated in the context of a host' immune responsiveness to HTLV-I and its genetic segregation.

Genetic Polymorphism in HLA Alleles and T-cell Mediated Immune Response against HTLV-I

The immunocompromised state of ATL (*22*) and the hyper-immunity of HAM/TSP are characteristic immunopathologies to distinguish the two clinical conditions. Two distinct immunopathologies caused by one etiologic agent (HTLV-I) suggest that genetic segregation of host factors may determine susceptibility to either ATL or HAM/TSP whose T-cell mediated immunity individually differ by the genetic make-

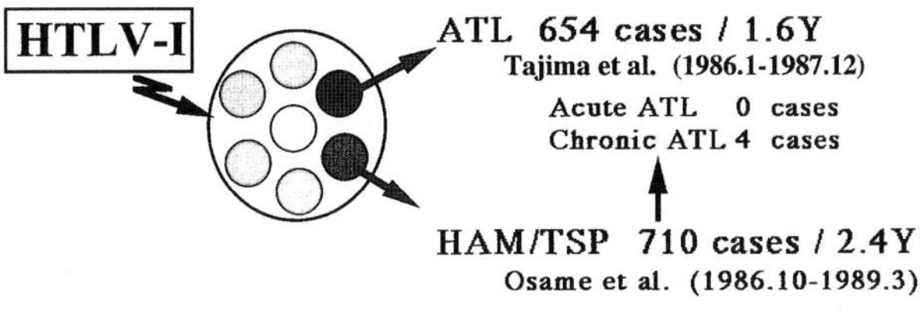

FIG. 2. Nationwide surveillance of ATL and HAM/TSP in Japan
ATL patients were registered for 2 years (January 1986–December 1987), and HAM/TSP patients for 2.5 years (October 1986–March 1989). No HAM/TSP case from ATL patients was found while 4 cases of chronic ATL patients from 710 patients with HAM/TSP were recognized (modified from Refs. *19* and *28*).

up of HLA alleles and immune response genes. HLA alleles are highly polymorphic and produce a huge variety of combinations, so that the genetic makeup of HLA used to be different among individuals and populations.

Individual HLA allele encodes its specific protein (HLA molecule) to bind a peptide composed of 9–16 amino acids that confers a unique immunoepitope to be recognized by T-cells. The T-cell epitopes are presented *in vivo* by dendritic cells and macrophage/monocytes in lymphoid organs and also by virally-infected cells. HTLV-I-infected T-cells are known to be potent antigen-presenting cells which express HTLV-I epitopes in the context of polymorphic HLA molecules (Fig. 3) (*26*). HTLV-I epitopes (Env, Gag, Pol, Tax) are presented in the complex with either class I or class II HLA molecules to generate CD4+ T- and CD8+ T-cell effectors (*10, 11, 13*). HTLV-I envelope antigens are effectively presented by HLA-DRB*0101,

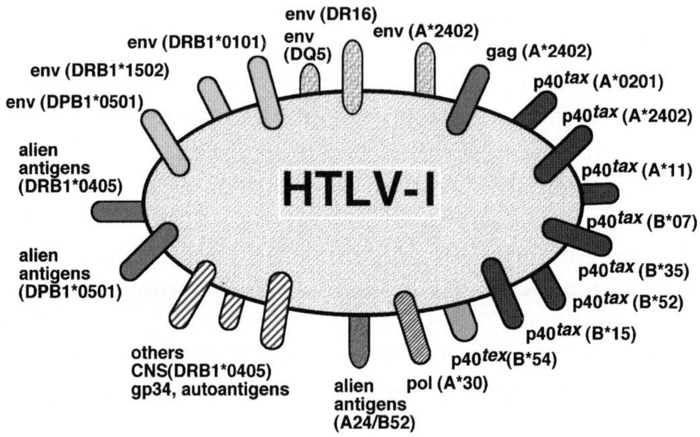

FIG. 3. T-cell epitopes on HTLV-I-infected cells
HTLV-I virion, p40 tax and alien (altered self) antigens were recognized in the context of specific HLA alleles (modified from Ref. *26*).

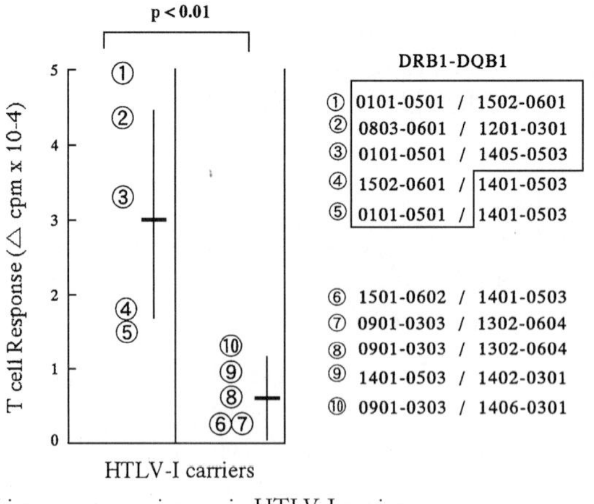

FIG. 4. HTLV-I immune responsiveness in HTLV-I carriers
High and low T-cell immune responses were segregated with two categories of HLA-DRB1-DQB1 haplotypes.

B*1502 and B*0803 which induced a high proliferative response of T-cells while HLA-DR B*1501, B*0901, B*1401 and B*1302 were less responsive (Fig. 4) (*26*). CD8+ T-cell effectors (CTL) for HTLV-I Tax epitopes are known to prevent HTLV-I carriers from ATL development. HLA-A*24, A*02, B*07, B*35, for example, are able to recognize Tax epitopes to generate anti-Tax CTL while HLA-A*26, B*4002, B*4006 and B*4801 were found to be poor in recognition of HTLV-I Tax epitopes. The individuals inheriting of these HLA class I alleles were thus unable to generate anti-Tax CTL and eventually became a risk for ATL development as described below.

Distribution of HLA Class I Alleles among ATL, HAM/TSP, Asymptomatic HTLV-I Carriers and Healthy Controls

We identified 45 distinct alleles of HLA-A, HLA-C and HLA-B in ATL patients, HAM/TSP patients, asymptomatic HTLV-I carriers and healthy controls from the southern Japanese population (Table I). ATL patients showed significantly increased frequencies of HLA-A*26, Cw*08, B*4002, B*4006 and B*4801 as compared with HAM/TSP patients, asymptomatic HTLV-I carriers or healthy controls. HAM/TSP patients showed significantly increased frequencies of HLA-A*24, Cw*07 and B*07 as compared with ATL patients, asymptomatic HTLV-I carriers or healthy controls. The two categories of HLA class I alleles were mutually segregated between ATL and HAM/TSP patients. Asymptomatic HTLV-I carriers showed an intermediate frequency of these particular HLA class I alleles associated with ATL or HAM/TSP.

A recently published report from Jeffrey *et al.* (*12*) prompted us to examine whether or not HLA-A*02 is relevant to a protective allele in HAM/TSP. An allele frequency of HLA-A*02 in HAM/TSP patients was found to be 19.9%, which was not significantly different from the frequency in asymptomatic HTLV-I carriers (23.7%) or healthy controls (21.1%) (Table I, A*02).

HTLV-I Tax and Env Epitopes and Anchor Motifs for HLA Class I Molecules Relevant to ATL

The immunological significance of HLA class I molecules relevant to ATL was determined by assigning HTLV-I Tax and Env epitopes recognized by CD8+ CTL from asymptomatic HTLV-I carriers possessing various HLA alleles. HLA-A*26, B*4002, B*4006 and B*4801 molecules were able to recognize 8 distinct epitopes of HTLV-I Env peptides with 5 anchor motifs (A*26, -V------L, -V------V; B*4002, -E-------A; B*4006, -E-------V; B*4801, -K------L). However, the three HLA class I (HLA-A*26, B*4002 and B*4006) molecules completely lacked anchor motifs on HTLV-I Tax peptides to generate anti-HTLV-I Tax CD8+ CTL, except one anchor motif for the B*4801 molecule (Table II, A*26, B*4002, B*4006 and B*4801). The lack of HTLV-I Tax peptide anchor motifs binding to HLA-A*26, B*4002 and B*4006 molecules was verified by "reverse immunogenetics" (*21*) of the anchor motifs of other peptides (*4, 7, 26*). Meanwhile, HLA-A*24, B*07 and B*52 molecules recognized 8–15 epitopes of HTLV-I Tax and Env with 1–5 anchor motifs. It is remarkable that HLA molecules relevant to HAM/TSP are highly potent to generate anti-HTLV-I Tax and Env CTL.

TABLE I. HLA Alleles and Haplotypes

HLA alleles	Frequency (%)			
	ATL	HAM/TSP	Asymptomatic HTLV-I carriers	Healthy controls
A*02	0.18	0.19	0.24	0.21
A*11	0.08	0.11	0.06	0.13
A*24	0.30	0.43*	0.39	0.37
A*26	0.26**	0.13	0.15	0.12
A*31	0.13	0.10	0.09	0.08
A*33	0.03	0.04	0.07	0.08
Cw*01	0.21	0.26	0.23	0.19
Cw*03	0.18**	0.10	0.12	0.07
Cw*07	0.09	0.19*	0.10	0.09
Cw*12	0.54	0.09	0.10	0.10
B*07	0.03	0.14*	0.05	0.04
B*15	0.11	0.06	0.07	0.11
B*35	0.08	0.07	0.07	0.11
B*4002	0.12**	0.06	0.11	0.04
B*4006	0.10**	0.03	0.07	0.10
B*51	0.09	0.10	0.08	0.08
B*52	0.05	0.09	0.10	0.10
DRB1*0101	0.02	0.13*	0.05	0.03
DRB1*0405	0.15	0.17	0.16	0.15
DRB1*0901	0.20**	0.13	0.13	0.12
DRB1*1501	0.08	0.02*	0.08	0.10
DRB1*1502	0.07	0.08	0.11	0.11
DQB1*0303	0.21**	0.15	0.14	0.13
DQB1*0401	0.14	0.17	0.15	0.16
DQB1*0501	0.02	0.13*	0.05	0.04
DQB1*0601	0.15	0.20	0.20	0.20
DQB1*0602	0.08	0.02*	0.08	0.09

	Haplotypes
ATL	A*26 Cw*03 B*4002 DRB1*0901 DQB1*0303
	A*26 Cw*03 B*4006 DRB1*0901 DQB1*0303
	A*26 Cw*03 B*15 DRB1*1501 DQB1*0602
HAM/TSP	A*24 Cw*07 B*07 DRB1*0101 DQB1*0501
	A*24 Cw*12 B*52 DRB1*1502 DQB1*0601
	A*24 Cw*01 B*54 DRB1*0405 DQB1*0401

CD8⁺ CTL Epitopes on HTLV-I Tax and Env Peptides Estimated by the Anchor Motifs of 25 Distinct HLA-A and -B Molecules

The limited numbers of HTLV-I Tax epitopes and anchor motifs binding to HLA-A*26, B*4002, B*4006 and B*4801 molecules suggest that the genetic risk of ATL may be determined by low immune responsiveness of anti-HTLV-I Tax $CD8^+$ CTL. To test this hypothesis, we estimated the number of HTLV-I Tax and Env epitopes by reverse immunogenetics using anchor motifs of HLA-A and -B molecules relevant to ATL and HAM/TSP, as shown in Table I. A total of 226 HTLV-I Tax and Env epitopes were assigned by reverse immunogenetics of anchor motifs bound to 25 distinct HLA class I molecules prevalent among Japanese ATL patients

TABLE II. HLA Anchor Motifs and CD8⁺ CTL Epitopes for HTLV-I Tax and EnV Peptides

HLA alleles	HTLV-I Tax												HTLV-I Env											
	Tax peptides (aa position)		HLA anchor motifs										Env peptides (aa position)		HLA anchor motifs									
			1	2	3	4	5	6	7	8	9	10			1	2	3	4	5	6	7	8	9	10
A*26	none												FTQEVSRLNINLHFS	(141–155)	E	V	-	-	L	-	-	-	L	
													RRAVPVAVWLVSALA	(311–325)	A	V	-	-	-	V	-	-	V	
													LLHEVDKDISQLTQA	(346–360)	E	V	-	-	-	I	-	-	L	
													PILQERPPLENRVLT	(411–425)	E	R	-	-	L	-	-	-	V	
B*4002	none												LNTEPSQLPPTAPPL	(176–190)	-	E	-	-	-	-	-	-	-	A
													GLDLLFWEQGGLCKA	(381–395)	-	E	-	-	-	-	-	-	A	
B*4006	none												PILQERPPLENRVLT	(411–425)	-	E	-	-	-	-	-	-	-	V
B*48	FQTKAYHPSFLLSHG	(281–295)	-	K	-	-	-	-	-	-	L		MGKFLATLILFFQFC	(1–15)	-	K	-	-	-	-	-	-	L	
													NLHFSKCGFPFSLLV	(151–165)	-	K	-	-	-	-	-	-	L	
													WKSKLLTLVQLTLQS	(206–220)	-	K	-	-	-	-	-	-	L	
A*24	LLFGYPVYVFGDCVQ	(11–25)	-	F	-	-	-	-	-	-	F		FFQFCPLIFGDYSPS	(11–25)	-	F	-	-	-	-	-	-	F	
	VPYKRIEELLYKISL	(186–200)	-	Y	-	-	-	-	-	-	L		DYSPSCCTLTIGVSS	(21–35)	-	Y	-	-	-	-	-	-	L	
	YHPSFLLSHGLIQYS	(286–300)	-	F	-	-	-	-	-	-	I		NLHFSKCGFPFSLLV	(151–165)	-	F	-	-	-	-	-	-	F	
	LIQYSSFHSLHLLFE	(296–310)	-	F	-	-	-	-	-	-	F		FNWTHCFDPQIQAIV	(271-285)	-	F	-	-	-	-	-	-	I	
	HLLFEEYTNIPISLL	(306–320)	-	Y	-	-	-	-	-	-	L													
B*07	MAHFPGFGQSLLFGY	(1–15)	-	P	-	-	-	-	-	-	L		SQLPPTAPPLLPHSN	(181–195)	-	P	-	-	-	-	-	-	L	
	TPPITHTTPNIPPSF	(91–105)	-	P	-	-	-	-	-	-	-	I	TAPPLLPHSNLDHIL	(186–200)	-	P	-	-	-	-	-	-	L	
	AFLTNVPYKRIEELL	(181–195)	-	P	-	-	-	-	-	-	L		SVPSSSSTPLLYPSL	(246–260)	-	P	-	-	-	-	-	-	L	
	IILPEDCLPTTLFQP	(206–220)	-	P	-	-	-	-	-	-	-	L	IQAIVSSPCHNSLIL	(281–295)	-	P	-	-	-	-	-	-	L	
	TLFQPARAPVTLTAW	(216–230)	-	P	-	-	-	-	-	-	L		PILQERPPLENRVLT	(411–425)	-	P	-	-	-	-	-	-	L	
	HEPQISPGGLEPPSE	(331–345)	-	P	-	-	-	-	-	-	L													
B*52	GFGQSLLFGYPVYVF	(6–20)	-	L	-	-	-	-	-	-	V		QALQPPCPNLVSYSS	(61–75)	-	Q	-	-	-	-	-	-	V	-
	LLFGYPVYVFGDCVQ	(11–25)	-	L	-	-	-	-	-	-	V		QHDVNFTQEVSRLNI	(136–150)	-	Q	-	-	-	-	-	-	I	-
	SALQFLIPRLPSFPT	(66–80)	-	L	-	-	-	-	-	-	F		SRLNINLHFSKCGFP	(146–160)	-	L	-	-	-	-	-	-	F	-
	VVCMYLYQLSPPITW	(151–165)	-	L	-	-	-	-	-	-	I		LTLQSTNYTCIVCID	(216–230)	-	Q	-	-	-	-	-	-	I	-
	PGQLGAFLTNVPYKR	(176–190)	-	L	-	-	-	-	-	-	V		GITLVALLLLVILAG	(446–460)	-	L	-	-	-	-	-	-	V	-

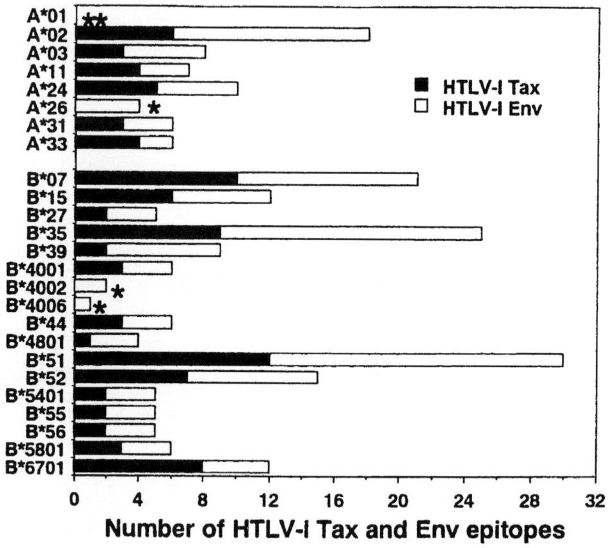

FIG. 5. CD8+ CTL epitopes on HTLV-I Tax and Env peptides estimated by the anchor motifs of 25 distinct HLA-A and HLA-B molecules

Numbers of HTLV-I Tax and Env epitopes were estimated by reverse immunogenetics of anchor motifs for HLA-A and HLA-B molecules, referring to our study and others. The HLA-01 molecules did not recognize both HTLV-I Tax and Env epitopes (**), and HLA-A*26, B*4002 and B*4006 did not recognize HTLV-I Tax epitopes (*) (modified from Ref. *33*).

(Fig. 5). HLA-A*26, B*4002 and B*4006 molecules recognized no epitope on HTLV-I Tax peptides but 7 epitopes on HTLV-I Env peptides. The HLA-B*4801 molecule recognized one HTLV-I Tax epitope and 3 HTLV-I Env epitopes, while HLA-A*01 recognized none of HTLV-I Tax and Env epitopes. HLA-A*02, B*07, B*35, B*51 and B*52 molecules recognized more than 15 epitopes on HTLV-I Tax and Env peptides. The other HLA class I molecules had intermediate numbers of HTLV-I Tax and Env epitopes, ranging from 5 to 12. These results indicated that HLA class I molecules of our ATL patients recognized more HTLV-I Env than HTLV-I Tax epitopes.

Relationship between Anti-HTLV-I Tax CD8+ CTL Epitopes and ATL Incidence

Correlation of HLA class I alleles and number of CD8+ CTL epitopes with risk of ATL was done by comparing the number of HTLV-I Tax and Env epitopes between ATL patients and asymptomatic HTLV-I carriers (Fig. 6). The number of Tax epitopes in ATL patients was significantly less than in asymptomatic carriers ($P<0.007$) while there was no significant difference in Env epitopes between the two groups ($P=0.321$). We further investigated the relationship between number of HLA binding epitopes and ages at ATL onset (Fig. 7). To this end, we compared two groups of ATL patients possessing HLA-A*26, B*4002, B*4006 and B*4801 alleles (group I) and those possessing other HLA class I alleles (group II) with respect to age of ATL at onset. Group I developed clinical symptoms of ATL at 52.2±12.3 years old and group II at 64.8±9.7, the age difference being statistically significant (Fig. 4,

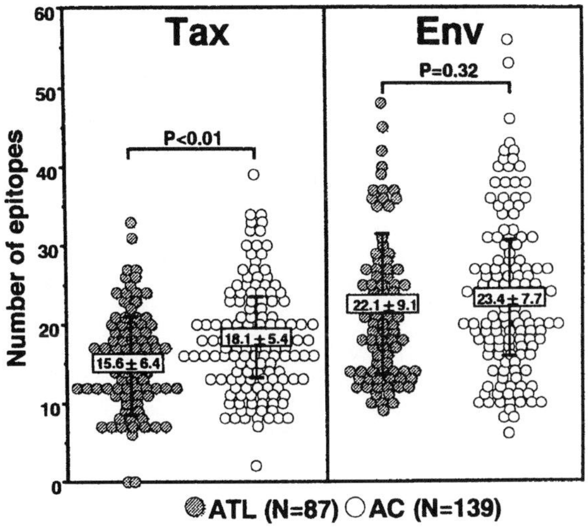

FIG. 6. Relationship between development of ATL and number of HTLV-I Tax epitopes recognized by CD8⁺ CTLs

HLA-A and HLA-B alleles of 87 ATL patients and 139 asymptomatic HTLV-I carriers were determined to enumerate HTLV-I Tax and Env epitopes. ATL patients recognized significantly fewer HTLV Tax epitopes than the asymptomatic carriers (modified from Ref. *33*).

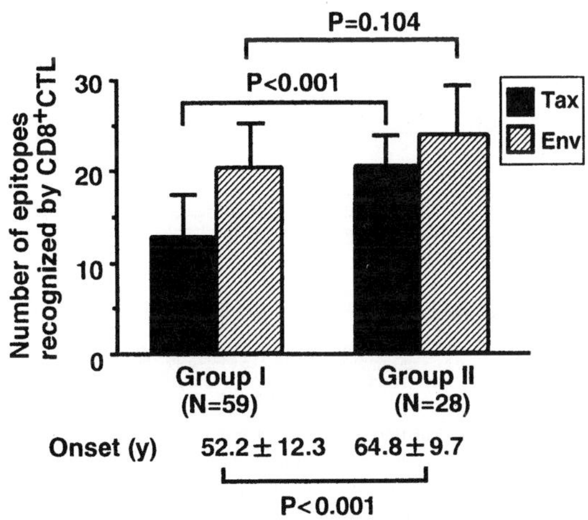

FIG. 7. Relationship between age at time of ATL onset and numbers of HTLV-I Tax epitopes recognized by CD8⁺ CTLs

Eighty-seven patients with ATL were divided into two groups, group I possessing HLA-A*26, B*4002, B*4006 and B*4801 and group II possessing other HLA class I alleles. The mean ages of group I and II at time of ATL onset were significantly different (modified from Ref. *33*).

P<0.001). Average numbers of HTLV-I Tax and Env epitopes recognized by the group I patients were 13.3±5.7 and 20.5±5.0, respectively, while the group II patients recognized 21.0±8.4 and 24.4±10.4. The difference in the average number of HTLV-I Tax epitopes between group I and group II ATL patients was significant

TABLE III. Heterozygosity of HLA Alleles Associated with ATL or HAM/TSP

HLA alleles	ATL (n=120)		HAM/TSP (n=135)	
	n	%	n	%
A loci				
A*26/A*26	8	6.7	0	0.0
A*26/A*24	19	15.8	21	15.6
A*26/A*02	10	8.3	8	5.9
A*26/A*11	7	5.8	2	1.5
A*26/A*31	7	5.8	3	2.2
A*26/A*33	3	2.5	1	0.7
A*24/A*24	13	10.8	25	18.5
A*24/A*26	19	15.8	21	15.6
A*24/A*02	14	11.7	18	13.3
A*24/A*11	4	3.3	13	9.6
A*24/A*31	8	6.7	10	7.4
A*24/A*33	2	1.7	4	3.3
B loci				
B*4002/B*4002	3	2.5	1	0.8
B*4006/B*4006	2	1.7	0	0.0
B*4002/B*4006	0	0.0	0	0.0
B*4801/B*4801	0	0.0	0	0.0
B*4002/B*4801	2	1.7	0	0.0
B*4006/B*4801	1	0.8	0	0.0
B*4002/6/48/B*07	2	1.7	4	3.3

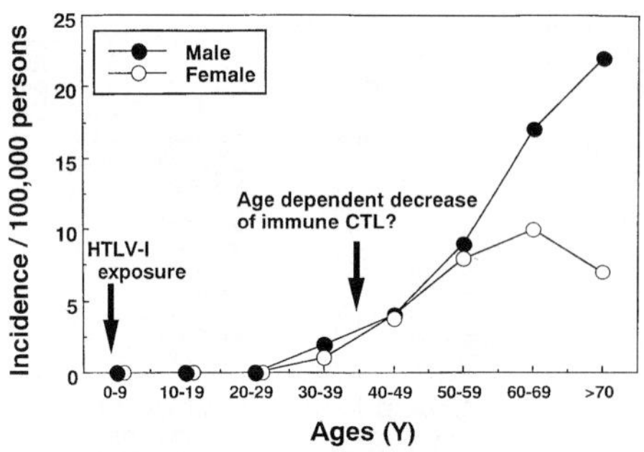

FIG. 8. ATL surveillance in Kagoshima (1991–1995)

Age-dependent morbidity of ATL was estimated by surveying ATL patients in Kagoshima population. The incidence of ATL patients increased exponentially after 30 years of age and reached a peak around 50–60 years in females but males went on beyond 70 years. Age-dependent decrease of immune CTL likely occurs at 30–40 years of age.

(P<0.001), whereas that for HTLV-I Env epitopes was not (P=0.104) (Fig. 5, Tax, Env).

Conclusion

Most HTLV-I-infected persons remain asymptomatic for 20–30 years after primary infection with HTLV-I (Fig. 8). During this latency period, the host's defense system exerts both humoral and cellular immune responses against HTLV-I-infected

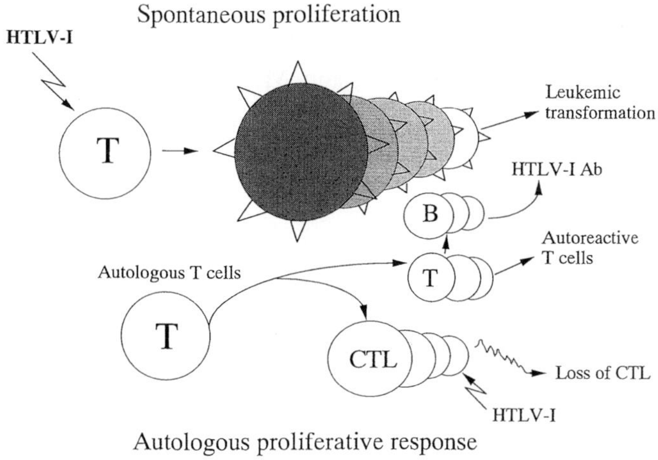

FIG. 9. HTLV-I-infected T-cells present HTLV-I-related antigens to elicit autologous T-cell proliferative response. $CD4^+$ T-cells may induce autoreactive T cells to cause neurological disorder like HAM/TSP. $CD8^+$ CTL may exclude HTLV-I-infected T-cells to prevent leukemic transformation and ATL development. The CTL may be re-infected with HTLV-I to lose CTL activity *in vivo*.

T-cells which present HTLV-I virion antigens as well as altered self antigens (Fig. 9). Some $CD4^+$ T-cell repertoires turn to autoreactive T-cells (*16*) and $CD8^+$ CTL proliferate to exclude HTLV-I-infected T-cells (*10, 11, 14*). Since most T-cell effectors cannot survive longer than several months so that anti-HTLV-I CTL must be constantly recruited to maintain effective immunity against ATL development. However, the hosts are immunologically compromised by aging and exogenous impact of deleterious effect on the immune system, presumably by impact of HTLV-I re-infection *in vivo* so as happened with human immunodeficiency virus (*5*). It thus is promising to develop a recombinant HTLV-I Tax vaccine for ATL by augmenting anti-HTLV-I Tax CTL.

REFERENCES

1. Cabrera, M. E. *et al.* HTLV-I positive adult T-cell leukemia/lymphoma (ATLL) in Chile. *Leukemia*, **8**, 1763–1767 (1994).
2. Cartier, L. *et al.* HTLV-I positive spastic paraparesis in a temperate zone. *Lancet*, **i**, 556–557 (1989).
3. Clerici, M., Lucey, D. R., Zajac, R. A. *et al.* Detection of cytotoxic T lymphocytes specific for synthetic peptides of gp160 in HIV-seropositive individuals. *J. Immunol.*, **146**, 2214–2219 (1991).
4. Davenport, M. P. and Hill, A. V. Reverse immunogenetics: from HLA-disease association to vaccine candidates. *Mol. Med. Today*, **2**, 38–45 (1996).
5. Flamand, L., Crowley, R. W., Lusso, P., Colombini-Hatch, S., Margolis, D. M., and Gallo, R. C. Activation of $CD8^+$ T lymphocytes through the T cell receptor turns on CD4 gene expression: implications for HIV pathogenesis. *Proc. Natl. Acad. Sci. USA*, **95**, 3111–3116 (1998).
6. Gotch, F. M., Nixon, D. F., Alp, N., McMichael, A. J., and Borysiewicz, L. K. High

frequency of memory and effector gag specific cytotoxic T lymphocytes in HIV seropositive individuals. *Int. Immunol.*, **2**, 707–712 (1990).
7. Goulder, P., Conlon, C., McIntyre, K., and McMichael, A. Identification of a novel human leukocyte antigen A26-restricted epitope in a conserved region of Gag. *AIDS*, **10**, 1441–1443 (1996).
8. Haniwara, K. Origin and affinities of Japanese viewed from cranial measurements. *Acta Anthropogenet.*, **8**, 149–158 (1984).
9. Ishida, T. and Hinuma, Y. The origin of Japanese HTLV-I. *Nature*, **322**, 564 (1986).
10. Jacobson, S., Shida, H., McFarlin, D. E., Fauci, A. S., and Koenig, S. Circulating $CD8^+$ cytotoxic T lymphocytes specific for HTLV-I pX in patients with HTLV-I associated neurological disease. *Nature*, **348**, 245–248 (1990).
11. Jacobson, S. *et al.* Induction of $CD4^+$ human T lymphotropic virus type-I-specific cytotoxic T lymphocytes from patients with HAM/TSP. Recognition of an immunologic region of the gp46 envelope glycoprotein of human T lymphotropic virus type-I. *J. Immunol.*, **146**, 1155–1162 (1991).
12. Jeffery, K. J. *et al.* HLA alleles determine human T-lymphotropic virus-I (HTLV-I) proviral load and the risk of HTLV-I-associated myelopathy. *Proc. Natl. Acad. Sci. USA*, **96**, 3848–3853 (1999).
13. Kannagi, M., Harada, S., Maruyama, I. *et al.* Predominant recognition of human T cell leukemia virus type I (HTLV-I) pX gene products by human $CD8^+$ cytotoxic T cells directed against HTLV-I-infected cells. *Int. Immunol.*, **3**, 761–767 (1991).
14. Katahira, Y., Yashiki, S., Fujiyoshi, T., Nomura, K., Tara, M., Mori, M., Setoyama, M., Kanzaki, T., Shida, H., and Sonoda, S. *In vitro* induction of cytotoxic T lymphocytes against HTLV-I-infected T-cells from adult T-cell leukemia patients, asymptomatic HTLV-I carriers and seronegative healthy donors. *Jpn. J. Cancer Res.*, **86**, 21–27 (1995).
15. Kayembe, K. *et al.* A cluster of HTLV-I-associated tropical spastic paraparesis in Equateur (Zaire); ethnic and familial distribution. *J. Neurol. Neurosurg. Psychiatry*, **53**, 4–10 (1990).
16. Nagai, M., Yashiki, S., Fujiyoshi, T. *et al.* Characterization of a unique T-cell clone established from a patient with HAM/TSP which recognized HTLV-I-infected T-cell antigens as well as spinal cord tissue antigens. *J. Neuroimmunol.*, **65**, 97–105 (1996).
17. Nelson, G. W., Kaslow, R., and Mann, D. L. Frequency of HLA allele-specific peptide motifs in HIV-1 proteins correlates with the allele's association with relative rates of disease progression after HIV-1 infection. *Proc. Natl. Acad. Sci. USA*, **94**, 9802–9807 (1997).
18. Osame, M. *et al.* HTLV-I-associated myelopathy (HAM), treatment trials, retrospective survey and clinical laboratory findings. *Hematol. Rev.*, **3**, 271–284 (1990).
19. Osame, M. *et al.* Nationwide survey of HTLV-I-associated myelopathy in Japan; association with blood transfusion. *Ann. Neurol.*, **28**, 51–56 (1990).
20. Parker, C. E., Nightingale, S., Taylor, G. P., Weber, J., and Bangham, C. R. Circulating anti-Tax cytotoxic T lymphocytes from human T-cell leukemia virus type I-infected people, with and without tropical spastic paraparesis, recognize multiple epitopes simultaneously. *J. Virol.*, **68**, 2860–2868 (1994).
21. Rammensee, H. G., Friede, T., and Stevanoviic, S. MHC ligands and peptide motifs: first listing. *Immunogenetics*, **41**, 178–228 (1995).

22. Shaw, G. *et al.* Human T-cell leukemia virus; its discovery and role in leukemogenesis and immunosuppression. *Adv. Intern. Med.*, **30**, 1–27 (1984).
23. Shoji, H. *et al.* HTLV-I-associated myelopathy and adult T-cell leukemia cases in a family. *Eur. J. Neurol.*, **29**, 33–35 (1989).
24. Sonoda, S., Fujiyoshi, T., and Yashiki, S. Immunogenetics of HTLV-I/II and associated diseases. *J. Acquir. Immune Defic. Syndr. Hum. Retrovirol.*, **13** (Suppl. 1), S119–S123 (1996).
25. Sonoda, S., Yashiki, S., Fujiyoshi, T. *et al.* Immunogenetic factors involved in the pathogenesis of adult T-cell leukemia and HTLV-I-associated myelopathy. *In* "Advances in Adult T-cell Leukemia and HTLV-I Research," ed. S. Takatsuki, Y. Hinuma, and M. Yoshida, Gann Monograph on Cancer Research No. 39, pp. 81–93 (1992). Japan Sci. Soc. Press, Tokyo.
26. Sonoda, S. *et al.* Immunogenetics of HTLV-I/II and associated diseases. *J. Acquir. Immune Defic. Syndr. Hum. Retrovirol.*, **13**, S119–S123 (1996).
27. Tajima, K. HTLV-I/II related disease with special reference to its distribution among mongoloids. *In* "Ethnoepidemiology of Cancer," ed. K. Tajima and S. Sonoda, Gann Monograph on Cancer Research No. 44, pp. 123–135 (1996). Japan Sci. Soc. Press, Tokyo.
28. Tajima, K. and the T- and B-cell Malignancy Study Group. The 4th nationwide study of adult T-cell leukemia/lymphoma (ATL) in Japan; estimation of risk of ATL and its geographical and clinical features. *Int. J. Cancer*, **45**, 237–243 (1990).
29. Tendler, C. L., Greenberg, S. J., Blattner, W. A. *et al.* Transactivation of interleukin 2 and its receptor induces immune activation in human T-cell lymphotropic virus type I-associated myelopathy: pathogenic implications and a rationale for immunotherapy. *Proc. Natl. Acad. Sci. USA*, **87**, 5218–5222 (1990).
30. Tomiyama, H., Miwa, K., Shiga, H. *et al.* Evidence of presentation of multiple HIV-1 cytotoxic T lymphocyte epitopes by HLA-B*3501 molecules that are associated with the accelerated progression of AIDS. *J. Immunol.*, **158**, 5026–5034 (1997).
31. Usuku, K., Sonoda, S., Osame, M. *et al.* HLA haplotype-linked high immune responsiveness against HTLV-I in HTLV-I associated myelopathy: comparison with adult T-cell leukemia/lymphoma. *Ann. Neurol.*, **23** (Suppl. 1), S143–S150 (1988).
32. Walker, B. D., Flexner, C., Paradis, T. J. *et al.* HIV-1 reverse transcriptase is a target for cytotoxic T lymphocytes in infected individuals. *Science*, **240**, 64–66 (1988).
33. Yashiki, S. *et al.* HLA-A*26, HLA-B*4002, HLA-B*4006, and HLA-B*4801 alleles predispose to adult T cell leukemia: the limited recognition of HTLV type 1 Tax peptide anchor motifs and epitopes to generate anti-HTLV type 1 Tax $CD8^+$ cytotoxic T lymphocytes. *AIDS Res. Hum. Retroviruses*, **17**, 1047–1061 (2001).
34. Yoshida, M., Osame, M., Kawai, H. *et al.* Increased replication of HTLV-I in HTLV-I-associated myelopathy. *Ann. Neurol.*, **26**, 331–335 (1989).
35. Yoshida, M., Osame, M., Usuku, K., Matsumoto, M., and Igata, A. Viruses detected in HTLV-I-associated myelopathy and adult T-cell leukaemia are identical on DNA blotting. *Lancet*, **i**, 1085–1086 (1987).

CELL-TO-CELL TRANSMISSION OF HTLV-I

Hiroshi SHIRAKI, Yasuko SAGARA, and Yukiko INOUE
*Department of Research, Fukuoka Red Cross Blood Center**

Cell-to-cell transmission of human T-cell leukemia virus type I (HTLV-I) is initiated with the interaction between viral envelope proteins and the host cell receptor, thus leading to HTLV-I cell entry via membrane fusion. Viral envelope attachment to the cellular receptor is mediated through receptor binding sites on amino acids 111 to 136 (gp46-111) and 197 to 216 (gp46-197) on the gp46 surface envelope glycoprotein, and 400 to 429 (gp21-400) on the gp21 transmembrane glycoprotein, respectively. Cellular receptors for cell-to-cell transmission of HTLV-I consist of a 71 kDa heat shock cognate protein (HSC70), β-actin and palmitoyl(16:0)-oleoyl(18:1)-phosphatidylglycerol (POPG) and these three receptor molecules form a receptor complex in target cell membrane, as a core molecule of HSC70. Infectivity-neutralizing antibodies bind to the sites surrounding the receptor-binding pockets, gp46-111 and gp46-197 on gp46, and structure of the complexes between neutralizing antibody and gp46 indicate a possible neutralizing mechanism. Thus, gp46 is the receptor binding and membrane fusion glycoprotein of HTLV-I and the target for infectivity-neutralizing antibodies. These findings provide a frame for developing therapeutic and preventative strategies to combat HTLV-I pathologenesis, as well as data on the basic biology of HTLV-I, including the selective tropism of virus and membrane fusion mechanisms governing virus entry.

Human T-cell leukemia virus type I (HTLV-I), a retrovirus with a wide geographic distribution, is associated with adult T-cell leukemia/lymphoma (ATLL) (*16, 32*) and HTLV-I-associated myelopathy/tropical spastic paraparesis (HAM/TSP) (*10, 29*). It has also been associated with a number of inflammatory diseases, such as pediatric infectious dermatitis (*23*) and uveitis (*24*). HTLV-I is a lifelong chronic infection, which may result in ATLL in 2% to 3% of asymptomatic carriers 40 to 50 years after the initial infection (*42*). Despite a vigorous host immune response, HTLV-I persists in the infected host, which means that the virus may be unable to evade host immune response.

HTLV-I primarily infects $CD4^+$ T-lymphocytes but, various established mammalian cell lines can be infected (*2, 12, 17, 19*). HTLV-I is spread more efficiently via cell-cell contact, and unlike human immunodeficiency virus type I (HIV-I), free virions are not so infectious (*28*). Hence, cell-to-cell infection via membrane fusion is considered the general mode of HTLV-I transmission. In endemic regions, HTLV-I is transmitted primarily from mother to child during breast-feeding and sexually between adults (*14, 27*). In the Western world, the principal routes of infection are

* Fukuoka 815-8588, Japan

parenteral (transfusion and needle sharing among intravenous drug users) and sexual contacts (*25, 28, 43*). Mother-to-child transmission is prevented by discouraging breast-feeding (*15*), but this has proved to be impossible in developing countries. Thus, an acquisition of therapeutic and preventative agents for HTLV-I is crucial.

To better comprehend HTLV-I-related infection, molecular events concerning HTLV-I entry into a cell have to be understood. This article focuses on cell-to-cell transmission of HTLV-I via membrane fusion between infected and target cells, and what is known of molecular events in HTLV-I entry into a cell, based on the observations in syncytium formation induced by HTLV-I-bearing cells, is discussed.

In vitro Cell-to-Cell Transmission of HTLV-I

To determine retrovirus infection *in vitro*, syncytium formation assay is useful to assess virus entry into a cell. This assay is based on formation of multinucleated cells (syncytia) by fusion between viral envelope proteins-expressing cells and receptor-bearing cells (*18, 26*). In retrovirus entry, this process is initiated by interaction between envelope proteins and cellular receptors, conformational changes of the envelope protein follow and lead to the membrane fusion required to deliver the viral genome to the cell cytoplasm. This assay has been extensively used to study cellular and molecular determinants that contribute to membrane fusion (*4, 13, 30, 44*), cell-to-cell viral transfer (*7*), and to host-encoded factors that inhibit viral entry into cells (*8, 45*). The assay system we used consists of T-cell line KT252 established from an ATL patient as HTLV-I-infected cells, and T-cell line MOLT-4 was used for uninfected cells. gp46 surface glycoprotein (gp46) and the gp21 transmembrane glycoprotein (gp21) at low concentration both inhibit in a linear manner syncytium formation, and the inhibitory effect of the mixture of gp46 and gp21 was approximately the sum of these glycoproteins alone (*34*). These observations suggest that gp46 and gp21 independently have a critical role in the syncytium formation induced by cell-to-cell contact.

When HTLV-I envelope protein is aligned to the murine or the feline retrovirus envelope, analogy to the attachment site for cell recognition is found in a region between amino acids 138 to 202 of gp46, and characteristics of a fusion peptide region are found in a 29 amino acid-long hydrophobic stretch at N-terminal (amino acids 313 to 331) of gp21 (*11*). Studies using site-directed mutagenesis revealed that the region corresponding to amino acids 75 to 101 and 195 and 205 on gp46 is critical for syncytia formation and mutations in N-terminal (amino acids 360 to 407) and C-terminal (amino acids 419 to 440) of the gp21 ectodomain affected syncytium formation and the potential for virus entry of HTLV-I (*6*).

Receptor-binding Sites in gp46 and gp21

Native HTLV-I envelope protein initially synthesized as a 61 kDa precursor (gp61), is subsequently cleaved into two mature products (gp46 and gp21). These two subunits are associated through noncovalent interaction (*31*) and are anchored to the cell surface by gp21. In retrovirus entry into a cell, the surface envelope subunit interacts with receptor-bearing cell membrane and induces transmembrane protein-mediated fusion between the viral envelope/infected cell membrane and the target cellular

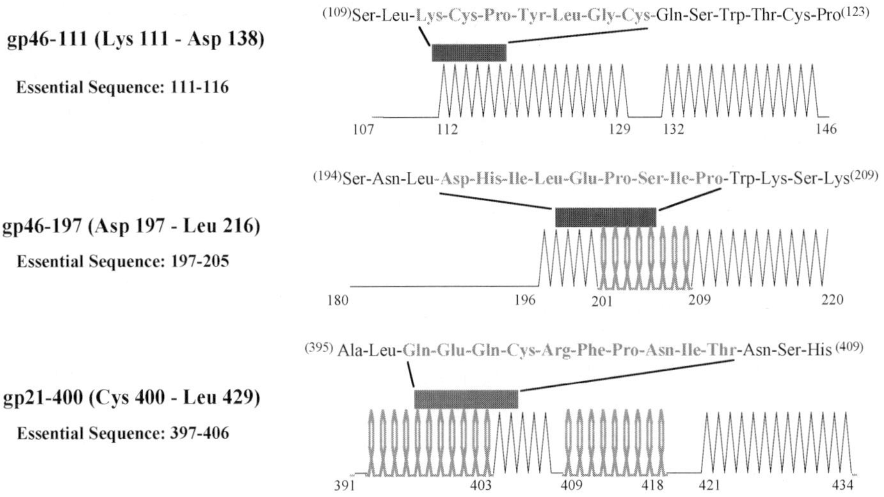

FIG. 1. Schematic representation of essential receptor-binding sites on HTLV-I envelope proteins Zigzag line, β-sheet structure; loop-line, α-helix structure; straight line, unordered structure. The numbers represent the amino acid position from N-terminal Met of the precursor gp62 envelope protein.

membrane. Thus, the retrovirus envelope protein has a role in receptor binding and membrane fusion glycoprotein in the process of virus entry.

To identify all potential binding sites in gp46 and gp21 to cellular receptors on target cells, we use a peptide library representing the complete envelope protein sequence, which delineates the linear epitopes recognized by sera from HTLV-I-infected blood donors (*21, 46*). This library consists of 23 different synthetic peptides containing 16 to 37 amino acids segments, overlapped by 2 to 5 amino acid in each peptide of the entire envelope protein sequence. In preparing this peptide library, the peptides for synthesis primarily selected hydrophilic regions including the β-turn structure and bilateral termini by analysis of the hydropathy (*22*) and the predicted secondary structure of the envelope protein (*3*). To locate the functional regions required for syncytium formation, each of the library peptides was added to a co-culture consisting of KT252 and MOLT-4 and we determined the inhibitory potential for syncytium formation as induced by HTLV-I-bearing cells. We found that peptide segments corresponding to amino acids 111 to 138 (gp46-111) and 197 to 216 (gp46-197) of gp46, and 400 to 429 (gp21-400) (numbering in gp62) of gp21 had inhibitory effects on syncytium formation, which meant these three peptide regions likely contribute as receptor-binding sites in envelope protein-mediated virus attachment to cellular receptor on the target cell membrane (*34, 35*). Essential receptor-binding sites in these three peptide segments were estimated to be amino acid sequences Lys111–Cys117 in gp46-111, Asp197–Pro205 in gp46-197 and Gln397–Thr406 in gp21, respectively, determined using a series of overlapping 10-mer and 12-mer peptides.

Structure of Receptor-binding Sites

Figure 1 shows the location, the essential receptor-binding site and predicted secondary structure of receptor-binding peptide segments. The essential receptor-binding site (Lys111 to Cys106) in the gp46-111 peptide locates at a highly potential area to form a β-sheet structure with highly hydrophobic residues. The essential binding site (Asp197 to Pro205) in the gp46-197 peptide locates at a highly ordered structure with β-sheet and α-helices with hydrophobic and hydrophilic residues. If modeled as α-helices, this peptide displayed amphipathic helices with a hydrophilic and hydrophobic face on opposite sides of the helical. The essential binding site in the gp21-400 peptide locates at a highly α-helices structure comprised of hydrophilic residues (*13*). Recently, X-ray crystallography and nuclear magnetic resonance studies have revealed the structure of the entire gp21 ectodomain (residues 313–445) that crystallized as a maltose-binding protein chimera. In this study, the gp21 ectodomain structure, like HIV gp21 and hemagglutinin of influenza virus, is constructed of long rods comprising an N-terminal central triple-stranded coiled-coil, a disulfide-bonded loop associated with a chain reversal at the base of the rod, and a structurally diverse C-terminal ectodomain segment that packs against the exterior of the central coiled-coil in an antiparallel manner (*47*). These structures are considered to be a fusion activated or postfusion conformation of gp21, referred to as helical hairpins, thus implying that the hydrophobic N-terminal fusion peptide and C-terminal transmembrane sequencing are juxtaposed at one end of the rod. In this structure, the peptide region gp21-400 located at the N-terminal of antiparallel C-terminal ectodomain segment is packed on the outside of the coiled coil. The idea that gp21 hairpins correspond to a fusion-activated conformation support the observations that gp21-400 is a potent inhibitor of envelope-mediated fusion and viral entry, like the C-terminal helical segment of HIV (*1, 9, 33*). In the HIV case, the gp41 C-terminal peptide appears to inhibit membrane fusion by binding to a CD4-induced site in the central coiled-coil of a gp120-gp41 fusion intermediate, blocking formation of the gp41 hairpin. Taken together, these results indicate that receptor-binding sites in these three peptide segments are qualitatively different in both character of amino acid residues and structural features, thereby implying the functional specificity of each peptide segment in the syncytium formation induced by HTLV-I-bearing cells.

Identification of Cellular Receptor

To isolate cellular components required for syncytium formation, we used materials bound to a Sepharose 4B affinity column coupled with each peptide of a library, including receptor-binding peptides (*35, 36*). After applying the MOLT-4 lysate to this column, followed by washing, the materials bound to the column were assayed for syncytium formation. We found that the bound materials on the affinity column with peptides gp46-111, gp46-197 and gp21-400, respectively, and had distinctly inhibitory activities for syncytium formation, while no activities were observed in column-binding materials from other peptides. These three materials apparently contain cellular components critical for the syncytium formation induced by HTLV-I-bearing cells. Figure 2 shows SDS-PAGE of these three bound-materials. With materials from the gp46-111 and the gp46-197 columns, three bands stained with Coomassie-brilliant blue were observed at positions corresponding to molecular masses of approximately 73, 43 and 34 kDa, while in the gp21-400 materials, there was only one band at 34 kDa (Fig. 2B). All of these bands showed inhibitory activities against

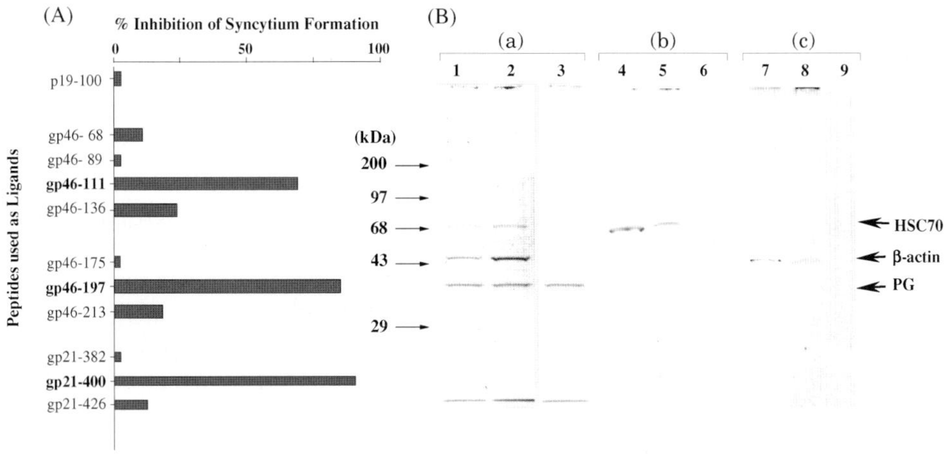

FIG. 2. Analysis of eluates from peptide-affinity chromatography

A) Inhibitory activity for syncytium formation. Data were estimated in comparison with the number of syncytia in co-culture with medium alone and that with the eluate. Values represent averages of triplicate determinations. The horizontal axis indicates the percent ratio of inhibition for syncytium formation and the vertical axis indicates eluates from affinity columns coupled with each peptide segment. B) SDS-PAGE pattern of eluates. (a) Staining with Coomassie-brilliant blue. (b) Immunostaining with anti-HSC monoclonal antibody. (c) Immunostaining with anti-β-actin monoclonal antibody. Lanes 1, 4, and 7, eluate from the gp46-111 column; lanes 2, 5, and 8, eluate from the gp46-197 column; lanes 3, 6, and 9, eluate from the gp21-400 column.

syncytium formation. Taken together these observations indicate that these three bound-materials likely function as receptor molecule attachment of the viral envelope attachment to the target cell membrane. Cellular components at 73 and 43 kDa were assigned to a 71 kDa heat shock cognate protein (HSC70) and β-actin, respectively, through amino acid sequencing using electron spray ionization tandem mass spectrometry (ESI-MS) and immunochemical staining using antibodies to HSC70 and β-actin, respectively (Fig. 2B, lanes 4–9).

The 34 kDa component is not a protein, rather this component was assigned to a phosphatidyl glycerol (PG) determined by thin-layer chromatography (TLC) and ESI-MS (*37*). Interestingly, all ESI-MS spectra of the 34 kDa band from three bound-materials is identical to the three column-bound materials identified as the PG molecule composed of palmitoic acid (16:0) and oleic acid (18:1) (POPG). In the ESI-MS spectrum, the 34 kDa band showed a unique fragment ion peak at *m/z* 255 and 281 corresponding to palmitoic ester and oleic ester and characteristic ion peaks of PG (ion peak *m/z* 491: PG lacking palmitoic acid and *m/z* 465: PG lacking oleic acid). Ion peaks derived from other fatty acids were nil in the spectra in these three 34 kDa components.

Regarding cellular distribution of the receptor, flow cytomeric analysis indicated that an anti-HSC70 monoclonal antibody reacted with 70% of MOLT-4 cells, while only 15% of the L-M (TK-) cells resistant to HTLV-I infection (*41*), scored positive. Furthermore, an anti-HSC70 antibody inhibited syncytium formation by HTLV-I, whereas no inhibition was seen using an anti-β-actin antibody. These observations indicate that the HSC70 molecule is expressed on surfaces of HTLV-I-permissive MOLT-4 cells.

Complex Formation of Cellular Receptor Molecules

In SDS-PAGE analysis, we observed that eluates from gp46-111 and the gp46-197 columns had three identical bands corresponding to HSC70, β-actin and POPG, respectively, while that from the gp21-400 column had only POPG. HSC70 is considered to be attached to the cell surface, as there is a lack of a transmembrane motif in the amino acid sequence. This means that for HSC70 to locate on the intact cell surface, cellular constituents are needed to anchor the membrane. We propose that the receptor apparatus comprises HSC70, β-actin and POPG which likely form a complex in MOLT-4 cell membranes.

To clarify molecular interactions in the receptor complex, we estimated the cross-linking among HSC70, β-actin and POPG using plastic plate wells coated with each receptor molecule (*38*). HSC70 bound with both β-actin and POPG while β-actin showed no binding with POPG, and vice versa. These observations support the view

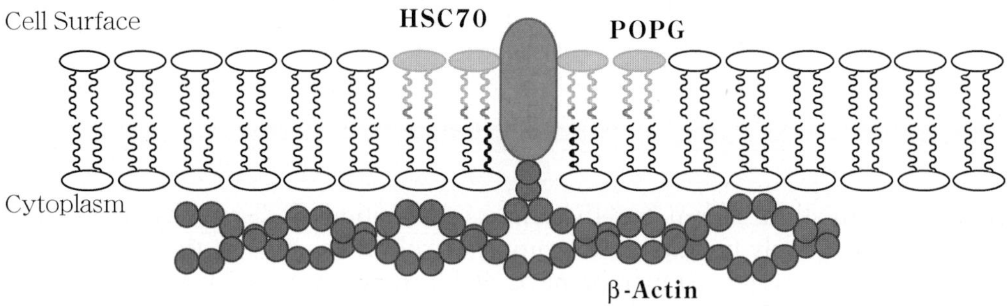

FIG. 3. Model for receptor complex for HTLV-I infection
The receptor complex in target cell membrane is formed by interaction with phosphatidyl glycerol and HSC70 with β-actin.

TABLE I. Cross-interaction among Each Molecule of the Receptor Complex

Receptor molecule	Interactive activity to		
	HSC70	β-Actin	PG
HSC70	n.t.[a]	+	+
β-Actin	+	n.t.	–
PG	+	n.t.	n.t.
PE	–	–	n.t.
PC	–	–	n.t.
PS	–	–	n.t.
PI	–	–	n.t.
PA	–	–	n.t.
Lyso PC	–	–	n.t.
Lyso PE	–	–	n.t.
Cardiolipin	–	–	n.t.

[a] n.t.: not tested.
Each receptor component was added to wells coated with one of the other components, and the supernatant was examined for remaining inhibitory activity for syncytium formation.

that the HSC70, as a core component, forms a receptor complex with β-actin and POPG (Fig. 3). In this model, both interactions of HSC70-β-actin and HSC70-POPG were significant, while that of β-actin-PG was sparse.

Interestingly, the phospholipid species required for receptor complex formation is to be specified PG. As shown in Table I, HSC70 binds with only PG, whereas binding was nil in other phospholipids (phosphatidyl ethanolamine PE and its lyso-form, phosphatidyl choline PC and its lyso-form, phosphatidyl serine PS, phosphatidyl inositol PI, phosphatidic acid PA and cardiolipin). These suggest that a receptor apparatus produced by translocation of cytosolic HSC70 molecules is likely to be a unique domain in the target cell membrane by cross-linking with POPG and β-actin, thereby leading to HTLV-I entry into a cell via membrane fusion.

Envelope Attachment to the Cellular Receptor

For HTLV-I entry into a cell, gp46 plays a key role in recognition of the cellular receptor for anchoring of a virus-producing cell to target cell membrane. Table II summarizes results on the inter-molecular interactions between viral envelope proteins and receptor molecules. HSC70 and β-actin bind to gp46 and gp46-197 but not to gp21, thus indicating that HSC70 and β-actin interact only with gp46 via the peptide region gp46-197. It is noteworthy that gp46 and gp46-197 bind to all three receptor molecules while gp21 binds only to PG.

In retrovirus entry into a cell, the functional role of receptor-binding envelope protein is to induce conformational change in the transmembrane subunit which is correlated with the membrane fusion activity. In HIV entry into a cell, gp120 first binds to the primary receptor CD4 (*5, 20*), the result being creation of a high affinity binding site for co-receptors (CXCR4 and/or CCR5). These binding events trigger a conformational change in gp41 (*9*). In HTLV-I entry, it is unclear whether gp46 attachment to the target cells evokes conformational changes of the envelope protein that facilitate the fusion activity on gp21. However, despite the lack of interaction between gp46-111 and HSC70/β-actin, we found that the eluate from the gp46-111 column contained HSC70 and β-actin with PG. On the other hand, the gp21-400, like gp46-111, binds only with PG, but the eluate from the gp21-400 column contained only PG without HSC70/β-actin. These findings imply that the structural

TABLE II. Interaction between the Envelope Protein and the Cellular Receptor

Ligand	Receptors isolated on the peptide column	Interactive activity to		
		HSC70	β-Actin	PG
BSA	n.t.[a]	–	–	–
gp46	n.t.	+	+	+
gp21	n.t.	–	–	+
pl9-100		–	–	–
gp46-197	HSC70, β-actin, POPG	+	+	+
gp46-111	HSC70, β-actin, POPG	–	–	+
gp21-400	POPG	–	–	+

[a] n.t.: not tested.

Receptor component was added to wells coated with viral envelope protein or peptide segments. After 3 h of incubation, the supernatants were examined regarding the remaining inhibitory activity for syncytium formation.

changes accompanying the envelope protein attachment to the cellular receptor including dissociation of the gp46-gp21 complex exposing the gp21-400 on the C-terminal portion of the ectodomain gp21 and dissociation of the interaction between HSC70/β-actin and PG.

Infectivity-neutralizing Antibody

Antibody neutralization in retrovirus infection is considered to be due to inhibition of virus attachment to the cellular receptor and/or processes following to binding, such as membrane fusion. Accordingly, molecular determinants for neutralization of virus infection by an antibody suggest new strategies for developing preventive and therapeutic agents to combat the HTLV-I endemic, and useful information on envelope functions in virus entry into cells.

To map the binding sites of infectivity-neutralizing antibodies on envelope proteins, we utilized a peptide-reactive antibody library (10 against gp46 and 6 against gp21) prepared from plasma of HTLV-I-infected blood donors, using an affinity column coupled with synthetic peptides identified as linear antigenic peptides (*39*). This library of 16 different antibodies consists of 10 antibodies through the entire gp46 and 6 antibodies through the entire ectodomain of gp21. These peptide-reactive antibodies exhibited overwhelming potency to the putative peptide segment used for purification of each antibody, except for a faint binding seen in some antibodies derived from gp21. Of the 16 antibodies reactive to the peptide segments on the envelope protein, six (Ab gp46-20, Ab gp46-89, Ab gp46-175, Ab gp46-197, Ab gp46-213 and Ab gp46-288) had potent infectivity-neutralizing ability for HTLV-I. It is noteworthy that all infectivity-neutralizing antibodies recognize only gp46. No neutralizing antibody binding site was not seen on gp21, despite the inclusion of antibody against receptor-binding site gp21-400 on gp21. Therefore, gp46 is the target for infectivity-neutralizing antibodies as well as the receptor-binding protein of HTLV-I, and the receptor-binding site gp21-400 likely resides inside of native gp21 envelope protein.

Table III summarizes results on antibody blocking of interactions between envelope protein and receptor molecules determined using plastic plate wells coated with

TABLE III. Effect of Infectivity-neutralizing Antibody on the Interaction among gp46 and Cellular Receptors for Syncytium Formation

Antibody	Antibody-binding site	Blocking effect for gp46 binding to	
		HSC70	PG
None[a]		–	–
Ab gp46-20	gp46 20-49	–	–
Ab gp46-89	gp46 89-115	–	+
Ab gp46-175	gp46 175-199	+	+
Ab gp46-197	gp46 197-216	+	+
Ab gp46-213	gp46 213-236	+	+
Ab gp46-277	gp46 277-292	–	–
LAT27	gp46 191-196	+	+

[a] Control experiments were as follows: HSC70 or PG was incubated on a plastic plate coated with gp46 for 3 h at 37°C without antibodies, then each supernatant was used to determine the activity for syncytium formation.

gp46 or gp21. Ab gp46-175, Ab gp46-197 and Ab gp46-213 preferentially blocked the interaction between HSC70 and gp46, which is equivalent to the inhibition of monoclonal antibody, LAT27, which recognized amino acid 191 to 196 of gp46 protein (*43*). On the other hand, Ab gp46-89, Ab gp46-175, Ab gp46-197 and Ab gp46-213 blocked the interaction between gp46 and POPG. Therefore, neutralization by Ab gp46-175 Ab gp46-197 and Ab gp46-213 is due to a block of gp46 binding to HSC70 and POPG, respectively, whereas neutralization by Ab gp46-89 is due to a block of the interaction between PG and gp46. Neutralizing antibody-binding sites surround the receptor-binding pocket on the receptor-binding protein gp46, and structures of the complexes between neutralizing antibody and gp46 indicate possible neutralization mechanisms. On the other hand, neutralization by the antibody recognizing the C-terminus (gp46-20) and N-terminus (gp46-277) is probably independent of the direct interaction between gp46 and cellular receptor and may suppress the flexible migration of envelope protein accompanying fusion-activation.

HSC70-mediated Necrosis

Biochemical studies indicate that HSC70 acts as a cellular receptor in the cell-to-cell transmission of HTLV-I through membrane fusion. The gp46 binding to HSC70 induced cell lysis due to pore formation on the surface of target cell MOLT-4 and cell death followed (*29*). This cytolysis increased in a linear manner with a gp46 concentration over 0.1 μM and was completed with 0.3 μM of gp46. This activity of gp46 was blocked when it co-existed with HSC70, whereas β-actin and PG did not do so, hence this cytolysis is likely to be mediated by HSC70 on the cell surface of target cells. This view is supported by findings that gp46-197, but not gp46-111 or gp21-400, also had cytotoxic activity to the target cell MOLT-4. Since there was no evidence of annexin V binding, the cytotoxic activities of gp46 and gp46-197 are due to necrosis, not to apoptosis.

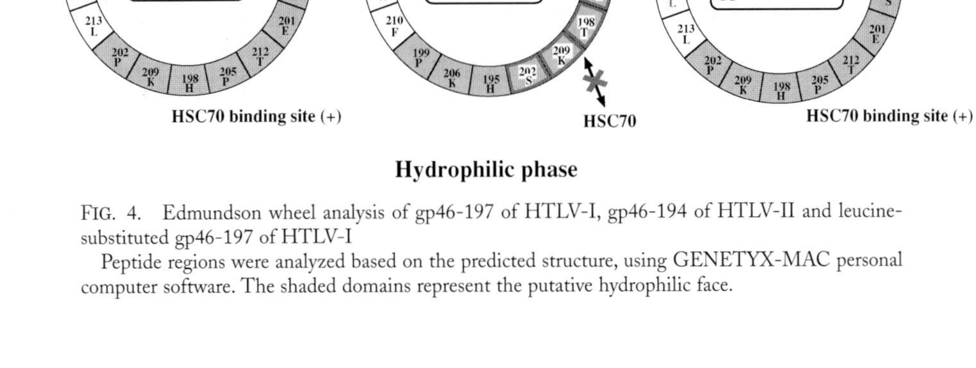

FIG. 4. Edmundson wheel analysis of gp46-197 of HTLV-I, gp46-194 of HTLV-II and leucine-substituted gp46-197 of HTLV-I

Peptide regions were analyzed based on the predicted structure, using GENETYX-MAC personal computer software. The shaded domains represent the putative hydrophilic face.

To clarify the functional role of gp46 in cytolysis through HSC70, we prepared HTLV-II peptide (gp46-194) corresponding to the HTLV-I gp46-197 and synthetic peptide (gp46-197SS) by leucine-substitution of the hydrophobic residue in the hydrophobic phase of gp46-197. HTLV-II also has a gp46 envelope protein, and the gp46-194 peptide segment (Glu194 to Leu213) is approximately 60% homologous to HTLV-I gp46-197 in amino acid sequence (*40*). Both HTLV-I gp46-197 and HTLV-II gp46-194 have high potential to form an α-helical amphipathic structure (Fig. 4). Interestingly, amino acid alteration in HTLV-I gp46-194 dominantly occurred in the hydrophilic phase of the HTLV-I gp46-197, thereby implying a lack of the HSC70 binding activity in gp46-194. In fact, the HTLV-II gp46-194 peptide has no binding activity to HSC70 or β-actin, and also no cytotoxic activity against MOLT-4. Furthermore, leucine substitution in the hydrophobic phase sequence of gp46-197 (gp46-197SS), which has binding activity to HSC70/β-actin but not to POPG, is a moderately decreased cytotoxic activity. These observations indicate that the amino acid sequence in the hydrophilic phase of gp46-197 contributes to interactions with HSC70 and the hydrophobic face to PG, and also that multiple residues in gp46-197, mainly the hydrophilic HSC70 binding phase, play a critical role in cytolysis via HSC70/PG binding.

Concluding Comments

HTLV-I virions, in contrast to HIV, are not very infectious, and are seen to transmit mainly through cell-to-cell contact. HTLV-I induces a lifelong chronic infection, which may result in ATLL in 2 to 3% of carriers 40 to 50 years after infection. These characteristics of HTLV-I infection have been attributed to low titer of viral particles released from infected cells and to a restricted cellular tropism. In addition, maintenance of a large pool of longlasting infected T-cells carrying the provirus in infected individuals has likely been instrumental in not achieving the extinction of HTLV-I.

Here, we have shown that HTLV-I, like HIV, represents a retrovirus whose principal receptor is comprised of more than one component. This raises the possibilities that the cell lacks one or more of these components and that cells resistant to virus infection are not necessarily receptor-negative. HTLV-I has a very wide host-range, which would suggest that functional homologues of the receptor components are widely distributed on cells from various species. For the HIV showing CD4-dependent entry, CD4 utilization is conserved despite co-receptor flexibility. This suggests that the principal binding site of receptor CD4 and not components of the receptor determine the observed receptor interference among these viruses. HTLV-I entry into a cell is a critical initial event in the cycle. All these findings provide a new frame for developing preventive and therapeutic strategies to combat HTLV-I pathogenesis and critical features of the basic biology of HTLV-I, including the selective tropism of virus and the membrane fusion mechanism governing virus entry.

Acknowledgments

We wish to thank Y. Maeda, Director of the Fukuoka Red Cross Blood Center for his continuous support.

REFERENCES

1. Chan, D. C., Chutkowski, C. T., and Kim, P. S. Evidence that a prominent cavity in the coiled coil of HIV type 1 gp41 is an attractive drug target. *Proc. Natl. Acad. Sci. USA*, **95**, 15613–15617 (1998).
2. Chen, I. S., McLaughlin, J., Gasson, J. C., Clark, S. C., and Golde, D. W. Molecular characterization of genome of a novel human T-cell leukemia virus. *Nature*, **305**, 502–505 (1983).
3. Chou, P. Y. and Fasman, G. D. Prediction of protein conformation. *Biochemistry*, **13**, 222–244 (1974).
4. Daenke, S., McCracken, S. A., and Booth, S. Human T-cell leukaemia/lymphoma virus type 1 syncytium formation is regulated in a cell-specific manner by ICAM-1, ICAM-3 and VCAM-1 and can be inhibited by antibodies to integrin beta2 or beta7. *J. Gen. Virol.*, **80**, 1429–1436 (1999).
5. Dalgleish, A. G., Beverly, P. C. L., Clapham, P. R., Crawford, D. H., Greaves, M. F., and Weiss, R. A. The CD4 (T4) antigen is an essential component of the receptor for the AIDS retrovirus., *Nature*, **312**, 763–767 (1984).
6. Delamarre, L., Pique, C., Pham, D., Tursz, T., and Dokhelar, M. C. Identification of functional regions in the human T-cell leukemia virus type I SU glycoprotein. *J. Virol.*, **68**, 3544–3549 (1994).
7. Delamarre, L., Rosenberg, A. R., Pique, C., Pham, D., and Dokhelar, M. C. A novel human T-leukemia virus type 1 cell-to-cell transmission assay permits definition of SU glycoprotein amino acids important for infectivity. *J. Virol.*, **71**, 259–266 (1997).
8. Desgranges, C., Souche, S., Vernant, J. C., Smadja, D., Vahlne, A., and Horal, P. Identification of novel neutralization-inducing regions of the human T-cell lymphotropic virus type I envelope glycoproteins with human HTLV-1-seropositive sera. *AIDS Res. Hum. Retroviruses*, **10**, 163–173 (1994).
9. Furuta, R. A., Wild, C. T., Weng, Y., and Weiss, C. D. Capture of an early fusion active conformation of HIV gp41. *Nat. Struct. Biol.*, **5**, 276–279 (1998).
10. Gessain, A., Barin, F., Vernent, J. C., Gout, O., Maurs, L., Calender, A., and de-The, G. Antibodies to human T-lymphotropic virus type I in patients with tropical spastic paraparesis. *Lancet*, **ii**, 407–410 (1985).
11. Haseltine, W. A., Sodroski, J. G., and Patarca, R. Structure and function of the genome of HTLV. *Curr. Top. Microbiol. Immunol.*, **115**, 177–209 (1985).
12. Hattori, T., Uchiyama, T., Toibana, T., Takatsuki, K., and Uchino, H. Surface phenotype of Japanese adult T-cell leukemia cells characterized by monoclonal antibodies. *Blood*, **58**, 645–647 (1981).
13. Hildreth, J. E., Subramanium, A., and Hampton, R. A. Human T-cell lymphotropic virus type 1 (HTLV-1)-induced syncytium formation mediated by vascular cell adhesion molecule-1: evidence for involvement of cell adhesion molecules in HTLV-1 biology. *J. Virol.*, **71**, 1173–1180 (1997).
14. Hino, S., Yamaguchi, K., Katamine, S., Sugiyama, H., Amagasaki, T., Kinoshita, K., Yoshida, Y., Doi, H., Tsuji, Y., and Miyamoto, T. Mother-to-child transmission of human T-cell leukemia virus type-I. *Jpn. J. Cancer Res.*, **76**, 474–480 (1985).
15. Hino, S., Katamine, S., Miyata, H., Tsuji, Y., Yamabe, T., and Miyamoto, T. Primary prevention of HTLV-1 in Japan. *Leukemia*, **11**, 57–59 (1997).

16. Hinuma, Y., Nagata, K., Hanaoka, M., Matsumoti, T., Kinoshita, K., Shirakawa, S., and Miyoshi, I. Adult T-cell leukemia: antigen in an ATL cell line and detection of antibodies to the antigen in human sera. *Proc. Natl. Acad. Sci. USA*, **78**, 6476–6480 (1981).
17. Ho, D. D., Rota, T. R., and Hirsch, M. S. Infection of human endothelial cells by human T-lymphotropic virus type I. *Proc. Natl. Acad. Sci. USA*, **81**, 7588–7590 (1984).
18. Hoshino, H., Shimoyama, M., Miwa, M., and Sugimura, T. Detection of lymphocytes producing a human retrovirus associated with adult T-cell leukemia by syncytia induction assay. *Proc. Natl. Acad. Sci. USA*, **80**, 7337–7341 (1983).
19. Hoxie, J. A., Matthews, D. W., and Cines, D. B. Infection of human endothelial cells by human T-cell leukemia virus type 1. *Proc. Natl. Acad. Sci. USA*, **81**, 7591–7595 (1984).
20. Klatzmann, D., Champagne, E., Chamaret, S., Gruest, J., Guetard, D., Hercend, T., Glickman, J.-C., and Montanier, L. T-lymphocyte T4 molecule behaves as the receptor for human retrovirus LAV. *Nature*, **312**, 767–768 (1984).
21. Kuroda, N., Washitani, Y., Shiraki, H., Kiyokawa, H., Ohno, M., Sato, H., and Maeda, Y. Detection of antibodies to human T-lymphotropic virus type 1 by using synthetic peptides. *Int. J. Cancer*, **45**, 865–868 (1990).
22. Kyte, J. and Doolittle, R. F. A simple method for displaying the hydropathic character of a protein. *J. Mol. Biol.*, **157**, 105–132 (1982).
23. Lagrenade, L., Hanchard, B., Flecher, V., Cranton, B., and Blattner, B. Infective dermatitis in Jamaica children—a marker for HTLV-1 infection. *Lancet*, **336**, 1345–1347 (1990).
24. Mochizuki, M., Watanabe, T., Yamaguchi, K., Tajima, K., Yoshikura, I., Nakashima, S., Shirao, M., Araki, S., Miyota, N., Mori, S., and Takatsuki, K. Uveitis associated with human T-lymphotropic virus type I : seroepidemiologic, clinical and virologic studies. *J. Infect. Dis.*, **166**, 943–944 (1992).
25. Murphy, E. L., Figueroa, J. P., and Gibbs, W. N. Sexual transmission of human T-lymphotropic virus type-I. *Ann. Intern. Med.*, **111**, 555–560 (1989).
26. Nagy, K., Clapham, P., Cheingsong-Popov, R., and Weiss, R. Human T-cell leukemia virus type I: induction of syncytia and inhibition by patient's sera. *Int. J. Cancer*, **32**, 321–328 (1983).
27. Nakano, S., Ando, T., Saito, K., Moriyama, I., Ichijo, M., Toyama, T., Sugamura, K., Imai, J., and Hinuma, Y. Primary infection of Japanese infants with adult T-cell leukemia-associated retrovirus (ATLV): evidence for virus transmission from mother to children. *J. Infect. Dis.*, **12**, 205–212 (1986).
28. Okochi, K., Sato, H., and Hinuma, Y. A retrospective study on transmission of adult T-cell leukemia virus by blood transfusion: seroconversion in recipients. *Vox Sang.*, **46**, 245–253 (1984).
29. Osame, M., Usuku, K., Izumo, S., Ijichi, N., Amitani, H., Igata, A., Matsumoto, M., and Tara, M. HTLV-I associated myelopathy, a new clinical entity. *Lancet*, **i**, 1031–1032 (1986).
30. Pique, C., Tursz, T., and Dokhelar, M. C. Mutations introduced along the HTLV-I envelope gene result in a non-functional protein: a basis for envelope conservation? *EMBO J.*, **9**, 4243–4248 (1990).

31. Pique, C., Pham, D., Tursz, T., and Dokhelar, M. C. Human T-cell leukemia virus type I envelope protein maturation process: requirements for syncytium formation. *J. Virol.*, **66**, 906–913 (1992).
32. Poiesz, B. J., Ruscetti, F. W., Gazdar, A. F., Bunn, P. A., Minna, J. D., and Gallo, R. C. Detection and isolation of type C retrovirus particles from fresh and cultured lymphocytes of a patient with cutaneous T-cell lymphoma. *Proc. Natl. Acad. Sci. USA*, **77**, 7415–7419 (1980).
33. Rimsky, L. T., Shugar, D. C., and Matthews, T. J. Determinants of human immunodeficiency virus type 1 resistance to gp41-derived inhibitory peptides. *J. Virol.*, **72**, 986–993 (1998).
34. Sagara, Y., Inoue, Y., Shiraki, H., Jinno, A., Hoshino, H., and Maeda, Y. Identification and mapping of functional domains on human T-cell lymphotropic virus type 1 envelope proteins by using synthetic peptides. *J. Virol.*, **70**, 1564–1569 (1996).
35. Sagara, Y., Ishida, C., Inoue, Y., Shiraki, H., and Maeda, Y. Trypsin-sensitive and -resistant components in human T-cell membranes required for syncytium formation by human T-cell lymphotropic virus type 1-bearing cells. *J. Virol.*, **71**, 601–607 (1997).
36. Sagara, Y., Ishida, C., Inoue, Y., Shiraki, H., and Maeda, Y. 71-kilodalton heat shock cognate protein acts as a cellular receptor for syncytium formation induced by human T-cell lymphotropic virus type 1. *J. Virol.*, **72**, 535–541 (1998).
37. Sagara, Y., Inoue, Y., Kojima, E., Shiraki, H., and Maeda, Y. HTLV type 1 envelope glycoprotein gp46 evokes necrosis by binding to receptor complex. *AIDS Res. Hum. Retrovirus*, **16**, 1701–1704 (2000).
38. Sagara, Y., Inoue, Y., Kojima, E., Ishida, C., Shiraki, H., and Maeda, Y. Phosphatidylglycerol participates in syncytium formation induced by HTLV type 1-bearing cells. *AIDS Res. Hum. Retrovirus*, **17**, 125–135 (2001).
39. Sato, H., Hirata, J., Furukawa, M., Kuroda, N., Shiraki, H., Maeda, Y., and Okochi, K. Identification of the region including the epitope for a monoclonal antibody which can neutralize human Parvovirus B19. *J. Virol.*, **65**, 1667–1672 (1991).
40. Shimotohno, K., Takahashi, Y., Shimizu, N., Gojobori, T., Golde, D. W., Chen, I. S. K., Miwa, M., and Sugimura, T. Complete nucleotide sequence of an infectious clone of human T-cell leukemia virus type II: an open reading frame for the protease gene. *Proc. Natl. Acad. Sci. USA*, **82**, 3101–3105 (1985).
41. Sommerfelt, M. A., Williams, B. P., Clapham, P. R., Solomon, E., Goodfellow, P. N., and Weiss, R. A. Human T-cell leukemia viruses use a receptor determined by human chromosome 17. *Science*, **242**, 1557–1559 (1988).
42. Tajima, K. The T and B Cell Malignancy Study Group, and co-authors. The 4th nationwide study of adult T cell leukemia/lymphoma (ATL) in Japan: estimates of risk of ATL and its geographical and clinical features. *Int. J. Cancer*, **45**, 237–243 (1990).
43. Tajima, K., Tominaga, T., Suchi, T., Kawagoe, T., Tomoda, H., Hinuma, Y., Oda, T., and Fujita, K. Epidemiological analysis of the distribution of antibody to adult T-cell leukemia virus. *Jpn. J. Cancer Res.*, **73**, 893–901 (1982).
44. Tanaka, Y., Zeng, L., Shiraki, H., Shida, H., and Tozawa, H. Identification of a neutralization epitope on the envelope gp46 antigen of human T cell leukemia virus type 1 and induction of neutralizing antibody by peptide immunization. *J. Immunol.*, **147**, 354–360 (1991).

45. Trejo, S. R. and Ratner, L. The HTLV receptor is a widely expressed protein. *Virology*, **268**, 41–48 (2000).
46. Washitani, Y., Kuroda, N., Shiraki, H., Sato, H., Ohshima, K., Kiyokawa, H., and Maeda, Y. Linear antigenic regions of the structural proteins of human T-cell lymphotropic virus type I by enzyme-linked immunosorbent assay using synthetic peptides as antigens. *J. Clin. Microbiol.*, **30**, 287–290 (1992).
47. Wilson, K. A., Maerz, A. L., and Poumbourios, P. Evidence that the transmembrane domain proximal region of the human T-cell leukemia virus type 1 fusion glycoprotein gp21 has distinct roles in the prefusion and fusion-activated states. *J. Biol. Chem.*, **276**, 49466–49475 (2001).

PHYLOGENETIC ANALYSIS OF HTLV-1 AND STLV

The Past Movement of HTLV-1-Carriers and the Origin of PTLV

Sadayuki OHKURA, Tomoyuki MIURA, Masahiro YAMASHITA, Kentaro IBUKI, Taichiro TAKEMURA, and Masanori HAYAMI

*Laboratory of Viral Pathogenesis, Institute for Virus Research, Kyoto University**

Two types of human T-cell leukemia virus (HTLV-1 and -2) and three types of simian T-cell leukemia virus (STLV-1, -2, and -3) have been described to date. Seroepidemiological studies have shown that all types of HTLV and STLV prevail in Africa, whereas no STLV-related retrovirus has been confirmed among New World monkeys. Phylogenetic studies of HTLV-1 revealed that the virus is divided into three distinct genotypes (the Cosmopolitan, Central African, and Australo-Melanesian groups) and that the Cosmopolitan group is further subdivided into five subgroups (Transcontinental, Japanese, West African, North African, and Peruvian Black subgroups). Since these phylogenetic lineages are well linked with geographic origins and the anthropological background of carrier populations, phylogenetic analysis of HTLV-1 allows us to investigate ancient and recent human movements that might have resulted in the unique current distribution of HTLV-1 in the world. Molecular characterization of HTLV-1 and STLV-1 revealed that Central African HTLV-1 is interspersed within lineages of African STLV-1, indicating multiple episodes of interspecies transmission between humans and nonhuman primates in Africa. In contrast, Melanesian HTLV-1 and Asian STLV-1 are much more diverged than African viruses, suggesting that the virus has evolved over a long period in Asia. In this review article, we discuss recent views on the past movement of human populations that carried HTLV-1, and on the origin of HTLV-1/STLV-1. Recent findings of additional STLV-3 infection not only in East Africa, but in Central Africa are also discussed.

Since the discovery of human T-cell leukemia virus type 1 (HTLV-1) as the first human retrovirus (*30, 64*), HTLV-1 infections have been found among various ethnic groups in the world. This oncoretrovirus causes two separate diseases, adult T-cell leukemia (ATL) and HTLV-1-associated myelopathy/tropical spastic paraparesis (HAM/TSP) (*16, 30, 63, 64, 77*). The risk of development of HAM/TSP has been associated with some variants of *pX* gene products, such as p12 protein which is encoded by open reading frame (ORF) I (*38*) and tax (*12*). However, there is no evidence of a strong correlation between the pathogenesis of HTLV-1 and specific nucleotide mutations on the HTLV-1 proviral genome. On the other hand, accumulated sequence data have shown that the phylogenetic lineages of HTLV-1 are corre-

* 53 Kawahara-cho, Shogoin, Sakyo-ku, Kyoto 606-8501, Japan

lated with the geographic origins and ethnic backgrounds of HTLV-1-carriers (*19, 53, 59*). Because the major route of viral transmission is from mother to child via breast-feeding (*28, 29*), this correlation allows us to investigate past dissemination episodes of the retrovirus and to gain insight into the past movements of virus-carrying populations.

To clarify the evolutionary history and origin of HTLV, it has been helpful to examine its simian counterpart, simian T-cell leukemia virus (STLV). Because HTLV-1 and STLV-1 have similar genome organizations and biological features, these two viruses are referred to as 'primate T-cell leukemia virus type 1' (PTLV-1). PTLV is of three types: PTLV-1 (HTLV-1/STLV-1), PTLV-2 (HTLV-2/STLV-2), and PTLV-3 (STLV-3). PTLV-2 has been detected in both humans and nonhuman primates. PTLV-3 has so far been found only in nonhuman primates. This review article examines the phylogeny of PTLV in the world. Part 1 discusses the phylogeny of PTLV-1, including interspecies transmissions and origins of this genotype. Part 2 discusses PTLV-2 and PTLV-3, with an emphasis on STLV-3, including our most recent findings.

Part 1. Type 1 Viruses (HTLV-1 and STLV-1)

1-1. Phylogeny of HTLV-1/STLV-1

Three groups of HTLV-1 in the world

A comparison of nucleotide sequences between type 1 (HTLV-1) and type 2 (HTLV-2) indicates that the highest variability occurs in the long terminal repeats (LTR) region (*72*). This indicates that the LTR region of HTLV provirus has much more evolutionary information than the protein-encoding regions. Phylogenetic trees based on informative genomic regions such as the LTR region show similar topologies with respect to the classification of HTLV-1s into three major groups: the Cosmopolitan (HTLV-1a), Central African (HTLV-1b), and Australo-Melanesian (HTLV-1c) groups (*19, 53, 100*) (Fig. 1).

(1) The Cosmopolitan group (HTLV-1a) and its five subgroups

HTLV-1s of the Cosmopolitan group (HTLV-1a) are distributed all over the world. HTLV-1a is further divided into four subgroups based on the LTR region. These are: the Transcontinental (A), Japanese (B), West African (C) and North African (D) subgroups (*15, 53, 81, 85, 94*) (Fig. 1). Recently, an additional subgroup was reported in the Black population of Peru, designated as subgroup E (*83*). As the names of the Japanese, North African and West African phylogenetic subgroups imply, these subgroups are well linked with the geographic origins of carriers. On the other hand, the Transcontinental subgroup (subgroup A) contains closely related viral strains from many parts of the world including South Africa, India, the Middle East, South America, the Caribbean basin and Japan (*53, 95–98*). Although subgroup A of HTLV-1a has a worldwide distribution, viruses in this subgroup tend to form clusters that are associated with the geographic origins or ethnic backgrounds of the carriers. Examples are the "Middle Eastern cluster" and the "Latin American cluster" (*61, 83*) (Fig. 1). These specific clusters provide important clues for tracing the dissemination of certain HTLV-1 strains. For instance, an HTLV-1 strain found in Europe was suggested to have originated in South America by showing a direct link between these two areas.

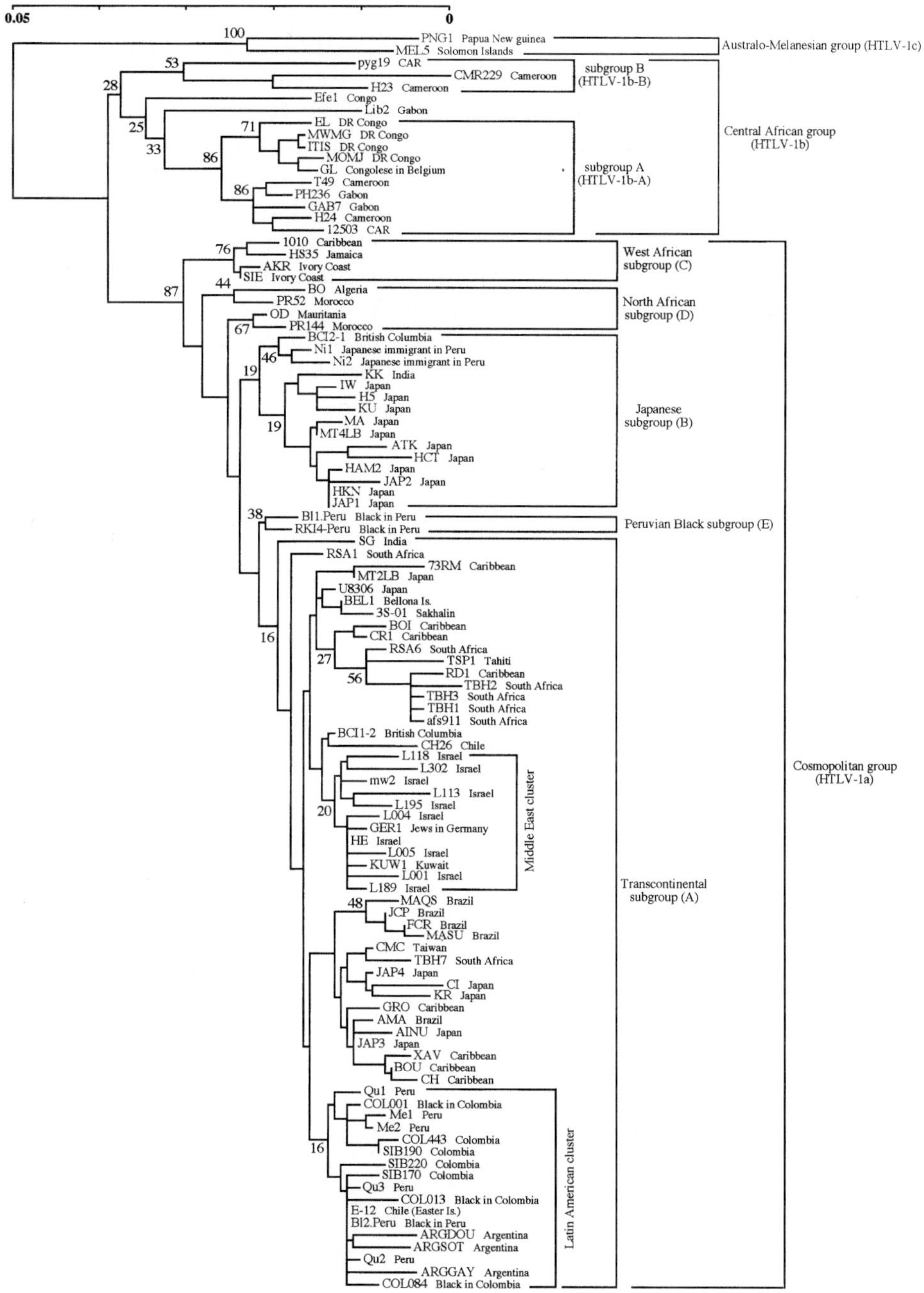

FIG. 1. Phylogenetic tree of HTLV-1s based on a partial LTR region (about 507 bp long, which corresponds to nucleotide positions 122 to 628 in ATK, a prototypic Japanese strain of HTLV-1a)

The tree was constructed by the Neighbor-Joining (NJ) method with a prototypic HTLV-2 strain, MoT (*70*), as a root of the tree. The scale at the top of the tree is the number of nucleotide substitutions per site estimated by Kimura's two parameter method, and the horizontal branch length indicates the genetic distance whereas the vertical length is for clarity only. Bootstrap values of 100 replications are represented at the nodes. The detailed procedure for the construction of the tree was described previously (*61*).

(2) The Central African group (HTLV-1b)

The Central African group of HTLV-1 (HTLV-1b) has been isolated exclusively from inhabitants in Central Africa: the Democratic Republic of Congo (DR Congo; formerly Zaire), Gabon, Cameroon, and Central African Republic (CAR). The carriers include both indigenous pygmies as well as Bantu-speakers. The nucleotide sequence of HTLV-1b differs from that of HTLV-1a by about 5%. HTLV-1b has at least two phylogenetic lineages. Although these lineages have been referred to as separate groups of HTLV-1 (HTLV-1b and HTLV-1d) (*47*), in this review article, we refer to them as the subgroups (A and B) of the Central African group (HTLV-1b) (Fig. 1). 'Subgroup A' of HTLV-1b (HTLV-1b-A) consists of the former HTLV-1b strains, and 'subgroup B' of HTLV-1b (HTLV-1b-B) consists of the diverged strains that have been referred to as HTLV-1d (*5, 47, 50, 56*). In addition, Salemi *et al.* reported the two diverged strains (Efe1 and Lib2) (*68*) (Fig. 1). At the moment, the precise grouping of these two strains is not clear.

(3) The Australo-Melanesian group (HTLV-1c)

HTLV-1c is the most diverged, and is found among native inhabitants of the Solomon Islands, Papua New Guinea, and Australia (*4, 17, 19, 60*). For example, the nucleotide sequence of a Papua New Guinean isolate, PNG-1 (*66*), differed by about 10% from ATK, a prototypic Japanese strain of HTLV-1a (*69*), based on the LTR

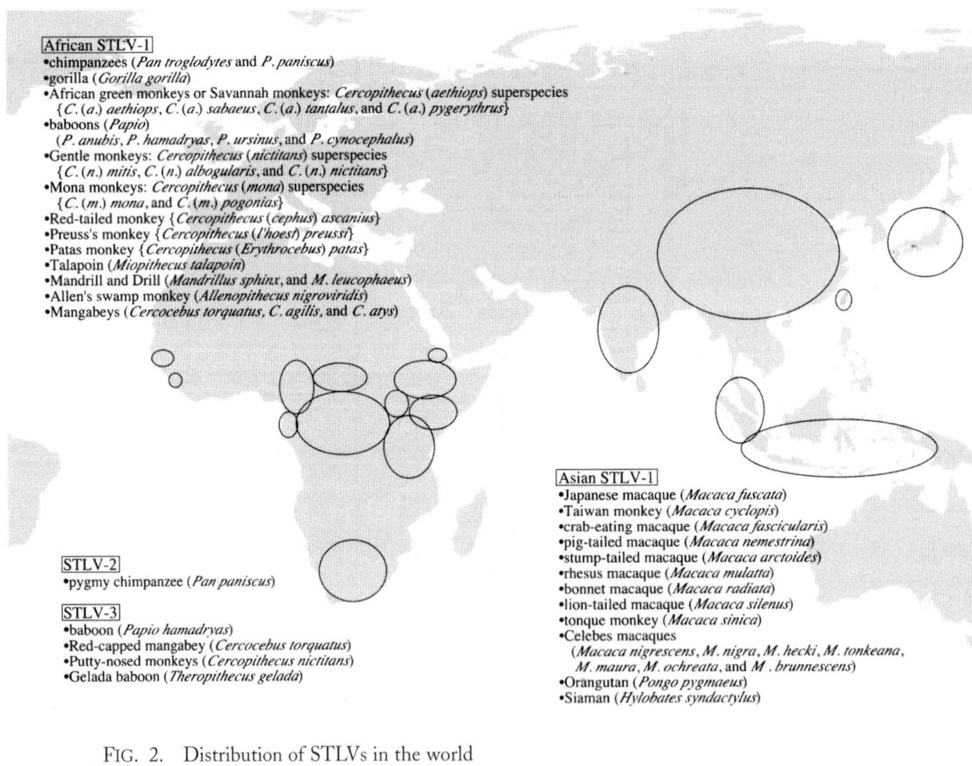

FIG. 2. Distribution of STLVs in the world

Circles represent the geographic areas where STLVs have been identified so far. For some strains, the geographic and species origins of their host nonhuman primates are not defined. Species names of STLV-carrying nonhuman primates are shown on the map. There has been no confirmed case of STLV infection in New World monkeys.

region. As shown in Fig. 1, HTLV-1c formed a distinct monophyletic cluster.

Epidemiology of STLV-1 in the world

Extensive serological surveys have revealed that many species of Old World nonhuman primates are naturally infected with STLV-1 (see Fig. 2). These nonhuman primates include chimpanzees, gorillas, monkeys of *Cercopithecus* species, baboons, mangabeys, and mandrills in Africa, and orangutans and macaque monkeys in Asia (Fig. 2). In marked contrast to Old World nonhuman primates, New World monkeys and prosimians do not appear to have antibodies to STLV-1 (*27, 31, 32*).

Whereas the seroprevalence of STLV-1 remains unknown among natural colonies or troops of each nonhuman primate, a few studies have shown that it increases with age (*32, 55*) and that STLV-1 infections were more frequent in females in the monkey colonies that were studied (*32, 45, 55*). As in the case of HTLV-1 infection in human, it has been suggested that viral transmissions occurred vertically from mothers to offspring, probably via breast-feeding, and horizontally from males to females via sexual contact (reviewed in Ref. *3, 18*).

Phylogeny of African STLV-1

Phylogenetic lineages of African STLV-1 have been shown to be closely associated with the geographic origins of their host monkeys, rather than with the species origins of the hosts (Fig. 3). This raised the possibility that interspecies transmission might have occurred among different species of African nonhuman primates in the past, and such episodes might still be going on. For example, the STLV-1s of eastern African monkeys, including African green monkeys, mitis monkeys, and baboons, resemble each other (Fig. 3). Similarly, analysis of a part of the LTR region showed that the STLV-1 strains of African green monkeys in southern Africa are similar to those of Chacma baboons in this geographic area (Fig. 3A). Since some of the above-mentioned phylogenetic lineages were not tightly clustered, further confirmatory investigations are needed for the precise grouping.

Phylogeny of Asian STLV-1

In contrast to the large amount of nucleotide sequence data on African STLV-1, little sequence data is available on Asian STLV-1. Nevertheless, extensive phylogenetic analyses have shown that, in general, Asian STLV-1 is much more diverged than African STLV-1 and HTLV-1 of the Cosmopolitan and Central African groups (HTLV-1a and -1b, respectively) (see Fig. 3). Many macaques living in Japan, Taiwan, Indonesia, China, and India, have been shown to have antibodies to STLV-1 (Fig. 4). In addition, our research group as well as other groups have found STLV-1 infections in orangutans, which live in Asia. Phylogenetic analyses based on the LTR region (*33*) and *pX* gene (*88*) have shown that STLV-1s of orangutans were interspersed within Asian STLV-1 lineages, and that these STLV-1s made a single cluster (supported by a bootstrap value of 100%) within the Asian STLV-1 lineages (data not shown) (*33*). Based on a partial sequence of the *env* gene, the STLV-1s from tonkeana monkeys (*32*) made a single cluster (Fig. 3B).

1-2. HTLV-1 of the Cosmopolitan group (HTLV-1a)

Dissemination of Cosmopolitan HTLV-1 in the world

Since West and North African Cosmopolitan HTLV-1 (HTLV-1a) (subgroups C and D, respectively) branched out at the node of the group (Fig. 1), HTLV-1a might have arisen in Africa (*44, 94*). On the other hand, since HTLV-1a did not

A.

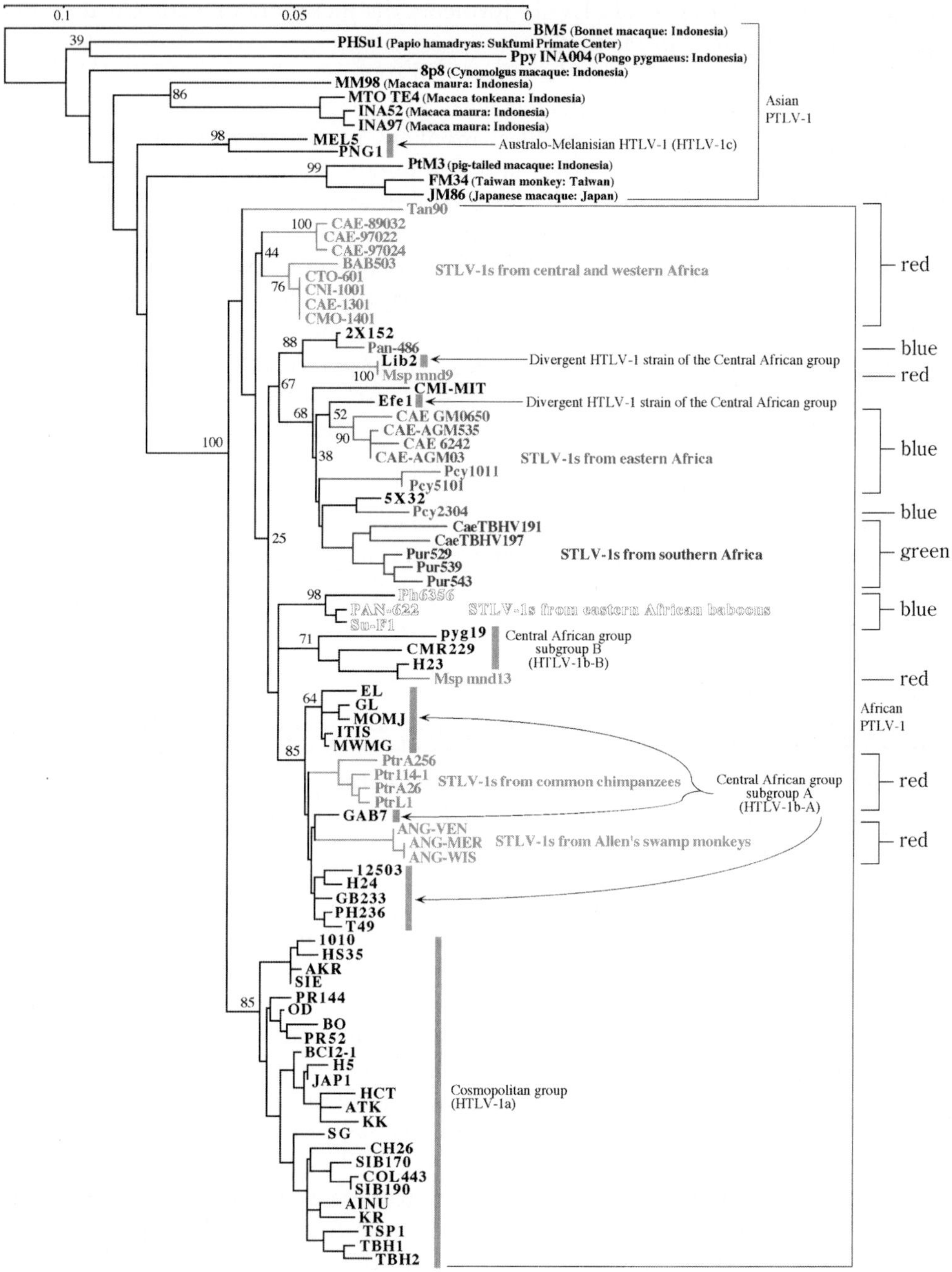

FIG. 3. Phylogenetic trees of STLV-1s including HTLV-1s, showing the evolutional relationship of STLV-1 and HTLV-1

(A) Phylogenetic tree based on a partial LTR region (about 500 bp long, corresponding to nucleotide positions 125 to 624 in HTLV-1 ATK). (B) Phylogenetic tree based on the 522 bp env fragment which encodes the 3′ end of gp46 and the 5′ end of gp21 proteins (corresponding to nucleotide positions 6068 to 6589 in ATK). The two trees have similar, but not identical, topologies. Phylogenetic lineages that are specific to the geographic origins of the hosts' nonhuman primates are indicated with colored letters. Blue indicates strains of eastern African monkeys, green indicates strains from south-

B.

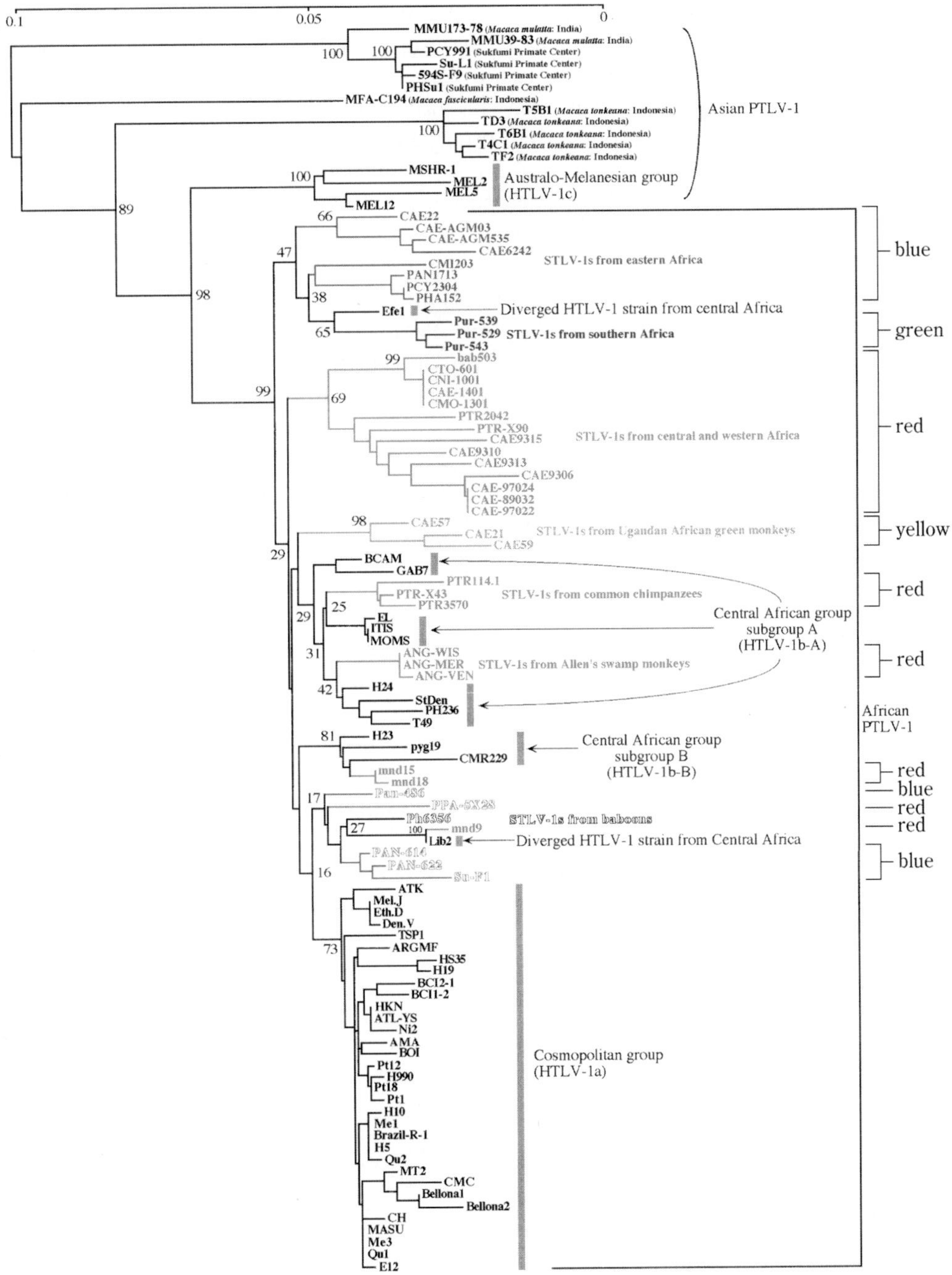

ern Africa, red indicates strains of central and western African monkeys, yellow indicates strains of Ugandan African green monkeys, and black indicates strains whose geographic origins are unknown. Since nucleotide sequences of the LTR region of the strains of Ugandan African green monkeys are not available, these strains are not shown in Fig. 4A. The tree was constructed by the NJ method with a prototypic HTLV-2 strain, MoT, as the root of the tree. Additional details are provided in the legend to Fig. 1. The LTR and env sequences used to construct the trees were obtained from studies cited in this paper (*8, 32, 34, 39, 42, 45, 46, 48, 52, 58, 65, 74, 79, 85, 86, 89–91, 93, 94*).

include any available STLV-1 (Fig. 3), dissemination of HTLV-1a in the world might have been accompanied by virus-carrying human movements. For example, in southern Africa, interspecies transmission from monkeys to humans seems unlikely, because HTLV-1a strains in this area did not resemble southern African STLV-1 strains. This observation implies that emergence of South African HTLV-1 was accompanied by recent human movements probably between South Africa and India (*98*). Nevertheless, the available phylogenetic data do not necessarily exclude the possibility that Asian HTLV-1a arose from Asian STLV-1, because there might exist a new Asian STLV-1 which is more closely affiliated with HTLV-1a.

In regard to world wide dissemination of HTLV-1a, especially the introduction of this genotype of HTLV-1 into the New World, the prevailing hypothesis is that African black slaves who were brought by European traders via the trans-Atlantic slave trade during the 16th to 18th centuries introduced this virus into the New World (*13, 14, 19, 39, 73*). This hypothesis is supported by phylogenetic studies of HTLV-1 based on a fragment of the *env* gene that encompasses most of the gp21-encoding region (522 bp long). These studies showed that HTLV-1 isolates from both western Africans and blacks living in the Caribbean basin and South America belong to the same Cosmopolitan group (HTLV-1a). However, since this env region is less informative than the LTR region (*72*), it cannot be used to further subdivide HTLV-1a.

On the other hand, based on the LTR region, HTLV-1 strains from western Africa, from where great numbers of slaves were sent to the New World, were clearly distinguished from those of South American inhabitants, including Native Americans (Fig. 1). Whereas only a few HTLV-1a strains from Caribbean blacks resembled West African HTLV-1a (subgroup C) viruses, the majority of South American and Caribbean HTLV-1 belonged to the Transcontinental subgroup (subgroup A) which did not include any available western African HTLV-1 (Fig. 1). Therefore, although there undoubtedly exist HTLV-1s that originated from African slaves in the New World, the trans-Atlantic slave trade does not appear to be the main route of intro-

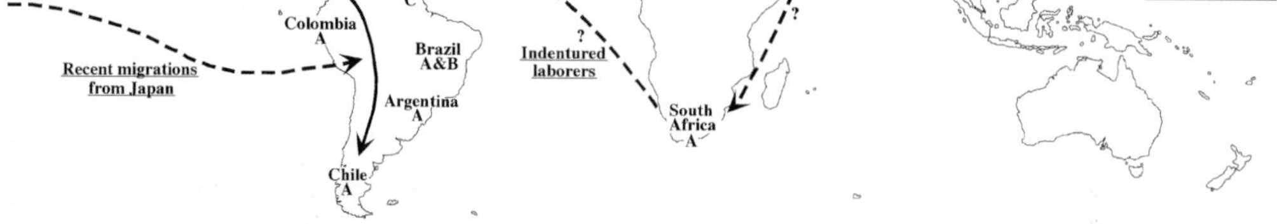

FIG. 4. Possible movements of human populations carrying the Cosmopolitan group of HTLV-1 (HTLV-1a) around the world, deduced from the topology of the phylogenetic tree and other anthropological data

Solid and broken lines indicate ancient and recent human movements, respectively.

duction of HTLV-1 on the continent. Rather, HTLV-1a seems to have been introduced to the New World by ancient human movements into the Americas from Asia, which might have followed human movements from Africa to East Asia (Fig. 4) (*53, 94*).

From Africa to Asia, and to the New World

As described above, the origin of HTLV-1a may be Africa (*44*) (Fig. 1). At the moment, little is known of the populations that contributed to the introduction of HTLV-1 to Asia. HTLV-1 in the Middle East and India may be a vestige of the movement of such virus-carrying populations from Africa to Asia.

Although when and how HTLV-1a was brought from Africa to Far East Asia is still cloaked in mystery, the spread of Cosmopolitan HTLV-1 (HTLV-1a) to Far East Asia might have resulted from the movement of Mongoloid peoples. The dissemination of HTLV-1a in Far East Asia, especially in Japan, is highly complicated for two reasons. First, there is little or no seroprevalence in East Asia other than Japan and Taiwan (*76*), and second, two separate subgroups (A and B) of HTLV-1a exist in Japan (*62, 80*). In particular, infection with HTLV-1 of the Japanese subgroup (subgroup B) mainly occurs in Japan and in Japanese immigrants in Brazil (*99*). Furthermore, subgroup B HTLV-1 mainly prevails among the Wajin (the mainland Japanese population living on the centrally located islands of Honshu, Shikoku, and Kyushu), whereas HTLV-1 of the Transcontinental subgroup (subgroup A) prevails among the Ryukyuans and the Ainu who live in the southernmost and northernmost parts of Japan, respectively (*62*). These findings suggest several independent migrations of virus-carrying Mongoloid lineages into Japan (*36, 37*). Therefore, elucidation of HTLV-1 dissemination in East Asia needs further analysis. The epidemiology of Japanese HTLV-1 will be discussed in another chapter (Tajima *et al.*).

Recent phylogenetic analyses of HTLV-1 have suggested that Mongoloid populations in East Asia further introduced the virus into the New World (*53, 94*). Phylogenetic studies based on the LTR region so far have demonstrated the presence of closely related HTLV-1 strains both among native Mongoloid lineages in Far East Asia, such as the Ainu in Japan (*53, 95*) and Nivkhi in Sakhalin (*97*), and among Native Americans. Native Americans are descended from Mongoloid peoples who first settled the Americas at least 10,000 years ago. In addition, ancient provirus of HTLV-1 was detected in an Andean mummy (*40*), and the ancient HTLV-1 was closely affiliated to HTLV-1 of the Ainu in Japan (strain AINU) as well as that of Native Americans in Colombia (strain SIB170) (*40*). These findings raise the possibility that HTLV-1 is an indigenous virus in South America. To summarize, a likely and simple scenario for this ancient origin of HTLV-1 in South America is as follows: the Asian antecedent of Native Americans carried HTLV-1 of the Transcontinental subgroup (subgroup A) from Far East Asia to the New World at least 10,000 years ago via the Bering land bridge, resulting in the existing presence of the same genotype in both areas. And later, HTLV-1 infections were spread to the new arrivals of the New World.

The history of HTLV-1 transmission in the New World is much more complicated than had previously been assumed. Recent studies have grouped some black South American HTLV-1s (strains COL001, COL013, COL084, and Bl2.Peru) (*54, 83*) with Native American isolates in subgroup A (Fig. 1). It seems likely that those black South American HTLV-1s were of Native American origin. Although many

Natives were annihilated by a combination of events including warfare, slavery, and epidemics, recent ethnogenetic studies reveal that the black South American populations that were investigated have preserved some of the genes from the original indigenous populations (*7*). For example, HLA analysis demonstrated a Native American contribution to the gene pool of the Colombian blacks, in spite of the fact that the investigated Colombian blacks were morphologically indistinguishable from African blacks (*11, 54*). From an evolutionary perspective, Native American HTLV-1 was successful, in that it spread to new arrivals on the continent.

1-3. *HTLV-1 of the Central African group (HTLV-1b) and African STLV-1*

HTLV-1 of the Central African group (HTLV-1b) is more diverged than that of the Cosmopolitan group (HTLV-1a) which prevails in western, northern, and southern Africa (Fig. 1). Molecular characterization of African STLV-1 revealed its close similarity to African HTLV-1 in genomic sequences. Indeed, HTLV-1b strains were interspersed within African STLV-1s (Fig. 3). For example, HTLV-1 strains of subgroup A of HTLV-1b (HTLV-1b-A) were closely related to STLV-1s from chimpanzees (*44*) and Allen's swamp monkeys (*52*), both of which live in Central Africa, with a statistically significant bootstrap value (85%) based on the partial (Fig. 3A) as well as the full-length (data not shown) LTR region. Strains of subgroup B of HTLV-1b (HTLV-1b-B) resembled STLV-1 of mandrills (*45*) with high bootstrap values of 71% and 81% based on the LTR region and the *gp21-env* gene, respectively (Fig. 3, A and B).

The observed close relatedness between Central African HTLV-1 and STLV-1 (Fig. 3, A and B) raises the possibility that interspecies transmission from nonhuman primates to humans occurred in Central Africa. Furthermore, such interspecies transmissions might have occurred several times because each of four types of HTLV-1 (subgroups HTLV-1b-A and HTLV-1b-B, and strains Efe1 and Lib2) made separate distinct clusters with African STLV-1 strains. Several other studies have also suggested that there have been multiple instances of interspecies transmission (*39, 44, 45, 47, 86, 91*). This seems reasonable because humans hunt and eat small monkeys in Africa, which allows hunters to be frequently exposed to, and therefore to have a high risk of acquiring simian viruses via blood contacts (*45*). Consistently, four of the five HTLV-1b-B strains described above were isolated from hunters (*45*).

Furthermore, the variability and high nucleotide homologies among African PTLV-1s support the African origin of PTLV-1, which implies that Asian STLV-1 was transmitted by movement of monkeys from Africa to Asia. For macaques, this migration appears to have occurred about 2,000,000 years ago (*9, 67*). Since PTLV-1 appears to have emerged at least 93,000±8,000 years ago (*67*), it is at least possible that it emerged before this migration.

1-4. *HTLV-1 of the Australo-Melanesian group (HTLV-1c) and Asian STLV-1*

As described above, the Australo-Melanesian group (HTLV-1c) is highly divergent and distinct from the other HTLV-1 genotypes. Recent phylogenetic analysis of Asian STLV-1s revealed that HTLV-1c strains were interspersed within Asian STLV-1 lineages (Fig. 3). The relation of the HTLV-1c to STLV-1 in Southeast Asia is quite different from that in Africa. That is, Asian STLV-1 was as divergent as

HTLV-1c, and no reported Asian STLV-1 resembled HTLV-1c strains as closely as the similarities between African HTLV-1 and STLV-1 (Fig. 3).

The observed genetic relations and high divergence of Asian PTLV-1 imply that interspecies transmissions occurred separately in Asia, raising the possibility of an Asian origin of PTLV-1 (*49, 66, 87*). Since Asian STLV-1 branched out before African STLV-1, the virus may have evolved for a longer time in Asian nonhuman primates than in African nonhuman primates. Because there is no evidence that monkeys migrated from Asia to Africa, all of these findings suggest that virus evolution in primates occurred in Asia independently from Africa.

On the other hand, interspecies transmission from nonhuman primates to humans might have taken place in Asia in the distinct past, and thereafter HTLV-1c strains might have evolved independently from the Asian STLV-1 (*39, 74*). This zoonosis might have occurred before the antecedent populations of the present-day Melanesians settled into the Solomon Islands and Papua New Guinea, neither of which has any indigenous nonhuman primates (*32, 44, 87, 94*). This is because, at present, Asian monkeys live along the route that ancestral settlers of Melanesia must have traveled in ancient times.

Taken together, the present analyses reasonably suggest that virus evolution and interspecies transmissions occurred on different continents (Africa and Asia). However, the available data are not sufficient to specify which continent is the place of origin of PTLV-1.

Part 2. Other Types of PTLV

2-1. PTLV-2

Only a brief description of the phylogeny of HTLV-2 will be given here as this subject has been covered by several other reviews (*18, 35, 67, 72*). In brief, HTLV-2 consists of two subgroups, HTLV-2a and -2b (*25*), and HTLV-2b seems to consist of several clusters (*35*). Both subgroups (HTLV-2a and -2b) are found in Amerindians and in Africans, as well as in intravenous drug users in the United States and Europe. Of particular interest is the presence of HTLV-2 among isolated pygmies who have inhabited Central Africa for a long time (*10, 20, 22, 24*). Moreover, HTLV-2 was isolated from Mongolia, an inland country on the Asian continent (*26*), but not in Far East Asia (*57*).

On the other hand, the distribution of STLV-2 is totally unknown because only a small number of STLV-2 infections have been identified. So far, two groups have independently identified STLV-2 infections in captive bonobo chimpanzees in DR Congo (*21, 43*). In contrast to the strong affiliation between HTLV-1 and STLV-1, the similarity between HTLV-2 and STLV-2 is somewhat limited. In spite of the limited homologies, STLV-2 is clearly related to HTLV-2 (see Fig. 5). For example, STLV-2PP1664 differed from HTLV-1ATK by 37%, whereas it differed from HTLV-2MoT by only 25% (*82*). Although an additional STLV-2 strain was isolated from spider monkeys in Panama (*6*), the presence of STLV-2 in the New World needs to be confirmed by further investigation. Therefore, there has not yet been any confirmed case of infection with an STLV-like retrovirus in New World monkeys.

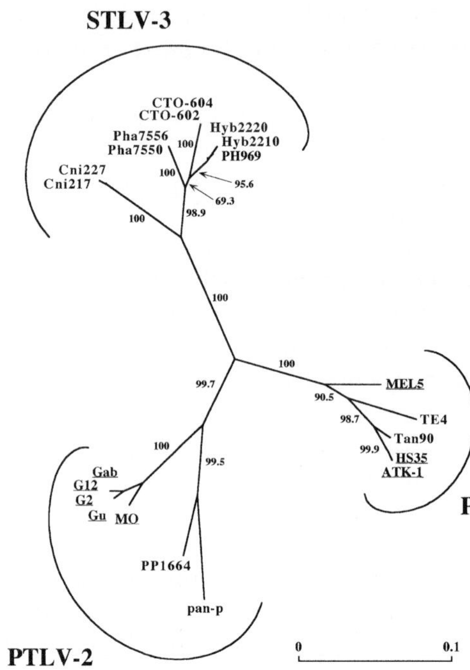

FIG. 5. Unrooted phylogenetic tree of PTLV including all of the three genotypes based on a short fragment of the *rex*-encoding gene, showing the phylogenetic relations of STLV-3 strains with PTLV-1 and -2 strains

The fragment is about 180 bp long, and corresponds to nucleotide positions 7491 to 7670 in PH969, the prototypic STLV-3 strain. The HTLV-1 and -2 strains are underlined in the tree. The tree was constructed by the NJ method. Bootstrap values of 1,000 replications are shown at the nodes. For details, see the legend to Fig. 1. The rex sequences used to construct the tree were obtained from references (*51, 78, 84*).

2-2. *Phylogenetic analysis of STLV-3*

In 1994, Goubau *et al.* (*23*) identified a third type of STLV (STLV-3; formerly PTLV-L) from a captive hamadryas baboon that was brought from Eritrea to a zoo in Belgium. This new STLV (PH969) is highly diverged and is equidistant from PTLV-1s and PTLV-2s in the phylogenetic tree (Fig. 5). In addition, the splicing pattern of the *pX* gene of PH969 has been shown to be slightly different from those of PTLV-1 and -2 (*82*).

Since its discovery, PH969 has been the sole representative of a monophyletic cluster. Recently, our research group conducted a serological survey of STLV infection among wild-caught monkeys in Ethiopia, a neighboring country of Eritrea. We found a high prevalence of STLV-3 among hamadryas baboons (50%) as well as among hybrids between hamadryas and anubis baboons (38%) (*78*). All baboons studied were from natural colonies in Ethiopia, and hamadryas and anubis baboons are pure populations as determined by genetic studies (*71*). In addition, Switzer *et al.* found two cases of STLV-3 infection in Ethiopian gelada baboons kept in captivity in US zoos (*75*). Collectively, our study as well as the previous ones indicated that STLV-3 is highly prevalent among hamadryas baboons in eastern Africa.

At the same time as our study, two European research groups identified STLV-3 in Cameroonian monkeys. Van Dooren *et al.* (*84*) reported two cases of STLV-3 infection among putty-nosed monkeys (2/122) (Cni217 and Cni227), and Meertens *et al.* (*51*) reported two cases of STLV-3 infection among wild-caught red-capped mangabeys (2/65) (CTO-602 and CTO-604). These findings suggest that STLV-3 is more widely distributed among various monkey species in Africa than was previously assumed.

A phylogenetic analysis of our STLV-3 strains from hybrid baboons based on an approximately 180-bp-long region of the *rex*-encoding gene (corresponding to nucleotide positions 7491 to 7670 in PH969) showed that they resembled PH969, and that they also made a group together with STLV-3 of pure hamadryas baboons and red-capped mangabeys (Fig. 5). This genetic lineage was supported statistically with a bootstrap value of 98.9%. On the other hand, the STLV-3 of putty-nosed monkeys was slightly different from the other STLV-3 strains, making a separate cluster with a bootstrap value of 100% (Fig. 5). This result implied the existence of at least two subgroups within STLV-3, although the precise subgrouping needs to be confirmed by analyses of other genomic regions.

The broad distribution and high prevalence of STLV-3 imply that interspecies transmission of STLV-3 from feral baboons to humans has occurred and will continue through accidental physical contact. This implication seems reasonable in view of the finding by several studies that multiple episodes of retroviral zoonoses, not only of STLV but also of simian immunodeficiency virus (SIV), have occurred in Africa. Furthermore, a western blotting assay indicated that a small percentage (0.43%) of Ethiopians exhibit HTLV-2-like sero-reactivity (*92*). Although the occurrence of HTLV-2 in Ethiopia is not necessarily related to the occurrence of STLV-3 in the baboons, in view of the high prevalence of STLV-3 in Ethiopia, it is possible that HTLV-2-like sero-reactivity among Ethiopians is due to infection with an STLV-3-like retrovirus.

Concluding Remarks and Future Perspectives

In the past two decades, much information on the molecular phylogeny and epidemiology of the PTLV retrovirus has become available. Such information has been useful to elucidate the past movement of HTLV carriers with the help of anthropological, genetic, and archeological knowledge. Thus, researchers in this field have proposed several hypotheses concerning ancient and recent human movements that might have contributed to the unique current distribution of HTLV. For example, the presence of the Transcontinental subgroup (subgroup A) of the Cosmopolitan group (HTLV-1a) among Native Americans could be interpreted to be a result of Mongoloid movement from Asia to the New World. In many cases, however, we are still unable to offer any explanation for the existence of certain isolates. For example, past dissemination episodes of subgroup A in many parts of the world cannot simply be explained by the available phylogenetic data and historical information. In addition, the present phylogenies do not necessarily reflect the precise evolutionary relationships of PTLV. Two potential sources of misestimation are the cross-contamination of monkey species by having different species share the same facilities (*90*), and the possibility that the evolutionary rate of PTLV proviruses differs among different primate species including humans. These differences may be due to different evolutionary rates of the host species or to differences in the host immune system or immune status (*41, 44, 67*). Therefore, at this moment, it would still be premature to draw a conclusive map of PTLV dissemination. Nevertheless, the phylogenetic information that has accumulated so far can offer new insights on the interdisciplinary field of ethnoepidemiology, and more extended phylogenetic analyses of PTLV would help to elucidate the evolutionary histories of the oncoretroviruses.

The high prevalence of STLV-3 among baboons should be a cause of concern because baboons have been considered as potential donors for xenotransplantation. In this case, investigations of the exogenous and endogenous retroviruses that the donor animals carry are essential to prevent the transmission of new infectious diseases to humans (*1, 2*). In fact, the US Food and Drug Administration recommends further research before proceeding with nonhuman primates xenotransplantation. Thus, development of systems to specifically screen certain retroviral pathogens and development of animal models to clarify their disease manifestations will be needed. Such animal models should also serve as useful tools for vaccine development as well as clarifying the molecular basis of PTLV pathogenesis.

Acknowledgments

We thank all of the members of our laboratory for helpful comments on this review article. This work was supported by Grants-in-Aid from the Japanese Ministry of Education, Culture, Sports, Science and Technology and by Research Fellowships of the Japan Society for the Promotion of Science for Young Scientists.

REFERENCES

1. Allan, J. S. Xenotransplantation at a crossroads: prevention versus progress. *Nat. Med.*, **2**, 18–21 (1996).
2. Allan, J. S. The risk of using baboons as transplant donors. Exogenous and endogenous viruses. *Ann. NY Acad. Sci.*, **862**, 87–99 (1998).
3. Allan, J. S., Leland, M., Broussard, S., Mone, J., and Hubbard, G. Simian T-cell lymphotropic viruses (STLVs) and lymphomas in African nonhuman primates. *Cancer Invest.*, **19**, 383–395 (2001).
4. Bastian, I., Gardner, J., Webb, D., and Gardner, I. Isolation of a human T-lymphotropic virus type I strain from Australian aboriginals. *J. Virol.*, **67**, 843–851 (1993).
5. Chen, J., Zekeng, L., Yamashita, M., Takehisa, J., Miura, T., Ido, E., Mboudjeka, I., Tsague, J. M., Hayami, M., and Kaptue, L. HTLV-I isolated from a Pygmy in Cameroon is related to but distinct from the known central African type. *AIDS Res. Hum. Retroviruses*, **11**, 1529–1531 (1995).
6. Chen, Y. M., Jang, Y. J., Kanki, P. J., Yu, Q. C., Wang, J. J., Montali, R. J., Samuel, K. P., and Papas, T. S. Isolation and characterization of simian T-cell leukemia virus type II from New World monkeys. *J. Virol.*, **68**, 1149–1157 (1994).
7. Crawford, M. "The Origins of Native Americans: Evidence from Anthropological Genetics" (1998). Cambridge University Press, UK.
8. Englebrecht, S., van Rensburg, E. J., and Robson, B. A. Sequence variation and subtyping of human and simian T-cell lymphotropic virus type I strains from South Africa. *J. Acquir. Immune Defic. Syndr. Hum. Retrovirol.*, **12**, 298–302 (1996).

9. Fa, J. A. The genus *Macaca*: a review of taxonomy and evolution. *Mamm. Rev.*, **19**, 45–81 (1989).
10. Froment, A., Delaporte, E., Dazza, M., and Larouze, B. HTLV-II among pygmies from Cameroon. *AIDS Res. Hum. Retroviruses*, **9**, 707 (1993).
11. Fujiyoshi, T., Yashiki, S., Fujiyama, C., Kuwayama, M., Miyashita, H., Onishi, H., Blank, M., Zaninovic, V., Blank, A., Cartier, L., Bynes, J. J., Harrington, J. W. J., Miura, T., Hayami, M., and Sonoda, S. Ethnic segregation of HTLV-I and HTLV-II carriers among South American native indians. *Int. J. Cancer*, **63**, 510–515 (1995).
12. Furukawa, Y., Yamashita, M., Usuku, K., Izumo, S., Nakagawa, M., and Osame, M. Phylogenetic subgroups of human T cell lymphotropic virus (HTLV) type I in the tax gene and their association with different risks for HTLV-I-associated myelopathy/ tropical spastic paraparesis. *J. Infect. Dis.*, **182**, 1343–1349 (2000).
13. Gallo, R. C. The first human retrovirus. *Sci. Am.*, **255**, 78–88 (1986).
14. Gallo, R. C., Sliski, A., and Wong-Staal, F. Origin of human T-cell leukemia-lymphoma virus. *Lancet*, **ii**, 962–963 (1983).
15. Gasmi, M., Farouqi, B., d'Incan, M., and Desgranges, C. Long terminal repeat sequence analysis of HTLV type I molecular variants identified in four north African patients. *AIDS Res. Hum. Retroviruses*, **10**, 1313–1315 (1994).
16. Gessain, A., Barin, F., Vernant, J., Gout, O., Maurs, L., Calender, A., and de The, G. Antibodies to human T-lymphotropic virus type-I in patients with tropical spastic paraparesis. *Lancet*, **ii**, 407–409 (1985).
17. Gessain, A., Boeri, E., Yanagihara, R., Gallo, R., and Franchini, G. Complete nucleotide sequence of a highly divergent human T-cell leukemia (lymphotropic) virus type I (HTLV-I) variant from Melanesia: genetic and phylogenetic relationship to HTLV-I strains from other geographical regions. *J. Virol.*, **67**, 1015–1023 (1993).
18. Gessain, A. and de The, G. Geographic and molecular epidemiology of primate T lymphotropic retroviruses: HTLV-I, HTLV-II, STLV-I, STLV-PP and PTLV-L. *Adv. Virus Res.*, **47**, 377–426 (1996).
19. Gessain, A., Gallo, R. C., and Franchini, G. Low degree of human T-cell leukemia/ lymphoma virus type I genetic drift *in vivo* as a means of monitoring viral transmission and movement of ancient human populations. *J. Virol.*, **66**, 2288–2295 (1992).
20. Gessain, A., Mauclere, P., Froment, A., Biglione, M., Le Hesran, J., Tekaia, F., Millan, J., and de The, G. Isolation and molecular characterization of a human T-cell lymphotropic virus type II (HTLV-II), subtype B, from a healthy Pygmy living in a remote area of Cameroon: an ancient origin for HTLV-II in Africa. *Proc. Natl. Acad. Sci. USA*, **92**, 4041–4045 (1995).
21. Giri, A., Markham, P., Digilio, L., Hurteau, G., Gallo, R. C., and Franchini, G. Isolation of a novel simian T-cell lymphotropic virus from *Pan paniscus* that is distantly related to the human T-cell leukemia/lymphotropic virus types I and II. *J. Virol.*, **68**, 8392–8395 (1994).
22. Goubau, P., Liu, H., De Lange, G. G., Vandamme, A. M., and Desmyter, J. HTLV-II seroprevalence in pygmies across Africa since 1970. *AIDS Res. Hum. Retroviruses*, **9**, 709–713 (1993).
23. Goubau, P., Van Brussel, M., Vandamme, A. M., Liu, H. F., and Desmyter, J. A primate T-lymphotropic virus, PTLV-L, different from human T-lymphotropic viruses

types I and II, in a wild-caught baboon (*Papio hamadryas*). *Proc. Natl. Acad. Sci. USA*, **91**, 2848–2852 (1994).

24. Goubau, P., Vandamme, A., Beuselinck, K., and Desmyter, J. Proviral HTLV-I and HTLV-II in the Efe pygmies of northeastern Zaire. *J. Acquir. Immune Defic. Syndr. Hum. Retrovirol.*, **12**, 208–209 (1996).
25. Hall, W., Takahashi, H., Liu, C., Kaplan, M., Scheewind, O., Ijichi, S., Nagashima, K., and Gallo, R. Multiple isolates and characteristics of human T-cell leukemia virus type II. *J. Virol.*, **66**, 2456–2463 (1992).
26. Hall, W., Zhu, S., Horal, P., Furuta, Y., Zagaamy, G., and Vahlne, A. HTLV-II infection in Mongolia. *AIDS Res. Hum. Retroviruses*, **10**, 443 (1994).
27. Hayami, M., Komuro, A., Nozawa, K., Shotake, K., Ishikawa, K., Yamamoto, T., Ishida, T., Honjo, S., and Hinuma, Y. Prevalence of antibodies to adult T-cell leukemia virus-associated antigens (ATLA) in Japanese monkeys and other non-human primates. *Int. J. Cancer*, **33**, 179–183 (1984).
28. Hino, S., Katamine, S., Miyata, H., Tsuji, Y., Yamabe, T., and Miyamoto, T. Primary prevention of HTLV-1 in Japan. *Leukemia*, **11** (Suppl. 3), 57–59 (1997).
29. Hino, S., Yamaguchi, K., Katamine, S., Sugiyama, H., Amagasaki, T., Kinoshita, K., Yoshida, Y., Doi, H., Tsuji, Y., and Miyamoto, T. Mother-to-child transmission of human T-cell leukemia virus type-I. *Jpn. J. Cancer Res.*, **76**, 474–480 (1985).
30. Hinuma, Y., Nagata, K., Hanaoka, M., Nakai, M., Matsumoto, T., Kinoshita, K., Shirakawa, S., and Miyoshi, I. Adult T-cell leukemia: antigen in an ATL cell line and detection of antibodies to the antigen in human sera. *Proc. Natl. Acad. Sci. USA*, **78**, 6476–6480 (1981).
31. Hunsmann, G., Schneider, J., Schmit, J., and Yamamoto, N. Detection of serum antibodies to adult T-cell leukemia virus in non-human primates and in people from Africa. *Int. J. Cancer*, **32**, 329–332 (1983).
32. Ibrahim, F., de The, G., and Gessain, A. Isolation and characterization of a new simian T-cell leukemia virus type 1 from naturally infected celebes macaques (*Macaca tonkeana*): complete nucleotide sequence and phylogenetic relationship with the Australo-Melanesian human T-cell leukemia virus type 1. *J. Virol.*, **69**, 6980–6993 (1995).
33. Ibuki, K. Phylogenetic analysis of simian T-cell leukemia virus type I from Asian apes including orangutan (*Pongo Pygmaeus*). *Ninth International Conference on Human Retrovirology: HTLV and Related Viruses, April 5–9, Kagoshima, Japan*, P194 (1999).
34. Ibuki, K., Ido, E., Setiyaningsih, S., Yamashita, M., Agus, L. R. P., Takehisa, J., Miura, T., Dondin, S., and Hayami, M. Isolation of STLV-I from orangutan, a great ape species in Southeast Asia, and its relation to other HTLV-Is/STLV-Is. *Jpn. J. Cancer Res.*, **88**, 1–4 (1997).
35. Ido, E., Yamashita, M., and Hayami, M. Ethnic variation of HTLV in the world. *Gann Monogr. Cancer Res.*, **44**, 107–120 (1996).
36. Ishida, T. and Hinuma, Y. The origin of Japanese HTLV-I. *Nature*, **322**, 504 (1986).
37. Ishida, T., Yamamoto, K., Omoto, K., Iwanaga, M., Osato, T., and Hinuma, Y. The prevalence of human retroviruses in native Japanese: evidence for a possible ancient origin. *J. Infect.*, **ii**, 153–157 (1985).
38. Johnson, J. M., Harrod, R., and Franchini, G. Molecular biology and pathogenesis of

the human T-cell leukaemia/lymphotropic virus Type-1 (HTLV-1). *Int. J. Exp. Pathol.*, **82**, 135–147 (2001).

39. Koralnik, I. J., Boeri, E., Saxinger, W. C., Monico, A. L., Fullen, J., Gessain, A., Guo, H. G., Gallo, R. C., Markham, P., Kalyanaraman, V., Hirsh, V., Allan, J., Murthy, K., Alford, P., Slattery, J. P., O'brien, S. J., and Franchini, G. Phylogenetic associations of human and simian T-cell leukemia/lymphotropic virus type I strains: evidence for interspecies transmission. *J. Virol.*, **68**, 2693–2707 (1994).
40. Li, H. C., Fujiyoshi, T., Lou, H., Yashiki, S., Sonoda, S., Cartier, L., Nunez, L., Munoz, I., Horai, S., and Tajima, K. The presence of ancient human T-cell lymphotropic virus type I provirus DNA in an Andean mummy. *Nat. Med.*, **5**, 1428–1432 (1999).
41. Li, W. H. and Tanimura, M. The molecular clock runs more slowly in man than in apes and monkeys. *Nature*, **326**, 93–96 (1987).
42. Liu, H., Goubou, P., van Brussel, M., Desmyter, J., and Vandamme, A. Phylogenetic analysis of a simian T lymphotropic virus type I from a Hamadryas baboon. *AIDS Res. Hum. Retroviruses*, **13**, 1545–1548 (1997).
43. Liu, H., Vandamme, A., Brussel, M., Desmyter, J., and Goubau, P. New retroviruses in human and simian T-lymphotropic viruses. *Lancet*, **344**, 265–266 (1994).
44. Liu, H. F., Goubau, P., Van Brussel, M., Van Laethem, K., Chen, Y. C., Desmyter, J., and Vandamme, A. M. The three human T-lymphotropic virus type I subtypes arose from three geographically distinct simian reservoirs. *J. Gen. Virol.*, **77**, 359–368 (1996).
45. Mahieux, R., Chappey, C., Georges-Courbot, M. C., Dubreuil, G., Mauclere, P., Georges, A., and Gessain, A. Simian T-cell lymphotropic virus type 1 from *Mandrillus sphinx* as a simian counterpart of human T-cell lymphotropic virus type 1 subtype D. *J. Virol.*, **72**, 10316–10322 (1998).
46. Mahieux, R., Chappey, C., Meertens, L., Mauclere, P., Lewis, J., and Gessain, A. Molecular characterization and phylogenetic analyses of a new simian T cell lymphotropic virus type 1 in a wild-caught African baboon (*Papio anubis*) with an indeterminate STLV type 2-like serology. *AIDS Res. Hum. Retroviruses*, **16**, 2043–2048 (2000).
47. Mahieux, R., Ibrahim, F., Mauclere, P., Herve, V., Michel, P., Tekaia, F., Chappey, C., Garin, B., Van Der Ryst, E., Guillemain, B., Ledru, E., Delaporte, E., de The, G., and Gessain, A. Molecular epidemiology of 58 new African human T-cell leukemia virus type 1 (HTLV-1) strains: identification of a new and distinct HTLV-1 molecular subtype in Central Africa and in Pygmies. *J. Virol.*, **71**, 1317–1333 (1997).
48. Mahieux, R., Pecon-Slattery, J., Chen, G. M., and Gessain, A. Evolutionary inferences of novel simian T lymphotropic virus type 1 from wild-caught chacma (*Papio ursinus*) and olive baboons (*Papio anubis*). *Virology*, **251**, 71–84 (1998).
49. Mahieux, R., Pecon-Slattery, J., and Gessain, A. Molecular characterization and phylogenetic analyses of a new, highly divergent simian T-cell lymphotropic virus type 1 (STLV-1marc1) in *Macaca arctoides*. *J. Virol.*, **71**, 6253–6258 (1997).
50. Mboudjeka, I., Zekeng, L., Yamashita, M., Takehisa, J., Ido, E., Miura, T., Ohkura, S., Ikeda, M., Kaptue, L., and Hayami, M. Prevalence and phylogenetic analysis of HTLV-I isolates in Cameroon, including those of the Baka Pygmy. *Jpn. J. Cancer Res.*, **88**, 619–624 (1997).
51. Meertens, L., Mahieux, R., Mauclere, P., Lewis, J., and Gessain, A. Complete sequence of a novel highly divergent simian T-cell lymphotropic virus from wild-

caught red-capped mangabeys (*Cercocebus torquatus*) from Cameroon: a new primate T-lymphotropic virus type 3 subtype. *J. Virol.*, **76**, 259–268 (2002).

52. Meertens, L., Rigoulet, J., Mauclere, P., Van Beveren, M., Chen, G. M., Diop, O., Dubreuil, G., Georges-Goubot, M. C., Berthier, J. L., Lewis, J., and Gessain, A. Molecular and phylogenetic analyses of 16 novel simian T cell leukemia virus type 1 from Africa: close relationship of STLV-1 from *Allenopithecus nigroviridis* to HTLV-1 subtype B strains. *Virology*, **287**, 275–285 (2001).
53. Miura, T., Fukunaga, T., Igarashi, T., Yamashita, M., Ido, E., Funahashi, S., Ishida, T., Washio, K., Ueda, S., Hashimoto, K., Yoshida, M., Osame, M., Singhal, B. S., Zaninovic, V., Cartier, L., Sonoda, S., Tajima, K., Ina, Y., Gojobori, T., and Hayami, M. Phylogenetic subtypes of human T-lymphotropic virus type I and their relations to the anthropological background. *Proc. Natl. Acad. Sci. USA*, **91**, 1124–1127 (1994).
54. Miura, T., Yamashita, M., Zaninovic, V., Cartier, L., Takehisa, J., Igarashi, T., Ido, E., Fujiyoshi, T., Sonoda, S., Tajima, K., and Hayami, M. Molecular phylogeny of human T-cell leukemia virus type I and II of Amerindians in Colombia and Chile. *J. Mol. Evol.*, **44** (Suppl. 1), S76–S82 (1997).
55. Mone, J., Whitehead, E., Leland, M., Hubbard, G., and Allan, J. S. Simian T-cell leukemia virus type I infection in captive baboons. *AIDS Res. Hum. Retroviruses*, **8**, 1653–1661 (1992).
56. Moynet, D., Cosnefroy, J. Y., Bedjabaga, I., Roelants, G., George-Courbot, M. C., and Guillemain, B. Identification of new genetic subtypes of human T cell leukemia virus type I in Gabon from encoding sequence of surface envelope glycoprotein. *AIDS Res. Hum. Retroviruses*, **11**, 1407–1411 (1995).
57. Neel, J. V., Biggar, R. J., and Sukernik, R. I. Virologic and genetic studies relate Amerind origins to the indigenous people of the Mongolia/Manchuria/southeastern Siberia region. *Proc. Natl. Acad. Sci. USA*, **91**, 10737–10741 (1994).
58. Nerrienet, E., Amouretti, X., Muller-Trutwin, M. C., Poaty-Mavoungou, V., Bedjebaga, I., Nguyen, H. T., Dubreuil, G., Corbet, S., Wickings, E. J., Barre-Sinoussi, F., Georges, A. J., and Georges-Courbot, M. C. Phylogenetic analysis of SIV and STLV type I in mandrills (*Mandrillus sphinx*): indications that intracolony transmissions are predominantly the result of male-to-male aggressive contacts. *AIDS Res. Hum. Retroviruses*, **14**, 785–796 (1998).
59. Nerurkar, V., Song, K., Bastian, I., Garin, B., Franchini, G., and Yanagihara, R. Genotyping of human T cell lymphotropic virus type I using Australo-Melanesian topotype-specific oligonucleotide primer-based polymerase chain reaction: insights into viral evolution and dissemination. *J. Infect. Dis.*, **170**, 1353–1360 (1994).
60. Nerurkar, V., Song, K., Saitou, N., Melland, R., and Yanagihara, R. Interfamilial and intrafamilial genomic diversity and molecular phylogeny of human T-cell lymphotropic virus type I from Papua New Guinea and the Solomon Islands. *Virology*, **196**, 505–513 (1993).
61. Ohkura, S., Yamashita, M., Cartier, L., Tanabe, D. G., Hayami, M., Sonoda, S., and Tajima, K. Identification and phylogenetic characterization of a human T-cell leukaemia virus type I isolate from a native inhabitant (Rapa Nui) of Easter Island. *J. Gen. Virol.*, **80**, 1995–2001 (1999).
62. Ono, A., Miura, T., Araki, S., Yamaguchi, K., Takatsuki, K., Mori, S., Hayami, M.,

Mochizuki, M., and Watanabe, T. Subtype analysis of HTLV-1 in patients with HTLV-1 uveitis. *Jpn. J. Cancer Res.*, **85**, 767–770 (1994).
63. Osame, M., Usuku, K., Izumo, S., Ijichi, N., Amitani, H., and Igata, A. HTLV-I associated myelopathy, a new clinical entity. *Lancet*, **i**, 1031–1032 (1986).
64. Poiesz, B., Ruscetti, F., Gazdar, A., Bunn, P., Minna, J., and Gallo, R. Detection and isolation of type C retrovirus particles from fresh and cultured lymphocytes of a patient with cutaneous T-cell lymphoma. *Proc. Natl. Acad. Sci. USA*, **77**, 7415–7419 (1980).
65. Saksena, N. K., Herve, V., Sherman, M. P., Durand, J. P., Mathiot, C., Muller, M., Love, J. L., LeGuenno, B., Sinoussi, F. B., Dube, D. K. *et al.* Sequence and phylogenetic analyses of a new STLV-I from a naturally infected tantalus monkey from Central Africa. *Virology*, **192**, 312–320 (1993).
66. Saksena, N. K., Sherman, M. P., Yanagihara, R., Dube, D. K., and Poiesz, B. J. LTR sequence and phylogenetic analyses of a newly discovered variant of HTLV-I isolated from the Hagahai of Papua New Guinea. *Virology*, **189**, 1–9 (1992).
67. Salemi, M., Desmyter, J., and Vandamme, A. M. Tempo and mode of human and simian T-lymphotropic virus (HTLV/STLV) evolution revealed by analyses of full-genome sequences. *Mol. Biol. Evol.*, **17**, 374–386 (2000).
68. Salemi, M., Van Dooren, S., Audenaert, E., Delaporte, E., Goubau, P., Desmyter, J., and Vandamme, A. M. Two new human T-lymphotropic virus type I phylogenetic subtypes in seroindeterminates, a Mbuti pygmy and a Gabonese, have closest relatives among African STLV-I strains. *Virology*, **246**, 277–287 (1998).
69. Seiki, M., Hattori, S., Hirayama, Y., and Yoshida, M. Human adult T-cell leukemia virus: complete nucleotide sequence of the provirus genome integrated in leukemia cell DNA. *Proc. Natl. Acad. Sci. USA*, **80**, 3618–3622 (1983).
70. Shimotohno, K., Takahashi, Y., Shimizu, N., Gojobori, T., Golde, D., Chen, I., and Sugimura, T. Complete nucleotide sequence of an infectious clone of human T-cell leukemia type II: an open reading frame for the protease gene. *Proc. Natl. Acad. Sci. USA*, **82**, 3101–3105 (1985).
71. Shotake, T. Population genetical study of natural hybridization between *Papio anubis* and *P. hamadryas*. *Primates*, **22**, 285–308 (1981).
72. Slattery, J. P., Franchini, G., and Gessain, A. Genomic evolution, patterns of global dissemination, and interspecies transmission of human and simian T-cell leukemia/lymphotropic viruses. *Genome Res.*, **9**, 525–540 (1999).
73. Song, K. J., Nerurkar, V. R., Pereira-Cortez, A. J., Yamamoto, M., Taguchi, H., Miyoshi, I., and Yanagihara, R. Sequence and phylogenetic analyses of human T cell lymphotropic virus type I from a Brazilian woman with adult T cell leukemia: comparison with virus strains from South America and the Caribbean basin. *Am. J. Trop. Med. Hyg.*, **52**, 101–108 (1995).
74. Song, K. J., Nerurkar, V. R., Saitou, N., Lazo, A., Blakeslee, J. R., Miyoshi, I., and Yanagihara, R. Genetic analysis and molecular phylogeny of simian T-cell lymphotropic virus type I: evidence for independent virus evolution in Asia and Africa. *Virology*, **199**, 56–66 (1994).
75. Switzer, W. M., Shanmugam, V., Van Dooren, S., Vandamme, A.-M., Bhullar, V., Parekh, B., and Heneine, W. Identification of a distinct STLV in an Ethiopian Gelada baboon (*Theropithicus gelada*). *AIDS Res. Hum. Retrovirol.*, **17**, S-11 (2001).

76. Tajima, K. and Hinuma, Y. Epidemiology of HTLV-I/II in Japan and the world. *Gann Monogr. Cancer Res.*, **39**, 129–149 (1992).
77. Takatsuki, K., Uchiyama, T., Sagawa, K., and Yodoi, J. *In* "Topics in Hematology," ed. S. Seno, F. Takaku, and S. Irino, pp. 73–77 (1977). Excerpta Medica Amsterdam, Amsterdam.
78. Takemura, T., Yamashita, M., Shimada, M. K., Ohkura, S., Shotake, T., Ikeda, M., Miura, T., and Hayami, M. High prevalence of simian T-lymphotropic virus type L in wild Ethiopian baboons. *J. Virol.*, **76**, 1642–1648 (2002).
79. Tsujimoto, H., Komuro, A., Iijima, K., Miyamoto, J., Ishikawa, K., and Hayami, M. Isolation of simian retroviruses closely related to human T-cell leukemia virus by establishment of lymphoid cell lines from various non-human primates. *Int. J. Cancer*, **35**, 377–384 (1985).
80. Ureta Vidal, A., Gessain, A., Yoshida, M., Mahieux, B., Nishioka, K., Tekaia, F., Rosen, L., and de Thé, G. Molecular epidemiology of HTLV type I in Japan: evidence for two distinct ancestral lineages with a particular geographical distribution. *AIDS Res. Hum. Retroviruses*, **10**, 1557–1566 (1994).
81. Ureta Vidal, A., Gessain, A., Yoshida, M., Tekaia, F., Garin, B., Guillemain, B., Schulz, T., Farid, R., and de Thé, G. Phylogenetic classification of human T cell leukemia/lymphoma virus I genotypes in five major molecular and geographical subtypes. *J. Gen. Virol.*, **75**, 3655–3666 (1994).
82. Van Brussel, M., Salemi, M., Liu, H. F., Gabriels, J., Goubau, P., Desmyter, J., and Vandamme, A. M. The simian T-lymphotropic virus STLV-PP1664 from *Pan paniscus* is distinctly related to HTLV-2 but differs in genomic organization. *Virology*, **243**, 366–379 (1998).
83. Van Dooren, S., Gotuzzo, E., Salemi, M., Watts, D., Audenaert, E., Duwe, S., Ellerbrok, H., Grassmann, R., Hagelberg, E., Desmyter, J., and Vandamme, A. M. Evidence for a post-Columbian introduction of human T-cell lymphotropic virus [type I] [corrected] in Latin America. *J. Gen. Virol.*, **79**, 2695–2708 (1998).
84. Van Dooren, S., Salemi, M., Pourrut, X., Peeters, M., Delaporte, E., Van Ranst, M., and Vandamme, A.-M. Evidence for a second simian T-cell lymphotropic virus type 3 in *Cercopithecus nictitans* from Cameroon. *J. Virol.*, **75**, 11939–11941 (2001).
85. Vandamme, A. M., Liu, H. F., Goubau, P., and Desmyter, J. Primate T-lymphotropic virus type I LTR sequence variation and its phylogenetic analysis: compatibility with an African origin of PTLV-I. *Virology*, **202**, 212–223 (1994).
86. Vandamme, A. M., Liu, H. F., Van Brussel, M., De Meurichy, W., Desmyter, J., and Goubau, P. The presence of a divergent T-lymphotropic virus in a wild-caught pygmy chimpanzee (*Pan paniscus*) supports an African origin for the human T-lymphotropic/simian T-lymphotropic group of viruses. *J. Gen. Virol.*, **77**, 1089–1099 (1996).
87. Vandamme, A. M., Salemi, M., and Desmyter, J. The simian origins of the pathogenic human T-cell lymphotropic virus type I. *Trends Microbiol.*, **6**, 477–483 (1998).
88. Verschoor, E. J., Warren, K. S., Niphuis, H., Heriyanto, Swan, R. A., and Heeney, J. L. Characterization of a simian T-lymphotropic virus from a wild-caught orang-utan (*Pongo pygmaeus*) from Kalimantan, Indonesia. *J. Gen. Virol.*, **79**, 51–55 (1998).
89. Voevodin, A., Miura, T., Samilchuk, E., and Schatzl, H. Phylogenetic characterization of simian T lymphotropic virus type I (STLV-I) from the Ethiopian sacred baboon

(*Papio hamadryas*). *AIDS Res. Hum. Retroviruses*, **12**, 255–258 (1996).

90. Voevodin, A., Samilchuk, E., Allan, J., Rogers, J., and Broussard, S. Simian T-lymphotropic virus type 1 (STLV-1) infection in wild yellow baboons (*Papio hamadryas cynocephalus*) from Mikumi National Park, Tanzania. *Virology*, **228**, 350–359 (1997).
91. Voevodin, A. F., Johnson, B. K., Samilchuk, E. I., Stone, G. A., Druilhet, R., Greer, W. J., and Gibbs, C. J., Jr. Phylogenetic analysis of simian T-lymphotropic virus type I (STLV-I) in common chimpanzees (*Pan troglodytes*): evidence for interspecies transmission of the virus between chimpanzees and humans in Central Africa. *Virology*, **238**, 212–220 (1997).
92. Vrielink, H., Sisay, Y., Reesink, H. W., Woerdeman, M., Winkel, C., de Leeuw, S. J., Lelie, P. N., and van der Poel, C. L. Evaluation of a combined lysate/recombinant antigen anti-HTLV-I/II ELISA in high and low endemic areas of HTLV-I/II infection. *Transfus. Med.*, **5**, 135–137 (1995).
93. Watanabe, T., Seiki, M., Tsujimoto, H., Miyoshi, I., Hayami, M., and Yoshida, M. Sequence homology of the simian retrovirus genome with human T-cell leukemia virus type I. *Virology*, **144**, 59–65 (1985).
94. Yamashita, M., Ido, E., Miura, T., and Hayami, M. Molecular epidemiology of HTLV-I in the world. *J. Acquir. Immune Defic. Syndr. Hum. Retrovirol.*, **13**, S124–S131 (1996).
95. Yamashita, M., Ishida, T., Ohkura, S., Miura, T., and Hayami, M. Phylogenetic characterization of a new HTLV type 1 from the Ainu in Japan. *AIDS Res. Hum. Retroviruses*, **17**, 783–787 (2001).
96. Yamashita, M., Kitze, B., Miura, M., Weber, T., Fujiyoshi, T., Takehisa, J., Chen, J., Sonoda, S., and Hayami, M. The phylogenetic relationship of HTLV-I from non-Mashhadi Iranians to that from Mashhadi Jews. *AIDS Res. Hum. Retroviruses*, **11**, 1533–1535 (1995).
97. Yamashita, M., Senyuta, N., Pavlish, O., Miura, T., Syrtsev, A., Stepina, V., Ibuki, K., Susova, O., Schrbak, L., Schatzl, H., Shih, J., Yakovleva, L., Weber, J., Gessain, A., de Thé, G., Hayami, M., and Gurtsevitch, V. Prevalence and phylogenetic characterization of HTLV-I isolates from Far Eastern Russia. *Mol. Biol.*, **29**, 1172–1182 (1995).
98. Yamashita, M., Takehisa, J., Miura, T., Ido, E., Becker, W. B., Robson, B. A., Becker, M. L., and Hayami, M. Presence of the widespread subtype of HTLV-I in South Africa. *AIDS Res. Hum. Retroviruses*, **11**, 645–647 (1995).
99. Yamashita, M., Veronesi, R., Menna-Barreto, M., Harrington, W. J., Jr., Sampio, C., Brites, C., Badaro, R., Andrade-Filho, A. S., Okhura, S., Igarashi, T., Takehisa, J., Miura, T., Chamone, D., Bianchini, O., Jardim, C., Sonoda, S., and Hayami, M. Molecular epidemiology of human T-cell leukemia virus type I (HTLV-1) Brazil: the predominant HTLV-1s in South America differ from HTLV-ls of Japan and Africa, as well as those of Japanese immigrants and their relatives in Brazil. *Virology*, **261**, 59–69 (1999).
100. Yanagihara, R., Nerurkar, V. R., Garruto, R. M., Miller, M. A., Leonmonzon, M. E., Jenkins, C. L., Sanders, R. C., Liberski, P. P., Alpers, M. P., and Gajdusek, D. C. Characterization of a variant of human T-lymphotropic virus type I isolated from a healthy member of a remote, recently contacted group in Papua New Guinea. *Proc. Natl. Acad. Sci. USA*, **88**, 1446–1450 (1991).

HTLV-I WORLDWIDE PATTERNS AND DISEASE ASSOCIATIONS

Elizabeth M. MALONEY[*1] and William A. BLATTNER[*2]

*Viral Epidemiology Branch, Division of Cancer Epidemiology and Genetics, National Cancer Institute[*1] and Institute of Human Virology, University of Maryland[*2]*

The 1980 publication reporting the discovery of the first human retrovirus, human T-lymphotropic virus type I (HTLV-I), culminated a decade-long search for the human homologue to cancer-causing retroviruses initially identified in the first decade of the 20th century. HTLV-II, the second human retrovirus, was discovered in 1982. This review will focus on HTLV-I. Since the discovery of HTLV-I, extensive progress has been made in characterizing the worldwide distribution and disease associations of this virus. The geographic distribution of HTLV-I varies based on the three types, termed Cosmopolitan, Central African, and Australo-Melanesian. Of these three types, the most widespread worldwide is the Cosmopolitan strain, having four subtypes: subtype A (Transcontinental), subtype B (Japanese), subtype C (West African), and subtype D (North African). Japan and the Caribbean are major areas of HTLV-I endemicity where prevalence ranges from 3 to 15% and is characterized by an age-dependent rise commencing in adolescence, with higher rates in females compared to males. HTLV-I is transmitted from mother to child usually via breastfeeding, parenterally by transfusion of infected cellular blood products or exposure to contaminated needles, and sexually from male to female, female to male and male to male. Heightened transmission is associated with the presence of high antibody titer and elevated proviral load, while elevated *tax*-specific antibody is linked to decreased transmission. A history of sexually transmitted infection, particularly those causing an ulcerative genital lesion, is also associated with increased risk for HTLV-I transmission. Diseases associated with HTLV-I share an inability to down-modulate viral expression. Some conditions such as adult T-cell leukemia/lymphoma, originally detected in Japan in 1977, result from a monoclonal expansion of HTLV-I infected T-cells. The ability of HTLV-I and particularly the Tax protein to disrupt cell cycle regulatory pathways provides a model for understanding the pathogenesis of HTLV-associated disease.

In 1977, Takatsuki characterized adult T-cell leukemia/lymphoma (ATL) as a new disease that clustered epidemiologically in the southern islands of Japan, leading to the speculation that an infectious agent might account for this pattern (*128*). The 1980 discovery of HTLV-I by Gallo linked the first human retrovirus to this lymphoid cancer (*102, 103*). The longstanding search for a human homologue to cancer-

[*1] Bethesda, Maryland, USA
[*2] Baltimore, Maryland, USA

causing retroviruses in animals, first discovered at the beginning of the 20th century, ended at a time when most researchers had abandoned the quest and focused instead on viral transforming genes that occur as oncogenes in human tumors. Techniques employed in the isolation of this virus and human T-cell lymphotropic virus type II (HTLV-II) two years later were applied to the isolation of the human immunodeficiency virus (HIV-1) in 1983, which was shown to be the cause of AIDS in 1984 (*7, 30, 32*). These initial breakthroughs were followed by the molecular characterization of both human *Retroviridae, Deltaretrovirus* (formerly termed oncornavirus) (HTLV) and *Lentivirus* (HIV) (*31*).

Geographic Distribution of HTLV-I

HTLV-I strains are classified into three types that exhibit up to 7.5% sequence variation and are termed the Cosmopolitan, Central African and the Australo-Melanesian types (*35*). The Cosmopolitan strain has the widest distribution worldwide and consists of four subtypes that vary by 0–3% (*78, 151*). Subtype A (Transcontinental) occurs in Japan, the Caribbean, Colombia, Chile, India, Iran and South Africa. Subtype B (Japanese) is found in Japan and India. Subtype C (West African) is found in the Caribbean and West Africa and subtype D (North African) is found in North Africa. There are serotypes that correspond to these genetically defined types.

HTLV-I has infected an estimated 15–20 million people worldwide, with the majority experiencing no disease (*22*). Over the lifetime of the infected carriers, an estimated 3–5% will develop an aggressive T-cell malignancy, ATL, and another 2% will develop a chronic inflammatory disease, HTLV-I associated myelopathy/tropical spastic paraparesis (HAM/TSP) (*6, 21, 68*).

In all endemic areas, HTLV-I prevalence increases with age and is approximately twice as high among females ≥40 years compared to males of the same age (*83, 86*). The distribution of HTLV-I prevalence is characterized by microgeographic clustering resulting in areas of high prevalence located adjacent to regions where infection is not found. Areas of high HTLV-I prevalence are usually geographically isolated areas where distinct ethnic groups experience little immigration or emigration, allowing the persistence of viral transmission (*119*).

Japan is an area of high HTLV-I prevalence. HTLV-I is present among the three ethnically distinct populations of Japan: the Ainu of the northern island of Japan (Hokkaido), the Wajin of the central islands of Honshu, Shikoku and Kyushu, and the Ryukyuans of the most southern islands (Okinawa). However, population prevalence rates range from 0–1% in Hokkaido to 15% in Okinawa, increasing steadily along a north to south gradient (*37, 48, 108*). In the Goto Islands and the village of Miyazaki in southern Japan, approximately 30% of residents ≥40 years of age are HTLV-I seropositive (*119, 124*). The high prevalence rates in southern Japan are likely due to the geographic isolation of the populations of the islands (*84*).

The Caribbean is another major area of HTLV-I endemicity. Seroprevalence rates among blood donors or pregnant women in Haiti, Dominican Republic, Martinique, Barbados, St. Lucia, Jamaica, Trinidad and Tobago range from 2.2 to 11% (*1, 15, 73, 86, 107, 109*). The Central American country of Honduras has a low prevalence (0.3%), although a high prevalence (17.7%) was detected among persons of African descent who resided on the Caribbean coast of Honduras (*114, 140*). Panama also has a low prevalence (5%) of HTLV-I (*106*).

Pockets of endemicity in South America include areas of Brazil, Colombia, Venezuela, Argentina, and Peru where prevalence rates among blood donors and pregnant women range from <1.0 to 6.7% (*13, 38, 63*). Several native ethnic peoples residing in the Andes highlands have HTLV-I prevalence rates ranging from 0.8 to 6.0% (*4, 38, 155*). However, prevalence rates of 1.0% and 2.8% were reported among the Embera Amerindians and persons of African descent living in Tumaco, along the Pacific coast of Colombia (*4, 134*). The highest HTLV-I prevalence (6.7%) in South America was reported among the population of Maripasoula, French Guiana (*98*).

Areas of Central and Western Africa have prevalence rates ranging from <1.0 to 10.9% based on studies among pregnant women or the general population. Affected countries include Mozambique, Nigeria, Zaire, Guinea-Bissau, Gabon, Cameroon, and Equatorial Guinea (*2, 12, 24, 61, 75, 146*). The Seychelles have a prevalence of 4.9% (*23, 62*)

Melanesia is endemic for a variant strain of HTLV-I (*152*). Prevalence in Papua New Guinea ranges from 0 to 14.6% with seropositive persons found in both coastal and hinterland regions (*126*). HTLV-I prevalence in aboriginal communities in Australia is as high as 14% (*11*). However, none of the healthy volunteers or blood donors in Jakarta or Irian Jaya of neighboring Indonesia were seropositive for HTLV-I (*126, 129*).

In the Middle East, HTLV-I prevalence is limited to the Iranian Jews originating from the city of Mashad in Khurusan, Northeastern Iran. The prevalence among a small group of these Iranian Jewish residents in a home for the elderly was reported as 12%, and later determined to be 20% in a population-based sample (*76, 77*). Zero to 1.3% of non-Mashad Israelis from Iran were HTLV-I seropositive (*76, 77*).

The US and Europe have low prevalence rates of HTLV-I. Less than 1% (0.025%) of blood donors from several US centers were HTLV-I seropositive (*148*). Similarly, in western European countries, prevalence rates among blood donors ranged from 0 to 0.007% (*130*). Most seropositive blood donors in the US and Europe were born in an endemic country or had sexual relations with persons from endemic countries (*113, 130*). The European countries of Russia, Latvia and Romania have prevalence rates ranging from 0.1 to 0.6% (*130*).

Low prevalence of HTLV-I was found among blood donors in Korea (0.13%) and no seropositives were detected among blood donors in China (0%), although blood donors in Taiwan had a very low prevalence (0.07%) (*19, 33, 54*). However, higher prevalence rates were reported for Fujian Province, China (2.0%) and Cheju Island, Korea (0.76%), located adjacent to endemic areas of Japan (*53, 154*). Zero prevalence was found in Thailand (*66*).

HTLV-I prevalence is associated with race and ethnicity. Throughout the world, the areas of highest prevalence are inhabited by persons of Japanese or African descent. Striking examples of this effect include Trinidad, where 41% of the population are of African descent and 40% are of East Indian descent, and prevalence of HTLV-I among persons of African descent is 3.2% compared to 0.2% among persons of Asian descent (*15*). In the Seychelles, HTLV-I prevalence was significantly higher among persons of African descent than Caucasians or persons of mixed race (*62*). In a population-based study of a village in French Guiana, all of the HTLV-I seropositive persons belonged to three of the six ethnic groups included in the study, and varied from 8% among the Noir-Marron who are descendants of African slaves to 4.1% and 3.6% among the Creoles and mixed Noir Marron/other ethnicity, respectively (*98*). Among blood donors in Sao Paulo, Brazil, HTLV-I prevalence was associated with Asian ancestry (*26*).

HTLV-I prevalence is also associated with markers of low socioeconomic status. Examples of this effect include a population of older (≥ age 55) persons in Guinea-Bissau and US blood donors for whom HTLV-I prevalence was significantly higher among those with a low level of education (*74, 113*). In the Seychelles and Jamaica, HTLV-I prevalence was significantly associated with low level of occupation (*62, 86*). In Trinidad, HTLV-I was associated with poor sanitation, a marker of low socioeconomic status (*15*).

HTLV-I Transmission

Mother-to-child transmission

HTLV-I is transmitted from infected mothers to their offspring via breastfeeding, parenterally by transfusion of infected cellular blood products or exposure to contaminated needles, and sexually. Maternal-child transmission occurs in 18–26% of breastfed offspring between 4 and 27 months of age in Jamaica and 6 to 24 months in Kagoshima, Japan (*3, 43, 55, 57, 125, 145*). Maternal antibody (HTLV-I IgG, but not IgM) is evident in virtually all cord blood specimens from children born to HTLV-I seropositive mothers, but wanes after 6–18 months (*43, 125, 135, 145*). While higher maternal-antibody level has been associated with decreased maternal-child transmission of HTLV-I in bottle-fed babies and babies breast-fed for <12 months, breastfeeding duration beyond the waning of maternal antibody is believed to result in maternal-child transmission (*52, 135, 145*) (Hisada, M., personal communication). In Jamaica, the maternal-child transmission rate in children who breastfed for <12 months was 9% compared to 32% among children who breastfed for ≥12 months (*145*). The effect of breastfeeding duration is supported in all studies that examined this association, and appears to be the strongest risk factor for maternal-child transmission (*3, 43, 55, 57, 125, 135*). However, seroconversion occurs in 3–6% of bottle-fed children, indicating that breast milk is not the exclusive vehicle of maternal-child transmission (*125, 135*).

Maternal HTLV-I antibody titer, maternal HTLV-I proviral load, and maternal *tax*-specific antibody have been shown to be risk factors for maternal-child transmission (*44, 111, 139, 145*). However, in a study that mutually adjusted for all three viral markers, only maternal antibody and proviral load remained independent predictors (Hisada, M. *et al.*, personal communication). Other risk factors for maternal-child transmission include low socioeconomic status and long (>4 h) duration of ruptured membranes (*145*). Additionally, a segregation analysis applied to a family study in French Guiana supported a role for a dominant gene predisposing to HTLV-I infection in children (*99*).

Sexual transmission

Sexual transmission of HTLV-I is a major route of infection, with male-to-female transmission considered to be four times more likely than the reverse (*118*).

Risk of male-to-female sexual transmission among couples is 60% over a 10-year period of sexual exposure and is associated with duration of sexual relationship, high HTLV-I proviral load, high antibody titer and presence of *tax*-specific antibody in the transmitting partner (*47, 49, 121, 124*). However, female-to-male sexual transmission was the only discernible risk factor for three US Marines involved in marital or prolonged non-marital sexual relationships in Okinawa, Japan (*16*). Male-to-male sexual transmission was also supported by the higher prevalence (15%) of HTLV-I among homosexual men in Trinidad compared to the general population prevalence (2.4%) (*10*). However, there was no significant difference in the HTLV-I prevalence between homosexual men and the population of food handlers in Jamaica (*87*).

Among food handlers in Jamaica, blood donors in the US, and homosexual males in Trinidad, a high number of sexual partners was a significant risk factor for seropositivity (*10, 90, 113*). Current diagnosis of syphilis among males and females and a history of penile ulcers in males were associated with HTLV-I seropositivity among Jamaican food handlers, while treponemal antibody seropositivity was associated with HTLV-I in a historical survey in Barbados (*85, 107*). History of syphilis and diagnosis of either syphilis or gonorrhea were associated with infection among female attendees at sexually transmitted infection (STI) clinics in Jamaica and Kyushu, Japan (*27, 92*). Among male STI clinic patients in Kyushu, HTLV-I seropositivity was associated with diagnosis of non-gonococcal urethritis (*92*). Among male and female STI clinic attendees in Jamaica, bruising during sex was associated with HTLV-I infection (*27*). It is likely that in the absence of prolonged exposure to a seropositive sexual partner, sexual transmission requires exposure to multiple partners and the presence of an inflammatory lesion or otherwise disrupted genital epithelium.

Parenteral Transmission

Parenteral transmission by transfusions or drug injection is another means of acquiring infection. Transfusion-transmission of HTLV-I is highly efficient. Risk of seroconversion is 44% within 2 months of transfusion with an HTLV-I seropositive cellular unit (*72*). Risk factors for transfusion-transmission include short shelf-life of blood unit (<1 week), male gender and current use of immunosuppressive medications (i.e. corticosteroids) (*72*). Risk associated with immunosuppressive medications implies that a blunted cellular immune response to HTLV-I following exposure increases susceptibility. Screening of the blood supply has been implemented in many countries where HTLV-I has been reported, including Jamaica, Japan, United Kingdom, France, Portugal, Netherlands, Denmark, Sweden, Australia and Israel, thereby eliminating transfusion-transmission of HTLV-I in many areas.

Parenteral drug abuse has been associated with transmission of HTLV-I and HTLV-II, a retrovirus which shares approximately 65% homology to HTLV-I. In the US most HTLV-positive injecting drug users (IDUs) are HTLV-II infected, whereas in Salvador, Brazil, 72% of HTLV-positive IDUs are HTLV-I infected (*25, 142*). Risk factors for HTLV-II seroconversion in the US include sharing drug abuse paraphernalia and "booting," which involves blood exposure (notably the use of an eyedropper rather than a syringe), and receiving money for sex (*142*). In Salvador, Brazil, HTLV-I prevalence among IDUs is twice as high in females compared to males (*25*), and duration of injecting drug use and diagnosis of syphilis are risk factors for

HTLV-I infection in both males and females (*25*). Within the setting of injection drug use, similar risk factors appear to be operating for HTLV-I and HTLV-II transmission.

Parenteral drug use of prescription medications using non-disposable syringes was associated with a two-fold increase in HTLV-I prevalence among the S. Pacific coastal population of Guapi, Colombia (*134*). In pregnant women in Quillabamba, Peru, surgery was a significant risk factor for HTLV-I seropositivity (*155*). Other means of parenteral transmission are rare. While Schreiber *et al.* described accidental exposure to needle sticks or sharp instruments as a significant risk factor for HTLV-I in a study of US blood donors in 1991–1992, there were no instances of HTLV-I seropositivity among the 1,000 persons in the survey who reported prior needle stick exposure in the US (*113*).

HTLV-I Associated Diseases

Adults

ATL is characterized by hypercalcemia, hepatomegaly, splenomegaly, lymphadenopathy, skin involvement, and presence of abnormal lymphocytes which are differentially distributed by subtype (*127*). ATL includes four subtypes: acute (57%), lymphoma (19%), chronic (19%), and smoldering (5%) (*115*). All subtypes have a monoclonal integration of HTLV-I proviral DNA within leukemic cells and a proliferation of CD4+ CD25+ cells (*137*). Acute and lymphoma subtypes are associated with a rapidly progressive course with survival time measured in months. The chronic subtype is associated with longer survival (2 years) and the smoldering subtype has a more indolent course (*40, 115*). In endemic regions, such as the Caribbean and Japan, the lifetime risk of an infected individual developing ATL is 3 to 5% (*6, 21, 101*). In Jamaica, the incidence rate is estimated as 2 per 10^5 general population and 16 per 10^5 HTLV-I carriers per year (*70*). The incidence of ATL in the general population of Japan is the same as in Jamaica (*127*).

HTLV-I is the major single cause of lymphoma accounting for 50–59% of total non-Hodgkin's lymphoma incidence in Jamaica and Japan (*6, 70*). ATL incidence is higher (75 per 10^5 carriers per year) among the Jamaican population infected in childhood, and 70% of lymphomas diagnosed in persons $<$ age 40 years in this population are attributed to HTLV-I infection (*21, 70*). HTLV-I seropositivity has been documented in 97–100% of mothers of ATL patients (*9, 147*). Coupled with the decline in the HTLV-I associated attributable risk for ATL after age 40, these factors strongly support exposure to HTLV-I in early life as an important risk factor for this lymphoid malignancy with a latency of decades (*70*).

The second HTLV-I-associated disease, HAM/TSP, was linked to HTLV-I in 1985 (*34*). HAM/TSP is a chronic, degenerative, neurologic disease characterized by weakness and hyperreflexia of the lower limbs, bladder dysfunction, and impotence. The incidence of HAM/TSP in Jamaica is 1.8 per 10^5 general population and approximately 20 per 10^5 HTLV-I-positive population (*68*). The lifetime risk of developing HAM/TSP among HTLV-I infected persons is 1.9% (*68*). The mean age of diagnosis is approximately 45 years in Jamaica and Japan, and females are three times more likely to develop HAM/TSP compared to males (*56, 68, 97*). In Jamaica, this excess in females was suggested to be due to markers of sexual transmission, as HAM/TSP cases were significantly more likely to have initiated sex at an early age and to have had more sexual partners (*56*). History of receiving a blood transfusion was a significant risk factor for HAM/TSP in Japan (*97*). HAM/TSP has been diagnosed within 3 years following receipt of a transfusion of an HTLV-I-positive blood unit (*39, 72, 112*). While rare cases have been diagnosed in childhood, HAM/TSP is generally associated with adult acquired infection via transfusion or sexual transmission.

HTLV-I-associated uveitis (HAU) is the third disease to be linked to HTLV-I infection in 1992 (*80*). Uveitis is an inflammatory disease of the eye that responds to corticosteroid therapy, but relapses post-treatment at intervals of 2 months to 2 years (*80*). Forty percent of uveitis patients previously considered to be idiopathic were associated with HTLV-I (*79, 80*). The mean age of patients is 43 years, although risk of uveitis associated with HTLV-I was significantly greater among younger patients (20 to 40 years) (*79*). The incidence of uveitis in Japan is similar to that of ATL; however, the incidence of uveitis in Jamaica is unknown.

Other HTLV-I-associated diseases among adults include Sjögren's syndrome, facial nerve palsy, arthropathy, polymyositis, and lymphadenitis (*8, 82, 93, 117, 120, 132*). Subclinical conditions have been associated with HTLV-I seropositivity in studies of carriers. HTLV-I was associated with detection of mild anemia among Japanese men and severe anemia among Jamaican food handlers (*47, 90, 120*). A significantly decreased eosinophil percent was associated with HTLV-I in Jamaica and Japan (*90, 144*). A high prevalence of abnormal lymphocytes or 'flower cells' was associated with HTLV-I among US blood donors and older Japanese adults, and in the Japanese study abnormal lymphocytes were associated with high HTLV-I proviral load (*83, 110, 114*). Lymphopenia was associated with HTLV-I in elderly male Japanese emigrants to Hawaii, while lymphocytosis was associated with HTLV-I carriers in Japan with abnormal lymphocytes (*47, 144*). Other HTLV-I-associated conditions include asthma among males and abnormal ECG, supporting an earlier report of an association of HTLV-I with history of cardiac disease (*83, 120*). HTLV-I was associated with lymphadenopathy and gait abnormality among US blood donors (*88*). Among the US blood donors HTLV-I was also associated with occurrence of bladder or kidney infections, which supports the HTLV-I association with the history of kidney disease in the Miyazaki, Japan cohort (*89, 120*).

Children

Infective dermatitis (ID) in children was first linked with HTLV-I infection in Jamaica in 1990, although it was reported as a unique dermatitis as early as 1966 and 1967 in Jamaica (*58, 123, 143*). ID is a severe, exudative eczema of the scalp, axillae, groin, nasal folds, eyelids, retroauricular areas and neck, with onset in childhood (*59*). Concurrent infections with *Staphylococcus aureus* and/or B-hemolytic streptococci of the skin and anterior nares, generalized lymphadenopathy, anemia, hyperimmunoglobulinemia, and elevated CD4 cell count and erythrocyte sedimentation rate are common features (*59*). The age at diagnosis ranges from 2 years to 16 years, but commonly occurs at 2 years of age (*17*; LaGrenade, L., personal communication). HTLV-I is the only known risk factor for this disease, although immune dysregulation and host genetic factors may play a role (*58, 59*). ID has been reported in other HTLV-I endemic areas including Trinidad, Dominican Republic, Barbados, Colombia, and rarely in Japan (*67, 105, 122, 136*).

HTLV-I infection in children has recently been significantly associated with other skin diseases, seborrheic dermatitis, and eczema, as well as subclinical conditions including severe anemia and persistent hyperreflexia in Jamaican children (Maloney, E. M., personal communication). Infected children with either seborrheic dermatitis or severe anemia had a two-fold higher HTLV-I proviral load preceding their diagnoses. HTLV-I was also associated with lymphadenopathy, presence of abnormal lymphocytes, and low eosinophil count, which were all of borderline statistical significance (Maloney, E. M., personal communication). It appears that there is a range of skin diseases associated with HTLV-I infection in children, while other subclinical conditions more frequently detected in infected children are similar to what is reported for adults.

Although ATL and HAM/TSP rarely occur in childhood, six cases have been reported in children who ranged in age from 1 year to 16 years. Skin disease was noted at presentation in five children, three of whom also had a history of skin disease (*36, 64, 65, 141, 149*). Additionally, five adults with either ATL or HAM/TSP had documented childhood diagnoses of infective dermatitis (*41, 60*). These data raise the speculation that infective dermatitis of childhood may be a harbinger of risk for ATL or HAM/TSP (*41, 60*).

Supporting this speculation are reports of childhood cases of ATL and HAM/TSP with histories of skin diseases. Ten cases of pediatric ATL clinically similar to adult ATL were described in Brazilian children ages 2 to 18 years for whom vertical transmission was documented in eight of the ten children (*104*). HAM/TSP has been reported in five Brazilian children ages 11 to 17 years over an 11-year period (*5*). All but one of these children were breastfed by HTLV-I-positive mothers, and were reported to have a history of dermatitis at the time of diagnosis (*5*). These data provide support for ID as a harbinger of risk for ATL and HAM/TSP.

HTLV-I Viral Markers

The natural history of viral markers has been studied in two adult cohorts of HTLV-I carriers. In a prospective study of Jamaican transfusion recipients of HTLV-I-positive blood with known dates of infection, the level of antibody titer established early in infection was linked to the level of proviral DNA load at time of infection and remained at that level (*71*). This same steady state of both proviral load and antibody titer characterized carriers from Miyazaki, Japan, who were followed for 11 years after infection (*95*). In cross-sectional studies of carriers, HTLV-I antibody titer and proviral load were significantly correlated among adults and children (r=0.58 for both) (*81*; Maloney, E. M., personal communication). These studies suggest that although proviral load and antibody titer varied between individuals, they remained at a steady level over time, indicating that in a healthy state proviral load is under immune control.

HTLV-I proviral DNA level is significantly higher in ATL and HAM/TSP patients compared to HTLV-I seropositive, asymptomatic carriers, as measured by copies of HTLV-I proviral DNA per peripheral blood mononuclear cells (PBMC) (*71, 91*), *in situ* PCR (*42*), reverse transcriptase (RT)-PCR (*28*), inverse PCR (*18*) percent integration of proviral DNA using Southern blot (*153*), and nested PCR (*133*). Proviral DNA load was as much as 16-fold higher in HAM/TSP patients compared to carriers (*71, 91*). Proviral load correlated highly with *tax* messenger RNA (mRNA), and *tax* mRNA correlated with both the frequency of *tax*-specific CD8+ T-cells and disease severity in HAM/TSP patients (*150*). These data suggest that *tax* mRNA is a marker of disease progression in HAM/TSP (*150*). Because of the high correlation between proviral DNA and *tax* mRNA levels, it is likely that both play a role in disease development. It is unknown whether HTLV-I proviral load is significantly higher in patients with ID, although these patients do have elevated levels of HTLV-I antibody titer and *tax*-specific antibody similar to what is reported for HAM/TSP patients (LaGrenade, L., personal communication). Because these two viral markers are highly correlated, ID patients are expected to have similarly elevated proviral DNA levels. Extrapolating from these data, HTLV-I proviral load and *tax* mRNA level may serve as markers of disease risk.

HTLV-I antibody titer levels are significantly higher in HAM/TSP patients compared to HTLV-I carriers (*71*), even when examined within strata of HTLV-I proviral load (*116*). ATL patients' median antibody titer levels are also significantly higher than those of carriers, but not different from patients with HAM/TSP (*71*). However, *tax*-specific antibody is detected in almost all (94%) HAM/TSP patients compared to approximately 60% of ATL patients and 48% of asymptomatic carriers (*51, 116*). While HTLV-I antibody titer may be another important marker of risk for disease development, *tax*-specific antibody may provide a tool for discriminating risk of developing ATL from risk of developing HAM/TSP. Therefore, additional study of this marker in the context of disease development is warranted.

The rare occurrence of ATL and HAM/TSP coupled with latency of disease development makes it difficult to study pre-diagnostic levels of viral markers in order to assess them as markers of disease development. In one study of three ATL patients prospectively studied for 6 years preceding ATL diagnosis, the median pre-diagnostic proviral load of 4.9 copies per 100 PBMC 6 years preceding development of ATL, was significantly higher than the median proviral load of 0.8 copies per 100 PBMC in age- and sex-matched carrier controls (*96*). A prospective study of twenty HTLV-I carriers at high risk of disease documented elevated HTLV-I proviral load an average of 10 months preceding development of HTLV-I-associated inflammatory disease in five carriers (*131*). The single child who developed ID in a Jamaican cohort study had an HTLV-I proviral DNA level at the time of infection in the range of that of adult ATL and HAM/TSP patients, suggesting that proviral load may be an important marker of risk of ID development in children (*69*). In one study of five ATL patients who were prospectively studied for 10 years preceding their diagnosis of ATL, risk of developing ATL was associated with a high HTLV-I antibody titer 10 years preceding disease development (*45*). Proviral load testing was not conducted on these specimens but due to the expected high correlation of these markers it is likely that proviral load accounts for the association of risk with antibody titer (*45*). Collectively, these sparse but important data suggest that HTLV-I proviral load is a useful marker of disease risk. The availability of a test to measure *tax* mRNA provides another potential marker of disease risk.

Comparative Epidemiology: the Caribbean and Japan

There are several differences in the epidemiology of HTLV-I and associated diseases between the Caribbean and Japan. In Jamaica, prevalence rises from 1.7% (10–19 years of age) to 9.1% (≥70 years) in males, and from 1.9% (10–19 years of age) to 17.4% (≥70 years) in females, with the initial female excess in seroprevalence in the third decade (*86*). Among Japanese adults, the increased seroprevalence in females compared to males occurs after the sixth decade (*84*).

The increased prevalence with age and the approximately two-fold increase in prevalence among adult women compared to adult men has been interpreted in different ways. One explanation is that more efficient sexual transmission of HTLV-I from males to females results in a relative increase in HTLV-I infections among females throughout their adulthood. In a cohort study of persons ≥40 years of age in Miyazaki, Japan, nine seroconversions occurred over a 10-year period. Twenty-seven point three percent of the women married to seropositive men seroconverted compared to 6.7% of the men married to seropositive women, where older age of the husband was found to be a risk factor (*118*). The authors speculated that post-menopausal females might be at increased risk of infection due to increased fragility of the vaginal epithelium, increased infectivity of older men, or long duration of exposure to an infected husband (*83*). Additionally, the 6.0% annual risk of male-to-female transmission among married couples over age 40 years (*121*) may explain the later age of rise in HTLV-I prevalence among women in Japan where long-term marital unions are the norm. In Jamaica, risk factors for HTLV-I seropositivity among female food handlers included high number of lifetime sexual partners, number of different men fathering a child by the woman and use of an intrauterine device (IUD) (*90*). These factors, coupled with the infrequent use of condoms by Jamaican women compared to the prevalent use of condoms by Japanese women, may explain the rise in prevalence among Jamaican females in the early years of their sexual activity (*14, 90*).

There may be additional differences in the dynamics influencing the age-prevalence curves in the Caribbean and Japan. In Japan, there is data supporting declining prevalence by birth-cohort. Serial surveys conducted between 1966 and 1988 and between 1990 and 1999 on women in their 20's and 30's who lived in Okinawa and first time blood donors in Nagasaki, showed declines in prevalence within age-groups which were significant among the blood donor group (*20, 138*). The authors speculated that the widespread supplanting of breastfeeding with bottle-feeding in the 1950s and improvements in socioeconomic conditions over time could account for this change, and discounted any effect of blood donor screening, as seropositive persons were not informed of their status until the late 1990s (*20*). The projected decrease in HTLV-I prevalence in Japan due to cultural and socioeconomic changes and public health interventions was estimated to be 50% per decade for both sexes, suggesting that HTLV-I will be non-endemic within a few generations (*20*). A decrease in prevalence from one generation (11.4%) to the next (1.3%) was reported among adult Okinawan immigrants in Hawaii that could be due to changes in cultural practices (i.e. increasing marriage outside of the Okinawan community by the younger generation) (*47*).

In the Caribbean, there is no data to support a cohort effect. The prevalence among pregnant Jamaican women was 3.8% in 1989–1990 and approximately 3.5% in 1996–2000 indicating that seroprevalence in women with a mean age of 27 years remained stable over this 11-year period (*145*; Maloney, E. M., personal communication). Additional studies are needed to corroborate this finding. However, it is likely that endemic countries which continue to practice breastfeeding for health, socioeconomic, and cultural reasons will not experience a cohort effect in seroprevalence.

An additional difference in HTLV-I epidemiology between the Caribbean and Japan is the difference in age of diagnosis for ATL patients. ATL patients from Jamaica, Trinidad, and Brazil are younger at diagnosis (43 years) than in Japan (54 years) (*97*). Additionally, in Jamaica males and females have the same incidence of diagnosis, while in Japan there is a male excess such that 2.5 male cases are diagnosed for each female case (*6, 70*). An explanation for geographic differences in age of diagnosis may be increased exposure to other infectious agents resulting in increased antigenic stimulation among Caribbean carriers, resulting in a shorter latency among Caribbean patients. In a study of ATL patients in Martinique, those with *Strongyloides stercoralis* (Ss) infection had a significantly earlier age of disease onset (39 years) compared to those without Ss infection (70 years), supporting a role for other infections in earlier age of diagnosis (*100*). In a comparison of HTLV-I proviral load in carriers with and without Ss infection, proviral load was more than five times higher in those with Ss infection and the increase was associated with oligoclonal expansion of infected cells (*29*). These data provide a potential mechanism to explain how other infectious agents may effect a shortened latency in some ATL patients (*29*).

Currently, there is inadequate data to explain the gender difference in ATL incidence in Japan, which is not observed in the Caribbean. Compared to females, male carriers in Japan have a higher percent of abnormal lymphocytes, which were associated with elevated HTLV-I proviral load, both of which are potential markers for ATL risk (*83*). These factors coupled with increased mortality among male patients with ATL lend support to the concept of a gender difference in ATL incidence in Japan (*46*). Further study of the differences in these factors between carriers in the Caribbean and Japan is needed to address the geographic differences.

Another difference in the epidemiology of HTLV-I-associated diseases between these two endemic areas pertains to the incidence of HAM/TSP. In Jamaica and Trinidad, the incidence of HAM/TSP is approximately 2 per 10^5 general population and 22 per 10^5 infected individuals (*68*). This incidence rate per infected individuals is much higher than that reported for Japan (3.1 per 10^5) (*50*). Risk factors for HAM/TSP differ between these two areas, with sexual risk factors predominating in the Caribbean, and blood transfusion predominating in Japan (*56*).

REFERENCES

1. Allain, J. P., Hodges, W., Einstein, M. H., Geisler, J., Neilly, C., Delaney, S., Hodges, B., and Lee, H. Antibody to HIV-1, HTLV-I, and HCV in three populations of rural Haitians. *J. Acquir. Immune Defic. Syndr.*, **5**, 1230–1236 (1992).
2. Analo, H. I., Akanmu, A. S., Akinsete, I., Njoku, O. S., and Okany, C. C. Seroprevalence study of HTLV-1 and HIV infection in blood donors and patients with lymphoid malignancies in Lagos, Nigeria. *Cent. Afr. J. Med.*, **44**, 130–134 (1998).
3. Ando, Y., Nakano, S., Saito, K., Shimamoto, I., Ichijo, M., Toyama, T., and Hinuma, Y. Transmission of adult T-cell leukemia retrovirus (HTLV-I) from mother to child: comparison of bottle- with breast-fed babies. *Jpn. J. Cancer Res.* (*Gann*), **78**, 322–324

(1987).

4. Arango, C., Maloney, E., Rugeles, M. T., Bernal, E., Bernal, C., Borrero, I., Herrera, S., Restrepo, A., Espinal, A., and Blattner, W. A. HTLV-I and HTLV-II coexist among the Embera and Inga Amerindians of Colombia. *J. Acquir. Immune Defic. Syndr. Hum. Retrovirol.*, **20**, 102–103 (1999).
5. Araujo, A., Fontenelle, L., Padua, P., and Maia, F. Juvenile HTLV-I myelopathy (Abstract). *AIDS Res. Hum. Retroviruses*, **17** (Suppl. 1), S20 (2001).
6. Arisawa, K., Soda, M., Endo, S., Kurokawa, K., Katamine, S., Shimokawa, I., Koba, T., Takahashi, T., Saito, H., Doi, H., and Shirahama, S. Evaluation of adult T-cell leukemia/lymphoma incidence and its impact on non-Hodgkin lymphoma incidence in southwestern Japan. *Int. J. Cancer*, **85**, 319–324 (2000).
7. Barre-Sinoussi, F., Chermann, J.-C., Rey, F., Nugeyre, M. T., Chamaret, S., Gruest, J., Dauguet, C., Axler-Blin, C., Vezinet-Brun, F., Rouzioux, C., Rozenbaum, W., and Montagnier, L. Isolation of a T-lymphotropic retrovirus from a patient at risk for acquired immune deficiency syndrome (AIDS). *Science*, **220**, 868–871 (1983).
8. Bartholomew, C., Cleghorn, F., Jack, N., Edwards, J., and Blattner, W. Human T-cell lymphotropic virus type I-associated facial nerve palsy in Trinidad and Tobago. *Ann. Neurol.*, **41**, 806–809 (1997).
9. Bartholomew, C., Jack, N., Edwards, J., Charles, W., Corbin, D., Cleghorn, F. R., and Blattner, W. A. HTLV-I serostatus of mothers of patients with adult T-cell leukemia and HTLV-I-associated myelopathy/tropical spastic paraparesis. *J. Hum. Virol.*, **1**, 302–305 (1998).
10. Bartholomew, C., Saxinger, W. C., Clark, J. W., Gail, M., Dudgeon, A., Mahabir, B., Hull-Drysdale, B., Cleghorn, F., Gallo, R. C., and Blattner, W. A. Transmission of HTLV-I and HIV among homosexual men in Trinidad. *JAMA*, **257**, 2604–2608 (1987).
11. Bastian, I., Hinuma, Y., and Doherty, R. R. HTLV-I among Northern Territory aborigines. *Med. J. Aust.*, **159**, 12–16 (1993).
12. Bertherat, E., Makuwa, M., Renaut, A., Nabias, R., and Georges-Courbot, M. C. HIV-1, HTLV-I, and HTLV-II in a semiurban population in East Gabon. *J. Acquir. Immune Defic. Syndr. Hum. Retrovirol.*, **19**, 430–432 (1998).
13. Biglione, M., Pizarro, M., Crespo, O., Severich, I., Martinez, P. L., Libonatti, O., Mercedes, A. M., and Astarloa, L. High prevalence of HTLV-I infection in Argentinian blood donors: a new HTLV-I-endemic area? *J. Acquir. Immune Defic. Syndr. Hum. Retrovirol.*, **20**, 101–102 (1999).
14. Blattner, W. A. Epidemiology of HTLV-I associated diseases. *In* "Human Retrovirology: HTLV," ed. W. A. Blattner, pp. 251–265 (1990). Raven Press, Ltd., New York.
15. Blattner, W. A., Saxinger, C., Riedel, D., Hull, B., Taylor, G., Cleghorn, F., Gallo, R., Blumberg, B., and Bartholomew, C. A study of HTLV-I and its associated risk factors in Trinidad and Tobago. *J. Acquir. Immune Defic. Syndr.*, **3**, 1102–1108 (1990).
16. Brodine, S. K., Oldfield, E. C., III, Corwin, A. L., Thomas, R. J., Ryan, A. B., Holmberg, J., Molgaard, C. A., Golbeck, A. L., Ryden, L. A., Benenson, A. S. *et al.* HTLV-I among U.S. Marines stationed in a hyperendemic area: evidence for female-to-male sexual transmission. *J. Acquir. Immune Defic. Syndr.*, **5**, 158–162 (1992).
17. Carberry, C., LaGrenade, L., Fletcher, V., Hanchard, B., Cranston, B., and Blattner,

W. A. The evolving natural history of infective dermatitis in Jamaica (Abstract). *West Indian Med. J.*, **41** (Suppl. 1), 44 (1992).

18. Cavrois, M., Gessain, A., Wain-Hobson, S., and Wattel, E. Proliferation of HTLV-1 infected circulating cells *in vivo* in all asymptomatic carriers and patients with TSP/HAM. *Oncogene*, **12**, 2419–2423 (1996).
19. Chang, C. S., Wu, Y. W., Pan, Y. C., Chen, Z. Y., Hsu, A., and Wang, C. S. Low prevalence of human T-cell lymphotropic virus type I infection among volunteer blood donors in Taiwan. *Transfusion*, **38**, 1118–1119 (1998).
20. Chiyoda, S., Kinoshita, K., Egawa, S., Inoue, J., Watanabe, K., and Ifuku, M. Decline in the positive rate of human T-lymphotropic virus type-1 (HTLV-1) antibodies among blood donors in Nagasaki. *Intern. Med.*, **40**, 14–17 (2001).
21. Cleghorn, F. R., Manns, A., Falk, R., Hartge, P., Hanchard, B., Jack, N., Williams, E., Jaffe, E., White, F., Bartholomew, C. *et al.* Effect of human T-lymphotropic virus type I infection on non-Hodgkin's lymphoma incidence. *J. Natl. Cancer Inst.*, **87**, 1009–1014 (1995).
22. De The, G. and Bomford, R. An HTLV-I vaccine: why, how, for whom? *AIDS Res. Hum. Retroviruses*, **9**, 381–386 (1993).
23. Delaporte, E., Buve, A., Nzila, N., Goeman, J., Dazza, M. C., Henzel, D., Heyward, W., St Louis, M., Piot, P., and Laga, M. HTLV-I infection among prostitutes and pregnant women in Kinshasa, Zaire: how important is high-risk sexual behavior? *J. Acquir. Immune Defic. Syndr. Hum. Retrovirol.*, **8**, 511–515 (1995).
24. Delaporte, E., Peeters, M., Durand, J. P., Dupont, A., Schrijvers, D., Bedjabaga, L., Honore, C., Ossari, S., Trebucq, A., Josse, R., and Merlin, M. Seroepidemiological survey of HTLV-I infection among randomized populations of western central African countries. *J. Acquir. Immune Defic. Syndr.*, **2**, 410–413 (1989).
25. Dourado, I., Andrade, T., Carpenter, C. L., and Galvao-Castro, B. Risk factors for human T cell lymphotropic virus type I among injecting drug users in northeast Brazil: possibly greater efficiency of male to female transmission. *Mem. Inst. Oswaldo Cruz*, **94**, 13–18 (1999).
26. Ferreira Junior, O. C., Vaz, R. S., Carvalho, M. B., Guerra, C., Fabron, A. L., Rosemblit, J., and Hamerschlak, N. Human T-lymphotropic virus type I and type II infections and correlation with risk factors in blood donors from Sao Paulo, Brazil. *Transfusion*, **35**, 258–263 (1995).
27. Figueroa, J. P., Morris, J., Brathwaite, A., Ward, E., Peruga, A., Hayes, R., Vermund, S. H., and Blattner, W. Risk factors for HTLV-I among heterosexual STD clinic attenders. *J. Acquir. Immune Defic. Syndr. Hum. Retrovirol.*, **9**, 81–88 (1995).
28. Furukawa, Y., Osame, M., Kubota, R., Tara, M., and Yoshida, M. Human T-cell leukemia virus type-1 (HTLV-1) Tax is expressed at the same level in infected cells of HTLV-1-associated myelopathy or tropical spastic paraparesis patients as in asymptomatic carriers but at a lower level in adult T-cell leukemia cells. *Blood*, **85**, 1865–1870 (1995).
29. Gabet, A. S., Mortreux, F., Talarmin, A., Plumelle, Y., Leclercq, I., Leroy, A., Gessain, A., Clity, E., Joubert, M., and Wattel, E. High circulating proviral load with oligoclonal expansion of HTLV-1 bearing T cells in HTLV-1 carriers with strongyloidiasis. *Oncogene*, **19**, 4954–4960 (2000).

30. Gallo, R., Salahuddin, S. Z., Popovic, M., Shearer, G., Kaplan, M., Haynes, B. F., Palker, T. J., Redfield, R. R., Oleske, J., Safai, B., White, F., Foster, P., and Markham, P. D. Frequent detection and isolation of cytopathic retroviruses (HTLV-III) from patients with AIDS and at risk for AIDS. *Science*, **224**, 500–503 (1984).
31. Gallo, R., Wong-Staal, F., Montagnier, L., Haseltine, W. A., and Yoshida, M. HIV/HTLV gene nomenclature. *Nature*, **333**, 504 (1988).
32. Gallo, R. C., Sarin, P. S., Gelmann, E. P., Robert-Guroff, M., Richardson, E., Kalyanaraman, V. S., Mann, D., Sidhu, G. D., Stahl, R. E., Zolla-Pazner, S., Leibowitch, J., and Popovic, M. Isolation of human T-cell leukemia virus in acquired immune deficiency syndrome (AIDS). *Science*, **220**, 865–867 (1983).
33. Geng, L., Zai, N., Xiao, Y., Song, F., Tanaka, A., Lou, H., Sonoda, S., and Kanzaki, T. Search for human T-lymphotropic virus type I carriers among northeastern Chinese. *J. Dermatol. Sci.*, **18**, 30–34 (1998).
34. Gessain, A., Barin, F., Vernant, J. C., Gout, O., Maurs, L., Calender, A., and De The, G. Antibodies to human T-lymphotropic virus type-I in patients with tropical spastic paraparesis. *Lancet*, **ii**, 407–410 (1985).
35. Gessain, A. and De The, G. Geographic and molecular epidemiology of primate T lymphotropic retroviruses: HTLV-I, HTLV-II, STLV-I, STLV-PP, and PTLV-L. *Adv. Virus Res.*, **47**, 377–426 (1996).
36. Gill, P. S., Harrington, W., Kaplan, M. H., Ribeiro, R. C., Bennett, J. M., Liebman, H. A., Bernstein-Singer, M., Espina, B. M., Cabral, L., Allen, S., Kornblau, S., Pike, M. C., and Levine, A. M. Treatment of adult T-cell leukemia-lymphoma with a combination of interferon alfa and zidovudine. *N. Engl. J. Med.*, **332**, 1744–1748 (1995).
37. Goto, K., Sato, K., Kurita, M., Masuhara, N., Iijima, Y., Saeki, K., and Ohno, S. The seroprevalence of HTLV-I in patients with ocular diseases, pregnant women and healthy volunteers in the Kanto district, central Japan. *Scand. J. Infect. Dis.*, **29**, 219–221 (1997).
38. Gotuzzo, E., Arango, C., Queiroz-Campos, A., and Isturiz, R. E. Human T-cell lymphotropic virus-I in Latin America. *Infect. Dis. Clin. North Am.*, **14**, 211, 235–239 (2000).
39. Gout, O., Baulac, M., Gessain, A., Semah, F., Saal, F., Peries, J., Cabrol, C., Foucault-Fretz, C., Laplane, D., Sigaux, F., and De The, G. Rapid development of myelopathy after HTLV-I infection acquired by transfusion during cardiac transplantation. *N. Engl. J. Med.*, **322**, 383–388 (1990).
40. Hanchard, B. Adult T-cell leukemia/lymphoma in Jamaica: 1986-1995. *J. Acquir. Immune Defic. Syndr. Hum. Retrovirol.*, **13** (Suppl. 1), S20–S25 (1996).
41. Hanchard, B., LaGrenade, L., Carberry, C., Fletcher, V., Williams, E., Cranston, B., Blattner, W. A., and Manns, A. Childhood infective dermatitis evolving into adult T-cell leukaemia after 17 years. *Lancet*, **338**, 1593–1594 (1991).
42. Hashimoto, K., Higuchi, I., Osame, M., and Izumo, S. Quantitative *in situ* PCR assay of HTLV-1 infected cells in peripheral blood lymphocytes of patients with ATL, HAM/TSP and asymptomatic carriers. *J. Neurol. Sci.*, **159**, 67–72 (1998).
43. Hino, S., Yamaguchi, K., Katamine, S., Sugiyama, H., Amagasaki, T., Kinoshita, K., Yoshida, Y., Doi, H., Tsuji, Y., and Miyamoto, T. Mother-to-child transmission of human T-cell leukemia virus type-I. *Jpn. J. Cancer Res.*, **76**, 474–480 (1985).

44. Hirata, M., Hayashi, J., Noguchi, A., Nakashima, K., Kajiyama, W., Kashiwagi, S., and Sawada, T. The effects of breastfeeding and presence of antibody to p40tax protein of human T cell lymphotropic virus type-I on mother to child transmission. *Int. J. Epidemiol.*, **21**, 989–994 (1992).
45. Hisada, M., Okayama, A., Shioiri, S., Spiegelman, D. L., Stuver, S. O., and Mueller, N. E. Risk factors for adult T-cell leukemia among carriers of human T-lymphotropic virus type I. *Blood*, **92**, 3557–3561 (1998).
46. Hisada, M., Okayama, A., Spiegelman, D., Mueller, N. E., and Stuver, S. O. Sex-specific mortality from adult T-cell leukemia among carriers of human T-lymphotropic virus type I. *Int. J. Cancer*, **91**, 497–499 (2001).
47. Ho, G. Y., Nomura, A. M., Nelson, K., Lee, H., Polk, B. F., and Blattner, W. A. Declining seroprevalence and transmission of HTLV-I in Japanese families who immigrated to Hawaii. *Am. J. Epidemiol.*, **134**, 981–987 (1991).
48. Kajiyama, W., Kashiwagi, S., Nomura, H., Ikematsu, H., Hayashi, J., and Ikematsu, W. Seroepidemiologic study of antibody to adult T-cell leukemia virus in Okinawa, Japan. *Am. J. Epidemiol.*, **123**, 41–47 (1986).
49. Kaplan, J. E., Khabbaz, R. F., Murphy, E. L., Hermansen, S., Roberts, C., Lal, R., Heneine, W., Wright, D., Matijas, L., Thomson, R., Rudolph, D., Switzer, W. M., Kleinman, S., Busch, M., and Schreiber, G. B. Male-to-female transmission of human T-cell lymphotropic virus types I and II: association with viral load. The Retrovirus Epidemiology Donor Study Group. *J. Acquir. Immune Defic. Syndr. Hum. Retrovirol.*, **12**, 193–201 (1996).
50. Kaplan, J. E., Osame, M., Kubota, H., Igata, A., Nishitani, H., Maeda, Y., Khabbaz, R. F., and Janssen, R. S. The risk of development of HTLV-I-associated myelopathy/tropical spastic paraparesis among persons infected with HTLV-I. *J. Acquir. Immune Defic. Syndr.*, **3**, 1096–1101 (1990).
51. Kashiwagi, S., Kajiyama, W., Hayashi, J., Noguchi, A., Nakashima, K., Nomura, H., Ikematsu, H., Sawada, T., Kida, S., and Koide, A. Antibody to p40tax protein of human T cell leukemia virus 1 and infectivity. *J. Infect. Dis.*, **161**, 426–429 (1990).
52. Katamine, S., Moriuchi, R., Yamamoto, T., Terada, K., Eguchi, K., Tsuji, Y., Yamabe, T., Miyamoto, T., and Hino, S. HTLV-I proviral DNA in umbilical cord blood of babies born to carrier mothers. *Lancet*, **343**, 1326–1327 (1994).
53. Kim, J. M., Chang, K. H., Choi, Y. H., Song, Y. G., Kang, S. M., Yoon, T. Y., Choi, J. M., Park, S. Y., and Lew, D. J. Seroprevalence of HTLV-I in Cheju Island, a Korean island adjacent to the endemic area of Japan. *J. Acquir. Immune Defic. Syndr.*, **22**, 409–412 (1999).
54. Kim, J. M., Song, Y. G., Oho, Y. C., Park, H. C., Kwon, K. H., Kim, E., Lee, S. H., and Kim, K. H. Antibodies to human T-cell lymphotropic virus type I (HTLV-I) by particle agglutination (PA) test in Korean blood donors. *Yonsei Med. J.*, **40**, 173–177 (1999).
55. Kinoshita, K., Amagasaki, T., Hino, S., Doi, H., Yamanouchi, K., Ban, N., Momita, S., Ikeda, S., Kamihira, S., Ichimaru, M., Katamine, S., Miyamoto, T., Tsuji, Y., Ishimaru, T., Yamabe, T., Ito, M., Kamura, S., and Tsuda, T. Milk-borne transmission of HTLV-I from carrier mothers to their children. *Jpn. J. Cancer Res.* (Gann), **78**, 674–680 (1987).
56. Kramer, A., Maloney, E. M., Morgan, O. St. C., Rodgers-Johnson, P., Manns, A.,

Murphy, E. L., Larsen, S., Cranston, B., Murphy, J., Benichou, J., and Blattner, W. A. Risk factors and cofactors for human T-cell lymphotropic virus type I (HTLV-I)-associated myelopathy/tropical spastic paraparesis (HAM/TSP) in Jamaica. *Am. J. Epidemiol.*, **142**, 1212–1220 (1995).

57. Kusuhara, K., Sonoda, S., Takahashi, K., Tokugawa, K., Fukushige, J., and Ueda, K. Mother-to-child transmission of human T-cell leukemia virus type I (HTLV-I): a fifteen-year follow-up study in Okinawa, Japan. *Int. J. Cancer*, **40**, 755–757 (1987).
58. LaGrenade, L., Hanchard, B., Fletcher, V., Cranston, B., and Blattner, W. Infective dermatitis of Jamaican children: a marker for HTLV-I infection. *Lancet*, **336**, 1345–1347 (1990).
59. LaGrenade, L., Manns, A., Fletcher, V., Derm, D., Carberry, C., Hanchard, B., Maloney, E. M., Cranston, B., Williams, N. P., Wilks, R., Kang, E. C., and Blattner, W. A. Clinical pathologic, and immunologic features of human T-lymphotropic virus type I-associated infective dermatitis in children [See comments]. *J. Hum. Virol.*, **1**, 293–298 (1998).
60. LaGrenade, L., Morgan, C., Carberry, C., Hanchard, B., Fletcher, V., Gray, R., Cranston, B., Rodgers-Johnson, P., and Manns, A. Tropical spastic paraparesis occurring in HTLV-1 associated infective dermatitis. Report of two cases. *West Indian Med. J.*, **44**, 34–35 (1995).
61. Larsen, O., Andersson, S., da Silva, Z., Hedegaard, K., Sandstrom, A., Naucler, A., Dias, F., Melbye, M., and Aaby, P. Prevalences of HTLV-1 infection and associated risk determinants in an urban population in Guinea-Bissau, West Africa. *J. Acquir. Immune Defic. Syndr.*, **25**, 157–163 (2000).
62. Lavanchy, D., Bovet, P., Hollanda, J., Shamlaye, C. F., Burczak, J. D., and Lee, H. High seroprevalence of HTLV-I in the Seychelles. *Lancet*, **337**, 248–249 (1991).
63. Lee, H., Anderson, E., Allain, J. P., and Gonzaga, A. HTLV-1 infection in Brazil. *Blood*, **73**, 1742 (1989).
64. Lewis, J. M., Vasef, M. A., and Seabury, S. M. HTLV-I-associated granulomatous T-cell lymphoma in a child. *J. Am. Acad. Dermatol.*, **44**, 525–529 (2001).
65. Lin, B. T., Musset, M., Szekely, A. M., Alexandre, J., Fraitag, S., Bodemer, C., Charpentier, A., Frenoy, N., Misset, J. L., Medeiros, L. J., and Rappaport, H. Human T-cell lymphotropic virus-1-positive T-cell leukemia/lymphoma in a child. Report of a case and review of the literature. *Arch. Pathol. Lab. Med.*, **121**, 1282–1286 (1997).
66. Louisiriotchanakul, S., Thongput, A., Khamboonruang, C., Taylor, G. P., Kunstadter, P., and Wasi, C. No evidence of HTLV-I or HTLV-II infection among the Hmong people of northern Thailand or injecting drug users in Bangkok. *J. Acquir. Immune Defic. Syndr.*, **23**, 441–442 (2000).
67. Mahe, A., Chollet-Martin, S., and Gessain, A. HTLV-I-associated infective dermatitis. *Lancet*, **354**, 1386 (1999).
68. Maloney, E. M., Cleghorn, F. R., Morgan, O. S., Rodgers-Johnson, P., Cranston, B., Jack, N., Blattner, W. A., Bartholomew, C., and Manns, A. Incidence of HTLV-I-associated myelopathy/tropical spastic paraparesis (HAM/TSP) in Jamaica and Trinidad. *J. Acquir. Immune Defic. Syndr. Hum. Retrovirol.*, **17**, 167–170 (1998).
69. Maloney, E. M., Hisada, M., Palmer, P., Brooks, K., Pate, E., Wiktor, S. Z., LaGrenade, L., and Manns, A. Human T cell lymphotropic virus type I-associated infective

dermatitis in Jamaica: a case report of clinical and biologic correlates. *Pediatr. Infect. Dis. J.*, **19**, 560–565 (2000).

70. Manns, A., Cleghorn, F. R., Falk, R. T., Hanchard, B., Jaffe, E. S., Bartholomew, C., Hartge, P., Benichou, J., and Blattner, W. A. Role of HTLV-I in development of non-Hodgkin lymphoma in Jamaica and Trinidad and Tobago. The HTLV Lymphoma Study Group. *Lancet*, **342**, 1447–1450 (1993).
71. Manns, A., Miley, W. J., Wilks, R. J., Morgan, O. St. C., Hanchard, B., Wharfe, G., Cranston, B., Maloney, E., Welles, S. L., Blattner, W. A., and Waters, D. Quantitative proviral DNA and antibody levels in the natural history of HTLV-I infection. *J. Infect. Dis.*, **180**, 1487–1493 (1999).
72. Manns, A., Wilks, R. J., Murphy, E. L., Haynes, G., Figueroa, J. P., Barnett, M., Hanchard, B., and Blattner, W. A. A prospective study of transmission by transfusion of HTLV-I and risk factors associated with seroconversion. *Int. J. Cancer*, **51**, 886–891 (1992).
73. Mansuy, J. M., Schlegel, L., Villeneuve, L., Mengelle, C., and Magnaval, J. F. Seroprevalence of retroviral infections among pregnant women in Martinique (French West Indies). *Am. J. Trop. Med. Hyg.*, **61**, 598–599 (1999).
74. Melbye, M., Poulsen, A. G., Gallo, D., Pedersen, J. B., Biggar, R. J., Larsen, O., Dias, F., and Aaby, P. HTLV-1 infection in a population-based cohort of older persons in Guinea-Bissau, West Africa: risk factors and impact on survival. *Int. J. Cancer*, **76**, 293–298 (1998).
75. Melo, J., Beby-Defaux, A., Faria, C., Guiraud, G., Folgosa, E., Barreto, A., and Agius, G. HIV and HTLV prevalences among women seen for sexually transmitted diseases or pregnancy follow-up in Maputo, Mozambique. *J. Acquir. Immune Defic. Syndr.*, **23**, 203–204 (2000).
76. Meytes, D., Schochat, B., Lee, H., Nadel, G., Sidi, Y., Cerney, M., Swanson, P., Shaklai, M., Kilim, Y., Elgat, M. *et al.* Serological and molecular survey for HTLV-I infection in a high-risk Middle Eastern group. *Lancet*, **336**, 1533–1535 (1990).
77. Miller, M., Achiron, A., Shaklai, M., Stark, P., Maayan, S., Hannig, H., Hunsmann, G., Bodemer, W., and Shohat, B. Ethnic cluster of HTLV-I infection in Israel among the Mashhadi Jewish population. *J. Med. Virol.*, **56**, 269–274 (1998).
78. Miura, T., Fukunaga, T., Igarashi, T., Yamashita, M., Ido, E., Funahashi, S., Ishida, T., Washio, K., Ueda, S., and Hashimoto, K. Phylogenetic subtypes of human T-lymphotropic virus type I and their relations to the anthropological background. *Proc. Natl. Acad. Sci. USA*, **91**, 1124–1127 (1994).
79. Mochizuki, M., Watanabe, T., Yamaguchi, K., Tajima, K., Yoshimura, K., Nakashima, S., Shirao, M., Araki, S., Miyata, N., Mori, S. *et al.* Uveitis associated with human T lymphotropic virus type I: seroepidemiologic, clinical, and virologic studies. *J. Infect. Dis.*, **166**, 943–944 (1992).
80. Mochizuki, M., Watanabe, T., Yamaguchi, K., Yoshimura, K., Nakashima, S., Shirao, M., Araki, S., Takatsuki, K., Mori, S., and Miyata, N. Uveitis associated with human T-cell lymphotropic virus type I. *Am. J. Ophthalmol.*, **114**, 123–129 (1992).
81. Morand-Joubert, L., Mariotti, M., Reed, D., Petit, J. C., and Lefrere, J. J. Correlation between viral DNA load and serum anti p19 antibody concentration in symptomless human T-lymphotropic virus type-I (HTLV-I)-infected individuals. *Int. J. Cancer*, **60**,

156–159 (1995).

82. Morgan, O. S., Rodgers-Johnson, P., Mora, C., and Char, G. HTLV-1 and polymyositis in Jamaica. *Lancet*, **ii**, 1184–1187 (1989).
83. Mueller, N., Okayama, A., Stuver, S., and Tachibana, N. Findings from the Miyazaki Cohort Study. *J. Acquir. Immune Defic. Syndr. Hum. Retrovirol.*, **13** (Suppl. 1), S2–S7 (1996).
84. Mueller, N., Tachibana, N., Stuver, S. O., Okayama, A., Ishizaki, J., Shishime, E., Murai, K., Shioiri, S., and Tsuda, K. Epidemiologic perspectives of HTLV-I. *In* "Human Retrovirology: HTLV," ed. W. A. Blattner, pp. 281–293 (1990). Raven Press, Ltd., New York.
85. Murphy, E. L., Figueroa, J. P., Gibbs, W. N., Brathwaite, A., Holding-Cobham, M., Waters, D., Cranston, B., Hanchard, B., and Blattner, W. A. Sexual transmission of human T-lymphotropic virus type I (HTLV-I). *Ann. Intern. Med.*, **111**, 555–560 (1989).
86. Murphy, E. L., Figueroa, J. P., Gibbs, W. N., Holding-Cobham, M., Cranston, B., Malley, K., Bodner, A. J., Alexander, S. S., and Blattner, W. A. Human T-lymphotropic virus type I (HTLV-I) seroprevalence in Jamaica: I. Demographic determinants. *Am. J. Epidemiol.*, **133**, 1114–1124 (1991).
87. Murphy, E. L., Gibbs, W. N., Figueroa, J. P., Bain, B., LaGrenade, L., Cranston, B., and Blattner, W. A. Human immunodeficiency virus and human T-lymphotropic virus type I infection among homosexual men in Kingston, Jamaica. *J. Acquir. Immune Defic. Syndr.*, **1**, 143–149 (1988).
88. Murphy, E. L., Glynn, S. A., Fridey, J., Sacher, R. A., Smith, J. W., Wright, D. J., Newman, B., Gibble, J. W., Ameti, D. I., Nass, C. C., Schreiber, G. B., and Nemo, G. J. Increased prevalence of infectious diseases and other adverse outcomes in human T lymphotropic virus types I- and II-infected blood donors. Retrovirus Epidemiology Donor Study (REDS) Study Group. *J. Infect. Dis.*, **176**, 1468–1475 (1997).
89. Murphy, E. L., Glynn, S. A., Fridey, J., Smith, J. W., Sacher, R. A., Nass, C. C., Ownby, H. E., Wright, D. J., and Nemo, G. J. Increased incidence of infectious diseases during prospective follow-up of human T-lymphotropic virus type II- and I-infected blood donors. Retrovirus Epidemiology Donor Study. *Arch. Intern. Med.*, **159**, 1485–1491 (1999).
90. Murphy, E. L., Wilks, R., Hanchard, B., Cranston, B., Figuero, J. P., Gibbs, W. N., Murphy, J., and Blattner, W. A case-control study of risk factors for seropositivity to human T-lymphotropic virus type I (HTLV-I) in Jamaica. *Int. J. Epidemiol.*, **25**, 1083–1089 (1996).
91. Nagai, M., Usuku, K., Matsumoto, W., Kodama, D., Takenouchi, N., Moritoyo, T., Hashiguchi, S., Ichinose, M., Bangham, C. R., Izumo, S., and Osame, M. Analysis of HTLV-I proviral load in 202 HAM/TSP patients and 243 asymptomatic HTLV-I carriers: high proviral load strongly predisposes to HAM/TSP. *J. Neurovirol.*, **4**, 586–593 (1998).
92. Nakashima, K., Kashiwagi, S., Kajiyama, W., Hirata, M., Hayashi, J., Noguchi, A., Urabe, K., Minami, K., and Maeda, Y. Sexual transmission of human T-lymphotropic virus type I among female prostitutes and among patients with sexually transmitted diseases in Fukuoka, Kyushu, Japan. *Am. J. Epidemiol.*, **141**, 305–311 (1995).

93. Nishioka, K., Maruyama, I., Sato, K., Kitajima, I., Nakajima, Y., and Osame, M. Chronic inflammatory arthropathy associated with HTLV-I. *Lancet*, **i**, 441–442 (1989).
94. Ohshima, K., Kikuchi, M., Masuda, Y., Sumiyoshi, Y., Eguchi, F., Mohtai, H., Takeshita, M., and Kimura, N. Human T-cell leukemia virus type I associated lymphadenitis. *Cancer*, **69**, 239–248 (1992).
95. Okayama, A., Stuver, S., Iga, M., Okamoto, M., Mueller, N., Matsuoka, M., Yamaguchi, K., Tachibana, N., and Tsubouchi, H. Sequential change of virus markers in seroconverters with community-acquired infection of human T lymphotropic virus type I. *J. Infect. Dis.*, **183**, 1031–1037 (2001).
96. Okayama, A., Stuver, S. O., Ishizaki, J., Murai, K., Suzuki, M., Kubuki, Y., Tachibana, N., Matsuoka, M., Yamaguchi, K., Etoh, K., Mueller, N. E., and Tsubouchi, H. HTLV-I proviral DNA load in carriers as a risk factor for adult T cell leukemia (Abstract). *AIDS Res. Hum. Retroviruses*, **17** (Suppl. 1), S40 (2001).
97. Osame, M., Janssen, R., Kubota, H., Nishitani, H., Igata, A., Nagataki, S., Mori, M., Goto, I., Shimabukuro, H., Khabbaz, R., and Kaplan, J. Nationwide survey of HTLV-I-associated myelopathy in Japan: association with blood transfusion. *Ann. Neurol.*, **28**, 50–56 (1990).
98. Plancoulaine, S., Buigues, R. P., Murphy, E. L., van Beveren, M., Pouliquen, J. F., Joubert, M., Remy, F., Tuppin, P., Tortevoye, P., De The, G., Moreau, J. P., and Gessain, A. Demographic and familial characteristics of HTLV-1 infection among an isolated, highly endemic population of African origin in French Guiana. *Int. J. Cancer*, **76**, 331–336 (1998).
99. Plancoulaine, S., Gessain, A., Joubert, M., Tortevoye, P., Jeanne, I., Talarmin, A., De The, G., and Abel, L. Detection of a major gene predisposing to human T lymphotropic virus type I infection in children among an endemic population of African origin. *J. Infect. Dis.*, **182**, 405–412 (2000).
100. Plumelle, Y., Gonin, C., Edouard, A., Bucher, B. J., Thomas, L., Brebion, A., and Panelatti, G. Effect of *Strongyloides stercoralis* infection and eosinophilia on age at onset and prognosis of adult T-cell leukemia. *Am. J. Clin. Pathol.*, **107**, 81–87 (1997).
101. Poiesz, B. J., Papsidero, L. D., Ehrlich, G., Sherman, M., Dube, S., Poiesz, M., Dillon, K., Ruscetti, F. W., Slamon, D., Fang, C., Williams, A., Duggan, D., Glaser, J., Gottlieb, A., Goldberg, J., Ratner, L., Phillips, P., Han, T., Friedman-Kien, A., Siegal, F., Rai, K., Sawitsky, A., Sheremata, L. W., Dosik, H., Cunningham, C., and Montagna, R. Prevalence of HTLV-I-associated T-cell lymphoma. *Am. J. Hematol.*, **66**, 32–38 (2001).
102. Poiesz, B. J., Ruscetti, F. W., Gazdar, A. F., Bunn, P. A., Minna, J. D., and Gallo, R. C. Detection and isolation of type C retrovirus particles from fresh and cultured lymphocytes of a patient with cutaneous T-cell lymphoma. *Proc. Natl. Acad. Sci. USA*, **77**, 7415–7419 (1980).
103. Poiesz, B. J., Ruscetti, F. W., Reitz, M. S., Kalyanaraman, V. S., and Gallo, R. C. Isolation of a new type C retrovirus (HTLV) in primary uncultured cells of a patient with Sezary T-cell leukaemia. *Nature*, **294**, 268–271 (1981).
104. Pombo-de-Oliveira, M. S., Moellman, A., and Dobbin, J. HTLV-I associated T-cell leukemia in pediatric patients (oral presentation). *AIDS Res. Hum. Retroviruses*, **17**

(Suppl. 1), S-30 (2001).

105. Rabkin, C. S., Corbin, D. O. C., Felton, S., Barker, H., Davison, D., Dearden, C., Blattner, W. A., and Evans, A. S. Human T-cell lymphotropic virus type I infection in Barbados: results of a 20-year follow-up study. *J. Acquir. Immune Defic. Syndr. Hum. Retrovirol.*, **12**, 519–522 (1996).

106. Reeves, W. C., Saxinger, C., Brenes, M. M., Quiroz, E., Clark, J. W., Hoh, M. W., and Blattner, W. A. Human T-cell lymphotropic virus type I (HTLV-I) seroepidemiology and risk factors in metropolitan Panama. *Am. J. Epidemiol.*, **127**, 532–539 (1988).

107. Riedel, D. A., Evans, A. S., Saxinger, C., and Blattner, W. A. A historical study of human T lymphotropic virus type I transmission in Barbados. *J. Infect. Dis.*, **159**, 603–609 (1989).

108. Robert-Guroff, M., Shüpbach, J., Blayney, D. W., Kalyanaraman, V. S., Merino, F., Sarngadharan, M. G., Clark, J. W., Saxinger, W. C., Blattner, W. A., and Gallo, R. C. "Seroepidemiologic Studies of Human T-cell Leukemia/lymphoma Virus, Type I. Human T-cell Leukemia/lymphoma Viruses," pp. 285–289 (1984). Cold Spring Harbor Laboratory, New York.

109. Rodriguez, E. M., de Moya, E. A., Guerrero, E., Monterroso, E. R., Quinn, T. C., Puello, E., de Quinones, M. R., Thorington, B., Glasner, P. D., Zacarias, F. *et al.* HIV-1 and HTLV-I in sexually transmitted disease clinics in the Dominican Republic. *J. Acquir. Immune Defic. Syndr.*, **6**, 313–318 (1993).

110. Sacher, R. A., Luban, N. L., Ameti, D. I., Friend, S., Schreiber, G. B., and Murphy, E. L. Low prevalence of flower cells in U.S.A. blood donors infected with human T-lymphotrophic virus types I and II. *Br. J. Haematol.*, **105**, 758–763 (1999).

111. Sawada, T., Tohmatsu, J., Obara, T., Koide, A., Kamihira, S., Ichimaru, M., Kashiwagi, S., Kajiyama, W., Matsumura, N., Kinoshita, K., Yano, M., Yamaguchi, K., Kiyokawa, T., Takatsuki, K., Taguchi, H., and Miyoshi, I. High risk of mother-to-child transmission of HTLV-I in p40^{tax} antibody-positive mothers. *Jpn. J. Cancer Res.*, **80**, 506–508 (1989).

112. Saxton, E. H., Lee, H., Swanson, P., Chen, I. S., Ruland, C., Chin, E., Aboulafia, D., Delamarter, R., and Rosenblatt, J. D. Detection of human T-cell leukemia/lymphoma virus type I in a transfusion recipient with chronic myelopathy. *Neurology*, **39**, 841–844 (1989).

113. Schreiber, G. B., Murphy, E. L., Horton, J. A., Wright, D. J., Garfein, R., Chien, H. C., and Nass, C. C. Risk factors for human T-cell lymphotropic virus types I and II (HTLV-I and -II) in blood donors: the Retrovirus Epidemiology Donor Study. NHLBI Retrovirus Epidemiology Donor Study. *J. Acquir. Immune Defic. Syndr. Hum. Retrovirol.*, **14**, 263–271 (1997).

114. Segurado, A., Granade, T., Parekh, B., Nunez, C. A., Meza, R., Amador, L., Terrell, S., George, J. R., and Lal, R. B. Presence of HTLV-I and HTLV-II infection in Honduras [Letter]. *J. Acquir. Immune Defic. Syndr. Hum. Retrovirol.*, **16**, 308 (1997).

115. Shimoyama, M. Diagnostic criteria and classification of clinical subtypes of adult T-cell leukaemia-lymphoma. A report from the Lymphoma Study Group (1984–87). *Br. J. Haematol.*, **79**, 428–437 (1991).

116. Shinzato, O., Kamihira, S., Ikeda, S., Kondo, H., Kanda, T., Nagata, Y., Nakayama,

E., and Shiku, H. Relationship between the anti-HTLV-1 antibody level, the number of abnormal lymphocytes and the viral-genome dose in HTLV-1-infected individuals. *Int. J. Cancer*, **54**, 208–212 (1993).
117. Smadja, D., Bellance, R., Cabre, P., Arfi, S., and Vernant, J. C. Clinical characteristics of HTLV-1 associated dermato-polymyositis. Seven cases from Martinique. *Acta Neurol. Scand.*, **92**, 206–212 (1995).
118. Stuver, S. O. and Mueller, N. E. Re: "Sexual transmission of human T-lymphotropic virus type I among female prostitutes and among patients with sexually transmitted diseases in Fukuoka, Kyushu, Japan." *Am. J. Epidemiol.*, **142**, 1247–1248 (1995).
119. Stuver, S. O., Tachibana, N., and Mueller, N. Case-control study of factors associated with human T-cell leukemia virus type I infection in southern Miyazaki, Japan. *J. Natl. Cancer Inst.*, **84**, 867–872 (1992).
120. Stuver, S. O., Tachibana, N., Okayama, A., and Mueller, N. E. Evaluation of morbidity among human T lymphotropic virus type I carriers in Miyazaki, Japan. *J. Infect. Dis.*, **173**, 584–591 (1996).
121. Stuver, S. O., Tachibana, N., Okayama, A., Shioiri, S., Tsunetoshi, Y., Tsuda, K., and Mueller, N. E. Heterosexual transmission of human T cell leukemia/lymphoma virus type I among married couples in southwestern Japan: an initial report from the Miyazaki Cohort Study. *J. Infect. Dis.*, **167**, 57–65 (1993).
122. Suite, M., Jack, N., Basdeo-Maharaj, K., Edwards, J., White, F., Blattner, W. A., and Bartholomew, C. Infective dermatitis in Trinidad and Tobago (Abstract). *AIDS Res. Hum. Retroviruses*, **10**, 447 (1994).
123. Sweet, R. D. A pattern of eczema in Jamaica. *Br. J. Dermatol.*, **78**, 93–100 (1966).
124. Tajima, K., Tominaga, S., Suchi, T., Kawagoe, T., Komoda, H., Hinuma, Y., Oda, T., and Fujita, K. Epidemiological analysis of the distribution of antibody to adult T-cell leukemia-virus-associated antigen: possible horizontal transmission of adult T-cell leukemia virus. *Gann*, **73**, 893–901 (1982).
125. Takahashi, K., Takezaki, T., Oki, T., Kawakami, K., Yashiki, S., Fujiyoshi, T., Usuku, K., Mueller, N., Osame, M., Miyata, K. *et al.* Inhibitory effect of maternal antibody on mother-to-child transmission of human T-lymphotropic virus type I. The Mother-to-Child Transmission Study Group. *Int. J. Cancer*, **49**, 673–677 (1991).
126. Takao, S., Ishida, T., Bhatia, K. K., Saha, N., Soemantri, A., and Kayame, O. W. Seroprevalence of human T-lymphotropic virus type 1 in Papua New Guinea and Irian Jaya measured using different western blot criteria. *J. Clin. Virol.*, **16**, 129–133 (2000).
127. Takatsuki, K., Matsuoka, M., and Yamaguchi, K. Adult T-cell leukemia in Japan. *J. Acquir. Immune Defic. Syndr. Hum. Retrovirol.*, **13** (Suppl. 1), S15–S19 (1996).
128. Takatsuki, K., Uchiyama, T., Sagawa, K., and Yodoi, J. Adult T-cell leukemia in Japan. *In* "Topics in Hematology," ed. S. Seno and F. Takaku, p. 73 (1977). Excerpta Medica, Amsterdam.
129. Tanggo, Y., Gultom, S. P., Simanjuntak, T., Sibuea, W. H., Matsuzaki, H., and Yamaguchi, K. Human T lymphotropic virus I in Indonesia. Very low seroprevalence in the Jakarta area: antibodies in healthy blood donors and in various nonhematological diseases. *Intervirology*, **43**, 77–79 (2000).
130. Taylor, G. P. The epidemiology of HTLV-I in Europe. *J. Acquir. Immune Defic. Syndr. Hum. Retrovirol.*, **13** (Suppl. 1), S8–S14 (1996).

131. Taylor, G. P., Tosswill, J. H., Matutes, E., Daenke, S., Hall, S., Bain, B. J., Davis, R., Thomas, D., Rossor, M., Bangham, C. R., and Weber, J. N. Prospective study of HTLV-I infection in an initially asymptomatic cohort. *J. Acquir. Immune Defic. Syndr.*, **22**, 92–100 (1999).
132. Terada, K., Katamine, S., Eguchi, K., Moriuchi, R., Kita, M., Shimada, H., Yamashita, I., Iwata, K., Tsuji, Y., Nagataki, S., and Miyamoto, T. Prevalence of serum and salivary antibodies to HTLV-1 in Sjogren's syndrome. *Lancet*, **344**, 1116-1119 (1994).
133. Tosswill, J. H., Taylor, G. P., Clewley, J. P., and Weber, J. N. Quantification of proviral DNA load in human T-cell leukaemia virus type I infections. *J. Virol. Methods*, **75**, 21–26 (1998).
134. Trujillo, J. M., Concha, M., Munoz, A., Bergonzoli, G., Mora, C., Borrero, I., Gibbs, C. J., Jr., and Arango, C. Seroprevalence and cofactors of HTLV-I infection in Tumaco, Colombia. *AIDS Res. Hum. Retroviruses*, **8**, 651–657 (1992).
135. Tsuji, Y., Doi, H., Yamabe, T., Ishimaru, T., Miyamoto, T., and Hino, S. Prevention of mother-to-child transmission of human T-lymphotropic virus type-I. *Pediatrics*, **86**, 11–17 (1990).
136. Tsukasaki, K., Yamada, Y., Ikeda, S., and Tomonaga, M. Infective dermatitis among patients with ATL in Japan. *Int. J. Cancer*, **57**, 293 (1994).
137. Uchiyama, T., Yodoi, J., Sagawa, K., Takatsuki, K., and Uchino, H. Adult T-cell leukemia: clinical and hematologic features of 16 cases. *Blood*, **50**, 481–492 (1977).
138. Ueda, K., Kusuhara, K., Tokugawa, K., Miyazaki, C., Yoshida, C., Tokumura, K., Sonoda, S., and Takahashi, K. Cohort effect on HTLV-I seroprevalence in southern Japan. *Lancet*, **ii**, 979 (1989).
139. Ureta-Vidal, A., Angelin-Duclos, C., Tortevoye, P., Murphy, E., Lepere, J. F., Buigues, R. P., Jolly, N., Joubert, M., Carles, G., Pouliquen, J. F., De The, G., Moreau, J. P., and Gessain, A. Mother-to-child transmission of human T-cell-leukemia/lymphoma virus type I: implication of high antiviral antibody titer and high proviral load in carrier mothers. *Int. J. Cancer*, **82**, 832–836 (1999).
140. Vallejo, A., Dubon, J. M., and Garcia-Saiz, A. Presence of human T-cell lymphotropic virus types I and II infections in Honduras [Letter]. *J. Acquir. Immune Defic. Syndr. Hum. Retrovirol.*, **12**, 29–30 (1996).
141. Vilmer, E., le Deist, F., Fischer, A., Griscelli, C., Nezelof, C., de Prost, Y., and Prieur, M. Smouldering T lymphoma related to HTLV-I in a Sicilian child. *Lancet*, **ii**, 1301 (1985).
142. Vlahov, D., Khabbaz, R. F., Cohn, S., Galai, N., Taylor, E., and Kaplan, J. E. Incidence and risk factors for human T-lymphotropic virus type II seroconversion among injecting drug users in Baltimore, Maryland, U.S.A. *J. Acquir. Immune Defic. Syndr. Hum. Retrovirol.*, **9**, 89–96 (1995).
143. Walshe, M. M. Infective dermatitis in Jamaican children. *Br. J. Dermatol.*, **79**, 229–236 (1967).
144. Welles, S. L., Tachibana, N., Orav, E. J., Okayama, A., Ishizaki, J., and Mueller, N. E. Changes in hematologic parameters among Japanese HTLV-I carriers. *J. Acquir. Immune Defic. Syndr.*, **7**, 92–97 (1994).
145. Wiktor, S. Z., Pate, E. J., Rosenberg, P. S., Barnett, M., Palmer, P., Medeiros, D., Maloney, E. M., and Blattner, W. A. Mother-to-child transmission of human T-cell

lymphotropic virus type I associated with prolonged breast-feeding. *J. Hum. Virol.*, **1**, 37–44 (1997).

146. Wiktor, S. Z., Piot, P., Mann, J. M., Nzilambi, N., Francis, H., Vercauteren, G., Blattner, W. A., and Quinn, T. C. Human T cell lymphotropic virus type I (HTLV-I) among female prostitutes in Kinshasa, Zaire. *J. Infect. Dis.*, **161**, 1073–1077 (1990).
147. Wilks, R., Hanchard, B., Morgan, O., Williams, E., Cranston, B., Smith, M. L., Rodgers-Johnson, P., and Manns, A. Patterns of HTLV-I infection among family members of patients with adult T-cell leukemia/lymphoma and HTLV-I associated myelopathy/tropical spastic paraparesis. *Int. J. Cancer*, **65**, 272–273 (1996).
148. Williams, A. E., Fang, C. T., Slamon, D. J., Poiesz, B. J., Sandler, S. G., Darr, W. F., Shulman, G., McGowan, E. I., Douglas, D. K., Bowman, R. J., Peetoom, F., Kleinman, S. H., Lenes, B., and Dodd, R. Y. Seroprevalence and epidemiological correlates of HTLV-I infection in U.S. blood donors. *Science*, **240**, 643–646 (1988).
149. Williams, C. K., Alexander, S. S., Bodner, A., Levine, A., Saxinger, C., Gallo, R. C., and Blattner, W. A. Frequency of adult T-cell leukaemia/lymphoma and HTLV-I in Ibadan, Nigeria. *Br. J. Cancer*, **67**, 783–786 (1993).
150. Yamano, Y., Nagai, M., Brennan, M., Mora, C. A., Soldan, S. S., Tomaru, U., Takenouchi, N., Izumo, S., Osame, M., and Jacobson, S. Correlation of human T-cell lymphotropic virus type 1 (HTLV-1) mRNA with proviral DNA load, virus-specific CD8(+) T cells, and disease severity in HTLV-1-associated myelopathy (HAM/TSP). *Blood*, **99**, 88–94 (2002).
151. Yamashita, M., Ido, E., Miura, T., and Hayami, M. Molecular epidemiology of HTLV-I in the world. *J. Acquir. Immune Defic. Syndr. Hum. Retrovirol.*, **13** (Suppl. 1), S124–S131 (1996).
152. Yanagihara, R., Nerukar, V. R., and Ajdukiewicz, A. B. Comparison between strains of human T lymphotropic virus type I isolated from inhabitants of the Solomon Islands and Papua New Guinea. *J. Infect. Dis.*, **164**, 443–449 (1991).
153. Yoshida, M., Osame, M., Kawai, H., Toita, M., Kuwasaki, N., Nishida, Y., Hiraki, Y., Takahashi, K., Nomura, K., Sonoda, S. *et al.* Increased replication of HTLV-I in HTLV-I-associated myelopathy. *Ann. Neurol.*, **26**, 331–335 (1989).
154. Zhuo, J., Yang, T., Zeng, Y., and Lu, L. Epidemiology of anti-human T-cell leukemia virus type I antibody and characteristics of adult T-cell leukemia in China. *Chin. Med. J.*, **108**, 902–906 (1995) (in English).
155. Zurita, S., Costa, C., Watts, D., Indacochea, S., Campos, P., Sanchez, J., and Gotuzzo, E. Prevalence of human retroviral infection in Quillabamba and Cuzco, Peru: a new endemic area for human T cell lymphotropic virus type 1. *Am. J. Trop. Med. Hyg.*, **56**, 561–565 (1997).

AUTHOR INDEX

SUBJECT INDEX